Plant Biology for Cultural Heritage

Biodeterioration and Conservation

Edited by
Giulia Caneva, Maria Pia Nugari, and Ornella Salvadori

Translated by
Helen Glanville

The Getty Conservation Institute
Los Angeles

The Getty Conservation Institute

Timothy P. Whalen, *Director*
Jeanne Marie Teutonico, *Associate Director, Programs*

The Getty Conservation Institute works internationally to advance conservation practice in the visual arts—broadly interpreted to include objects, collections, architecture, and sites. The Institute serves the conservation community through scientific research, education and training, model field projects, and the dissemination of the results of both its own work and the work of others in the field. In all its endeavors, the GCI focuses on the creation and delivery of knowledge that will benefit the professionals and organizations responsible for the conservation of the world's cultural heritage.

Originally published in Italy as *La Biologia Vegetale per I Beni Culturali, Vol. I: Biodeterioramento e Conservazione*
Italian edition © 2005 Nardini Editore — Firenze
info@nardinieditore.it
www.nardinieditore.it
www.nardinirestauro.it

Francesco Bertini, *Graphic Design*
Cristina Corazzi, *Editorial Coordinator*
Paola Bianchi, *Technical Coordinator*

English translation © 2008 The J. Paul Getty Trust

First published in the United States of America by the Getty Conservation Institute

Getty Publications
1200 Getty Center Drive, Suite 500
Los Angeles, California 90049-1682
www.getty.edu

Gregory M. Britton, *Publisher*
Mark Greenberg, *Editor in Chief*

Ann Lucke, *Managing Editor*
Tobi Levenberg Kaplan, *Editor*
Beatrice Hohenegger, *Manuscript Editor*
Pamela Heath, *Production Coordinator*
Hespenheide Design, *Designer*

Printed in Hong Kong

Library of Congress Cataloging-in-Publication Data

Biologia vegetale per i beni culturali. English.
   Plant biology for cultural heritage : biodeterioration and conservation / edited by Giulia Caneva, Maria Pia Nugari and Ornella Salvadori; translated by Helen Glanville.
      p. cm.
   "Originally published in Italy as La biologia vegetale per i beni culturali, vol. I: Biodeterioramento e conservazione."
   Includes bibliographical references and index.
   ISBN 978-0-89236-939-3 (pbk.)
1. Botany. 2. Biodegradation. 3. Cultural property—Conservation and restoration. 4. Cultural property—Protection. I. Caneva, Giulia. II. Nugari, M. P. III. Salvadori, O. IV. Getty Conservation Institute. V. Title.
   QK45.2.B5613 2009
   069'.53—dc22

                                     2008015043

The translation of this work has been funded by SEPS
Segretariato Europeo per le Pubblicazioni Scientifiche

Via Val d'Aposa 7 - 40123 Bologna - Italy
seps@alma.unibo.it - www.seps.it

Front cover (clockwise from top left):

Biodeterioration of mural paintings in the Crypt of the Original Sin in Matera, Italy. Photo G. Caneva.

Bacteria isolated from the same mural paintings, observed with a scanning electron microscope (see Fig. 1.2). Photo G. Caneva.

Statue of the Ogre in the Park of the Monsters of Bomarzo in Viterbo, Italy (see Plate 23c). Photo G. Caneva.

Algal community growing on stone, observed with an optical microscope. Photo M. L. Tomaselli.

CONTENTS

FOREWORD TO THE ENGLISH EDITION

by Timothy P. Whalen
Director, The Getty Conservation Institute

From nearly invisible microorganisms that can damage the surface of objects in museums to towering trees that can slowly break apart great monuments, plants and other biological agents pose serious challenges to the conservation of cultural heritage. Much of the work conservators do involves arresting, mitigating, and preventing the damage caused by bacteria, fungi, algae, lichens, and plants. *Plant Biology for Cultural Heritage: Biodeterioration and Conservation* provides a comprehensive examination of the science of plant biology and its practical applications for conservation. Through case studies and in-depth analyses, this book presents the mechanisms of biodeterioration, investigates the correlation between biodeterioration and environment, describes destructive organisms, and analyzes the effects of different materials, climatic conditions, and geographic settings on biodeterioration.

The Getty Conservation Institute undertook the publication of *Plant Biology for Cultural Heritage* in English at the urging of Giacomo Chiari, chief scientist at the GCI. He became acquainted with *La Biologia Vegetale per i Beni Culturali* through his longtime colleague Giulia Caneva, one of the volume editors. Giacomo recognized the book's value to students, specialists, and all those interested in the care and protection of cultural heritage, and he convinced us to make the book available in English.

We are grateful to Giacomo Chiari, to Giulia Caneva and her co-editors, Maria Pia Nugari and Ornella Salvadori, and to the book's many contributors for amassing such comprehensive, useful information on the wide range of issues related to the biological degradation and the conservation of art, architecture, and archaeological sites around the world. We would also like to thank Helen Glanville for translating the text from the original Italian; Joy Mazurek, assistant scientist at the Getty Conservation Institute, for her careful review of the English version; and Beatrice Hohenegger, freelance editor, for her engaged, intelligent, and skillful review of the translation and copyedit of the manuscript.

Conservators and conservation scientists often analyze and address the effects of biodeterioration, such as staining, surface changes, detachment, and insoluble deposits. By linking these effects to their origins, we hope that *Plant Biology for Cultural Heritage* will allow conservation professionals to understand such problems and learn new ways to address them effectively.

FOREWORD

by Carlo Blasi

President, Società Italiana Scienza della Vegetazione (Italian Society of Vegetation Science)

Plant Biology for Cultural Heritage is a work of great interest for a widely diverse readership; its contents reflect the need to bring together the different branches of knowledge involved in the investigation of the complexity of environmental phenomena.

Recent international guidelines in the area of environment and nature conservation have, for some time already, set up an interrelation of the various disciplines of physics, biology, the humanities, and economics. Even the agreements aimed at the Conservation of Biological Diversity do not limit themselves to giving indications that might be pertinent exclusively to the biologist, the ecologist, or the naturalist; instead, while addressing the subject of sustainable development and a fairer sharing of resources, they place the conservation of diversity within a wider context, integrating other spheres of competence such as economy, history, and the social, political, and anthropological disciplines.

Plant Biology for Cultural Heritage is a volume directed to specialists, students, and all those with a passion for the preservation of anything pertaining to the history of mankind and its articulated coexistence with the natural environment. For this reason, it is clear from the beginning that this book has required the involvement of a great many different specialists on themes and lines of investigations that were well defined by the editors.

While it is true that biodeterioration takes place by a number of different mechanisms (physical, mechanical, or chemical processes), it is also true that, in addition to understanding the substrate, it is essential to have knowledge of the organisms and the biocoenoses that can be encountered in the rich and varied field of cultural heritage. To better understand the relationships between physical and biological elements, it is useful to define the environmental conditions, always keeping in mind that there is no clear demarcation between the various causes since often there is a contemporaneous convergence of multiple factors instead. Even in the area of biodeterioration, it is possible to speak of nature, environment, and ecology. The link with ecology in particular allows us to refer to models of analysis and treatment that have already been amply codified on a territorial scale. In this sense, the present volume has succeeded in its main objective, which is to provide information about the scientific and technical instruments needed to operate within a field that is in constant expansion while at the same time extending the fundamental knowledge and the scope of the investigation to all those involved in environmental fields as well as in monitoring, prevention, and remediation.

Biodeterioration is analyzed and described in terms of models and dynamics, exactly as plants are analyzed: we start with the identification of the populations present and of the structural and functional models and proceed to the monitoring and the definition and planning of the maintenance operations.

Moreover, the considerable space devoted to the disciplines referring to systematic botany will be of value to any plant ecologist. There is much talk about biodiversity, but unfortunately the study of systematic botany has been removed from many college courses in biology—a catastrophic event unique to Italy. By contrast, in Germany, systematic botany and zoology are instead mandatory foundation subjects even for biotechnology courses with a molecular focus. Thus, highlighting the importance of the knowledge of the plant world from a taxonomic, functional, and ecological point of view when speaking of biodeterioration and cultural heritage is an effort that will be much appreciated.

The scope of this book, as delineated by the editors, also allows the reader to analyze a great wealth of environments, beginning with enclosed spaces (such as libraries, archives, churches and crypts, tombs, and catacombs) and including monuments and artifacts in urban environments, in parks, in rural or coastal contexts, in fountains and nymphaea, in loggias and porticoes, and finally in maritime rocky environments and lakes.

Of primary importance are the discussions of urban environments; it has been known for some time now that the behavior of energy fluctuations in urban ecosystems reflects that found in the natural ecosystem. It is essential that space be given to the problems connected to urban environments since in our country (Italy) urban ecological systems now cover almost ten percent of the national territory. In addition, monitoring and prevention operations in the area of cultural heritage offer extraordinarily positive effects in terms of quality of life in general: one need only think of the damaging effects of urban pollution on cultural heritage and on the health of the inhabitants.

The final part of the volume is dedicated to the subject of climate and biogeography, treated both on a global and on a national scale. Although Italy is a small country, its bioclimatic and biogeographical heterogeneity is unique in the world. The present volume has correctly highlighted the importance of geographical issues, as it becomes more and more pressing to place any project of bioindication and remediation on the scale of landscape as well.

Indeed, it is impossible to speak of cultural heritage without taking into account the general context of landscape, understood as the opportunity for integrating natural and anthropic components. If the new vision of landscape that emerged from the Florence Convention (2000)—no longer based solely on elements of an aesthetic and perceptual nature at a subjective level—succeeds in replacing the old concept with a more territorial vision in which perception becomes synonymous with the recognition of the identity of place, it will undoubtedly be easier to apply the analytical models and interventions proposed in this volume.

FOREWORD

by Caterina Bon Valsassina

Director, Istituto Centrale del Restauro (Central Conservation Institute)

With a few notable exceptions, one of the factors that has characterized the development of Western culture is the compartmentalization of knowledge; this has led to the isolation of various spheres of knowledge and thus to fewer opportunities for constructive interactions resulting from possible collaborations that could have been of benefit to the different fields of knowledge. In his essay *Les paroles et les choses*, Foucault pinpoints two precise moments in the history of Western civilization, in which this "fracture" is initiated (Foucault calls it "discontinuity") and which will lead to the eventual separation of knowledge (and specialization): the beginning of the seventeenth century (think of Galileo and Descartes), and the nineteenth century, which "marks the beginnings of our modernity" (M. Foucault, *Les paroles et les choses*, 1966, p. 12).

In contrast, in the great Eastern civilizations, the unity of knowledge—we might say, the "holistic" approach to knowledge—has remained intact over the centuries; that is why, today, we look to these civilizations as to a mythical golden age in which philosophy, religion, art, and science harmoniously merged to give rise to an immensely complex culture of the very highest quality. This view may represent a very efficient antidote for the defects resulting from the separation of the different fields of knowledge in our own culture, among which the most pernicious was (and, unfortunately, still is) a kind of irrational "jealousy" that pushes researchers and scholars in the most disparate fields to closely guard their own personal wealth of knowledge in a sort of egotistical private niche.

In the field of the conservation and restoration of cultural heritage, the knocking down of boundaries between the different disciplines has been an obligatory path in the direction of a "virtuous" choice: indeed, without constant interaction of historical, scientific, and technical research, it would have been impossible to achieve any result. From the outside, finding a practical and applicable harmony within the various fields and cultural backgrounds of art historians, archaeologists, architects, conservators, chemists, physicists, biologists, and geologists would seem a utopia, an impossible dream; and, indeed, it is not easy. Yet, it is not impossible, either. It was realized with the model—first epistemological, then organizational—conceived in 1939 by Argan and Brandi, of the Istituto Centrale del Restauro (Central Conservation Institute), which was founded on these very principles, constantly seeking to harmonize the various skills and specialties in the most constructive and creative way, and held together by that strongest of common bonds: the safeguard of our cultural heritage.

The text presented here is further proof of the strength of the interdisciplinary approach in our field; it also tackles a subject of unquestionable interest.

The knowledge of plant biology constitutes one of the essential elements for a correct approach for anyone working in the area of conservation and restoration of cultural heritage. The alterations induced by biological factors on works of art—be they works in stone, paintings on canvas, frescoes, or any other artifacts—are indeed extremely frequent and severe. Over the centuries and millennia, artistic creativity has induced artists to use the widest variety of materials; it is for this reason that today, when confronted with the necessity, or better yet, the obligation to conserve these works to the best of our ability, we must use what science puts at our disposal to attempt to prevent damage induced by biological agents on such a heterogeneous assortment of materials. However, within the field of restoration and conservation, biology must not be considered as a mere diagnostic necessity with subsequent treatment of the work of art: the deepest and most competent application of biology in the field of art must go beyond the restricted confines of biodeterioration. First of all, the "ecological" context of the work of art must be considered; and, for conservation purposes, this must be followed by a careful evaluation of the environmental conditions of that context. In this sense, therefore, taking on the hard responsibility to safeguard our cultural heritage means committing to acquiring the knowledge of the biological environmental conditions and, where possible, intervening on them to make them as compatible as possible with the requirements for the optimal conservation of the work.

It is precisely with the awareness of the importance of the delicate relationship between the work and the environment that the Ministero per i Beni e le Attività Culturali (the Department for Cultural Heritage and Activities) published in the Official Gazette n. 150 (October 2001) the *Atto di indirizzo sui criteri tecnico-scientifici e sugli standard di funzionamento e sviluppo dei musei* (Act addressing the technical and scientific criteria and the operating and development standards of museums),[1] in which a whole chapter is dedicated to defining a sort of checklist of the problems specifically connected with the relationship between the work of art and its environment.

From a wider perspective, it is also true that while the effects of biological agents on works of art are indeed more often than not negative (biodeterioration), they may also sometimes be positive and offer a resolution of certain forms of degradation (bioremediation).

Texts such are these are all the more welcome then—born as they are of the harmonious cooperation of so many experts with knowledge of the application of biology to works of art—because they offer a precious contribution not only to the conservation and restoration of our own heritage and that of other countries, but also, and more importantly, to a thorough and in-depth understanding of the complex relationship, in the broadest sense of the term, between the work of art and biology.

The organization of the text is deliberately didactic in its format and therefore aimed at young people who may wish to enter the fascinating world of the preservation of works of art. But this in no way lessens the volume's more general and deeper appeal, which can indeed constitute an invaluable aid to anyone—expert or not—who wants to explore the captivating world of plant biology and better understand its complex issues. In a way, it would be a little like rediscovering what Crollius described at the beginning of the seventeenth century: " . . . it really seems that plants speak . . . with their marks and signs . . . manifesting their inner virtues hidden by the silent veil of Nature" (first version in Latin in 1609, translated into French in 1624, cited in M. Foucault, ibid., p. 41).

The hope is that this text—stubbornly willed into being by its authors—will not be an isolated meteor but, rather, mark the beginning of a constellation of studies, researches, and publications capable of pointing in the direction of further openings and connections in the field of conservation and restoration, always guided by the interdisciplinary approach.

---

[1] Laws in Italy and the European Union affecting the conservation of cultural heritage are cited throughout this book. Readers may find that such laws are related to regulations in other countries, as well.

INTRODUCTION

by Giulia Caneva, Maria Pia Nugari, and Ornella Salvadori

Today, the relationship between the humanities and the scientific disciplines—especially in such fields as conservation—has reached a fruitful and productive consolidation, thanks to the ever-increasing initiatives that have led to a new dialogue between cultural contexts that are only seemingly removed from each other.

By the 1980s, a great deal of literature had begun to accumulate relating to scientific analysis and diagnosis of the degradation of works of cultural heritage and to the problems of improving their conservation. The past decade has seen a further development of applied research in this area as well as an increasing—and increasingly shared—attention focused on such problems.

However, as might be expected, published materials dealing with this specific subject are dispersed throughout a vast literature and are often difficult to find, or else they focus only on specific aspects of the problem. It seemed, therefore, more useful than ever to put together a publication offering a general, but also appropriately detailed, overview of the subject.

More specifically, the objective of this project (which is divided into two complementary volumes)[1] is to present an organic and articulated presentation of the different scientific contributions that the field of plant biology has made to the understanding, conservation, and display of our cultural heritage. As such, this book is intended as a contribution to an already well-developed discipline, one that can offer critical knowledge toward the solution of practical issues and applications.

The material has been organized to satisfy the requirements of the reforms of [Italian] university rulings on didactics; these place an increasing emphasis—both in the humanities and in the science departments—on the technical and scientific elements linked to the conservation of cultural heritage. Also, we gathered knowledge gleaned at an international level, with the belief that it is important to achieve—especially in this field—common objectives for the development of humanity in its core values.

Part One of the book focuses on understanding the problems linked to biological degradation of cultural heritage, first discussing general topics and then moving on to more specialized issues. Thus, the general mechanisms of biodeterioration are described and discussed within specific typologies, but reference is also made to the groups most frequently involved. This is followed by a discussion of the basic notions of ecology as applied to biodeterioration and as related to morphological, physiological, and ecological characteristics of the groups most frequently involved with the biological degradation of materials.

In treating the analysis of degradation of cultural heritage more specifically, we approach the problem from several points of view in order to provide clear answers to the different practical problems encountered in the world of conservation of materials.

First, we identify specific problems associated with each material, since the physical and chemical composition of a material is one of its most important variables. We then move on to treat specific problems of biodeterioration as related to various environmental typologies. Next, we present a broad overview of the most salient problems as related to different

---

[1] The second volume, *La Biologia Vegetale per i Beni Culturali, Vol. II: Conoscenza e Valorizzazione*, has not been translated into English.

geographical locations and, therefore, climate conditions.

In Part Two, we address the problems involved in conserving these materials, based on the latest scientific data, beginning with prevention of degradation, then addressing its control, and finally providing information on the diagnostic techniques to be employed.

Overall, we did not think it useful to provide minimal knowledge gleaned from within a generalized context, and we also avoided oversimplifying the issues; we did not want to encourage the false hope that any one individual might be able to single-handedly resolve all problems by using the best theoretical and conservation strategies. Rather, we tried to direct conservators toward identifying specific problems, for which they can then receive the necessary support from specialists in the field of plant biology. This text is therefore addressed to anyone interested in the problems of conservation of our historical, artistic, and archaeological heritage—to readers from both humanistic and scientific backgrounds. Clearly, a different reading of the text is advised for each of the two groups of readers, with the former omitting some of the more specialized sections and the latter taking a more comprehensive view.

We hope that this book will prove to be of value not only in a didactic context but also as a scientific instrument highlighting the most salient aspects of the knowledge acquired in this field to date.

CHAPTER 1

PROCESSES OF BIODETERIORATION: GENERAL MECHANISMS

by Daniela Pinna and Ornella Salvadori

H. J. Hueck (1965, 1968) was the first to define biodeterioration as "any undesirable change in the properties of a material caused by the vital activities of organisms." Alterations induced by such activities may vary to a large extent, from irreversible transformations and breakdown of the substrate to the simple, and unwelcome, presence of organisms without particular consequences for the materials involved (aesthetic damage). The term biodeteriogen refers to microorganisms or organisms that cause damage to materials, although the term is often loosely applied to include any organisms discovered on artifacts or objects, whether or not their capacity to damage the substrate has been established or is known.

Within the cultural heritage sector, the term biodeterioration, rather than biodegradation, is employed, as the latter usually refers to the biological processes through which organic macromolecules are decomposed, with particular reference to the transformation—by means of microorganisms—of toxic compounds (such as pesticides and pollutants) into other less toxic or harmless compounds. Therefore, while biodeterioration has negative connotations of damage, biodegradation generally implies a beneficial or positive process.

The objective of this chapter is to identify the main elements of the processes of biodeterioration and to contribute to an understanding of these processes by providing general answers to such essential questions as:

—What is biodeterioration?
—How does it occur?
—What are its causes?
—How does it manifest itself?

The general issues of when such processes of biodeterioration occur (in relation to environmental factors) and what the causal agents (biodeteriogens) are will be discussed in Chapters 2 and 3, before delving into more specific problems.

1.1 GENERAL PRINCIPLES

Different types of mechanisms result in the biodeterioration of materials: physical or mechanical processes leading to phenomena such as loss of cohesion, rupture, or disaggregation; and chemical processes that lead to transformation, degradation, or decomposition of the substrate. These processes usually occur simultaneously, but one type can predominate over the other depending on the kind of substrate and the biotic community, as well as environmental conditions. We must also keep in mind that there is no clear separation between the chemical and physical (abiotic) causes of deterioration on one side and the biological ones on the other, since any physical or chemical process can induce or influence the activity of organisms.

To understand the processes of biodeterioration, we must first and foremost remember that the biological colonization of a material may involve:

—the utilization of the substrate as a food source;
—the use of the material solely as a support for the colony's development.

It is therefore important to distinguish between autotrophs and heterotrophs: the first are organisms able to manufacture for them-

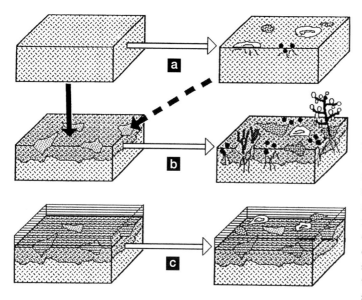

Fig. 1.1 *Primary (a), secondary (b), and tertiary (c) bioreceptivity in a stone material (from Guillitte 1995)*

enzymatically: rarely do auto-trophs colonize such materials (as, for example, when algae and lichens are found on wood kept in an outdoor environment). This is mainly due to the metabolic specificity of the attacking heterotrophs, which rapidly colonize and destroy the material they use as a food source, unlike the autotrophs, which grow more slowly and use the substrate for anchorage. The differing conditions in which objects of an organic nature are preserved, when compared with those of an inorganic nature, are also important factors: the former are usually kept in confined spaces where the environmental conditions (especially light and humidity) are, in general, limiting factors for auto-trophs (Chapter 2).

selves the organic materials necessary to their development, while the second must extract these organic materials from the substrate to survive. Thus, autotrophs (i.e., some bacteria, algae, lichens, mosses, and vascular plants, see Chapter 3) can colonize inorganic substrates, while heterotrophs (i.e., numerous bacteria and fungi as well as animal forms—the latter, however, not included in this context) are only able to grow on such substrates if organic substances are available to them from external sources. It is for this reason that any analysis of the processes of biodeterioration cannot ignore the chemical nature of the materials (in addition to taking into account the physical parameters and the surrounding environmental conditions), and it is useful to draw a distinction between the problems associated with organic substrates, whether of plant or of animal origin, and those associated with inorganic ones (Chapter 4).

The biodeterioration of works of art made up of organic materials (such as wooden sculptures, paintings on panel and canvas, books, parchment, etc.) most commonly occurs because the microorganisms involved use them as a source of nutrients, and therefore have the ability to decompose their organic constituents

Inorganic materials are generally used as a support, not for nutritional purposes (with the exception of the absorption of mineral salts). They are therefore more easily attacked by autotrophic organisms, though heterotrophs are often found alongside them. The biodeterioration of cultural heritage is rarely caused by only one group of organisms and depends rather on the complex interactions between different coexisting groups.

Aside from the environmental factors that trigger biodeterioration phenomena and the specific processes of deterioration of the materials—which will be analyzed in Chapters 2 and 4 respectively—we believe it would be useful to highlight the properties inherent to the materials before embarking on the analysis of individual processes.

The term bioreceptivity—introduced by Guillitte (1995) with reference to stone artifacts and subsequently expanded to cover all cultural heritage—refers to the aptitude of a material to being colonized by one or more groups of organisms, without this colonization automatically resulting in biodeterioration. The nature of the materials, i.e., their differing chemical composition, surface roughness, and porosity, as well as their state of preservation, are all important factors in the establishment and development of organisms that could, in various ways, lead to the deterioration of artworks. To identify the different stages of the colonization process, we distinguish among primary, secondary, and tertiary bioreceptivity. Primary bioreceptivity indicates the initial potential of a material to being colonized. Over time, as a result of the action of organisms and/or other exogenous factors, this becomes secondary receptivity. Finally, any human activity that interferes with the material (for example, consolidations, treatments with biocides, etc.), and therefore alters its characteristics, induces tertiary bioreceptivity (Fig. 1.1). The actual bioreceptivity of various materials is not known, but with appropriate laboratory tests using different groups of organisms it should be possible to develop indices that could offer information toward determining the degree of risk of colonization of a particular material, or the likely efficacy of specific treatments (Guillitte 1995).

1.2 PROCESSES OF BIODETERIORATION

It is not always possible to distinguish biological damage of a purely aesthetic nature from structural damage with irreversible consequences such as the degradation and decomposition of the materials. This is particularly true in the case of inorganic materials, where the microflora does not make direct nutritional use of the substrate, with the exception of mineral salts, which are more or less present depending on the nature of the constituent materials. Often, the simple fact of finding a biological species on an object of historical-artistic-archaeological interest has led a number of authors to postulate effects linked

to biodeterioration more on the basis of a hypothetical than a proven action. In some instances, the presence of organisms on a substrate does not seem to cause any discernible alteration, either in its chemical composition or in its physical characteristics (Realini et al. 1985; Pietrini et al. 1985), but much depends on the interval of time elapsed from the inception of biological growth. Indeed, the development of some of the organisms is exceedingly slow, and the damage may not be noticeable for many years, or even decades. Some authors have even spoken of an effect of bioprotection exercised by some organisms, as for instance in the case of lichens on porous stone in outdoor environments. This hypothesis is based on the observation that in some instances where lichens have not been present, alterations such as exfoliations, efflorescence of salts, disintegration, pulverization, and honeycombing are more widespread. These organisms would then be exercising a protective function toward the substrate, reducing the intensity of water exchanges and the effects of atmospheric agents of degradation such as wind, rain, pollutants, and saline aerosol (Lallemant and Deruelle 1978; Ariño et al. 1995; Wendler and Prasartet 1999). But this is a complex issue, and biodeteriogenic activity cannot be ignored, as we observe in Chapter 4 (section 3.1).

Fig. 1.2 *Image of a biofilm as seen under SEM (Crypt of the Original Sin, Matera)* – Photo G. Caneva

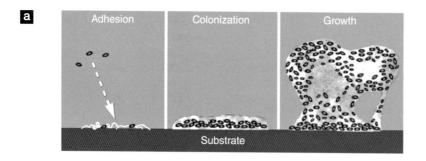

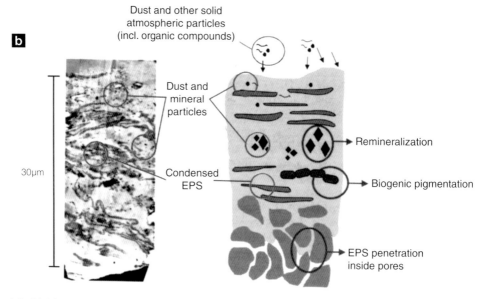

Fig. 1.3 *(a) Schematic representation of the various stages in the formation of a biofilm; (b)* left, *image under the* microscope, right, *schematic reconstruction of the same (modified from Hoppert and Schieweck 2004)*

Conversely, when organic materials, such as paper or wood, are attacked by heterotrophic cellulolytic microorganisms, the damage is often severe and becomes visible quickly, and the materials lose their mechanical characteristics.

As a general rule, in the case of degradation phenomena caused by the development of biological populations, it is necessary to take into account the objective damage, due to physical and chemical processes linked to biological growth, rather than the aesthetic damage. It should also be mentioned that although an initial biological colonization may cause little damage, sometimes considered only aesthetic,

it can nonetheless create favorable conditions for the establishment of more aggressive species.

The capacity of microorganisms and organisms to adhere to the substrate is very important: indeed, the ability to transform the substrate is strictly linked to good adhesion. While the more highly evolved and complex organisms (lichens, mosses, higher plants) adhere to, and penetrate, the substrate with specialized structures such as rhizines, rhizoids, or actual roots, microorganisms attach themselves to the substrate by forming a biofilm. This phenomenon, which is a precursor to possible physical and chemical damage, begins with nonspecific reversible reactions, which are primarily dependent on the physical and chemical properties of the cells and of the substrate. Once the microorganisms are in the vicinity of a surface, however, they are able to create stable interactions through the use of specific molecules and structures (lipopolysaccharides, membrane proteins, flagella). Moreover, organisms that adhere to a surface often secrete extracellular polymeric substances (EPS), the function of which is to "cement" them to the surface. This viscous, sticky layer of microorganisms, organized into complex associations and immersed in a polysaccharide matrix, is known as biofilm (Morton and Surman 1994). These matrices form hydrogels that contain about 98% water. The composition of the EPS in the biofilm (made up of carbohydrates, proteins, nucleic acids, lipids/phospholipids, and humic acids) varies, and in consequence so do its chemical and physical properties. The relative quantities of the different polysaccharides depend on the physiological state of the microorganisms and on the availability of nutrients. The presence of salts can affect the properties of the EPS as it reduces their viscosity.

The architecture of the biofilm consists of an intricate network within which microorganisms form microcolonies separated by interstitial voids (Figs. 1.2 and 1.3). From a biological point of view, this architecture allows the circulation of interstitial fluids between the empty spaces and also offers an optimal spatial arrangement for the circulation of nutrients even in the deepest layers of the biofilm. From an abiotic point of view, the formation of biofilm is of considerable importance for the activation and the development of alteration processes, as it is within the biofilm that a retention of fluids and an accumulation of aggressive metabolic compounds take place. This is why biofilms are able to maintain, at the interface with the surface, an environment that can be radically different from the surrounding environment in terms of pH and chemical composition. Biofilms can also absorb particles and corrosive atmospheric pollutants, thereby contributing to an increase in the reaction speed of the chemical corrosion processes (Warscheid and Braams 2000).

The mechanisms of microbial adhesion are complex and not yet fully understood. The initial attachment to the most varied materials and the initial phases of the formation of the biofilm depend on the surface typology, the kind of substrate, and the degree of its roughness. Irregularities in the surface are, in fact, preferential points of attachment and become veritable protective niches for the microorganisms (Roldàn et al. 2003). Well-developed biofilms can form in subaerial or aquatic environments, on both inorganic and organic artifacts (for example on fibroin, a protein found in silk) (Seves et al. 2000). An essential prerequisite in the breakdown of cellulose by cellulolytic anaerobic bacteria is their adhesion to the polysaccharide. In many bacteria, the proteins for adhesion and the enzymes required for the hydrolysis of cellulose are organized into a complex structure called "cellulosome."

The presence of biofilms on outdoor surfaces is very important even during the conservation phase; indeed, these structures can react most unsatisfactorily to treatments aimed at their removal, because the exopolysaccharides that surround the cells give them a certain level of protection from external factors, for instance from biocides and other toxic substances.

1.3 PROCESSES OF A PHYSICAL NATURE

These include all those mechanisms leading to a micro- or macroloss of cohesion in the substrate due to a mechanical action of the organisms, i.e., movement or growth. The fragments thus produced keep the same chemical makeup as the original material, but are easily detached from the substrate due to the pressure exercised by the growing organisms or by their parts (for example, fungal hyphae or the roots of plants). The fragmented substrate then offers a greater surface area to other agents of degradation, especially in outdoor environments, i.e., rain, wind, freeze-thaw cycles, pollutants, etc. The distinction between mechanical and chemical processes is useful from a didactic point of view, but it should be remembered that, in the great majority of cases, the two processes occur simultaneously, so that the action triggered mechanically is accompanied and facilitated by

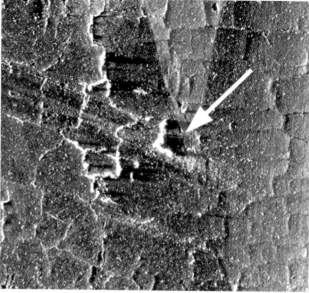

Fig. 1.4 *Detachment of the paint layer due to fungal attack (panel painting,* Enthroned Madonna with Saints, *by Mariotto di Nardo, fourteenth century, Church of Santa Margherita, Tosina, Borselli, Florence)* – Photo I. Tosini

the simultaneous chemical interaction between the substrate and the metabolites released by the organisms.

As a general rule, the damages provoked by organisms are more severe than those produced by microorganisms. This is due to the organisms' larger dimensions and, consequently, the more intense pressure produced—one need only think of the damage inflicted by roots on wall structures, for example. In certain instances, however, and especially when the substrate is a painted surface, even the loss of the tiniest fragments, caused by the action of microorganisms, can result in very serious damage; an example of this is the detaching of fragments of paint layer from the canvas support, due to the entirely mechanical action generated by the growth of fungal mycelium (Fig. 1.4). Analogous phenomena are encountered on frescoes as a result of the development of fungi and/or actinomycetes. The degradation resulting from the action of black fungi or microcolonial fungi (MCF) on stone in outdoor environments seems also due to the mechanical force exercised by hyphae,

as the production of acids by these microorganisms has yet to be proven (Dornieden and Gorbushina 2000; Urzì et al. 2000a). The mucilaginous sheaths produced by cyanobacteria and by some algae, and more generally speaking the biofilms, can induce mechanical-type damage on the substrate both because of their adhesive qualities and as a result of their changing dimensions.

Fig. 1.5 *Damage caused by root growth (archaeological site of Cobà, Mexico)* – Photo G. Caneva

In addition, the presence of colored patinas formed by microorganisms on stone in outdoor environments can be the source of physical stresses in the substrate, in that these patinas can induce a rise in temperature, a change in the thermohydric expansion phenomena, and an increase in water retention. In Israel, for example, the temperature was measured on some mortar walls covered with black patinas formed by cyanobacteria, and an increase of 8°C was found in comparison to noncolonized areas (Garty 1990).

The damage caused by ruderal plants on monuments or archaeological sites is well known and is linked either to particular biological forms (woody species, shrubs, or grasses, see Chapter 3, section 6) or to the characteristics of the root system (Caneva 1985, 1994). The roots tend to penetrate the areas offering the least resistance, such as mortars, and the damage is visible from a distance of many meters from the area of implantation (Fig. 1.5).

1.4 CHEMICAL PROCESSES

The mechanisms of chemical alteration are due to the effects of the metabolic processes of the organisms present. The chemical transformation of the substrate can be due to the excretion of intermediate metabolic products or of substances with an inhibitory or waste function, or else as the results of assimilatory processes and their subsequent production of extracellular enzymes, when the organisms are using the substrate for nutritional purposes. The processes of chemical alteration of the substrate due to the presence of microbial communities begin at the moment of adhesion and during the subsequent development of the biofilm, and take place within it (Ascaso et al. 2002). In comparison to plants and animals, bacteria, algae, and fungi have a high surface-to-volume ratio due to their small dimensions; this allows for the rapid diffusion of metabolic products from the cells to the surrounding environment.

The principal processes of chemical biodeterioration are the following:

—the production of inorganic and organic acids;

—the production of CO_2;
—the chelation of elements;
—the production of alkalis;
—selective mobilization and accumulation of elements;
—cationic exchanges;
—the production of enzymes;
—the production of pigments.

1.4.1 Acidolysis

The best known and most widely studied type of damage is the one associated with the release of acids (H^+ ions or protons) by biodeteriogens, especially in the case of inorganic materials (Warscheid and Braams 2000). This type of damage is caused by the release of inorganic acids (carbonic, nitric, or sulphuric acid) as well as organic ones (citric, succinic, glutamic, fumaric, malic, formic, acetic, lactic, gluconic acids) that do not possess, or only to a slight degree, a complexing action. The lytic action of acids (acidolysis) is due to their capacity to react directly with molecules from the substrate, giving rise to reaction products, mostly soluble salts. As an example, we shall use the reaction between calcium carbonate, the principal component of calcareous substrates, and the sulphuric acid produced by sulphur-oxidizing bacteria, which is representative of the damage mechanisms:

$$CaCO_3 + H_2SO_4 \rightarrow CaSO_4 + H_2CO_3$$

The release of strong inorganic acids leads to phenomena of corrosion, due to the interaction of these substances with the substrate, and also to the formation of more or less soluble salts as well as various other reaction products. In the case of stone, an increase in the quantity of salts can cause stress inside the pores of the material, which in turn leads to the formation of fissures; in addition, the salts can crystallize on the surface of the stone.

The quantity and the kind of acids produced vary according to the species. For example, both the strictly xerophilic and the facultatively xerophilic fungi produce fumaric, malic, lactic, and acetic acids, but the strictly xerophilic fungi produce 5–9 times more malic acid than the facultatively xerophilic species (Arai 2000). The production of acids, in particular citric acid, by fungi has long been exploited by the food industry.

Organic acids interact with materials either through the action of protons or through the chelation of metallic ions (see below). In addition to organic acids, many other molecules (such as amino acids and polysaccharides) containing ionic groups may be released (Sand 1997). The excretion of organic material is usually the result of a growth imbalance, as for instance the presence of an excess of nutrients or the deficiency of compounds containing nitrogen or phosphorus; other times, the organic compound is the final product of a metabolic process (acetic acid, for instance). Through fermentation, many microorganisms produce a variety of substances (such as acetic, formic, or butyric acids, and alcohols) that act as organic solvents on the substrates, causing them to swell and dissolve, either partially or totally.

Also, we should not forget to mention phenomena linked to respiration, which are responsible for the release of carbon dioxide; in an aqueous environment, the carbon dioxide produced during respiration by aerobic organisms does indeed transform into carbonic acid:

$$CO_2 + H_2O \rightleftarrows H_2CO_3$$

Carbonic acid gives rise to acidolysis phenomena on stone, resulting in the dissolution of the carbonates of calcium and magnesium present in calcareous stone, mortars, or plasters, and producing calcium and magnesium bicarbonates, which are highly soluble:

$$CaCO_3 + H_2CO_3 \rightleftarrows Ca(HCO_3)_2$$
$$MgCO_3 + H_2CO_3 \rightleftarrows Mg(HCO_3)_2$$

When the carbon dioxide has been consumed, for instance during photosynthesis, the equilibrium of the reaction moves to the left, and calcium carbonate is precipitated; in this way, cyanobacteria and algae can induce the formation of organogenic calcareous stone like travertine.

The production of acids is also important from the ecological point of view, in that it favors the development of acidophilic species that could not live in neutral or alkaline conditions. Thus, the colonization of a substrate follows a dynamic movement favoring acidophilic species in the secondary phases of colonization.

1.4.2 Complexolysis

Regarding the attack of stone or metals, the lytic processes due to chelation complexes or compounds are rather frequent and occur as the result of the stereochemical makeup of some molecules produced by the organisms, which allows them to bind a hydrogen or metal ion to two atoms of the same molecule. In particular, carbonyl oxygen and amino nitrogen (both electron donors) can provide another atom with a pair of electrons, such as the hydrogen of the hydroxy group or a metal (electron receptors). The bond thus formed is therefore the result of the donation of electrons by an atom to the free orbital of another atom, so that the pair of electrons is shared by both atoms. Once formed, these bonds are like covalent bonds and do not therefore result in dissociation. All molecules that have the power to effectuate a chelation (from the Greek χηλή = chele, pincers) are called chelating agents.

In chelation complexes, the metal ion coordinates with a polyfunctional organic base to form a stable ring compound called chelate. The factors that may influence chelate formation are:

—the basic strength of the chelating group;
—the electronegativity of the donor atoms of the basic group in the chelating agent;
—the ring size;
—the characteristics of the metal ion;
—the resonance;
—steric effects.

The ligands that contain two, three, four, or more donating atoms are called polidented (for example, bi-, tri-, tetra-dented) (Fig. 1.6).

Many of the organic compounds produced by microorganisms and organisms are able to

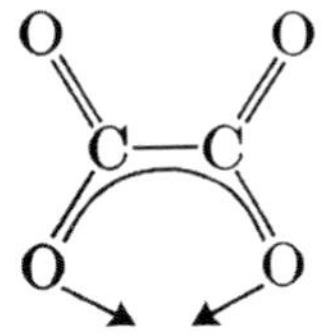

Fig. 1.6 *Example of a bidentate ligand (oxalate ion)*

complex or chelate metal ions from the substrate. Among the complex-forming agents are: simple organic acids (such as oxalic, citric, tartaric, 2-ketogluconic, fumaric, malic, malonic, and aspartic), phenols (for example, salicylic acid and 2-3 dihydroxybenzoic acid), and many other so-called lichenic substances. Many lichenic substances often contain donating polar groups in ortho-positions, i.e., adjacent, and it is precisely this that facilitates the complexing action and the formation of the ring (for example, –OH and –COOH in evernic and lobaric acids, –OH and –CHO in salazinic acid and atranorin) (Fig. 1.7).

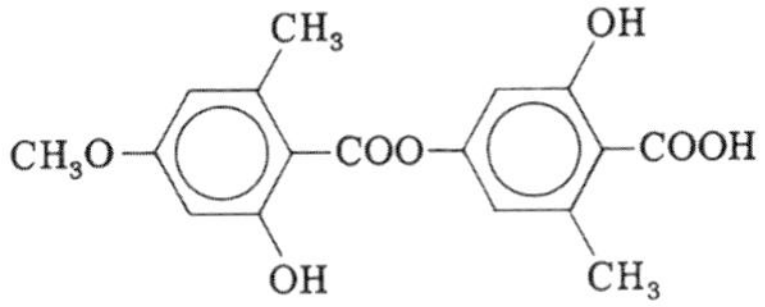

Fig. 1.7 *Structural formula of evernic acid with -OH and -COOH in the ortho-positions*

Even the exopolymers produced by microorganisms containing anionic groups such as amino acids, peptides, or sugar acids, can act as complexing agents (Sand 1997).

The complexing properties of some molecules are exploited positively in some activities. EDTA (ethylenediaminetetracetic acid, i.e., disodium and tetrasodium salts) for example, can be used during conservation for local removal of patinas containing calcium on stone artifacts (sulphatations, oxalates, calcium whitewashes), or of corrosion patinas on metal artifacts; it can even be used in gardening to remove the excess calcium and stop the soil from becoming too alkaline. Citric acid and its salts can be used in the cleaning of polychrome artifacts (Cremonesi 2001).

Oxalic acid, a strong complexing agent, is an intermediate product of the cellular metabolism of all organisms. There are several precursors to oxalic acid: oxaloacetate (one of the by-products of the Krebs cycle), glyoxylic acid, and ascorbic acid (Franceschi and Horner 1980). In comparison with other organic acids, oxalic acid has a high capacity for the degradation of minerals because of its complexing and acid properties. It can form complexes with metal ions such as calcium, magnesium, iron, copper, etc. There are numerous studies on the function of oxalic acid within an organism and also on the factors that induce its production by fungi, lichens, and other plants in relation to particular physiological and/or environmental situations. Several factors are involved and, at times, may coexist, without excluding one another.

The presence of calcium oxalate has often been noted on works of art of various natures (stone artifacts, frescoes, paintings on canvas, polychrome sculptures, mosaics, glass, wood,

Fig. 1.8 *Example of a calcium oxalate patina on marble monument (Temple of Antonino and Faustina, Rome)* – Photo G. Caneva

stuccowork) (Matteini and Moles 1986); it can be found in its monohydrate form (whewellite, $CaC_2O_4 \cdot H_2O$) or in the bihydrate form (weddellite, $CaC_2O_4 \cdot (2 + x)H_2O$). Often found on outdoor stone artifacts are colored patinas ranging from yellow to brown, which are primarily calcium oxalate (Fig. 1.8). On the subject of their formation, as well as on their possible protective role, there is a lively and wide-ranging debate within the scientific community; among the more accredited hypotheses we would like to mention: their derivation from intentional treatments carried out with a protective/decorative aim, or else from causes extraneous to human intervention (for instance, production by lichens) (AA.VV. 1989; Realini and Toniolo 1996).

Aside from the production of oxalates through chemical phenomena, such as the oxidation by atmospheric oxygen of possible organic substances present in the substrate, there is also frequent evidence of its biological production, in which it has been demonstrated that the production of oxalic acid is directly linked to the presence of calcium in the substrate. The organism tends to maintain its own ionic balance, and the excess of calcium is removed through the formation of calcium oxalate, a stable and nontoxic substance. For certain fungi, for example, a high concentration of calcium acetate added to the culture soil reduces the growth of the mycelium, but stimulates the production of oxalic acid. The formation of oxalic acid is therefore a process of biomineralization that has been developed by many organisms as a means of growth on calcium-rich substrates (Pinna 1993).

Many authors are in agreement that the fungi producing calcium oxalate contribute to the formation of the soil, and that oxalate crystals can also play a role in the retention of calcium in the soil and therefore in the retention and/or recycling of the elements required for the growth of plants. The production of oxalic acid in fungi is also dependent on the pH, maximum formation occurring in the interval of pH 3–5. An additional factor that induces the production of oxalates in fungi and lichens is a high intensity of light. This—in the same way as other environmental factors causing

oxidative stress to the organism (such as sulphur dioxide, for example)—induces a process of detoxification in the cells. Such stress situations increase the production of ascorbic acid (a precursor to oxalic acid) which, as an antioxidant, protects the organism from free radicals (Modenesi et al. 1998).

Finally, oxalic acid can be involved as a pathogenic element in the action of some plants' pathogenic fungi. Many studies (Havir and Anagnostakis 1983; Marciano et al. 1983; Punja et al. 1985; Stone and Armentrout 1985; Smith et al. 1986) have indeed demonstrated that oxalic acid:

—increases the permeability of the cytoplasmic plant membrane;
—reduces the pH to values favoring the growth of fungi and the activity of the fungal enzyme endopolygalacturonase;
—forms chelating bonds with the calcium of the plant cell wall modifying the calcium pectate to a form that can be easily hydrolyzed by the enzyme;
—is capable of inhibiting the action of the enzyme polyphenol oxydase, thus limiting the production of compounds derived from phenolic oxidation, compounds that protect the pectic substances in the walls of plant cells.

1.4.3 Alkaline Reactions

The production of alkaline substances by organisms is not frequent. It can be found mostly in the production of ammonia-based substances by certain microorganisms, especially ammonifying microorganisms that are not damaged by the accumulation of ammonia, which is normally a toxic substance. When the accumulation of such substances brings the pH to values higher than 9, the silica in silica-containing stone may be dissolved. The fixation of CO_2 during photosynthesis may also result in a weak alkalinization (Warscheid and Braams 2000). The raising of the pH may also lead to a variation in the microflora, inasmuch as it encourages the presence of alkalophilic organisms.

1.4.4 Selective Mobilization and Accumulation of Elements

The biogenic structure of biofilms plays an important role in the processes of calcium biomobilization by microorganisms, leading to the formation of microcavities in stone substrates. The anionic nature of biofilm exopolymers maintains a high level of hydration in the matrix and therefore has an antidehydrating effect on microorganisms; at the same time it absorbs cations, so that the polymers become sites for enucleation for the formation of minerals, as in the case of materials with high calcium content (Ascaso et al. 2002). The cells form extracellular carbonate deposits following the cell's contours.

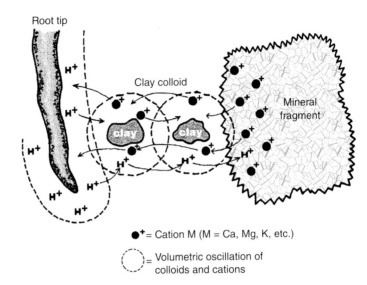

Fig. 1.9 *Extraction of metal ions by root tips through contact-exchange mechanism (modified from Keller and Frederickson 1952)*

Calcium ions are readily removed from the substrate and precipitate, in the form of calcium carbonate, onto the polysaccharide covering of the cyanobacteria (Albertano 2002). Among the calcifying species, *Scytonema julianum* and *Loriella osteophila* are considered the most damaging because of their capacity to mobilize calcium cations from calcareous substrates (Hernandez-Marine et al. 1999; Saiz-Jimenez 1999). Macronutrients such as nitrogen and phosphorus can also be mobilized from the substrate and metabolized or conserved inside the cells.

Moreover, many heterotrophic bacteria produce organic acids that may also cause the dissolution of stone through the mobilization of cations such as Ca^{2+}, Fe^{3+}, Mn^{2+}, Al^{3+}, and Si^{4+} (Kumar and Kumar 1999).

1.4.5 Cationic Exchange

Microorganisms and plants are able to absorb metal cations for their sustenance but also in amounts beyond that. The transformation of mica into vermiculite, carried out by fungi, takes place by a similar process, in which K^+ are exchanged with Na^+ ions (Weed et al. 1969).

The exopolymers contained in biofilms constitute an anionic medium that, in addition to keeping the fibrous matrix hydrated, also has high properties of cationic exchange entrapping aerosol, dust, nutrients, minerals, and complex organic compounds (Albertano 2002; May et al. 2002).

Processes of cationic exchange also take place in plants that utilize them in the processes of mineral nutrition, thanks to the chemical properties of the root tips. The hydrogen ions present on their surfaces can be exchanged with cations in solution following the lyotropic series (Ba^{2+} > Ca^{2+} > Mg^{2+} > Cs^+ > Rb^+ > NH_4^+ > K^+ > Na^+ > Li^+): the transfer of cations takes place through a net of colloidal particles by a mechanism of contact-exchange (Fig. 1.9) (Keller and Frederickson 1952). These processes explain the biochemical damage inflicted by ruderal plants and even grasses growing on a wall substrate; the damage is proportional to the acidity of the root tips, generally to be found between the values of pH 4–6 (Caneva and Altieri 1988).

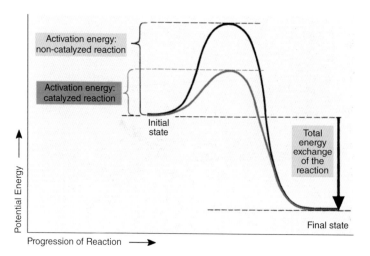

Fig. 1.10 *Effect of enzymes on the activation energy of a reaction (modified from Raven et al. 2002)*

1.4.6 Enzymatic Degradation

Enzymes are proteins, generally globular in shape, that catalyze chemical reactions without becoming part of the products of the reaction itself. They bond temporarily to one or more of the reagents (substrate) of the reaction they catalyze; by doing so, they lower the activation energy required and thus greatly increase the reaction speed (Fig. 1.10). Because enzymes, like all catalysts, remain unchanged by the reaction, they may be used repeatedly and are therefore required in very small quantities.

In order to carry out its action, the enzyme must bond with at least one of its reagents with forces that are not covalent (hydrogen bonds, ionic and hydrophobic interactions). Most of these forces are weak, and a three-dimensional complementarity is therefore required between the configuration of the substrate and that of the enzyme (active site)—like a key in a lock—for the substrate and the enzyme to bind (Fig. 1.11). Therefore, these are catalysts with a high level of specificity and generally catalyze only one chemical reaction or else reactions that involve substrates with very similar structures; they can act on millions of molecules every second. Many enzymes also require the presence of other nonproteinaceous cofactors, for instance some of the metal ions (Zn^{2+}, Cu^{2+}, Mn^{2+}, K^+, Na^+) or small organic molecules

called coenzymes (B vitamins are precursors of coenzymes, for example). Enzymatic action is strongly influenced both by pH and temperature; that is, every enzyme has a functional best at a specific pH and temperature for its activity, which is reduced the farther the conditions are from these values. This is because these factors can influence the tertiary structure of the enzyme as well as its noncovalent bonds, which are essential for it to remain bound to the substrate. Hydrogen bonds rupture readily with increases in temperature (Lehninger et al. 1994).

Enzymes are divided into six main groups depending on the kind of reaction they catalyze: 1. oxidoreductases, which catalyze oxidation-reduction reactions; 2. transferases, which transfer functional groups; 3. hydrolases, which operate reactions of hydrolysis; 4. lyases, which remove groups of atoms without hydrolysis; 5. isomerases, which rearrange functional groups; and 6. ligases, which join two molecules. The root of an enzyme's name originates either in the name of the reaction induced or in the substrate it interacts with, and the suffix -ase is added to the root (for example, cellulase, lipase, protease, tannase).

We differentiate between intracellular enzymes, which catalyze reactions inside cells, and extracellular enzymes, which catalyze chemical reactions outside of the cells. To the latter group belong the enzymes that degrade organic macromolecules (such as cellulose, lignin, proteins), reducing them to small molecules that are able to cross the cytoplasmic membrane. The prefixes endo- and exo- distinguish between those enzymes that degrade polymeric organic substances (cellulose, lignin, collagen, fibroin) at any point in the chain (endo) and those that do so only at the extremities (exo). In order to do this, the enzyme must be able to bind to fibers and to arrive at the

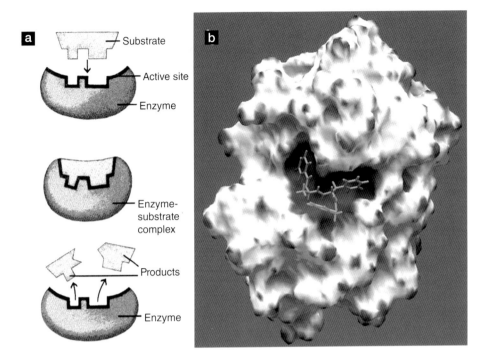

Fig. 1.11 *Enzymes: (a) example of complementarity between an active site of an enzyme and its substrate; (b) three-dimensional structure of an enzyme and its active site (image elaboration by F. Polticelli)*

centers that can be hydrolized or oxidized; because of this, the areas most easily attacked are those with a low degree of crystallization, i.e., amorphous areas.

Constitutive enzymes are always produced by cells, independently from the substrate on which they originate, while inductive or adaptive enzymes are produced in response to the presence of specific molecules in the substrate (for instance, the cellulase complex). Enzymatic activity is regulated by different mechanisms, such as the localization in the membranes (cytoplasmic, of mitochondria or chloroplasts), the synthesis of inactive precursors, feedback inhibition, and activation of precursors.

As we shall see in Chapter 4, fungi produce a wide range of enzymes that play a very important role in the degradation of organic materials, among which are cellulose and lignin. Many of these enzymes are now produced commercially and are finding an ever wider use in different industrial sectors (for example, asparaginase, amylase, catalase, cellulase, dextranase, β-glucanase, glucoamylase,

glucose-oxidase, hemicellulase, laccase, lipase, pectinase, protease, tannase). Fungal enzymes that degrade cellulose include cellulase, which hydrolyzes all types of cellulose; glucancellobiohydrolase, which degrades crystalline cellulose to a cellobiose; and glucanase, which hydrolyzes amorphous cellulose. All these enzymes are glycoproteins and are resistant to thermal degradation. Cellulase is also produced by certain bacteria that are, however, active in alkaline conditions and with very high values of a_w (see p. 39), conditions that occur only infrequently.

Lignin, which is a complex aromatic polymer, is degraded by extracellular oxidative enzymes produced by white rot fungi such as lignin peroxidase, manganese-dependent peroxidase, and laccase, a phenoloxidase containing copper. Ligninolytic enzymes are not highly selective; as a result, other aromatic substances (for instance, pentachlorophenol, dioxin, aromatic polycyclic hydrocarbons) can be oxidized and biodegraded by white rot fungi.

Although many proteins are water soluble, they are unable to pass through the cell membrane except in fragments of no more than 3–5 amino acids, which is why only the microorganisms that produce extracellular protease can assimilate proteins. There are two main kinds of protease: endohydrolase, which breaks protein chains into two smaller sections, and exohydrolase, which detaches individual amino acids from the extremities of the peptide chains.

Tannases are hydrolytic enzymes that break down the tannins (phenol polymers of plant origin) found in inks into sugars and phenolic acids.

Organic materials that are attacked enzymatically are prone to severe degradation, which can irreversibly compromise their conservation: paper becomes felt-like and so brittle that it tends to crumble away; parchment increases in porosity with numerous perforations; the sensitive layer of photographs is devastated; the mechanical properties of wood are impaired, and increasingly so the more the microorganisms develop (Gallo 1992; Florian 1997).

Precisely because of their high level of specificity, enzymes have been utilized in recent years in paper conservation, for the cleaning of polychrome sculptures, and in the removal of animal glue and aged acrylic resins, providing a valid alternative to traditionally used solvents, both from the point of view of toxicity for the operators and of the structural integrity of the component materials (Cremonesi 1999). Proteolytic enzymes have also been tested, with good results, in the cleaning of stone colonized by lichens (Chapter 8, section 5.2) (Capponi and Meucci 1987).

1.4.7 Production of Pigments

The pigments produced by cells belong to different chemical families and may take on different colorations. They can be found in the cytoplasm, the wall, the capsule, or they can be excreted onto the growth substrate. They are classified as endopigments, which are linked to the cellular structures and thus impart a color to the organism, and exopigments, which are excreted into the surrounding environment. Therefore, the color induced in the substrate is due either to the presence of colored cellular structures or to the excretion of pigment into the environment. The color is not necessarily correlated to the chemical nature of the pigment, because chemically dissimilar pigments can have the same color; also, the observed color may be due to the mixture of several pigments.

Plant organisms and cyanobacteria have fundamental photosynthetic pigments, which participate in the light reactions of photosynthesis (chlorophyll a, green in color), and accessory photosynthesizing pigments. The latter include other chlorophylls, different from chlorophyll a: carotenoids (carotene, xanthophyll, etc.), which are mostly yellow; and phycobilins (phycocyanin, phycoerythrin), which are blue and red. Accessory photosynthesizing pigments act as transfers of the captured light energy to the principal photosynthesizing pigment.

In addition to these, organisms (and included here are heterotrophic bacteria and fungi) synthesize other pigments, such as flavonoids (anthocyanins, flavones, and others) colored red, blue, and violet; quinones (naphthoquinone, anthraquinone, anthracyclinones, naphtodiantrones) variable in color, depending also on pH; and the brown-black melanins. The biological function of numerous bacterial and fungal pigments is not clear. Some of them seem to act as a filter against noxious light radiation, while others seem to be endowed with antibiotic properties. The tannins (phenolic compounds) and the products of their oxidation (quinones) are powerful antibiotics, which explains, for example, the particular resistance of the wood of certain tropical trees (for example teak, *Tectona grandis*) to the attack of fungi and insects.

The production of pigments is determined both by genetic and environmental factors. It is therefore a stable characteristic in some cases and can consequently be used to differentiate prokaryotic species at a taxonomic level (for instance, *Pseudomonas* sp., streptomycetes). Pigments do not seem to be essential; indeed, the loss of production capacity does not seem to damage the producing microorganisms (Kutzner 1981).

Table 1.1 Some types of pigments (quinone and naphthoquinone derivatives) produced by streptomycetes (modified from Karbowska-Berent and Strzelczyk 2000)

Pigments	Producers	Crystals	Acid Environment	Alkaline Environment	Antibiotic Activity
Granaticin	*Streptomyces thermoviolaceus, S. sp.*	Red	—	Blue	Antibiotics against gram-positive bacteria
Actinorhodin	*S. coelicor*	Red	—	Blue	
Phenocyclinone	*S. coelicor*	—	—	Blue	
Coelicolorin	*S. coelicor*	Purple	Red (at pH 6–7 violet)	Blue-green	
Ericamycin	*S. varius*	Deep red	Red	Purple	
Rubidin	*S. sp.*	Dark red	Red	Blue	
Julimycin	*S. shiodaensis*	Reddish orange	Reddish orange	Purplish blue	
Aquayamycin	*S. misawanensis*	Orangish yellow	Red orange	Violet blue	
Chrothiomycin	*S. pluricolorescens*	Dark purple	Purplish red	Blue	
Rhodomycin	*S. purpurascens*	Red	Red	Blue	Wide-spectrum antibiotics
Antibiotic S-583-B	*S. purpurascens*	Red	Red	Purplish blue	
Trypanomycin	*S. diastatochromogenes*	Reddish brown	Reddish orange	Violet blue	
Litmocidin	*Proactinomyces cyaneus, S. antibioticus*	—	Purplish red (violet at pH 7)	Blue	
Griseorhodin	*S. californicus*	Red	—	Violet	
Antibiotic U-12, 241	*S. bellus*	Dark red grains	Red	Blue	
Mycorhodin	*S. sp.*	Light red	Purplish red	Purple (dark blue at pH> 8.1)	

The production of pigments is specific to a species, a characteristic that is very relevant for streptomycetes (Table 1.1) and fungi (Fig. 1.12). The quantity of pigment produced is influenced by various factors: the availability of nutrients (in particular as a source of C and N); the presence of metals in the substrate; pH and buffering capacity of the substrate; temperature and light (Szczepanowska and Lovett 1992; Nyuksha 1994).

Regarding stone, the color change of the substrate is due mainly to the production of photosynthetic pigments and melanin (Warscheid and Braams 2000). The stains are therefore mostly linked to chlorophylls of cyanobacteria, green algae, and lichens, to the

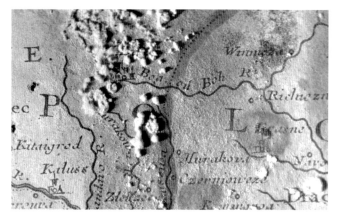

Fig. 1.12 *Stains produced by fungal development on a paper material (map)* – Photo I. Tosini

attention to the origin and function of the melanins (Urzì et al. 1994). Melanins are phenolic pigments, with color ranging from brown to black, and high molecular weight. They are secondary metabolites of many fungi (in particular dematiaceous ones), of actinomycetes, and of some bacteria and plants; they are not essential to growth, but in particular conditions they increase the capacity for survival and competitiveness, because they protect organisms against UV radiation, desiccation, changes in temperature, and hydrolytic enzymes (Florian 1997). Melanins are usually insoluble in water, acid solutions, and common organic solvents. This usually makes their removal difficult; however, melanines can be bleached with hydrogen peroxide.

On organic materials, on the other hand, the main damage is connected to the excretion of exopigments by bacteria and fungi. These can be water soluble and can quickly diffuse in the middle of the cells that produced them, and can also precipitate near those cells as amorphous or crystalline particles. It should also be emphasized that some pigments are colorless when synthesized and acquire a coloration only at the moment of excretion; others, also called indicators, change color, and often their solubility, according to the pH (for instance, red and insoluble in an acid environment and blue and soluble in an alkaline environment) (Table 1.1).

Alterations induced by pigments can have greater or lesser relevancy depending on the bonding established with the substrate. Yet, since we are dealing with artifacts of artistic interest, it is clear that, even when they do not cause structural damage, the presence of colored stains is displeasing and can seriously compromise a correct interpretation of the work of art. Indeed, one of the main

breakdown products of chlorophylls, and to other pigments such as phycobiliproteins and carotenoids. The changes of color in stone can be categorized as follows:

- —black coloration (melanin and melanoidine, degradation products of chlorophyll, iron and manganese minerals);
- —green and greenish colors (chlorophylls);
- —yellow, orange, brown coloration (carotene, carotenoids, and degradations products of chlorophyll, such as phycobiliproteins);
- —orange, pink, and red coloring (carotenoids).

The studies on the coloring of outdoor stone due to microorganisms have devoted particular

problems is the difficulty in identifying the correct solvent for the removal of pigments, especially from organic material. Countless pigments of different chemical nature have been identified (and most likely many more will be identified in the future); they can be grouped into three principal families: derivatives of toluquinone, of naphthoquinone, and of beta-methyl-quinone.

The rust-colored alterations, commonly known as foxing or fox-spots, frequently present on aged paper and parchment, have been the subject of much interest. Research has been carried out in two main directions: microbiological studies to verify the hypothesis of a biological origin for these stains (Arai 2000; Florian 2000; Montemartini Corte et al. 2003), and chemical studies to establish if they may rather be caused by the presence of iron within the paper (Chapter 4, section 1.2).

1.5 THE APPEARANCE OF BIODETERIORATION

The morphology of biodeterioration varies according to: the species present, their physiological conditions, the nature of the substrate, the climate, and the seasons. Macroflora (for example plants, mosses, and lichens) is more easily recognizable than bacteria, algae, and microfungi, because of larger dimensions: plants, mosses, or lichens are easily recognized with the naked eye. However, if a microbial colonization is well established thanks to favorable environmental and substrate conditions, the microflora can also be easily identified despite its microscopic dimensions. Indeed, since microorganisms are frequently the producers of pigments (such as chlorophylls, phycobilins, carotenoids, xanthophylls, melanines, etc.), it is generally this characteristic that makes their detection easier. When the organisms gain a nutritional advantage from attacking the substrates, the microflora is detectable from the colonies visible to the naked eye (for example, fungi on organic materials) (Fig. 1.13).

The appearance of biological colonizations also depends on the physiological conditions of the organisms, which are closely correlated both to internal, and species-specific, factors (i.e., the age of the population or its vital phase) and to external factors such as availability of nutrients and micro- and macroclimate. In unfavorable conditions, the organisms' morphology may indeed change, both in form and color (for example, algal patinas). The morphology of biological alterations is therefore termed either *typical* or *atypical*, indicating respectively alterations that are easily relatable to a biotic cause and alterations that are difficult to interpret (Giacobini 1974; Caneva et al. 1994a).

At times, microbial communities (or the deterioration caused by their growth) can be confused with damage of a purely chemical or physical nature (Fig. 1.14). Stains, for example, could be associated to chemical causes, resulting from oxidation processes or deposit of mineral salts, or to biological causes resulting from microbial attacks, which involve the utilization, the oxidation, or the reduction of ions, or the production of pigments. Examples of this may be the oxidation of pigments in mural paintings, the blackening of outdoor stone artifacts, whitish efflorescences, foxing on paper (Plates

Fig. 1.13 *Development of fungal colonies on the reverse of a painting* – Photo ICR

Fig. 1.14 *Roman marble group with "blackening of biological origin"; progressive magnifications of the image, demonstrating the biological origin of the alteration (Museo Nazionale Romano, Rome)* – Photo G. Caneva

12, 14, 20, etc.). An additional factor that makes it difficult to identify a biological attack is the growth of microorganisms deep inside the material, which can be missed even in the diagnostic stage if only surface sampling is carried out.

The presence of organisms can cause the so-called fouling or soiling, which refers to alterations in the external appearance of the work of art, such as chromatic variations, biological patinas, or visual impairment of the artifacts. Although these purely aesthetic damages are often the most emphasized and at times the only ones to receive attention, they are perhaps the most visible but often the least important aspect of the problem. Also, let us keep in mind that values related to aesthetic aspects are influenced by individual and cultural sensibilities, and can change in a relatively short time. In the nineteenth century, for example, the aesthetic value of ruins was highly regarded, and the effect of

plants climbing up monuments was much appreciated (Ruskin, in Martines 1983). Algal patinas, lichens, and plants were not only tolerated, but even prized because they signified the passage of time. Conversely, in the last decades the preference has been to remove biological patinas, incrustations, and plants covering ruins, not only for conservation reasons but also to achieve an effect of order and cleanliness (Fischer 1972). Despite this, biodiversity—in certain archaeological contexts—has been recently claimed as an "added value" to the intrinsic historical/artistic/archaeological merits, stressing the notion that the reduction or removal of organisms should be undertaken only when the damage to the substrate has been clearly identified (Chapter 4) (Nimis et al. 1992; Ariño and Saiz-Jimenez 1996a; Ceschin et al. 2003).

Providing a uniform terminology for the description of the alterations of materials has always been considered a priority by the

Table 1.2 Appearance of biological alterations of cultural heritage

	Wood	Paper	Textiles	Parchment, Leather	Stone	Glass	Metals
Autotrophic Bacteria	nd	nd	nd	nd	Black crusts, black patinas, exfoliations, powdering	Pitting, opacification, black spots, blackening of water-logged materials	Corrosion
Heterotrophic Bacteria	Changes in mechanical characteristics	Stains, changes in mechanical characteristics (felting and fragility)	Stains, discoloration, loss of strength	Staining, loss of tensile strength, softening	Black crusts, mucilaginous patinas, exfoliations, color change, stains	Same as above	Corrosion
Actinomycetes	Same as above	Same as above	Same as above	Stains, white patches, loss of tensile strength	Grayish-white powder and patinas, grayish-white efflorescences	nd	nd
Fungi	Stains, alterations in color, cracking, changes in mechanical characteristics	Same as above	Stains, alterations in color, loss of strength	Stains, loss of tensile strength, rigidity	Staining, exfoliations, pitting	Opacification, black spots	nd
Cyanobacteria and Algae	Patinas of different colors (especially green)	nd	patinas	nd	Patinas and films of varying color and consistency	nd	nd
Lichens	Incrustations, patches	nd	nd	nd	Incrustations, patches, pitting	Pitting, opacification	nd
Mosses and Liverworts	Greenish/ gray thalli and greenish stains in the initial stages	nd	nd	nd	Greenish/ gray thalli and greenish stains in the initial stages	nd	nd
Higher Plants	nd	nd	nd	nd	Grasses, shrubs, and woody species induce fractures, collapsing of structures, detachment of materials	nd	nd

Key: nd = not described

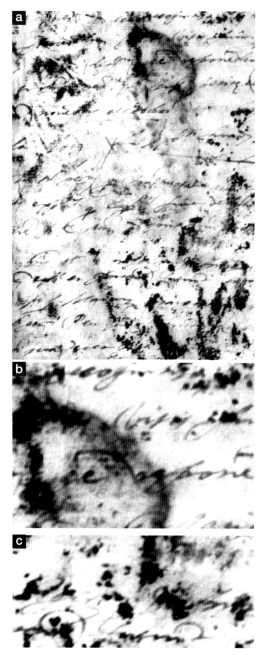

Fig. 1.15 *Powdery formations on paper support as a result of fungal attack (from Gallo 1992)*

operators in the field, in order to avoid that alterations with the same morphology be described with different terms. Italian NOR-MAL Recommendation 1/80 (1980)—later brought up to date and enriched with photographic material (NORMAL 1/88, 1990)—is the first glossary of alterations of natural and artificial stone, irrespective of their genesis. As far as wood is concerned, biological alterations are typical and have been described unambiguously (brown rot, white rot, soft rot, mildew) (Chapter 4, section 1.1b), although the terms have not been standardized yet.

The terms used at present to describe alterations on paper and on other materials are not unanimously accepted, and the literature contains scarce information. Gallo and Valenti (1999) proposed a form to be used for the collection of data on the state of paper conservation (books) and identified three different morphologies typical of damage caused by microorganisms: powdery or woolly accretions on bindings (Fig. 1.15); chromatic alterations of bindings and paper; fragility of the material. The descriptions of the main alterations of various materials are listed in Table 1.2.

CHAPTER 2
ECOLOGY OF BIODETERIORATION

by Giulia Caneva and Simona Ceschin

The biodeterioration of materials is closely correlated to the chemical and chemical-physical nature of the substrate as well as to the characteristics of the surrounding environment. In other words, a complex system of interrelationships among organisms, materials, and environment guides our understanding of the genesis, the dynamics, and the peculiarities of alteration phenomena in materials. As a result, it also directs our choices for the most effective strategies in preventing deterioration.

It is therefore critical for anyone working in conservation of cultural property to acquire some fundamental notions of ecology, especially in relation to defining the ecological requirements of various organisms and the influence that environmental parameters may have on them.

2.1 GENERAL PRINCIPLES

As ecology is the scientific study of the interrelationship between organisms and their environment, the ecology of biodeterioration deals with the relationships between organisms that can attack materials and the environmental factors that condition their development.

In the study of ecology the basic unit is the ecosystem, defined as biogeocoenosis, i.e., a community (koinosis) of organisms (bio) that interact among themselves and with the physical and edaphic (geo) environment. When it reaches maturity, such a system presents a well-defined trophic structure, a balanced biotic diversity, and cyclical processes of materials taking place within it.

Simply stated, an ecosystem is a "system defined through ecological parameters"—such a concept can be applied to many different contexts and, even though it is based on a concrete reality, it is the result of an abstraction.

To talk about a system as an ecosystem sets up the terms of the type of analysis, one that examines the functional relationships between the various components of an ecosystem and the processes that occur within it.

Figure 2.1 illustrates the ecological approach to the "system of a work of art" (the term "system" being used in the sense of a group of elements interacting with one another), showing the close interrelationships among the characterizing factors.

In every ecosystem there are cyclical processes of transformation of the material by chemical-physical and biological agents (termed biogeochemical cycles), which establish complex communities that are mutually interactive in a nutritional sense and that form trophic chains. Essentially, the nutritional position of an organism within a system can be categorized according to the following basic types:

—*Producers* are autotrophs that independently synthesize the required organic substances; various bacteria, algae, lichens, bryophytes, and vascular plants belong to this group. From a nutritional point of view, they do not need the substrate (in this case the work of art) except for those mineral elements that can provide nutrients. Biodeterioration from producers is generally not as serious as from other nutritional groups and any damage they inflict is indirect.
—*Destroyers*, which are heterotrophs, need organic substances to survive.

They include various types of bacteria and fungi (as well as insects and other animals) and are the most harmful colonizers for works of art because they are able to attack and metabolize specific organic components found in the materials (for example, protein, cellulose, etc.).

—*Consumers*, which are also heterotrophs, use the living matter of other organisms; they are not highly relevant for this sector, except for a few types that play an indirect role in controlling the growth of other populations that can damage materials (for example, micrograzing by gastropods, especially snails, that eat the lichens and the algae colonizing the stone). Other consumers, such as *Mycetophagus acari*, for example, may reduce the local extent of some biological

attacks, but may contribute to the spread of fungal colonization on the substrate.

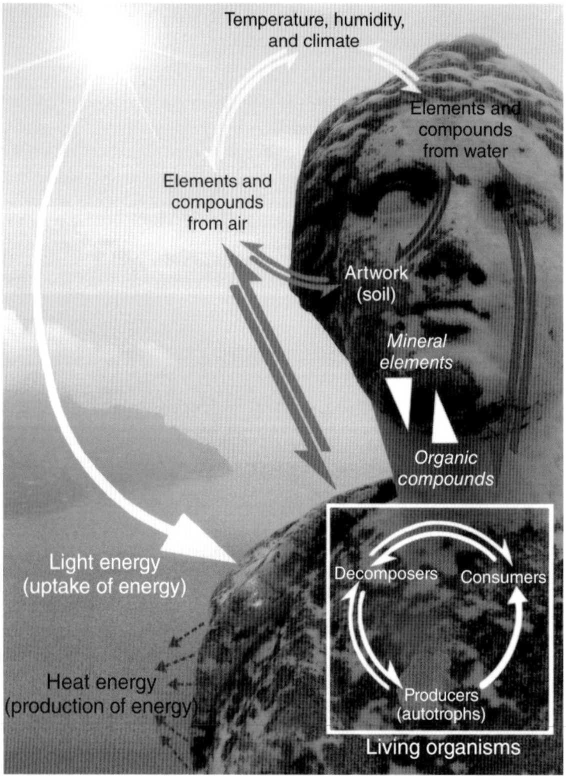

Populations present within a community can manifest two different types of growth: exponential growth and logistical growth. In the first, the initial increase in the number of individual organisms is slow and is followed by rapid explosion according to a logarithmic model—this type of growth is most typical in bacterial populations. In the second, on the other hand, an initial slow phase is followed by rapid growth, which continues until an equilibrium value is achieved (specific biological capacity, or K). After a series of oscillations around this value, the density of the population stabilizes.

Solar energy is the main driving factor for photoautotrophs, which use it in the processes of photosynthesis. This energy is unidirectional, and in every stage or step of the trophic chain there is a dispersion of energy that cannot be reused, thus determining an increment in entropy in accordance with the second law of thermodynamics.

Fig. 2.1 *Theoretical scheme of an artwork seen as an ecosystem (Villa Cimbrone, Ravello)* – Photo G. Caneva, elaboration by Caneva and A. Merante

2.2 ECOLOGICAL FACTORS AND THEIR RELATIONSHIPS WITH BIODETERIORATION

Ecological factors are defined as those physical, chemical, or biological factors in the environment (literally understood as "a space suitable for sustaining life processes") that can influence the life of an organism. As we shall see, the factors that can condition biological growth are numerous and therefore make the study of the ecological approach complex and multifaceted. Two laws, defined as the basic laws of ecology, clarify the general terms of these relationships.

Liebig's law (or law of the minimum, 1840) states that under conditions of stationary equilibrium, essential substances become limiting factors if their quantity is close to the minimum. In other words, the growth of an organism depends on the factor present in minimal quantity with respect to its needs and not by substances required in a greater quantity, as one might suppose. Additionally, even the use of the nondeficient elements is proportional to the scarcity of the element closest to the minimum. Water is often the main limiting factor for growth when there is a biological attack on materials—in quantitative terms, this is justified because water is the main component in biological structures. Wherever there is

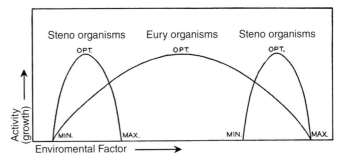

Fig. 2.2 *Possible types of ecological ranges of biological organisms: stenospecies with narrow ecological ranges (on the sides), and euryspecies with a wide range (in the center) (modified from Odum 1973)*

an abundance of water, such as in hypogea or in fountains, the concentration of certain salts or the quantity of light could be the limiting factors and would thus be the determining factors for biological growth (Plate 2). In theory, therefore, in order to prevent or control a biological attack on a material it would be enough (but not always easy) to identify the limiting factor that can be most easily modified and to reduce its values to a level below the limit necessary for growth.

Shelford's law (or the law of tolerance, 1913) states that organisms have not only an ecological minimum but also a maximum, which together define an interval representing the limits of tolerance. This range obviously contains the optimal value for each species (Fig. 2.2). This law is therefore an extension of the previous one, and it underscores that the same factor can be positive or negative depending on the values required by the organism. Thus, not only the minimum value but also the maximum value may be a critical factor. Consequently, any ecological factor can be a limiting factor based on the proximity of its value to the minimum or the maximum limits of tolerance, thus conditioning or inhibiting the presence of a biological species.

The threshold limit is the value above or below which the growth of an organism is inhibited. The knowledge of this value, which varies depending on the organisms, is clearly extremely useful in the prevention of biological deterioration (Chapter 7).

In ecology, different terms are used to indicate the ecological exigencies of a species; when we want to indicate their preference for

a specific ecological factor, we add the suffix –phile or -philic (literally, "friend", or "having a need of . . ."). For example, a species is termed hygrophile or is hygrophilic when it requires an elevated water content, while those species requiring dry conditions are termed xerophiles, i.e., they are xerophilic.

In order to show the relative amplitude of the range of tolerance with respect to a given factor, the prefixes steno- (narrow, small) or eury- (broad) are added to the ecological factor in question (for example: -hydric or -hygric for water or humidity, respectively; -thermic for temperature; -haline for salts, etc.). Species with a broad ecological valence, i.e., having a very broad range of tolerance with respect to numerous parameters, are called eurytopic, while those having narrow intervals are called stenotopic.

As far as application is concerned, it is important to remember that organisms with narrow ecological ranges for a given factor can be used as bioindicators (i.e., with their presence, and sometimes absence, they can be used to identify values for a given environmental parameter). In addition to using the Lichen Biodiversity Index (LBI) as an indicator of the level of air pollution (Nimis 1990), the ecological information connected to the presence of a given species can be used to understand some of the environmental parameters important to the conservation of cultural property, such as water content, salt concentration, and light and shade levels.

Various ecological indices that express preferentiality with respect to different environmental parameters have been developed for vascular plants (Landolt 1977; Ellenberg 1979) and for other groups of plant-like organisms. Specifically, various indices have been codified for lichens in relation to pH, nitrophytic, hygrophytic, and photophytic levels (Piervittori and Laccisaglia 1993; Wirth 1995). The latter were translated into numerical values and adapted to the characteristics of the Italian national territory (Nimis and Martellos 2001; Nimis 2003).

We should also point out that individual factors cannot be entirely separated from one another, given that their interaction produces different effects. Generally speaking, if the values of a given environmental parameter are not optimal, the limits of tolerance for the other factors are also narrowed—nonetheless, compensatory effects can occur (Fig. 2.3). Only a careful systematic-ecological analysis of the species present can provide useful information on the causes leading to the establishment of a given biological colonization and furnish precious indications on the best methods for controlling its presence (Caneva and Salvadori 1989).

When studying the distribution of a species it is important to keep in mind the following three distinct concepts: ecological niche, habitat, and chorology.

The ecological niche expresses an ideal ecological space, a "hypervolume" defined by all the factors necessary for the growth of an organism (for example, levels of temperature, humidity, and nutrients, as well as relationships with other biotic components of the system, etc.). This niche is synthetically—but perhaps somewhat hermetically—defined as "the functional role of an organism in the system within which it lives."

Habitat (literally, the "place where a species lives") is a type of environmental location where an organism finds the ideal conditions for its development (for example, lake or marine environments, savannas, tropical rain forests, etc.), without the definition including an exact geographical reality. Each habitat hosts species with different ecological niches that interact with each other in the system.

Chorology is the geographical distribution of a species and is the result of both its ecological requirements and of the historical events that characterize a geographical area over time (Chapter 6, section 1.1).

Generally, the most important environmental factors for the development of different organisms are water, temperature, light, and nutrients, because their values are, more often than for other factors, close to the minimum limit for the survival of the species. This helps to explain why climatic factors play a primary role in biodeterioration processes (Chapter 6, section 2).

Fig. 2.3 *Variations of the ecological space as a function of the relative development of two parameters influencing growth: examples of the interaction between these factors for three species of fungi (the black area indicates the optimal ecological space, while the broken lines indicate the possible extensions of the ranges) (modified from Deacon 2000)*

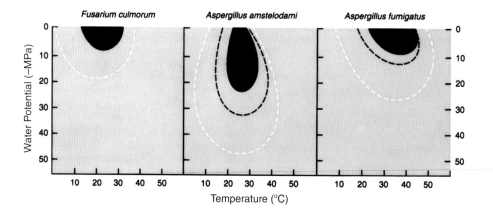

To facilitate this understanding, we shall first give an analytical description of the role of each single factor and then offer an integrated overview, introducing concepts on the spatial and temporal complexities.

2.2.1 Water

Water plays a fundamental role in life, both quantitatively and qualitatively. Water represents between 70 and 80% of the weight of a metabolically active organism, a level that is reduced to 8–12% in the physiologically quiescent forms, which some organisms take on strategically in order to have better resistance (for example, spores of bacteria or seeds).

The main reason for this high requirement for water is that it is indispensable to many of the metabolic processes in all living organisms (it is produced in many scission reactions and consumed in many biosynthesis reactions of complex polymeric substances) and also in order to maintain normal cellular functions, although its quantity varies more or less significantly according to the species. The role played by water is also important in the mechanisms of thermoregulation—often organisms protect themselves from excessively high temperatures by controlling the mechanisms of transpiration and evaporation.

At an ecological level, organisms are classified into aquatic, hydrophilic, mesophilic, and xerophilic (in the order of a decreasing requirement for water); a few organisms, termed poikilohydric (in particular, cyanobacteria and lichens), have a tolerance for very low levels of water and are able to suspend biochemical reactions in the wake of dehydration without incurring any damage, resuming metabolic functions as soon as conditions become favorable again.

The equilibrium moisture content (EMC, the amount of water in a given material at equilibrium) is calculated as a percentage of the weight loss of a material when it is dried to reach a constant weight. Generally, organic materials act as "buffering systems" in relation to variations in environmental relative humidity (RH), tending to absorb water with increases in the RH, or releasing water when

the RH decreases (see below in this chapter, hygroscopicity, section 2.4). For this reason, adsorbent materials (for example, ArtSorb, silica gel) are often used in museum display cases to mitigate the effects of changes in humidity on artworks or artifacts.

Water can be present in materials under three forms:

— water that is chemically bound with strong covalent bonds and that is part of the molecular structure of organic polymers; in this case, the water has no solvent properties and does not directly enter into chemical reactions;
— free multilayered water, bound with weak hydrophilic bonds or hydrogen to the surface of materials and present in capillary structures with a diameter less than 30μm;
— free condensed water that acts as a solvent and that can freeze and be rapidly exchanged with the exterior; this form clearly has the most influence on biological organisms (Florian 2002).

In biodeterioration, the quantity and especially the availability of water are considered the main factor determining the speed at which a surface is colonized. Because of this primary role, most treatments for the prevention of biodeterioration aim at reducing the water factor. In evaluating the actual amount of water available for an organism, the quantity of free water or water activity (a_w) must be considered. These values depend on characteristics within the substrate (specifically, porosity or hygroscopicity) and, in particular, on the sum of free multilayered water and free condensed water, but also on microclimate phenomena. Under normal conditions, only part of the total water content of a material can be exchanged with the surrounding environment—this is expressed as:

$$a_w = p/p_o$$

(p = partial pressure of the water vapor in the material; p_o = pressure of pure water vapor)

Water activity (a_w) is equal to 1 if the water is pure. By adding solutes, the vapor pressure of the aqueous solution decreases and, with it, the value of a_w. Essentially, this parameter indicates the quantity of water, in relation to the total content, that is available for use by microbic spores or seeds in order to germinate and grow. Every microbic species has the capacity to develop within a precise range of a_w values; below this range (generally 0.60–0.70) growth is no longer possible. Most organisms require a_w values between 0.75 and 0.99 (Florian 1988, 1997). Xerophilic fungi have the capacity to develop at a_w values between 0.60 and 0.80; bacteria generally require a_w values > 0.90, while halophilic bacteria can even grow in values > 0.75.

In confined spaces, the relative humidity in the air (see below in this chapter, section 2.6) and the surface temperatures are the main factors determining the amount of water available for microorganisms. Indoor RH values are influenced by exchanges with the outdoor environment and therefore follow any climate changes, but they are also conditioned by elements specific to the building, such as water content in the walls (due to phenomena of capillary dampness rising, percolation and/or condensation, broken pipes, etc.). Outdoors, the presence of water can come from atmospheric precipitation, condensation phenomena, and capillary dampness rising from the ground,

creating differentiated situations for biological growth (Fig. 2.4).

In the absence of water there is little chance for development of microflora, but once development has already taken place, many organisms have the capacity to withstand long periods of drought. This phenomenon is especially evident with black or greenish patinas from cyanobacteria and algae that develop in locations with water percolation or driving rain (Plate 1). As often occurs in relatively dry urban environments, these patinas can be dark in color, and they are often confused with "black crusts" from pollutants. Even without specific identifying analysis, biological patinas can be recognized by their distribution, which clearly follows increasing water gradients, while crusts and deposits resulting from pollution converge in areas that are never in contact with water (Fig. 2.5).

Parameters that define the nature of water from a qualitative standpoint include the chemical nature of aqueous solutions and, specifically, their acidity or alkalinity and the concentration of any solutes they may contain.

The pH value of the substrate is a fundamental parameter for the growth of biological species. This is due to the fact that most enzymatic reactions have an optimal pH (for example, cellulase complex has an optimum value between 6.4 and 8.2); therefore, each organism demonstrates a preference for specific acidic or basic values in accordance with its own specific enzymatic makeup.

Most organisms tend to live in chemically neutral conditions (neutrophilic species); nonetheless, some species thrive in various types of situations, as in the case of acidophilic species, or conversely, the alkalophiles. Most fungi, for example, develop on substrates with pH values between 4 and 7 (Florian 1997). Extreme pH values are damaging to the growth of organisms because of the chemical reactivity of the H^+ or OH^- ions in solution (consider the effects of acid rain on forest eco-

Fig. 2.4 *Differential growth of cyanobacteria on a marble capital, related to the effects of water percolation (Mausoleum of Boemondo, Canosa, Bari)*

▼ Pollutants

Overhang

Black
crusts

Percolation
lines

▼ Biological
Colonization

Patinas of
cyanobacteria
and algae

Ruderal plants

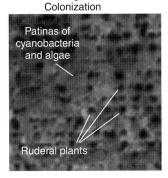

Fig. 2.5 *Section of the embankment walls of Lungotevere (Rome). Depending on the amount of water involved, it is possible to see: dark areas corresponding to biological colonization (wetter areas, on the right), black crusts resulting from pollution (dry areas because of the overhang, on the left), and light areas with low water content (middle areas) where there is no deposition of pollutants and no biological growth* – Photo G. Caneva

there is a flow of water from the solution with the lower concentration to the one with the higher concentration. If the solution surrounding the cell is the less concentrated of the two (hypotonic solution, irrespective of the nature of the particles), plant cells—as well as bacterial and fungal ones—will swell, but because of the presence of cell walls they will not break—unlike animal cells, which do not have walls. If the medium has a higher concentration (hypertonic solution), however, the water leaves the cells causing plasmolysis phenomena, which are obviously lethal. Microbic growth is inhibited when spores in the air cannot germinate on extremely hypertonic surfaces relative to their cytoplasmic values, and for this reason, salt has been used throughout history to preserve food and to prevent the putrefaction of organic materials.

Borderline situations occur with many bacteria, among these some *Streptomyces* isolated from hypogean environments (Krumbein 1988). It has indeed been observed that extreme halophiles can grow even in high salt concentrations (up to 30–40%). These organisms seem to find a particularly favorable habitat on the surface of frescoes in humid environments where there is an outcropping of many types of salts, such as carbonates, chlorides, nitrates, sulphates, etc. (Piñar et al. 2001).

systems and on plant and animal communities in general). The growth of microorganisms can modify the pH of a substrate through secretion of metabolites, thus favoring the establishment of acidophilic or alkalophilic species. In some cases, a substrate can have a buffering effect, as in the case of carbonatic materials with respect to acid pollutants.

On a chemical level, solution concentration plays an important role in osmotic pressure, a phenomenon occurring when a semipermeable membrane separates two solutions having different concentrations, as in the case of biological membranes. Indeed, when a cell is immersed in a liquid with different concentration values,

2.2.2 Light

Light is composed of the radiations of the electromagnetic spectrum with wavelengths of 380–750 nm, which the human eye perceives in the chromatic range between violet and red. Therefore, light represents only a small part of the broad range of radiations emitted by the sun, and the difference between light and dark—in terms of physics—is nothing but the difference of a few nanometers of wavelength or, in other words, of energy.

Most solar radiations reaching the Earth are made up of light radiation, because most of the inferior wavelengths—high in frequency and thus charged with greater energy—are absorbed by the upper strata of the atmosphere. High frequency radiations (for example, UV, gamma, and x-rays) can ionize or induce scission in chemical bonds and cannot be used in biological processes because of their inhibitory and damaging action (Chapter 8, section 3). Low frequency radiations (wavelengths greater than the visible spectrum) are energetically weaker (for example, infrared, radar, radio) and produce an increase in temperature—important from the biological point of view—but do not trigger changes in molecular configuration.

The utilization of light occurs thanks to the photosynthetic processes, which allow the transformation of light energy into chemical energy, producing carbohydrates (glucose) and oxygen from inorganic components such as carbon dioxide and water. The body of reactions that takes place during this process requires the presence of specific photosynthetic pigments, i.e., of molecules able to absorb light within a certain band of visible light that is useful to such a process. These are the chlorophylls, carotenoids, xanthophylls, phycoerythrins, phycocyanins, etc., which take on the color of the reflected wavelengths (for instance, green in the case of the chlorophylls, which in this band have an absorption deficiency). Accessory pigments, and in particular phycoerythrin and phycocyanin, present in cyanobacteria and red algae, play in part a protective role from photooxidation reactions, but mostly serve to widen the spectrum of absorption for chlorophyll, capturing light in complementary bands (Fig. 2.6).

From an ecological point of view, light represents the main source of growth energy in photosynthesizing organisms such as algae, lichens, mosses, and vascular plants—these are the main producers in an ecosystem and they require relatively high levels of light to carry out their metabolic activities. For chemosynthesizing organisms such as sulphobacteria or nitrobacteria, and for heterotrophic organisms such as fungi and actinomycetes, light is obviously not a limiting factor nutritionally, but in some cases it can influence metabolic phenomena such as development, spore germination, reproduction, pigment production, etc.

Organisms are divided into heliophiles—in need of high levels of solar illumination—and sciaphiles, preferring low light levels; the extremes in this spectrum have a different range of preferred values.

In the context of light, the most important parameters to evaluate are:

—quality (color);
—quantity (intensity);
—duration (over time).

Fig. 2.6 *Difference in the absorption of light by various photosynthesizing organisms, as a function of their array of pigments (modified from Stainer et al. 1970)*

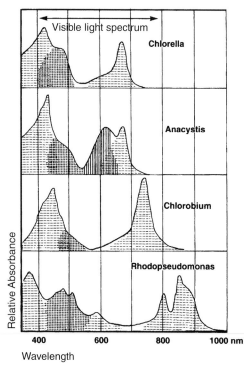

The quality of light is expressed by frequency (ν), or, conversely, by the wavelength (λ) of the radiations (λ = 1/ν); it is correlated to the energetic level of its photons and is visible as different chromatic types or colors. Biologically speaking, organisms can have preferential absorption in some visible bands, depending on the pigments available. Even though green light is the worst type of light for photosynthesis there are organisms adapted to use its energy—organisms are selected based on their ability to best make use of the energy of the various wavelengths of light radiations from one source or another (Fig. 2.6). There are substantial differences between sunlight and artificial illumination: the former is much richer in blue radiations when compared to incandescent light bulbs, which emit large amounts of radiations in the red and infrared lengths due to lower temperatures of the emission source. Fluorescent tubes are rich in blue radiations and poor in red, while low-pressure sodium vapor lamps emit a monochrome yellow light.

Light intensity is a quantitative parameter that expresses the density of incident photon flux over a given unit of time. In photometry, the unit of measurement for luminous flux, that is, the amount of luminous energy emitted from a source, is the lumen (lm). The density of incident photons on a given surface (illuminance) is expressed in lux (lm/m²). When light is related to the activity of photosynthesis, this unit of measure is usually substituted by the Einstein, a measurement of the photosynthetic active radiation (PAR) represented by Avogadro's number of photons (6.023×10^{23}). More generally, in radiometry, the radiant flux—or the quantity of radiant energy emitted from the source in a given unit of time—is expressed in Watts (Joules^{-1}).

Illumination duration influences the presence and the activity of organisms related both to photosynthesis and to photoperiodic phenomena, i.e., phenomena related to the alternating of light and dark.

Adequate intensity and duration of luminous flux is fundamental for efficient photosynthetic processes (Fig. 2.7).

In indoor environments, the intensity as well as the duration of illumination greatly affects the acceleration or inhibition of photosynthesizing organisms, limited in this context to the more sciaphilic and pioneer organisms (specifically, algae and cyanobacteria); this is also related to the increase in temperature caused by many light systems, which can be quite relevant (Camuffo 1998). Indeed, the presence of photoautotrophic microflora is often identified in those areas receiving sufficient luminous energy, while areas receiving weaker light do not show any growth phenomena (Fig. 2.8). The design of lighting systems must therefore take these concerns into account

Fig. 2.7 *Increase in the biomass of a vascular plant* (Abutilon) *as a result of the quantity of light and nutrients (modified from Bazzaz 1996)*

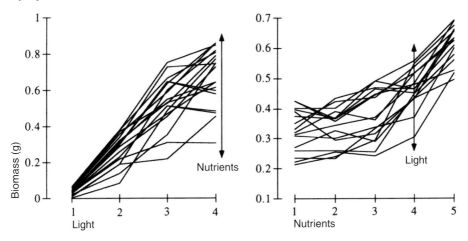

Fig. 2.8 *Colonization by algae (darker areas), on the ceiling of a hypogeum, limited to those areas that receive sufficient light (Tombe Latine, Rome)* – Photo ICR

in order to provide adequate preventive conservation (Chapter 7, section 1.1, and Plate 2).

Some microorganisms (for example, cyanobacteria or fungi) protect themselves from excessive light intensity or from the presence of ionizing radiations thanks to the presence of pigments, such as melanin, that provide dark pigmentation capable of blocking in-depth radiation penetration. Other organisms avoid direct exposure by growing beneath the surface, in an endolithic habitus, which can also be a way of adapting to very cold environments or of protecting against dehydration (Chapter 6) (Golubic et al. 1981).

2.2.3 Temperature

From a thermodynamic point of view, temperature represents a change in state describing the property of a material as allowing the spontaneous passage of energy (in the form of heat) from a body with a higher temperature to one with a lower temperature. Heat is not a material entity in itself, but represents one of the ways in which energy is transferred (in the same way as work), and temperature is a function of the average kinetic energy of the constituent molecules.

Ecologically speaking, temperature—along with atmospheric precipitation—is the primary descriptor of climatic phenomena. It has an impact on biological development not only in terms of hygrometric values but also because it significantly conditions metabolic activity.

The constraints that temperature exercises on vital phenomena are primarily related to its effects on the chemical-physical properties of water, which, as we said, is the main component of biological structures. Below 0°C, the crystallization of water into ice results in an increase in volume and, as a consequence, it ruptures the biological membranes causing the death of the cell. Life can continue at these temperatures only in the quiescent state of dehydrated cells. As the temperature increases, the loss of water from evaporation also increases, unless specific survival mechanisms and thickening of the cell walls are set in motion, but always in a quiescent state (for example, bacterial spores and seeds). Thus, even though the theoretical interval is relatively broad (but still small in absolute terms), in concrete terms, life exists in a narrower interval, for reasons tied to the metabolic reactions that occur in every organism.

Temperature influences the kinetics of chemical reactions. Every increase of 10°C doubles the rate of the chemical reaction. In biological organisms, however, there is an enzymatic kinetic that follows different rules. Each enzyme has an optimal value of activity at specific temperatures, and the preferentiality of a given organism is in relation to all those parameters (Fig. 2.9). The three-dimensional structure of an enzyme—and consequently the morphology of the active site (Fig. 1.11b)—is a result of the changes in temperature that favor or obstruct the formation of bonds (especially weak bonds such as H bonds, van der Waals forces). If the active site were to lose its specific configuration because of the breakage of these bonds, the catalyzed reactions would cease. The fact that the thermal optima are not at a balanced center, as in Figure 2.2, but closer to maximum values of tolerance, as in Figure 2.9, is due to the positive influence of temperature on the reaction kinetics. Increases in temperature greater than 50–60°C are in any case dangerous for organisms, both because of

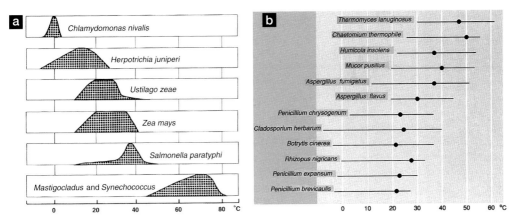

Fig. 2.9 *(a) Temperature tolerance range for various organisms (from Larcher 1975 in Bullini et al. 1998);
(b) fungal species (from Deacon 2000)*

the effects on enzymatic kinetics and because they result in an excessive fluidification of the biological membranes, thus modifying their permeability.

Temperature also influences the relative environmental humidity—the two parameters have an inverse relationship: at the same absolute humidity, the RH decreases with an increase in temperature and increases when the temperature falls (see below in this chapter, section 2.6).

Most organisms grow at an optimal temperature interval between 20°C and 30°C. Those organisms adapted to low temperatures, with a metabolism active between 0°C and 10°C, are called psychrophiles, while organisms adapted to high temperatures (between 35°C and 60°C) are called thermophiles. The survival of these organisms in extreme situations is due to specific adaptation that foresees a rapid replacement of damaged proteins. Nonetheless, microorganisms can tolerate a wide range of temperatures and are able to survive sudden and extreme variations, partially thanks to their ability to develop resistant structures.

Latitudinal variations in temperature determine the distribution of organisms and, therefore, also the biomes. Local variations in temperature based on altitude changes determine the altitudinal zoning of living organisms in relation to induced climatic effects. Similarly, temperature is a relevant factor in biodeterioration, even though outdoor values are the most

critical. The variations found in confined spaces, which are modified or planned to also comply with the requirements of human comfort, are generally of reduced proportions and relatively independent of the climatic context, with a few exceptions found mostly in non air-conditioned tropical environments. Normally, the temperatures inside buildings are more constant compared to values outside, and the day-night fluctuations are minimized; the average values vary—depending on the geographical locality, on the architectural typology of the building, and on the degree of insulation from external conditions—but usually do not constitute limiting factors. More critical conditions can develop in cooler and more humid environments, such as caves and hypogea, in which even small increases in temperature accelerate the colonization and growth of microflora as a result of getting closer to optimal thermal conditions for the greater majority of species—the effects of the decrease of humidity are not relevant, in any case, since the values are always very high in these environments (Chapter 5, section 1.4).

2.2.4 Characteristics of the Substrate

We already mentioned the bioreceptivity of materials in the previous chapter (Fig. 1.1). The chemical and physical properties of the edaphic substrate ("soil" in ecological terms) play a very important role in biodeterioration.

One of the physical parameters in biore-ceptivity is porosity, which can be defined as the proportion between the volume of empty spaces (adjoining or isolated) of a given amount of material and the total volume of that same amount of material. Porosity is therefore cor-related to the presence of intermolecular or intercrystalline spaces, but the most relevant type of porosity is the one with adjoining pores, because of the water passage or reten-tion. Thus, a substrate with high porosity is more susceptible to biodeterioration because the germination of spores and the development of microorganisms, or organisms, that may be present on the substrate can take place only in sufficiently hydrated conditions.

The rugosity, or roughness, factor of any surface is proportional to its deviation from lin-earity. As with porosity, the rugosity of a sub-strate can facilitate biodeterioration thanks to the indirect effect of surfaces absorbing water. An irregular and rough surface also facilitates the physical establishment of microorganisms.

Normally, organic materials are highly porous with marked surface roughness. Thus, from a physical standpoint, they are more sus-ceptible than inorganic materials to attacks from microflora. It is easy to see how, under the same exposure conditions, very porous stone materials, such as mortars, tufa, or sand-stones (with a porosity of about 30% or more), are colonized by microorganisms, while mate-rials with a porosity of less than 10%, such as marble, basalts, and compact rocks, have greater resistance to biological colonization (Plate 4) (Tomaselli et al. 2000b; Caneva et al. 2004). Additionally, the deterioration of mate-rials outdoors resulting from exposure to envi-ronmental agents (sunlight, freeze-thaw cycles, action from wind and water) causes an increase of these parameters to a variable degree. Even compact materials can undergo biological attack if there is an increase in the superficial microporosity and rugosity.

The hygroscopicity of a material is a mea-sure of its capacity to absorb water when the ambient RH (relative humidity) increases or to release water when the RH decreases. Every thermohydrometric value of the air, therefore, corresponds to a specific percentage of water content in materials such as paper and pro-teinaceous materials, wood, etc. The higher the RH, the more a hygroscopic material absorbs water; yet, at the same level of RH, different materials may have very different water con-tents at equilibrium (Fig. 2.10). For instance, only when the level of RH is higher than 65% does the water content of cotton or good qual-ity papers reach 10%, thus allowing micro-organisms to carry out their metabolic activity and hence degrade the support. On the other hand, at those same RH values, other materials such as collagen already exceed the water con-tent sufficient to trigger biodeterioration, and if they are not treated promptly with tannin, they are rapidly attacked.

Fig. 2.10 *Water content at equilibrium of various materials, at 20°C and 60% RH (light columns) and at 100% RH (dark columns) (modified from Florian 2002)*

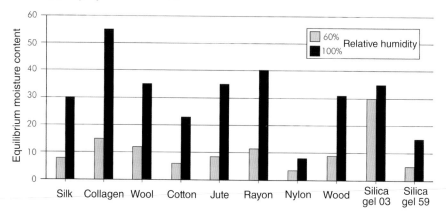

Hysteresis describes the phenomenon by which the values assumed by a physical quantity at the end of a process differ from the values of that same quantity before the process began (Fig. 2.11). Consequently, once some types of hygroscopic materials, such as paper and wood, take on large quantities of water they never return to exactly the same water content they had originally—the values always remain higher even if the same initial hygrometric conditions are reproduced. Thus, once equilibrium conditions are reached, a material that has already been exposed and "aged" can hold more water than "new" materials, even at the same RH— this is the result of the material's "microenvironmental history" and of any treatments (for example, humidification, freezing, etc.) it was subjected to in the past. This is an example of the often irreversible processes that occur in nature.

From the point of view of biodeterioration, the chemical characteristics of the substrate are a discriminating factor because the various chemical components can provide possible nutrients. The analysis of the dried components of the biomasses allows the identification of the types of qualitative and quantitative nutrients required. The primary functions of nutrients are structural; the nutrients required in the greatest quantities (concentrations of 1,000 mg/kg of dry matter, also termed macronutrients) are: carbon (C), oxygen (O), hydrogen (H), nitrogen (N), sulphur (S), phosphorus (P), in addition to magnesium (Mg), calcium (Ca), and potassium (K), which play a fundamental role in other essential functions (components of pigment molecules, enzymatic cofactors, etc.). Other elements are generally required in smaller quantities and are therefore termed micronutrients: chlorine (Cl), iron (Fe), boron (Bo), manganese (Mn), zinc (Zn), copper (Cu), nickel (Ni), and molybdenum (Mo).

Nutritional requirements differ from organism to organism depending on whether they are photoautotrophs, chemoautotrophs, photoheterotrophs (often overlooked because of their rarity), or chemoheterotrophs. Photoautotrophs require only inorganic nutrients (mineral salts), and sometimes, as in the case of cyanobacteria and some lichens, there is no need for

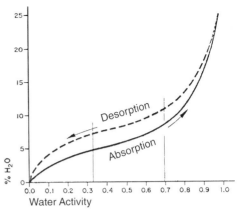

Fig. 2.11 *Example of hysteresis demonstrating how the absorption and desorption curves for water in a hygroscopic material do not match (modified from Florian 1997)*

nitrogen compounds to be present, because the organisms have the capacity to fix atmospheric nitrogen. Chemoautotrophs require some reduced inorganic compounds, i.e., sulphur, ammonium, and nitrites, which they employ in the oxidation-reduction reactions from which they obtain energy. Heterotrophic microorganisms, on the other hand, require organic compounds of various nature, such as carbohydrates and proteins and, depending on their enzymatic makeup, they can be specialized in the utilization of certain compounds, such as cellulose, lignin, keratin, collagen, etc. Normally, complex molecules or hydrogenated organic substances are attacked with greater difficulty (Fig. 2.12). This is the reason why the biological attack of inorganic materials (stone, metal, glass) is primarily, although not exclusively, carried out by autotrophs, while organic materials are mostly attacked by heterotrophic organisms, which use them as a source of nutrients. Moreover, the microbial infestation of these materials is conditioned by the intrinsic characteristics of the organism, first and foremost its oligotrophic capacity, which allows the initial colonization of the substrate without the use of the nutrients it contains, as well as its capacity of adhesion (Poindexter 1981; Urzì et al. 1991).

The accumulation of nutrients, termed eutrophication, favors the development of some species that are adapted to these conditions, in particular to higher levels of nitrates and phosphates; consequently, they are present wherever such an accumulation of substances occurs. This is the case with nitrophilic lichens such as, for instance, *Caloplaca citrina, Xanthoria parietina, X. calcicola*, with an orange thallus, which are mostly found on roofs, architectural moldings, and the horizontal parts of statues where there is an accumulation of nitrogenous substances carried by rain, or of bird droppings (ornithocoprophilic lichens) (Plate 3), or of fertilizer derivatives transported by the wind, as found on monuments in rural areas. On monuments on which the biological colonization has evolved further, we then observe the development of nitrophilic vascular plants such as *Sinapis alba, Chenopodium album, Ballota nigra, Solanum nigrum, Chelidonium majus, Hyosciamus albus*, or also of urophilic plants such as *Urtica* sp. pl., which can be considered as bioindicators for this particular environment. On the other hand, an accumulation of phosphates is sometimes found in fountains and water courses, where it partly develops as a strategy for the reduction of carbonate deposits (Chapter 5, section 2.4).

2.2.5 Chemical Characteristics of the Atmosphere

From a chemical standpoint, the atmosphere is composed of some inert gases (elementary nitrogen $N_2 = 78.08\%$, argon $Ar = 0.93\%$, and minute quantities of helium), oxygen ($O_2 = 20.95\%$), and carbon dioxide ($CO_2 = 0.03\%$).

Nitrogen and the other inert gases, being very stable and nonreactive chemically, are of little use from a biological point of view, their utilization being limited to the rare and energy-consuming processes of nitrogen fixation. Oxygen, however, because of its physical-chemical properties, has a strong affinity for electrons and is therefore a strong oxidizer (i.e., capable of extracting electrons) and therefore plays an important role in biological growth. Aerobic organisms are those that require O_2 in concentrations that are normal for respiration reactions; microaerophilic organisms favor concentrations between 5 and 20%; anaerobic organisms are inhibited by the presence of oxygen (in concentrations greater than > 5%) and therefore can only be found at variously significant depths inside materials or in the soil. Facultative aerobic and anaerobic organisms also exist: these can grow both in the presence and the absence of oxygen.

Pollutants are those substances that are not naturally part of the atmosphere (or indeed of water or soil), or that are present in concentrations greater than the norm. Atmospheric pollutants are normally divided into atmospheric particles (hydrocarbons, silicates, spores, pollens) and gaseous compounds (SO_2, H_2S, NO, NO_2, CO, CO_2, HF, HCl, O_3). Among the compounds that are naturally found in the composition of the atmosphere a brief mention should be made of carbon dioxide (CO_2), a gas that is commonly emitted in large quantities by industrial

Fig. 2.12 *Relationship between the complexity of organic molecules and their degradability by fungal microflora (modified from Deacon 2000)*

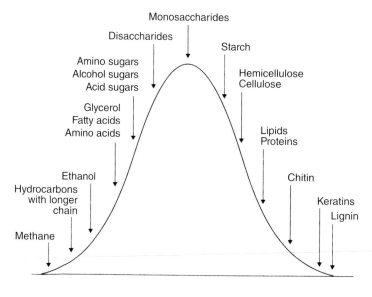

processes of oxidation as well as by normal combustion reactions; its present rise is considered to be responsible for the greenhouse effect, which has such grave consequences from a climatic point of view. Its introduction into an aqueous environment makes the medium acidic through the production of carbonic acid, which clearly produces a lythic effect.

While it is well known that pollutants associated with vehicular traffic, industrial emissions, and heating systems (in particular NO_x and SO_2 and the by-products of hydrocarbons) have a highly aggressive effect on the material constituents of cultural heritage, in the case of biological organisms these effects vary according to the nature of the pollutants and the type of organism involved. As far as the toxic effects of chemical pollutants are concerned, not all organisms are equally sensitive, and the species that are mostly harmed are the ones that do not have mechanisms of passive protection or excretion, such as briophytes and lichens in particular. For this reason, both these organisms are often used as bioindicators of pollution: their spread and density fall rapidly in relation to such a parameter, and a number of indices have been proposed for the purpose of correlating the reduction in density of sensitive species with the pollution factor (Nimis et al. 2002).

Acid pollutants in the atmosphere, such as NO_x, SO_2, CO_2, etc., severely reduce the pH of rainfall, which in turn can acidify the substrates. This phenomenon inhibits the growth of neutrophilic and alkalophilic populations, while both gaseous and particulate polluting substances, even in low concentrations, can be used as nutrients by many microbial groups (Krumbein 1988; Ortega-Calvo et al. 1995). In urban polluted areas (see earlier in this chapter, section 2.1), for example, the deposit on the surfaces of buildings and monuments of both saturated and unsaturated hydrocarbons, derived from the partial combustion of oil products, conditions the patterns of colonization favoring the growth of heterotrophic microorganisms that are able to utilize these substances as a source of carbon. In high-traffic areas, therefore, we see a prevalent growth of heterotrophic fungi and bacteria, rather than of autotrophic organisms (Saiz-Jimenez 1995a; Ortega-Calvo and Saiz-Jimenez 1997; Laiz et al. 2002).

2.2.6 Climatic Factors

Climate is defined as the resultant over time of the combined action of meteorological factors, primarily, temperature and precipitation. A more in-depth description of climate types includes other parameters, specifically, insolation and cloud cover, because of their effects on the energetic input that comes from light as well as wind intensity and direction.

Climatic phenomena vary over the course of time giving rise to the seasons; therefore, a description of climate must begin from its analysis over the full annual cycle. Conventionally, this period is measured in twelve months, and in this interval the most representative climatic parameters are precisely measured. These measurements include any daily fluctuations, at least in the characterizing seasons.

We use the terms macro-, meso-, and microclimate according to the scale at which we analyze climate. In this section we concentrate on climatic phenomena of a smaller scale—the influence of climate on a regional climatic and biogeographical scale is discussed in Chapter 6.

Wind cannot be overlooked as a climatic parameter. Wind is the movement of air masses generated by differences in atmospheric pressure and temperature. These are correlated to the term windiness, or the state of being hit by wind, and to the term ventilation, which refers to the movement of air due to the effect of the winds. The main effects of ventilation are a drop in surface temperature; a dehydration effect (in the case of dry winds); the transportation of spores, pollution, and other biological particulate (Chapter 7, section 3); and where they are deposited.

In outdoor environments, ventilation plays an important role in biodeterioration, especially in terms of rainfall angle—if rainfall occurs in windy conditions, the resulting angle can cause rainwater to strike even vertical surfaces, in what is called driving rain. Specific analyses were carried out on older buildings in the city of Rome to evaluate which environmental factors of exposure (light, temperature, exposure to dominant winds or to rain winds) were the most relevant in terms of biological colonization. The study showed that driving rain

Fig. 2.13 *Difference in the colonization of old plaster in relation to driving rain from the south-southeast on two buildings in the outskirts of Rome* – Photo G. Caneva

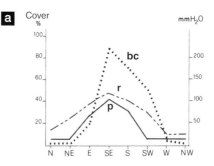

Fig. 2.14 *(a) Relationship between driving rain* (r) *in Rome, biological colonization of the wall surfaces* (bc), *and spread of pitting phenomena on the surfaces of the Trajan Column* (p); *(b) pitting phenomena on the surfaces of the Trajan Column as documented on a historical photograph from the Germanic Archaeological Institute (from Caneva et al. 1994b)*

was the single most determinant factor for biological colonization (Caneva et al. 1992a) (Fig. 2.13). This correlation was very useful in understanding the formation and development of pitting on the southern face of Trajan's Column in Rome (Fig. 2.14) (Caneva et al. 1994b).

Similarly, the type of microclimate resulting from exposure can cause a differentiation in biological colonization, as is seen on the city walls of the Tuscan city of Lucca, which have remained intact along the whole perimeter of the ancient city (Fig. 2.15).

In enclosed environments, ventilation takes on a double valence. Air currents can transport biological particles from the outside to the inside (spores, cells, thallus fragments, etc.), favoring their impact on the objects and consequently increasing the level of biological contamination on the surfaces. On the other hand, air currents can also bring about the dehydration of the substrate and of the cells during the initial phase of colonization, thus blocking growth, as happens, for example, to

fungi during the hydration phase and the germination of the spores. Artificial illumination can also cause air circulation, creating warmer air cells that move upwards, carrying heat, dust, etc.

It is unlikely that, under normal circumstances, the intensity of air movement alone is a limiting factor. If air movement influences growth, most likely it does so indirectly, by inducing changes in other environmental factors such as temperature and water availability (Scott 1994).

In grottoes and semien-closed environments, even under sufficient light conditions, there is a reduction in lichen colonization (not algae) in areas that are essentially defined by reduced ventilation. This situation could be tied to a reduction of the exchange of atmospheric gases, of the diffusion of biological aerosol, and of the dispersion of forms of vegetative or sexual propagation of plant-like organisms.

In ecology, microclimate is defined as the climate of a given biotope, that is, an area where physical-chemical phenomena are relatively uniform.

Fig. 2.15 *Difference in the colonization (white, to the left, with a predominance of lichens, dark gray with a predominance of algae and cyanobacteria) on the bricks of a city wall, related to the microclimate gradients resulting from the conditions of exposure (city walls of Lucca)* – Photo G. Caneva

More specifically, microclimate can be defined as "the synthesis of the physical conditions of the environment resulting both from physical variables in the atmosphere (for example, temperature, humidity, insolation, air velocity) and from exchange with other bodies (infrared radiation, heating, lighting) within a time period that is representative of the natural or artificial conditions typical of a given place" (Camuffo 1998).

Due to their importance in the prevention of deterioration, the fundamental parameters in describing microclimate are discussed and described in more detail in Chapter 7. In synthesis, these parameters are temperature (T), specific humidity (SH), absolute humidity (AH), and relative humidity (RH).

In addition to temperature, the most important microclimatic parameter for biological development is relative humidity (RH), which is the level of saturation of water vapor in the air and is expressed as a % ratio between the quantity of water vapor (q) present in the air and the maximum quantity of water vapor (Q) that could be contained in the air if it were saturated under the same conditions of temperature and pressure:

$$RH = (q / Q) \times 100$$

The quantity of water vapor that can be contained in a given volume of air varies in relation to temperature because temperature is correlated to molecular movement, i.e., to the kinetic energy of the particles—the higher the temperature, the faster water vapor moves in the air and, thus, the more particles can be contained in it before condensation (passage from a vapor state to a liquid state) occurs (Fig. 2.16 and psychrometric diagram in Fig. 7.10).

Other factors influencing the microclimatic conditions in indoor environments (for example, museums or churches) are: the presence of air-conditioning or heating systems and their operating cycles; the movement of visitors; the opening of doors and windows; light sources (especially incandescent lights); and exposure conditions. All these factors affect the values of microclimatic parameters, which vary according to the location and time of the investigation (Fig. 2.17 and Chapter 7, section 1).

The presence of visitors, especially, brings with it large amounts of water vapor emitted from respiratory phenomena. The estimate of the emission is around 50 mg/h and determines an increase in temperature that is more or less perceptible depending on the relation to the initial temperature and to the volumes of

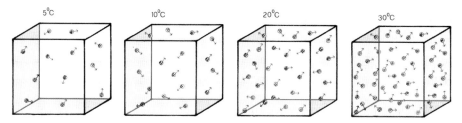

Fig. 2.16 *Different quantities of water vapor present in the air, at a constant RH, as a function of different thermal values*

the space in question. In addition, the presence of people carries with it a significant quantity of airborne microflora, and when visitor concentration is particularly high (as, for example, in the Sistine Chapel) water contents can reach levels between 150 kg/h to 250–300 kg/h.

For these reasons, limiting the number of visitors (not only in terms of daily absolutes, but also as maximum acceptable numbers per hour) in environments where the risk of altering the microclimate is high (for example, tombs, hypogea, grottoes, or museums with high visitor traffic) has become a standard preventive practice (Fig. 2.17).

The macroclimate is the average climate on a regional scale (macro, in the sense of referring to a large territorial extension). It is representative of median climatic values in a given territory, values that, at an ecosystem level, however, exhibit local diversifications (mesoclimates), including the more specific

characterization of different microhabitats (microclimates). Progressing up the scale, there is the mesoclimate, which is the result of modifications in the macroclimate in relation to topographical variations (different exposures due to hills or valleys), or of other significant factors (the presence of lakes, cities, forests, etc.).

The bioclimate is the result of the analysis of climate factors in relation to their capacity to influence biological phenomena. These factors must be analyzed in context because their effect can be accentuated or diminished as a function of other factors. In particular, the relationship between rainfall and temperature is of fundamental importance; suffice to think of how organisms are subjected to conditions of higher aqueous stress when low levels of rainfall occur at the same time as high temperatures, which is typical of Mediterranean areas during the summer months.

Fig. 2.17 *Isolines of temperature, RH, and SH in the Museum of Fine Arts in Antwerp (from Bernardi, in Castellano et al. 2002)*

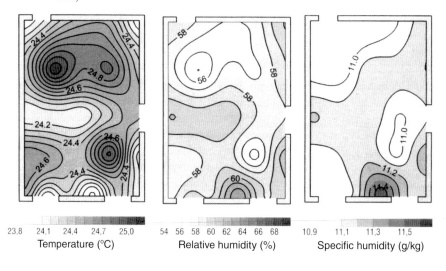

2.3 PLANT AND MICROBIAL COMMUNITIES: FUNCTIONAL AND DYNAMIC ASPECTS

2.3.1 The Concept of Community and Phytosociology

As we already mentioned, the ensemble of species that live together in a common environment or habitat form a biotic community, or biocoenosis, in which the living organisms that are part of it interact among each other in various ways. The cohabitation of the species is not limited to random spatial proximity; rather, it is determined by a complex series of relationships between the species, among which are trophic relationships, which form the basis of the structural and functional stability of the community itself. Such interactions can be of a negative nature—such as competition, parasitism, predation, and antibiosis, in which only some of the species benefit from the interaction—or of a positive nature, such as commensalism, cooperation, mutualism, and symbiosis, in which the advantages are reciprocal (Table 2.1).

Even within particular ecosystems such as the single monument or work of art, negative interactions can be established between the various populations, especially competition (for nutritional or spatial motives) and antibiosis, i.e., the release of substances that inhibit the growth of other species—as in the case of actinomycetes and fungi (from which derive antibiotic substances of well-known medical use, such as streptomycin and penicillin)—thus one species, especially in the early stages, prevails over other populations. In theory, species that have the same ecological requirements cannot live together; after a certain period of time the direct or indirect competition that necessarily occurs tends to eliminate one or the other of the species, unless—for the sake of survival—each one succeeds in developing its own specialized ecological niche. In reality, with the progress of time, the biodiversity of the initial populations increases and, slowly, other kinds of positive interactions tend to establish themselves, such as cooperation and mutualism, and up to the most extreme cases of symbiosis, as in the case of lichens.

The structure of a community is determined by the combination of different factors, such as the physical environment, which with its edapho-climatic characteristics directly selects the constituent species; the dimensions of the community; the typology of the species present; their longevity; and the way they increase. It was recently emphasized that even for microbial colonization it is necessary to analyze the phenomena by examining the community as a whole, observing that different species do not grow in isolation, but as part of a whole, which is also due to the presence of extracellular polymeric material (EPS) forming a biofilm (Chapter 1).

In the study of plant populations, it has been observed that in the presence of certain environmental conditions a certain kind of community will repeatedly establish itself; from this has emerged the science of phytosociology, i.e., the science that studies plant communities. It was first developed in the school of Zurich-Montpellier, thanks to the theories of Braun-Blanquet (1928) who proposed the floristic-statistical method, thus by him defined because it was founded on a detailed study of the flora and on a statistical sampling of the object under scrutiny.

The present-day approach could be described rather as floristic-ecological, considering the relevance of ecological information in the definition of the communities (Chapter 9, section 1.3). In this scientific approach, the plant association represents the basic unit. In order for an association to be recognized as such, it is necessary for it to be characterized by a well-typified body of flora and by characteristic species that must be more or less exclusive and that serve to distinguish the association under scrutiny, in terms of flora and ecology, from all other species present in the territory under consideration. The study of these communities from a phytosociological perspective is not limited to vascular plants, but includes other kinds of plant populations such as lichens, mosses, and algae. Since systematic knowledge of the world of fungi and bacteria is still limited, and the identification of bacteria and fungi still presents greater difficulty in comparison to higher plant organisms, and considering also the greater complexity and variability of

Table 2.1 Types of interaction between biological species

Types of Interaction	Species A to B	Species B to A	Description of the Interaction
Predation	+	–	Species A kills and eats species B
Parasitism	+	–	Species A grows at the expense of species B
Antibiosis	+	–	Species A releases substances toxic to species B
Commensalism	+	0	Species A benefits from the presence of species B, which is not affected
Amensalism	0	–	Species A harms species B, but does not benefit by doing so
Cooperation	+	+	The two species mutually benefit from the other's presence, but could survive alone
Mutualism	+	+	The two species mutually benefit from the other's presence and cannot live on their own
Symbiosis	+	+	The two species mutually benefit each other and form a single organism

Key: + = positive interaction; – = negative interaction; 0 = absence of any interaction

microbial communities, there is no parallel science of microbial communities, even though attempts at defining associations have been made in the "soil" category.

In a nonhomogenous environmental context, different communities develop contemporaneously but with precise spatial separation, giving rise to what may be defined ecological successions, i.e., groups of communities that follow one another in a particular

space as a result of a limiting ecological factor. This is a recurring case on monuments and in archaeological contexts where, depending on the level of humidity, of light, or even of the typology and concentration of mineral salts, substitution of communities takes place, in equilibrium with different environmental conditions. This correlation represents an

Fig. 2.18 *Different plant communities (various patinas and incrustations of algae, cyanobacteria, and lichens in the more sheltered areas, and higher plants in the more exposed areas) found in a rupestrian habitat according to various environmental factors, and in particular inclination and exposure (modified from Rizzi-Longo et al. 1980 and Poldini 1989)*

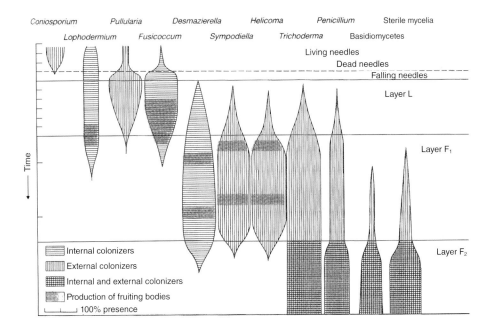

Fig. 2.19 *Example of ecological succession with time, in fungal populations in different layers of the rhizosphere (from Richards 1974 and Begon et al. 1989)*

important potential in the field of conservation diagnostics, in that a floristic-ecological approach makes it possible to employ species as bioindicators of environmental parameters.

As an example, we cite a sequence of algal and plant associations found in natural rocky environments, where we observe that the biological communities present undergo considerable changes according to the environmental gradients (in this instance mainly light and water) (Fig. 2.18).

2.3.2 Dynamism, Cycles, and Ecological Successions

In addition to an increase in time within the populations, there is also, especially in the case of inorganic materials, which are not directly utilized to a nutritional end, a more or less slow dynamic of the communities themselves, determined by the interactions among the species that define the communities and also among these and the various environmental parameters. In the deterioration processes of organic materials, there is a much more rapid

succession over time of microbial populations (Fig. 2.19), characterized by organisms that operate sequentially from a metabolic point of view (for instance, some cellulose-consuming organisms); or else, if the material does not seem to be chemically complex and difficult to metabolize, the populations that attack it first are the ones that are directly responsible for its destruction.

Dynamic phenomena can either proceed in a particular direction, or else exhibit more or less cyclical variations, without, however, ever returning the populations in question to the same original starting point. In fact, both the microbial communities as well as the phanerogamic ones normally exhibit more or less substantial seasonal fluctuations, which are due either to climatic variations or to the varying availability of nutrients. Although seasonal cycles are not always verifiable, in various cases they have been observed experimentally. Already in studies dating from the 1970s on the sulphur-bacteria found on various Roman monuments, a significant seasonal fluctuation had been observed, even though it was not found consistently in all the sites sampled

(Barcellona et al. 1973). The statistical analysis carried out on 255 samples from various stone monuments in Italy on which the presence of sulphur-bacteria was being investigated, had shown how seasonality, although playing a significant role, could not alone justify significant variations, since, depending on the sites sampled, there was evidence of both constant behavior and clear fluctuations (Caneva et al. 1989). The presence of seasonal fluctuations was also observed in heterotrophic bacteria on English monuments in sandstone (Tayler and May 1991); microbiological investigations on the façade of the Ca' d' Oro in Venice showed an increase in the values of bacteriological species between spring and summer from 10^3 to 10^5 units forming a colony per gram (UCF/g), while the fungal populations remained much more stable (Salvadori et al. 1994). This data has been related to the effects of the winter season, which in temperate climates is clearly a limiting factor for biological growth. An evident seasonal quality to the presence of microflora was noted several times in succession, for instance in the case of the microbiological charge in the caves of Altamira where, depending on the groups of organisms under study, considerable variations (both positive and negative) in the different seasons were observed, in relation to the most influential ecological factors (different requirements for water, nutrients, resistance to drought, etc.) (Arroyo and Arroyo 1996). As a further confirmation of the influence of seasonal changes, we should bear in mind that aerobiological investigations carried out both in museums and in outdoor environments showed significant variations, relating to the period of sampling as well as to local environmental conditions (Mandrioli and Caneva 1998). In the case of algae, seasonal fluctuations have been amply described and in some instances even accurately measured, both as color changes linked to the density of growth (Young and Urquhart 1996) and as qualitative/quantitative changes and of changes in cover benefiting more xerotolerant populations (Caneva et al. 2003).

In the domain of vascular plants, such a phenomenon has been described particularly for therophytes (ephemeral plants that form a coenosis and that can be classed as *Thero-Brachypodietea, Helianthemetea annuae*), which in Mediterranean climates have a maximum explosion in the spring or autumn and then disappear during the other seasons.

The contradictory data found in the literature may be the result of the different typology of the organisms studied, or of the methodologies adopted, or else of the initial conditions of the cases being studied. It is clear, for instance, that in environmental conditions that are always relatively favorable, seasonal fluctuations are less marked; also, in the instance of certain types of organism such as lichens and mosses, the phenomenon is less clearly visible in that they have a perennial thallus, on which the influence of seasonal changes is mainly on the intensity of metabolic activity.

Apart from the cyclical phenomena described above, we normally observe, over time and in a specific site, the progressive succession and substitution of biocoenoses, so that the best adapted communities can assert themselves at each phase of this evolving process. The reason for this succession is that, when a body of organisms grows on a substrate, it alters its intrinsic characteristics, for instance making its surface rougher, increasing its porosity, fracturing the material, adding organic substances such as primary or secondary metabolites, modifying the pH, etc.; as a result, other species that were previously unable to colonize the newly exposed surface, now find themselves able to establish themselves. If the incoming species are competitively dominant, they substitute the previous species giving rise to the process of succession.

An ecological succession does not carry on indefinitely over time; after the initial communities (pioneer communities) have been succeeded by increasingly more mature and structured ones, a final dynamic equilibrium is reached, characterized by biocoenoses that best respond to the climate and the conditions of the substrate (climax communities) (Fig. 2.20). Should outside limiting factors exist (strong inclination of the substrate, which does not allow the accumulation of soil, as is the case for all vertical surfaces in monuments; strong winds; high and low temperatures; etc.) that hinder the attainment of such climactic communities (i.e., those at the climax stage), only paraclimax or subclimax stages can be achieved, which are floristically and structurally more primitive.

If the development of a biological community begins on naked, never-before-colonized substrata, as is the case with exposed rocks, recent installations of works of art in stone, sand dunes, glacier moraines, or lava flows, the succession that occurs is defined as primary and develops extremely slowly, sometimes requiring millennia to reach the final stages of maturity. The succession will be much more rapid—a few centuries or even only decades—if it occurs instead in habitats that have already been colonized and where the soil has already been partly formed, but in which the plant elements have disappeared, perhaps in response to perturbations such as hurricanes, fires, deforestation, or clearance for cultivation—in these instances, the succession is defined as secondary. In the case of monuments that have undergone conservation interventions, in which the use of biocides has eliminated the more or less evolved dynamic stages that had been reached by the biological populations, the succession phenomena are secondary, so long as there are no other limiting factors.

The colonization of lava soils is a classic case of primary succession in which, on the basis of the different dates of the flows, it is possible to reconstruct the different evolutionary stages of its micro- and macrophytic populations. Various studies on both Mount Etna and Mount Vesuvius have shown how on the more recent lava, dating from about fifty years ago, a clear domination of the lichen populations can be observed, while the presence of higher organisms is negligible; on the more ancient lava, on the other hand, dating from a century ago, the reduction in the number of lichens is met with a considerable increase in phanerogamic species, and especially annual and perennial grasses (Ferro and Furnari 1968; Mazzoleni et al. 1989; Poli Marchese et al. 1995).

Obviously, the beginning date of the colonization of a substrate cannot be generalized, as it depends on the site's temperature and rainfall conditions as well as on the exposure conditions and the porosity of the substrate. In the case of tropical climates, the development times are definitely very short—in the order of a few months—while the time span can be of years or even decades in Mediterranean and temperate zones (in relation to rainfall and temperatures); even longer time spans can be observed in desert climates, or in the presence of protective strata such as oxalates.

Similarly, the development from pioneer stages to evolved ones may occur over only a few decades in a tropical climate—the literature showing how in less than a century the tropical forest is able to completely reclaim an abandoned archaeological site (Fig. 2.21)—or may take as long as a few millennia in difficult environmental conditions.

Fig. 2.20 *Example of ecological succession from the pioneer stages (left) through to maturity (right)—the spatial sequence from left to right corresponds to a time sequence (modified from Chapman and Reiss 1994)*

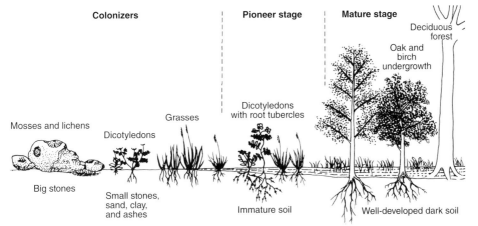

One element that emerges from the ecological analysis of successions that establish themselves after volcanic eruptions, which can be compared to other "primary" contexts such as that of the colonization of new monuments, is that if the soil is constituted by a pedogenetically primitive substrate, there is nothing onto which superior organisms can take root. First, the substrate has to be attacked by initial colonizers—usually bacteria (especially cyanobacteria), algae, and lichens—which disaggregate the substrate and form some humus, and only after this process can plants and other more advanced organisms attack the substrate.

Specific studies on the dynamics of colonization, conducted after the conservation interventions on some marble statues in the Boboli Gardens in Florence, have shown that processes of new colonization had already begun a year after the treatments and that the first species to colonize was the green alga *Coccomyxa*, which in only two years was able to form large expanses of green patinas. The success of this species, which is also found on the barks of surrounding trees, has been explained on the basis of its ecology and physiology, and in particular in relation to its oligotrophic capacity (optional), in addition to its ability to tolerate high solar radiation and to its efficient adhesion to the substrate, thanks to the hydrophobicity of its wall sheath and the excretion of EPS. Only subsequently did the cyanobacteria (also favored by the presence of exopolysaccharide sheaths and also by other well-known characteristics such as being able to fix nitrogen and tolerance to both desiccation and intense radiation) prove dominant in these still primary succession stages (Tomaselli et al. 2000c; Lamenti et al. 2000b).

Even though the autotrophs are without a doubt the most prevalent organisms, the role of heterotrophic species (such as some of the chemoorganotrophic bacteria and meristematic fungi) has been reevaluated in recent times.

Fig. 2.21 *(a) Final succession in a tropical forest (archaeological area of Cobá, Mexico)* – Photo G. Caneva; *(b) Archaeological site of Chichén Itzá, Mexico, in the final dynamic stages of succession, at the time of its discovery (late-nineteenth-century drawing) and (c) in its present state, with growth limited to pioneer populations to the exclusion of all others* – Photo G. Caneva

STRUCTURAL, FUNCTIONAL, AND ECOLOGICAL CHARACTERISTICS OF THE MAIN BIODETERIOGENS

The systematic and taxonomic knowledge of different living organisms, which forms the basis for any further physiological and ecological study, is still very limited, especially in the case of bacterial and fungal organisms (in addition to insects, which are not discussed here) (Fig. 3.1) This is because it is difficult to recognize certain species solely from their morphological characteristics; a more integrated systematic analysis has today been undertaken using biochemical and biomolecular techniques alongside microscopic ones.

From the initial Linnean separation of living organisms into two great kingdoms,

Animalia and Plantae, the plant kingdom has been further subdivided: first Bacteria, then Fungi, then Archaea were separated from the Eubacteria, leaving a large group of very heterogeneous eukaryotes in the Protista kingdom. As of today's knowledge, living organisms are therefore classified into six kingdoms, as shown in Table 3.1 (the Animalia kingdom is not included), but this organization must be considered temporary as certain groupings are artificial and will undoubtedly be reorganized at some time in the future.

Fig. 3.1 *Number of described species (light columns) and estimated species (dark columns) (modified from Hammond 1995)*

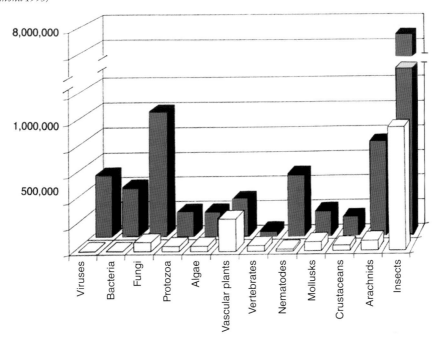

Table 3.1 Present systematic classification of the species under consideration*

PROKARYOTES
Kingdom Archaea
Kingdom Eubacteria 　　**Bacteria**, including **Cyanophyta** 　　and **Actinomycetales**
EUKARYOTES
Kingdom Protista (Algae) 　　**Division Bacillariophyta** 　　**Division Chlorophyta** 　　**Division Chrysophyta** 　　Division Rhodophyta
Kingdom Fungi 　　Division Chytridiomycota 　　Division Zygomycota 　　**Division Ascomycota** 　　**Division Basidiomycota** 　　**Division Deuteromycota**
Kingdom Plantae 　　**Division Bryophyta** 　　　　Class Anthocerotae 　　　　**Class Hepaticae** 　　　　**Class Musci** 　　Division Psilotophyta 　　Division Lycopodiophyta 　　Division Equisetophyta 　　**Division Polypodiophyta** 　　**Division Pinophyta** 　　　　Class Cycadopsida 　　　　Class Ginkgoopsida 　　　　**Class Pinopsida** 　　　　Class Gnetopsida 　　**Division Magnoliophyta** 　　　　**Class Magnoliopsida** 　　　　　　**(Dicotyledonae)** 　　　　**Class Liliopsida** 　　　　　　**(Monocotyledonae)**

*Included here are the groups traditionally classed as plants. The taxa in **bold** are those mostly responsible for the deterioration of works of art.

3.1 BACTERIA (EUBACTERIA AND ARCHAEA)

by Giancarlo Ranalli, Clara Urzì, and Claudia Sorlini

Bacteria are organisms with a wide range of metabolic functions and capacities and are mainly characterized by the prokaryotic organization of their cells. In the past, they were included within the plant kingdom, more because of the absence of characteristics typical of animal organisms than because of any true similarity with the other groups.

At a molecular level, though, bacteria have significant internal differences, and phylogenetic reconstruction has demonstrated that life on Earth evolved along three principal lines, in three distinct domains: Eukarya, which includes animals, plants, and microorganisms with nuclei enclosed by a membrane, and two prokaryotic domains, Bacteria and Archaea, including only unicellular organisms, in which the nuclear region is not surrounded by a membrane (Fig. 3.2).

The domain Bacteria is highly articulated and divided into twenty-three phyla (Boone et al. 2001); these include red and allied bacteria (Proteobacteria); Cyanobacteria; gram-positive bacteria; green sulphur bacteria and nonsulphur bacteria; Bacteroides; and the Flavobacteria.

The Archaea are characterized by a preference for unique ecosystems or extreme conditions, such as salinity, high temperature, or high acidity. It is thought that the Archaea, unlike the Eubacteria, evolved more slowly precisely because of the high specificity of their habitats (Polsinelli et al. 1993).

In the domain of Bacteria, around nine hundred genera and 5,100 species are known, in addition to seventy genera of Archaea (Salkinoja-Salonen et al. 2003). However, it is generally thought that microbial species existing in nature are very numerous and that only a very small fraction of them is known (1–5%). The majority of bacteria that are still unknown are such, either because it has proved impossible to cultivate them under laboratory conditions (viable but not cultivable [VBNC] microorganisms), or because we lack the knowledge of the environment in

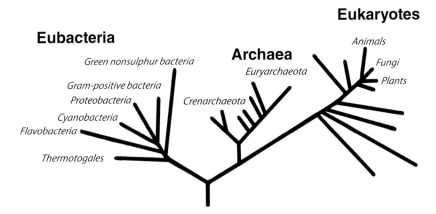

Fig. 3.2 *Simplified schematic representation of the universal phylogenetic tree obtained from the comparative analysis of RNA sequences (modified from Woese 2000)*

which they live. Indeed, it is interesting to note how the detailed study of various environments and monuments has allowed the isolation and description of several bacterial strains, which can be considered as new genera or species.

In this chapter, we will therefore be considering those groups (especially Eubacteria) that are most frequently associated with the deterioration of works of art. We will describe their main features, leaving the more specific details to the later treatment of degradation phenomena of the various materials. The description of Cyanobacteria can be found below in this chapter, in section 3.3 on algae.

3.1.1 Morphological and Structural Characteristics

The cell morphology of the various groups of prokaryotes (Eubacteria and Archaea) is heterogeneous (Plate 5). The cells can be spherical, rod-shaped, spiraled, lobed, disk-shaped, filamentous, irregular, or pleiomorphic (the same strain can assume different shapes at different stages of its cellular development). The diameter can vary between 0.1 and 1.5 μm, and the length can reach 10–15 μm (Fig. 3.3).

Fig. 3.3 *Main morphologies observed in Bacteria and Archaea*

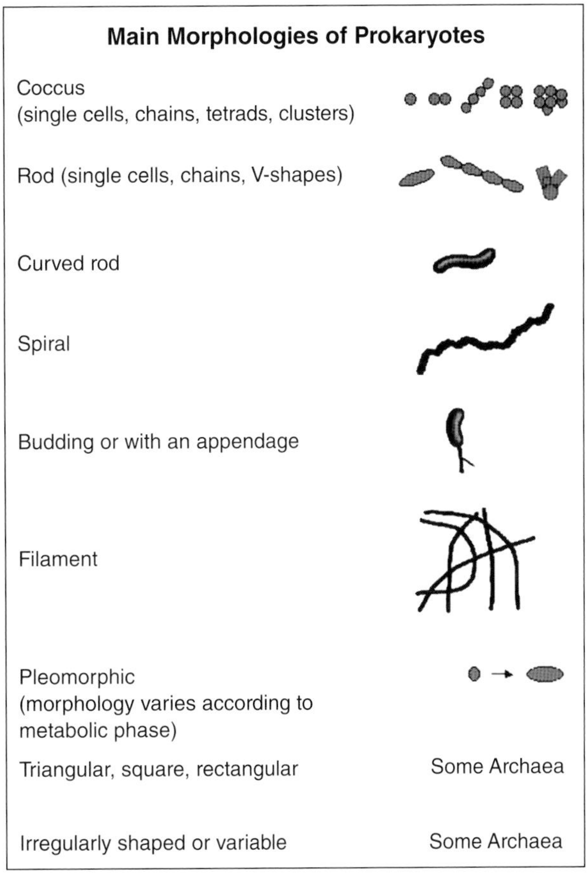

Main Morphologies of Prokaryotes	
Coccus (single cells, chains, tetrads, clusters)	
Rod (single cells, chains, V-shapes)	
Curved rod	
Spiral	
Budding or with an appendage	
Filament	
Pleomorphic (morphology varies according to metabolic phase)	
Triangular, square, rectangular	Some Archaea
Irregularly shaped or variable	Some Archaea

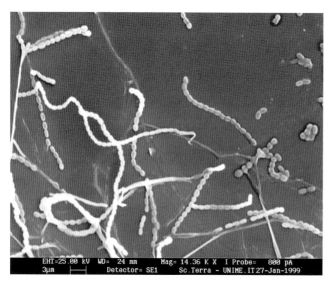

Fig. 3.4 Streptomyces *sp. strain seen in SEM* – Photo C. Urzì

The actinomycetes are gram-positive bacteria that are often filamentous in their structure, their hyphae forming a structure called mycelium, thus termed by analogy with the one formed by fungi (Fig. 3.4).

3.1.2 Physiological and Reproductive Characteristics

Cellular differentiation, also sometimes found in the bacterial world, includes the formation of a new structure inside the cell in the course of one and the same cellular cycle. Some bacteria form spores, while among the cyanobacteria some species differentiate some of their cells into heterocysts capable of fixing molecular nitrogen (see below in this chapter, section 3.3.1).

The main differences between the cellular structure of Eubacteria and that of Archaea lie in the composition of the cell wall (peptidoglycan in most Eubacteria; pseudomurein, polysaccharides, proteins, or glycoproteins in the Archaea) and in the RNA polymerase, which is made up of a greater number of subunits for the Archaea (between eight and twelve) as compared to the four found in Eubacteria.

Because of the difference in their cell wall structures, Eubacteria are subdivided into gram-positive and gram-negative, depending on the different behavior when stained with Gram's staining (discovered by Christian Gram in 1884): gram-positive bacteria will be colored in blue, while gram-negative are pink. The principal constituents of the cell wall are N-acetylglucosamine (NAG) and N–acetylmuramic acid (NAM) bound together with a beta 1,4 glucosidic bond. To this polymer, a chain of five amino acids is joined: L-alanine, D-alanine, D-glutamic acid, lysine, and/or diaminopimelic acid (DAP). In gram-positive bacteria, the peptidoglycan is multilayered and is the principal constituent of the wall, while in the gram-negative organisms it is found in one or more loose layers. In addition, the gram-negatives have an external lipopolysaccharide membrane (LPS).

The spore formed inside the vegetative cell is called endospore, to distinguish it from other forms of spores produced by other organisms, which are called exospores because they are formed outside of the cell. The bacterial spore appears highly refractile under phase contrast microscopy, because of its low water content and the thickness of the spore's external layers.

The prokaryotic cell does not possess mechanisms of sexual reproduction—although there are methods for the exchange of genetic material—and it duplicates itself by means of binary scission, some organisms through simple or multiple budding or by fragmentation (arthrospores).

Actinomycetes can produce spores asexually (conidia), which originate—after the formation of transversal septa—at the level of the sporophores (aerial filaments above the colony). Sporophores have a variety of appearances (straight, bent in a spiral, verticillate), the description of which can be useful diagnostically.

From the nutritional point of view, besides the requirements for macronutrients for metabolic needs, what should also be taken into consideration is the request for fundamental mineral elements, which, even though they are indispensable for numerous metabolic functions (for instance, enzymes), are usually sufficient even when present as trace elements (micronutrients). Such elements are often available even in what might appear to be the most inhospitable environments, such as stones, mortars, bricks, etc., and can be provided either by the substrate on which they grow, or by atmospheric dust, aerosols, or rainfall.

Listed below are the principal groups of bacteria that are biodeteriogenic for cultural property, classified according to their autotrophic or heterotrophic nutritional characteristics.

3.1.2a *Autotrophic Bacteria*
Sulphur-oxidizing Bacteria
These groups are often considered to be among the most dangerous for the conservation of stone material (chemolithotrophic), because they have the capacity to produce sulphuric acid (by the oxidation of hydrogen sulphide, elemental sulphur, or thiosulphate), which is an inorganic acid with a strong degrading action (Mishra et al. 1995a). Among these, the thiobacilli, which directly oxidize sulphides to sulphates (*Thiobacillus*, *Thiomicrospira*), are the most dangerous for artifacts because of their ecological characteristics and their greater production of sulphuric acid than other species. Less common—being a microaerophilic species—is the group of *Beggiatoa*, which normally are able to oxidize hydrogen sulphide to elemental sulphur and then, when the latter is no longer available, they oxidize sulphur to sulphate.

Nitrifying Bacteria
These are known to be among the earliest colonizers of stone surfaces and can play an important role in degradation, as a result of the production of nitric acid. They include ammonia-oxidizing or nitrosifying bacteria (*Nitrosomonas*, *Nitrosococcus*), which oxidize ammonia to nitrous acid, and the nitrite-oxidizing, or nitrifying, bacteria, which oxidize nitrous acid to nitric acid (*Nitrobacter*, *Nitrococcus*). The overall chemical process carried out by these bacteria is termed *nitrification*.

Hydrogen Bacteria
These are found in the soil and in water and use molecular hydrogen as an energy source, converting CO_2 to organic carbon. Their importance derives from the fact that they remain active even when environmental conditions are limiting, or if the primary source of energy is not continuously available. Almost all are facultative chemolithotropic organisms, having also the capacity to use organic compounds as a source of energy (chemoorganotrophs); they are active in both the presence and the absence of organic substances, so long as a high concentration of hydrogen is available. This group of hydrogen bacteria includes both gram-positive organisms (*Bacillus*) and gram-negative ones; among the latter, the most common and better known genera are *Pseudomonas*, *Paracoccus*, and *Alcaligenes*.

Iron Bacteria
These microorganisms obtain their energy from the aerobic oxidation of ferrous (Fe^{2+}) to ferric (Fe^{3+}) iron. In addition to being found on iron artifacts, they are found on stones containing pyrite (ferric sulphide), or on frescoes, wall paintings, etc., in which reduced iron compounds are present. In acid environments, *Thiobacillus ferrooxidans* is the most frequently found and is able to grow autotrophically, utilizing both ferrous ions and compounds reduced from sulphur as electron donors.

3.1.2b *Heterotrophic Bacteria*
A great number of heterotrophic bacteria (both Eubacteria and Archaea) have the capacity to colonize works of art. Heterotrophic bacteria carry out various types of enzymatic activity, which have a direct effect on the colonized material, especially when it is organic.

A great number of genera are responsible for the degradation of artifacts of organic nature. Among these we especially would like to mention the role of actinomycetes; these are able to utilize a great variety of different organic substrates and, in addition, to produce compounds with antibiotic properties, which limits the colonization of other strains that are sensitive to them.

Heterotrophic bacteria have also been frequently found on monuments and mural paintings; most are gram-positive bacteria belonging to the genera *Bacillus, Brevibacillus, Micrococcus, Kocuria, Clostridium, Frankia, Geodermatophilus, Blastococcus, Staphylococcus, Streptomyces,* and related genera (such as *Nocardia, Rhodococcus, Streptoverticillium, Micromonospora*), and gram-negative bacteria (*Pseudomonas, Acinetobacter*). Recently, a number of halophilic Archaea, such as *Halomonas* and *Halococcus*, were found on frescoes in Northern Europe. Depending on the nature of the enzymatic activity encountered, heterotrophic bacteria can be classified according to the following groups.

Proteolytic and Ammonifying Bacteria
These bacteria are able to hydrolyze proteinaceous substances into peptides, and then the peptides into amino acids by producing extracellular hydrolytic enzymes, called protease and peptidase; amino acids are then further broken down with the liberation of ammonia. Bacteria that are able to carry out this reaction are called ammonifying; in general, proteolytic bacteria are also ammonifiers. The most frequently identified species belong to the genera *Pseudomonas, Sarcina, Bacteroides,* and *Streptomyces*.

Cellulolytic Bacteria
These bacteria are divided into primary cellulolytic bacteria, which are specialized in the degradation of cellulose to the extent that they are unable to grow if this compound is not present (*Cytophaga, Sporocytophaga, Sporangium*), and facultative cellulolytic bacteria, which are able to utilize other organic compounds as well (*Vibrio, Cellvibrio, Cellfalcicula*). Cellulolytic bacteria often have the capacity to also break down lignin and other components of wood, such as resins, gums, dyes, tannic acid, waxes, and fats.

Amylolytic Bacteria
Many microorganisms, among which are certain species of *Bacillus* and *Clostridium*, are able to break down starch. The speed of this process varies according to the chemical composition of the different molecules that constitute the starch, in particular the percentage of amylopectin present—being a branched compound,

amylopectin represents the slowest component to be broken down by bacteria.

Lipolytic Bacteria
Relatively few genera are able to break down lipids through the production of lipase (specific esterase that hydrolyzes the ester-bonds between glycerol and the fatty acids); worth mentioning among these are *Bacillus, Alcaligenes, Staphylococcus, Clostridium*, etc. These microorganisms are able to degrade artifacts in which fatty substances are present, either as natural components (for instance, wood), or introduced during the manufacture of the artifact (pastels, for instance).

Denitrifying Bacteria
Phylogenetically, the great majority of denitrifying bacteria are part of the Proteobacteria. They are the only anaerobic bacteria that can be found relatively easily on the surface of artifacts, if organic substances are present. They are facultative anaerobes that, when in contact with oxygen, are able to metabolize the substrate by means of aerobic respiration, although they are able to reduce nitrates only in anaerobic conditions. They are also able to utilize other electron acceptors, such as the ferrous ion (Fe^{3+}). The biodeteriogenic activity of this group is, in all respects, similar to that of other heterotrophic bacteria, i.e., always tied to specific enzymatic activity (amylase, protease, esterase) and to the production of catabolites such as organic acids, whether they are metabolizing in aerobic or anaerobic conditions (*Bacillus denitrificans, Pseudomonas stutzeri, Achromobacter sewerinii*, etc.).

3.1.3 Ecological Characteristics

It is well known that in nature microorganisms play an important ecological role in the food chain and in biogeochemical cycles, as for instance in the carbon cycle (processes of transformation of organic compounds and mineralization), the nitrogen cycle (N_2-fixing, ammonification, nitrification, denitrification), and phosphorus and sulphur cycles (oxidation, reduction), etc.

The most influential environmental factors for bacterial growth are: the availability of water;

O_2; temperature; light (limited to photoautotrophs); and pH, although the great variability present in nature determines a wide range of adaptability to different conditions. For this reason, it is impossible to define the ecology of the group without specifically defining the species.

High concentrations of salts generally inhibit growth; however, many bacteria, including some *Streptomyces* found in hypogean environments (Urzì et al. 2002) as well as extreme halophilic species, are able to grow in high concentrations of salts (13% for bacteria, and up to 30–40% for halophiles). Halophiles have also been found on the surfaces of frescoes exposed to humid conditions, which presented salt efflorescences of different compositions, such as carbonates, chlorides, nitrates, sulphates, etc. (Rölleke et al. 1996, 1998; Piñar et al. 2003).

We should add that, in most environments, microorganisms are not found as "pure" isolated organisms, but as members within a community, often surrounded by a common polysaccharide matrix (biofilm or microbial mat) and that, as for other systematic groups, the homeostasis of the system is guaranteed by the whole community. Indeed, in a microbial community, the microorganism community as a whole can buffer the effects of negative environmental factors, for example the release of acids and/or bases. Changes in the composition of the community take place through processes of succession, which occur as the duration of stressful conditions increases.

Among the factors determining the selection during the succession are: the availability of nutrients released by previous species; the modifications in the concentration of inorganic nutrients; the formation of toxins produced by the original colonizers; the competition for limited resources (especially the organic carbon source); the predation; and the environmental factors.

The conditions that set in on the surfaces of works of art are often such that the colonization takes place by several physiological/metabolic groups, and the subsequent deterioration is caused by the whole community (Urzì and Krumbein 1994). It should further be noted that bacteria have the capacity to develop resistant structures, such as spores and cysts, that increase their chances of survival, even in unfavorable environments.

3.2 FUNGI

by Oriana Maggi, Anna Maria Persiani, Filomena De Leo, and Clara Urzì

Eumycota are heterotrophic eukaryotic organisms and are characterized by the presence of a rigid cell wall principally made out of chitin. They belong to the Fungi or "Mycota" kingdom; at present, roughly 80,000 are known species, but it is estimated that almost 1.5 million are in existence (Kirk et al. 2001).

Fungi are subdivided into four phyla: Chytridiomycota, Zygomycota, Ascomycota, and Basidiomycota, which have a sexual cycle (perfect fungi). In addition, there is the group of Deuteromycota, known as "mitosporic" or "imperfect fungi," since only an asexual reproductive cycle is known for them; at present, these fungi are defined as anamorphic fungi (Hawksworth et al. 1995; Kirk et al. 2001).

Fungi play an important role in the degradation of cultural property; it may be said that no material exists that can avoid being damaged by these organisms. Fungi exercise a direct action on organic materials utilized as nutrients as well as an indirect action on inorganic materials— the latter, even though they are not metabolized directly, can nevertheless contain organic elements able to support the fungal growth. Several species are indeed saprophytic, with strong degrading capacities for organic materials such as polysaccharides and proteins. In addition, fungi adapt to a variety of environmental conditions—environments where artifacts of historical and artistic interest are preserved, such as hypogean areas, churches, museums, libraries, and archives, can therefore provide ideal habitats for their development.

3.2.1 Morphological and Structural Characteristics

Fungi have different structures and vegetative systems and range from unicellular organisms (for example, yeast-like, spherical, or ovoid cells) to filamentous ones with relatively complex structures. Most of them present thin branched filaments called hyphae, which originate from the germination of the fungal spore. Hyphae are shaped like a narrow tube with a diameter of 2–12 μm. The whole body of hyphae forms the mycelium, which gives rise to the colony called thallus (Fig. 3.5 and Plate 6). Some fungi alternate the yeast-like phase with the hyphal phase (dimorphism). The hyphae also form fructiferous bodies (carpophores), especially characteristic of the ascomycetes and basidiomycetes (Fig. 3.6) (Rambelli and Pasqualetti 1996; Deacon 1997; Gow et al. 1999).

Unicellular forms appear as isolated round or oval cells. In certain conditions, some forms of yeasts may form a pseudomycelium consisting of a chain of more or less elongated cells that have remained attached after cell replication phase (Plate 6).

Hyphae have cross (transverse) septa (septate hyphae), except for the Zygomycota (coenocytic hyphae) (Fig. 3.6). In the septate hyphae, the cytoplasmic continuity and the migration of organelles between adjacent sections of the mycelium are guaranteed by the presence of one or more perforations at the level of the septum (septal pores).

The vegetative mycelium can also produce structures that guarantee the survival and the dispersal of the species, such as the sclerotia and the rhizomorphs (Rayner et al. 1995;

Deacon 1997; Jennings and Lysek 1999). Sclerotia are structures that are made up of aggregates of hyphae, with various morphologies and rich in nutrients; they remain dormant or quiescent in adverse environmental conditions, germinating when conditions improve. They have often been found on deteriorated materials, and their formation plays an important role in the processes of colonization and persistence on the substrates, contributing to the biodeterioration of works of art (Plate 6).

Rhizomorphs are structures that possess a morphology similar to that of roots; they are myceliar chains in which the tips of the individual hyphae grow parallel to one another. Their main function is to form organs of aggressive colonization, which penetrate into the substrate in search of nutrients and to set up reserves. These structures are produced especially by some of the basidiomycetes of wood rot, such as *Serpula lacrymans*. They have often been identified not only on wood materials kept in humid conditions, but also on plasters, wall paintings, and stuccowork, in hypogean environments, when temporary or permanent wooden structures are present (Singh 1993) (Plate 17).

Fungi have a rigid cell wall, composed mainly of polysaccharides such as chitin—a polymer of N-acetylglucosamine present also in arthropods—and (1-3) and (1-6)-β-D-glucan, to which are associated lipids, amino sugars, and proteins (among them glycoproteins) (Gooday 1995a; Carlile et al. 2001) (Table 3.2).

In fungi with a pigmented mycelium (for instance, Dematiaceae), melanines are present in the cell wall. These confer resistance against numerous chemical and physical agents, such as reducing substances, UV radiation, γ- and x-rays, and attacks by enzymes. This capacity is due to the chemical nature of melanin, a combination of quinones and hydroquinones that limits the damage caused by the formation of free radicals. Their presence in the cell wall also seems to play a fundamental role in the processes of biodeterioration of stone, favoring the penetration of the fungus into the substrate.

Fig. 3.5 *Development of a fungal mycelium, following the formation of spores by means of asexual reproduction, and their germination*

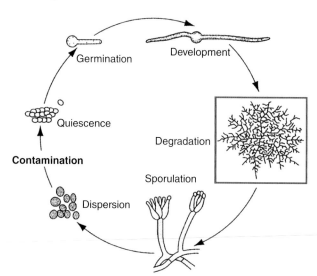

Germination

Development

Quiescence

Degradation

Contamination

Sporulation

Dispersion

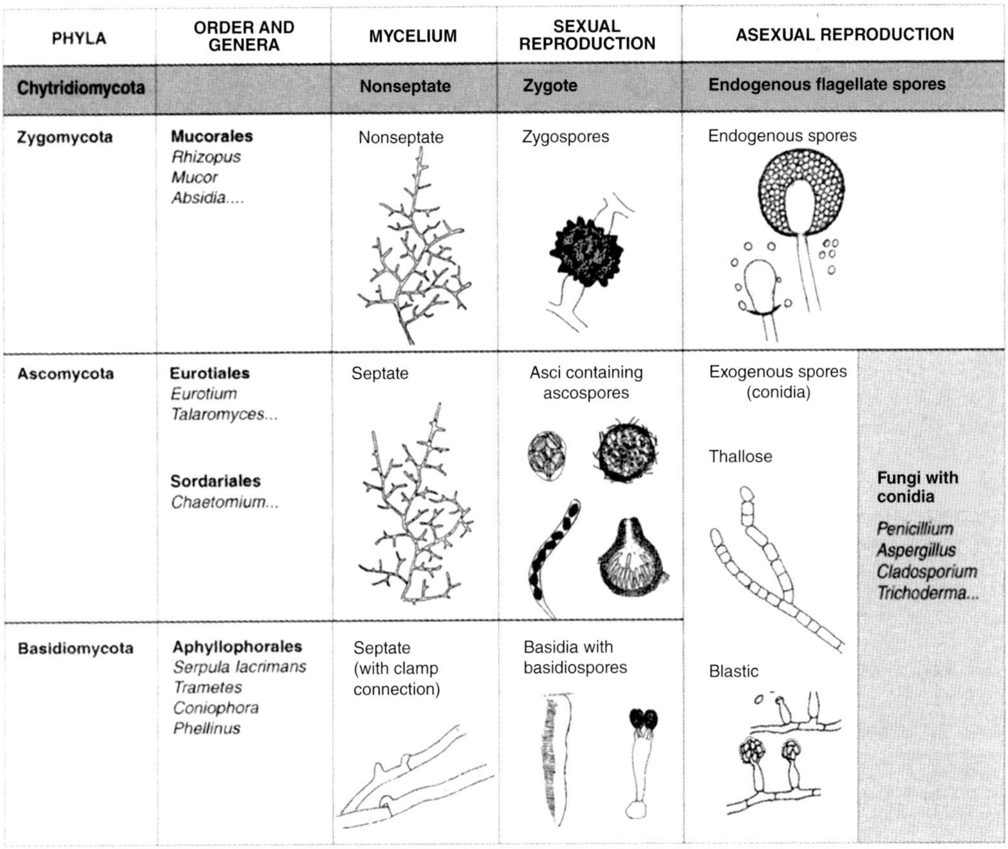

PHYLA	ORDER AND GENERA	MYCELIUM	SEXUAL REPRODUCTION	ASEXUAL REPRODUCTION
Chytridiomycota		Nonseptate	Zygote	Endogenous flagellate spores
Zygomycota	**Mucorales** *Rhizopus* *Mucor* *Absidia....*	Nonseptate	Zygospores	Endogenous spores
Ascomycota	**Eurotiales** *Eurotium* *Talaromyces...* **Sordariales** *Chaetomium...*	Septate	Asci containing ascospores	Exogenous spores (conidia) Thallose **Fungi with conidia** *Penicillium* *Aspergillus* *Cladosporium* *Trichoderma...*
Basidiomycota	**Aphyllophorales** *Serpula lacrimans* *Trametes* *Coniophora* *Phellinus*	Septate (with clamp connection)	Basidia with basidiospores	Blastic

Fig. 3.6 *Main phyla of Eumycota, with different diagnostic elements*

3.2.2 Physiological and Reproductive Characteristics

The majority of the species of fungi are saprophytic and able to utilize a large number of different organic substrates, thus fulfilling a fundamental role within the Earth's ecosystem. Fungi are chemoorganotrophic organisms, because they utilize organic substances both for the production of energy and for biosynthetic processes.

They have the capacity to use a wide variety of carbon sources such as hexoses, pentoses, organic acids, fatty acids and triglycerides, hydrocarbons, amino acids and proteins, as well as polysaccharides such as cellulose and hemicellulose (Markham and Bazin 1991) and lignin (Buswell 1991). Lignin is mainly used

Table 3.2 Main polymers present in the cellular walls of fungi (from Carlile et al. 2001)

Taxonomic groups	Fibrous polymers	Gelatinous polymers
Basidiomycetes, Ascomycetes, Deuteromycetes (Dematiaceae)	Chitin β (1→3), β (1→6)-glucan	Mannoprotein, α (1→3)-glucan
Zygomycetes	Chitosan, chitin	Polyglucuronic acid, mannoprotein, polyphosphates
Chytridiomycetes	Chitin, glucan	Glucan

by some species of basidiomycetes, which are responsible for white rot in wood.

Most fungi are able to use inorganic nitrogen as a source for their growth as well as organic nitrogen compounds (proteins and amino acids) (Dix and Webster 1995).

Many fungi, such as *Cladosporium*, have a wide metabolic diversity and are able to grow even in conditions of oligotrophy (scarcity of nutrients availability). Because of this capacity, they have often been identified as the fungi responsible for the degradation of those frescoes in which the only organic source available may have been dust particles deposited on the surface.

The degradation activity of microfungi, which in the case of artifacts of an organic origin coincides with the biodeterioration of the work of art itself, takes place through the production of intracellular and extracellular enzymes, specific to each substance (see Chapter 1, section 4.6). In addition, the development of the mycelium on a material constituting the work of art may cause physical and mechanical damage due to hyphae penetration.

Fungi reproduce and are dispersed by means of spores, which are produced either by sexual or asexual reproduction by specialized structures (Fig. 3.6). The two methods of reproduction may occur at different stages of the life cycle of the fungi (Carlile et al. 2001). When the two reproductive forms have a different morphology, we speak of pleomorphism. In pleomorphism, sexual and asexual generation can take place as independent means of propagation (Elliot 1994). Sexual reproduction is termed teleomorphic, while asexual reproduction is anamorphic.

Fungal spores have two main functions: the dispersion and the continuity of the species. There is great variety in their structure and morphology; indeed, they can be unicellular or multicellular, smooth or highly ornamented, immobile or mobile by means of flagella such as in the chytridiomycetes. Independently from the type of reproduction that generates them, spores share some characteristics, such as: a thick and sometimes pigmented cell wall; a low metabolic rate; reduced water content; and high levels of reserve materials.

The life cycle of a spore can be divided into a series of events that do not necessarily occur in the life cycle of every spore. After formation and maturity, most spores enter into a period of quiescence, called dormancy, which is followed by germination, in some cases only after a phase of activation (Griffin 1996). In germination, therefore, some spores germinate as soon as they find themselves in favorable environmental conditions, while others require a period of "maturation" or a specific activation stimulus before they can germinate.

Thermal shocks and some chemical substances function as activators (Cotter 1981). Among the activating chemical compounds are, at various concentrations: detergents, organic acids, alcohol, and other solvents, all more or less commonly used in conservation. This means that dangerous and rapid colonizations by fungi can occur as a result of contamination that takes place before or during the conservation phase of the work of art (Florian 1993). This underlines the importance of selecting suitable conservation materials in environments where the risk of biodeterioration is high, in addition to reducing the risks of spore contamination of the surfaces to be treated.

Spore germination consists of an irreversible series of events that are essentially: rehydration and swelling of the cell; resumption of metabolic activity, initially using the reserve of materials within the spore; new formation of the material for the cell wall; and emergence of the germinative tube, which will originate the hypha. The environmental conditions necessary for the occurrence of germination vary from species to species, and many fungi have a different optimum for the germination and growth phases—fundamental to all, however, is the availability of water.

3.2.2a *Asexual or Agamic Reproduction*

This is the most common type of reproduction and is usually carried out through the activity of specialized hyphae (conidiophores) (Fig. 3.7), which, at particular stages of the fungal development, produce spores by mitosis, called conidia (Plate 6). A particular type of asexual reproduction is by thallospores, which derive from the normal cells of the thallus through fragmentation of the hyphae (arthrospores), or—as in the case of unicellular fungi—

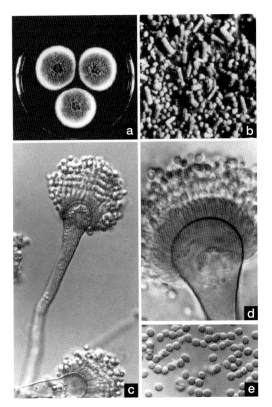

Fig. 3.7 Aspergillus fumigatus: *observation under the microscope, morphology of the colonies, and its reproductive structures: (a) colony; (b) conidial heads; (c–d) conidiophores; (e) conidia*

through a process of gemmation. A particular type of thallospores are the clamydospores; these are spherical cells, with a thick cell wall containing melanin, and with a greater diameter than the cells from which they originate. They are produced as resistant spores, both by filamentous and by yeast-like species, when environmental conditions are poor.

3.2.2b *Sexual Reproduction*

This kind of reproduction takes place as the result of the fusion of two cells of different sex (haploid gametes) or of specialized structures (gametangia of different polarities). This process gives rise to a diploid zygote, which usually immediately undergoes meiosis, originating a haploid generation, or else a more or less protracted dicaryotic phase, at the end of which spores are produced by meiosis (ascospores, basidiospores; Table 3.3 and Plate 6).

3.2.3 **Ecological Characteristics**

Saprophytic fungi are widely distributed, and many species are ubiquitous. The key element that determines the distribution of fungal species is the presence of a substrate able to support their growth (Wicklow and Carrol 1981; Dix and Webster 1995).

Environmental factors play a critical role in the development, persistence, and germination of fungal species.

As far as pH is concerned, most fungi are acidophiles, with optimal values for growth generally between 4 and 6; however, many species are able to grow in a wider range of pH values stretching from 2 to 9. An optimum pH is very important for all enzymatic activities— for example, the optimum for the cellulase complex lies in the interval between 6.4 and 8.2 (Deacon 1997).

Table 3.3 *Types of spores linked to sexual reproduction*

Zygospores	Spores of the Zygomycota; formed through the fusion of the isogamous gametanges (structurally identical), followed by nucleic fusion and meiosis. They have thick walls and contain reserve material.
Ascospores	Spores of the Ascomycota; formed after karyogamy and meiosis that take place in the sac-shaped cell that will contain the spores (ascus). The carpophores (ascoscarps) are: cleistothecia (closed structure), perithecia (structure with an opening), and apothecia (cup-shaped structure).
Basidiospores	Spores of the Basidiomycota; nuclear fusion and meiosis occur in the cell that will bear the spores (basidia). They can be pigmented or ornamented but with less variety than the ascospores. They are borne by the carpophore termed basidiocarp.

Almost all fungi are strict aerobes in that they require the presence of oxygen at least in one phase of their life cycle. Many imperfect fungi, such as *Fusarium oxysporum* and *Aspergillus fumigatus*, can be considered facultative aerobes, because they can also grow in the absence of oxygen by fermenting sugars. Those fungi that are able to develop at low pressures of oxygen (pO_2 = 0.18) are termed microaerophiles—many of the fungi responsible for the soft rots of water-logged woods or those that degrade organic materials that are partially buried (for example, archaeological objects) possess this characteristic.

The requirement for free water (activity of water, a_w) is lower for fungi than for bacteria; indeed, most species of bacteria require the equivalent of 0.98 a_w in the substrate on which they grow, while fungi defined as xerophiles can grow at values of less than 0.85 a_w (Pitt and Hocking 1985).

As far as the optimal temperature for growth is concerned, the majority of fungi are mesophilic (22–28°C); many species, however, also can grow at higher or lower temperatures. *Cladosporium herbarum*, for example, has been frequently found as a biodeteriogen of frescoes in crypts and hypogean environments (Fig. 5.5), where temperatures fluctuate around 10°C; its tolerance range, however, is very wide and extends from temperatures below 0° up to 40°C (Deacon 1997).

In general, the light as factor of growth has little effect on fungi, even though, in some instances, it has been shown that exposure to visible light can encourage the formation and germination of the spores.

On the other hand, prolonged and strong exposure to sunlight can have a germicidal effect thanks to the UV component and its oxidizing action. Fungi containing melanin in the cell wall, however, are not very susceptible to radiation in comparison to hyaline species.

grouped together through their ability to produce a melanin pigment, generally olive-black in color, and by their capacity to grow meristematically.

So far, twenty-five genera of black fungi with meristematic growth have been described. Although they are morphologically very similar to one another, they can be very different from a phylogenetic point of view. It has been shown (Hoog De et al. 1999; Sterflinger et al. 1999) that black fungi with meristematic growth can be classified into at least four different orders of ascomycetes: Chetothyriales, Dothideales, Capnodiales, and Pleosporales. However, for many species, their exact taxonomic position remains unclear.

Meristematic growth, characteristic of this group of fungi, is the result of the isodiametric growth of the cells, with a thickening of the wall into which the melanin is deposited, and of the internal conidiation (endoconidiation) (Fig. 3.8 and Plate 6). This has a protective effect on the cells during the germination phases, making them less vulnerable to environmental stresses (Wollenzien et al. 1995). This kind of growth is not very common in the kingdom of Fungi and is thought to be the response of the organisms to hostile environmental conditions. Both the production of melanin and the meristematic growth endow the cells with resistance to "extreme" environmental conditions such as strong sunlight or scarcity of water and of nutrients, all of which frequently occur on outdoor stone materials in Mediterranean regions (Urzì et al. 2000a).

In some of the stages of their development or in certain environmental conditions, species such as *Sarcynomyces petricola* (Wollenzien et al. 1997) may also have a unicellular yeast-like aspect; during this phase they propagate by budding. A more appropriate term for these species is black yeasts. But for other species (*Coniosporium* spp., *Trimmatostroma* spp.)

3.2.4 Meristematic Fungi

by Clara Urzì and Filomena De Leo

Black fungi with meristematic growth (or microcolonial fungi, MCF) are a broad and heterogeneous group of microorganisms,

Fig. 3.8 *Schematic representation of the formation of a meristematic cell through the thickening of the fungal cell and the formation of endoconidia (from Wollenzien et al. 1995)*

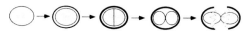

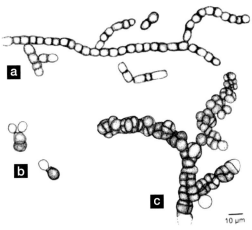

Fig. 3.9 *Fungi with meristematic growth in different stages of development: (a) chains of cells in the early stages of development; (b) budding cells; (c) formation of a thick mycelium containing more septa in the later stages of growth*

(Sterflinger et al. 1997; De Leo et al. 1999; De Leo et al. 2003), meristematic growth (Fig. 3.9) represents a type of growth that makes taxonomic identification of the microorganism very difficult, and at times impossible, if it is based solely on morphology.

Black fungi with meristematic growth form small black colonies (60–100 μm in diameter) on rocks and grow very slowly (1–2 months), both in natural conditions and in the usual media used for culture. Moreover, they are extremely versatile and able to grow on the most disparate substrates, even in oligotrophic conditions (Urzì et al. 2000a).

They are resistant to various types of environmental stresses, such as high temperatures (> 100°C) (Sterflinger and Krumbein 1995; Sterflinger 1998) as well as to prolonged periods of exposure to UV radiation and to osmotic stresses (up to 40%) (Zalar et al. 1999).

3.3 ALGAE AND CYANOBACTERIA

*by Maria Luisa Tomaselli
and Anna Maria Pietrini*

Algae and cyanobacteria are photoautotrophic organisms, devoid of tissues and organs, that demonstrate a remarkable diversity in their cellular organization and in their vegetative structures.

Algae, which are included in the kingdom Protista, are made up of eukaryotic cells; they exhibit a full range of pigments (chlorophylls, carotenoids, and phycobilins) as well as highly specific reserve materials and wall components, which are used as discriminating elements for their systematic classification (Van Den Hoek et al. 1995; John et al. 2003).

Cyanobacteria, on the other hand, are prokaryotic organisms, linked to the kingdom Eubacteria. In the past, they were considered algal organisms (blue-green algae) and belonged to the phylum Cyanophyta, because of their capacity to carry out photosynthesis in a similar fashion to higher plants. The traditional nomenclature of blue-green algae derives, similarly to the other classes of Algae, from the typical color of their cells due to the prevalence of the blue pigment, phycocyanin, over the green of chlorophyll.

Subaerial and lithophilic algae are those that commonly colonize archaeological remains, works of architecture, stuccowork, mosaics, frescoes, and wall paintings in general. Their appearance is indicated by a range of chromatic alterations, more often than not with powdery patinas and gelatinous layers of different colors: green, gray, black, orange, brown, and purplish-red, depending on the type of biocoenosis and on the particular combination of photosynthetic pigments present within the individual organisms (Plate 7). The establishment of cyanobacteria and algae is facilitated by the irregularity of the stone substrate. Depending on the type of relationship with the latter, such organisms can be distinguished into epilithic, living on the surface of the stone; chasmoendolithic, living in fissures and cavities in the stone but in contact with the surface; and endolithic, which live inside the stone. The latter category includes cryptoendolithic organisms, which colonize structural cavities inside the stone and develop in strata parallel to the stone surface, and euendolithic, which actively penetrate the stone, forming cavities with differing morphologies (Golubic et al. 1981; Hoffmann 1989). Algae and cyanobacteria give rise to more or less extensive colonizations that, in the case of works of art, not only modify their aesthetic appearance but also contribute to their deterioration, with variations depending on the form taken by their growth.

The effect of the epilithic forms, which are the ones most commonly found, can be either simply one of cover—so that the stone substrate underlying the biological growth is left intact and devoid of the presence of salts deriving from the interaction between the organisms and the substrate—or else it can be corrosive, in which case it manifests itself with pitting of the stone surface, often corresponding to the geometry of the microbial colonies. The deteriogenic action of chasmolithic organisms often results in the detachment and lifting of scales of the stone, whereas the effect of the euendolithic forms (which fortunately are somewhat rare) is a spoiling action, linked to the active perforating power of some of the species (Giaccone and Di Martino 1999).

The genera of algae that are most frequently referred to in studies on the biodeterioration of stone, belong mostly to the division Chlorophyta (green algae), with thirty-five species identified; less represented are the Bacillariophyta (diatoms), and even rarer the Rhodophyta (red algae).

As far as cyanobacteria are concerned, more than forty genera with unicellular and filamentous morphology are widely distributed on monuments and on works in outdoor environments or exposed to artificial light.

3.3.1 Structural and Morphological Characteristics

Algae include unicellular species—in which one single cell covers all the vital functions and behaves both as a vegetative organism and a reproductive organ—as well as multicellular species. In the latter, the organisms possess a vegetative body called thallus with a simple structure devoid of tissues and complex organs, but which, in the more evolved forms, attains a morphological differentiation similar to that found in higher plants. Unicellular organisms can live isolated or form colonies of various shapes and structures, and can have flagella. The thallus of the multicellular organisms can

be filamentous, with or without lateral ramifications, fronded, tubular, laminar or multilayered, and often attached to the substrate. Algae can be of widely varying dimensions: unicellular species are microscopic (< 10 μm), while multicellular organisms are for the most part macroscopic and may reach a length of several meters in the marine forms.

Chlorophyta (green algae) include organisms that are very varied in their structure, size, and the type of reproduction. There are coccoid forms, unicellular or colonial, mobile with flagella or immobile, multicellular with a filamentous thallus, laminar or tubular, uninucleate or multinucleate (Fig. 3.10 and Plate 7). The cell wall, which is layered and usually present, is made of cellulose, often with a large number of other polysaccharides on the outside, which confer a mucilaginous consistency to the organisms, while the rigidity is sometimes increased by incrustations of calcium carbonate. The chlorophyll pigments—chlorophylls a and b—are contained inside the chloroplasts, one or more of which are present in a variety of shapes (ribbon-, grain-, spiral-, or star-shaped). The carotenoid pigments, on the other hand, are accumulated on the outside of the chloroplast. Even the reserve materials, which consist primarily of starch, are deposited in the chloroplasts.

Reproduction can be both sexual and asexual, through the formation of zoospores and gametes.

Fig. 3.10 *Typical morphologies of the most frequently found green algae on stone monuments:* (a) Chlorococcum *sp.; (b)* Ulothrix *sp.;* (c) Tetracystis *sp.*

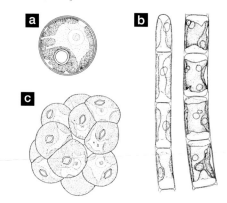

The Bacillariophyta, better known as diatoms, are widely distributed in nature and frequently found on works in stone that are permanently wet. They include unicellular organisms that can be isolated or gathered together in colonies and that can be either free or attached to the substrate by means of a peduncle. They contain chlorophylls a and c and a high percentage of carotenoids (carotenes and xanthophylls), which make them yellow-brown in color. The principal characteristic of diatoms is their cell wall—called frustule—which is rigid, impregnated with silica (SiO_2), and constituted of two thecae, which fit onto one another, like a box with a lid. The upper theca is called epitheca, while the smaller, lower one is termed hypotheca. The siliceous material deposited on the thecae forms an elegant and convoluted decoration, the distribution of which is used as a diagnostic characteristic. The ornamentation follows two basic designs that lead to the characterization of two subclasses: Centric and Pennate. In the centric diatoms, the design follows a radial pattern and is symmetrical from a central point, while in the pennate diatoms it is longitudinal and usually bilaterally symmetrical (Fig. 3.11).

Rhodophyta (red algae) owe their name to their characteristic coloring, which ranges from pink to purplish-red to brown, because of the prevalence of the pigment phycoerythrin over phycocyanin and over the chlorophylls a and d. They are rarely represented in the context of biodeteriogens of works in stone in subaerial environments and have been found only sporadically in monument fountains or in very humid hypogean environments; but they do play a not inconsiderable role in the case of submerged works. One of the rare examples was the presence of *Bangia atropurpurea* on the Fontana dell'Organo in Tivoli (Pietrini, unpublished). Another case was the polymorphous species *Phragmonema sordidum*, already noted as an alga found in caves (Friedmann 1956; Hoffmann 1989), and also found on the frescoes in the Colombario degli Scipioni in Rome (Ricci and Pietrini 2004).

Cyanobacteria are the prokaryotic organisms exhibiting the greatest morphological diversity, as we can see by the presence of unicellular, filamentous, branched, and non-

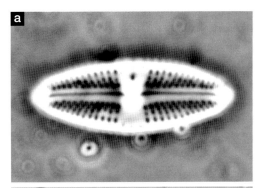
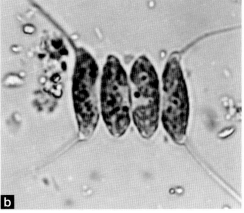

Fig. 3.11 *Morphology under the optical microscope of (a) diatom* Navicula; *(b) green algae (*Scenedesmus*)*

branched genera. They also vary tremendously in size: from unicellular organisms with a diameter of no more than 0.2 μm, to filamentous forms with a length of up to 100 μm (Fig. 3.12 and Plate 7). There are nonmotile forms, forms that move by gliding motility, and more rarely freely swimming forms.

At the systematic level, they have been traditionally considered algal organisms, assigned to the phylum Cyanophyta, classified as Cyanophyceae, and on the basis of their morphological characteristics subdivided into five orders: Chroococcales, Pleurocapsales, Oscillatoriales, Nostocales, and Stigonematales. However, because of their prokaryotic nature, in the past few decades they have been classified as bacteria (Eubacteriales); similarly to what occurs in bacterial taxonomy, their assignment is based on clonal and axenic cultures taking into account, in addition to morphological characteristics, also cytological, biochemical, physiological,

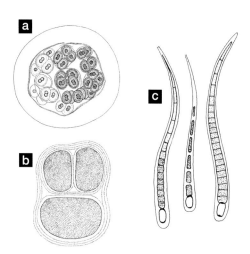

Fig. 3.12 *Morphology of cyanobacteria: (a)* Gloeocapsa *sp. and (b)* Chroococcus *sp., both unicellular forms; (c) filamentous (*Calothrix *sp.)*

and genetic ones. To the traditional five orders correspond, in the phylum Cyanophyta, as many subsections, distinguished by their various diagnostic traits (Castenholz 2001).

In addition to the common vegetative cells, in certain genera with a filamentous organization, two different kinds of specialized cells can be found: akinetes and heterocysts. The heterocysts are located at intervals along the filaments; they have a less dense cytoplasmic content than the vegetative cells and a thickened cell wall; at the points, in which they are connected with the adjacent cells, they have two polar nodules made of cyanophycin. The heterocysts are the site where elemental nitrogen is fixed; their differentiation along the filaments is favored by a deficiency of nitrogen compounds. Akinetes, or spores, are resistance organs produced during stressful conditions, for instance when there is a deficiency of light or of nutrients. They allow the organism to survive during periods of adversity, having thickened walls and a considerable quantity of reserve substances, characteristics that often give them a different pigmentation from that of vegetative cells.

The main characteristic of the cyanobacteria cell consists in the complex system of photosynthetic membranes, or thylakoids, which contain chlorophyll a and phycobiliprotein pigments (phycocyanin and phycoerythrin), the latter appearing as semispherical bodies— phycobilisomes—on the surface of the photosynthetic membranes. The cells are surrounded by a thin cytoplasmic membrane and by a multilayered cell wall composed of a layer of peptidoglycan and a further external layer of lipopolysaccharides. In addition to these layers, there may also be a gelatinous matrix, although it may only be present during certain phases of the development cycle. The matrix may form a structured envelope known as sheath, capsule, or calyx, or else simply be amorphous mucilage. It constitutes a fundamental characteristic for the colonization and the survival of the organism: it permits adhesion to the substrate; it functions as a reserve of water as it slowly absorbs it and then gradually releases it, allowing the organism to overcome periods of adversity; it inhibits predation; and, most importantly, it is the cement that holds together the microbial cells, thus giving rise to the formation of the biofilm. Investigations of stone monuments have shown that most cyanobacteria present are forms endowed with a gelatinous matrix (Lamenti et al. 2000a; Tomaselli et al. 2000a). The chemical nature of such a matrix, generally made up of negatively charged polysaccharides due to the presence of acid groups, can also contribute to the deterioration of the stone through complex-forming and leaching processes of constituent elements. Such a matrix also represents a source of organic material that supports the growth of heterotrophic microorganisms, which are nutritionally more demanding.

Often the sheath is variously pigmented due to the presence of compounds that mask the color of the cells and have a variety of functions, among which is the protection from high levels of light radiation and in particular UV radiation. Some species (*Scytonema julianum*, *Loriella osteophila*, *Geitleria calcarea*, *Hepyzonema pulverulentum*, *Leptolyngbya* sp., *Fischerella* sp.) are defined as calcifying because they are able to mobilize calcium ions from calcareous substrates and to precipitate calcium carbonate onto the polysaccharide sheaths (Pietrini and Ricci 1993; Hernandez-Marine and Canals 1994; Ariño et al. 1997b; Albertano 1998, 2003; Hernandez-Marine et al. 1999).

3.3.2 Physiological and Reproductive Characteristics

Most algae and cyanobacteria are obligate photoautotrophs, i.e, unable to grow in the absence of light. Some species, however, are able to grow chemoautotrophically, in the absence of light utilizing a variety of organic substrates, or mixotrophically, in the presence of light, photosynthesizing and at the same time utilizing chemical compounds. The various starch-like polysaccharides and oils that are produced by the photosynthetic reactions are also utilized to allow the continuation of normal metabolic functions in the dark for a limited period of time. Respiratory metabolism leads to the production of CO_2, which determines the formation of carbonic acid; this, along with some of the acid catabolites secreted by the cells, contributes to the deterioration of stone surfaces through processes of dissolution.

Nutritional requirements include some of the fundamental nutritional elements such as nitrogen, phosphorus, potassium, sulphur, calcium, iron, magnesium, and other elements required in small quantities. Nitrogen can be assimilated in various inorganic forms (NO_3^-, NO_2^-, NH_4^+) as well as organic ones (urea, amino acids); it can also be made by organic atmospheric nitrogen through the action of the enzyme nitrogenase, in the case of nitrogen-fixing species of cyanobacteria.

Reproduction occurs differently in each of the groups of algae, but it is always a form of the two principal types: asexual (agamic or vegetative), when there is no mixing of genetic material through fusion of the gametes, or sexual or gamic when genetic mixing takes place. In unicellular algae, asexual reproduction occurs through simple cell division or by the formation of various kinds of endogenous spores, which on germination give rise to new individuals. In multicellular algae and in colonial forms, reproduction may also occur by fragmentation of the thallus into several pieces, each one with the capacity to give rise to a new organism. In sexual reproduction there is fertilization, i.e., the cytoplasmic and nuclear union of the male and female gametes. The result of this union is a single cell, the zygote, which has twice the number of chromosomes compared to a single gamete. For the majority of green algae, the zygote is the only diploid cell of the life cycle; after meiosis, haploid spores are formed that, on germination, give rise to haploid individuals (gametophytes). In other algae, on the other hand, there is an alternation of haplo- and diplobiontic phases.

In the case of cyanobacteria, reproduction is primarily vegetative: in unicellular genera, it takes place through cell division by means of binary (Chroococcales) or multiple (Pleurocapsales) scission, with the release of motile cells called baeocytes. In the genus *Chamaesiphon* reproduction occurs by budding, while in the filamentous genera it takes place by fragmentation of the trichome and formation of hormogonia, which are short chains of cells, or by germination of the akinetes.

3.3.3 Ecological Characteristics

Algae and cyanobacteria are widely distributed throughout the world, from the poles to the tropics. They are particularly abundant in aquatic environments, both in fresh water and in brackish and marine waters, but there are also many terrestrial species that live on the ground. Many of these also have a subaerial life, growing on various substrates (plants, wood, rocks, walls, etc.), as long as these are sufficiently humid or covered by a temporary or permanent film of water. Mostly, they live as free organisms, but some also as symbionts (for example lichens). Generally, the distribution and abundance of these phototrophic organisms depend on: physical and chemical factors (light, temperature, salinity, rate of water flow, nutrients, etc.); biological factors (such as predation, competition, antagonism); and also on the interaction of all these factors.

In the context of groups that are relevant for biodeterioration, we note that most green algae are distributed worldwide and are prevalent in a fresh water habitat. The terrestrial and subaerial species are commonly found on the illuminated surfaces of buildings and works in stone, where they live either free or as phycobionts of the lichens.

Because their development depends on the presence of a considerable amount of water, diatoms are found in aqueous environments and on damp ground, but they can also be present on works in stone in the areas that are in contact with the ground—i.e., rich in humidity due to capillary rising action—and are also particularly frequent on the rims of fountains.

Red algae, found in abundance in warm seas and less frequently in fresh waters, consist of benthonic species, which live attached to rocks or to other substrates by means of adhesive disks or anchoring filaments. A small number of them live in different habitats from the typical ones, colonizing soils and caves.

In aqueous habitats, the main environmental parameter that influences the development of algae and cyanobacteria is light. In addition to having an effect on growth, the intensity and the quality of light are also responsible for considerable variations in the color of organisms. The presence of high levels of light intensity determines an increase in the synthesis of carotenoid pigments, which give cells a yellow-orange pigmentation. At the other extreme, low levels of light intensity lead to an accumulation of phycobiliproteic pigments in cyanobacteria, which gives the cells a very intense color and, in green algae, leads to an increase in chlorophyll.

Nutritional deficiencies, in particular of nitrogen-containing substances, determine—except in nitrogen-fixing cyanobacteria—the degradation of phycobiliprotein and chlorophyll pigments, which results in a greater prominence of the other pigments, and a green-yellowish-orange color. Vice versa, a saline environment can intensify the natural blue-green color in cyanobacteria. Indeed, on the weathered surfaces of stone monuments, where conditions of high salinity exist, populations of intensely pigmented cyanobacteria can be observed, if their development is not inhibited by other factors.

In terrestrial habitats in which light is not normally a limiting factor (with the exception of caves and other hypogean environments), the main conditioning environmental factor is the presence of water. Aside from species that live directly in water, as is the case with many diatoms, most algae and cyanobacte-

ria have an optimum growth at RH values close to 95–100%. However, it has been noted that most colonizing species also require a high level of humidity in the stone substrate on which they live in order to carry out their metabolic activities. In conditions of dryness, the cells undergo a process of shrinking, due to the loss of water, and often become cyst-like, with a thickening of the cell walls and of the external mucilage layer. Thus, they transform themselves into resistance cells, or spores, also taking on a different pigmentation that tends towards the yellow-brown. Particular conditions of drought can also be overcome thanks to the presence of the gelatinous sheaths, made up of hygroscopic polysaccharides, that envelop the cells.

Temperature is another important factor that regulates the development of cyanobacteria and algae; the range within which life is possible for these organisms is fairly wide, as there are thermophilic, mesophilic, and psicrophilic species in existence. Generally, the thermal optimum for the growth of mesophilic cyanobacteria is higher than that of the algae, as the first are favored by temperatures between 25 and 35°C, and the second by values between 20 and 30°C.

As far as pH is concerned, some species such as *Phormidium tenue*, *P. autumnale*, and *Microcoleus vaginatus* grow well on acid siliceous rocks (for example granites), while others, among which are *Chroococcidiopsis* and *Leptolyngbya*, develop on alkaline rocks, such as calcareous rocks and marble (Tomaselli et al. 2000b).

We cannot overestimate the importance of cyanobacteria and algae as pioneer organisms with a high colonizing capacity of particularly difficult environments, such as the surfaces of monuments or artifacts of artistic and/or historical interest, either in outdoor environments or in confined and illuminated spaces. It is they who set the process of colonization in motion; with their metabolic products and cell debris, they form a layer rich in nutrient substances that offers an ideal culture medium for the growth of heterotrophic bacteria and fungi, thus triggering ecological successions.

3.4 LICHENS

by Rosanna Piervittori, Pierluigi Nimis, and Mauro Tretiach

Lichens are among the most frequent organisms to be found on stone monuments in an outdoor environment. They enrich environments that are usually somewhat lacking in biodiversity, for example archaeological sites and urban centers, but they can also be agents of chromatic alterations and of biodeterioration of surfaces.

For a long time, lichens were considered autonomous organisms, without any evident links to other systematic groups, and were even assigned their own system of classification (Lichens). But they are a very heterogeneous group of higher fungi that have become specialized from a nutritional point of view, living symbiotically with one or more populations of the algae and/or cyanobacteria that are present within their vegetative body (thallus).

More correctly, lichens could be defined as small, self-sufficient ecosystems, formed by a fungus (mycobiont) that depends nutritionally on primary producers (photobionts), to which are often associated (although largely ignored) bacteria, and sometimes other saprophytic or parasitic fungi (Honegger 1998, 2001).

Conventionally, the name of the lichen corresponds to that of the mycobiont, although the product of the lichen symbiosis is generally very different from the individual symbionts, if these are indeed able to grow separately (Hawksworth 1994; Hawksworth et al. 1995).

Overall, more than 21% of all fungi (about 15–18,000 species) are lichenized. The phenomenon is particularly widespread among the ascomycetes (46%), much less so with the deuteromycetes (1.2%), and even less with the basidiomycetes (0.3%). Most species (85%) are associated to unicellular or filamentous green algae, while little more than 10% of the species live in symbiosis with cyanobacteria, and 3–4% live in symbiosis with both (Honegger 2001). However, only a relatively small number of algae and cyanobacteria seems adapted to live in symbiosis. So far, only forty or so genera have been identified, fifteen of which are cyanobacteria. Most of the photobionts belong to systematic groups that normally live in a free state, with a few notable exceptions (for example, the green algae of the genus *Trebouxia*, which are among the most widespread lichenic partners, but also extremely rare in the free state). It should be emphasized that the photobiont has been specifically identified in only a very small percentage of lichenic species (< 2%), and in some instances the identification has not even been attempted at the generic level (Galun 1988). The most recent molecular studies have also shown a genetic heterogeneity that was completely unsuspected a few years ago; therefore, the general picture may change yet again, and change considerably, in the near future.

In any case, the degree of selectivity of the mycobionts toward their photosynthetic partner seems to be quite variable, from highly selective species (when there is only one compatible mycobiont) to others that are only moderately selective (Tretiach 1996).

3.4.1 Structural and Morphological Characteristics

Morphologically speaking, more than 50% of lichens produce a largely undifferentiated thallus. In this case, the hyphae of the mycobiont multiply on the cells of the photobiont, eventually penetrating into the substrate and giving rise to crustose thalli (microlichens) (Plate 8). The crustose thalli adhere closely to the substrate and look like a crust that is attached to it. In these lichens there is no lower cortex, and the hyphae of the medulla penetrate to varying degrees into the substrate; for this reason, they cannot be removed without in part damaging their thallus. Within this category, morphological variation is significant: there are leprous formations (made up of pulverulous accumulations, usually water repellent); continuous or areolate ones (in which the surface is subdivided into numerous, more or less irregular areola); verrucose (with convex areola); peltate (the areola are fixed at a central point and have rising edges); and placodiomorphic (lobed in the most marginal areas). On carbonate rocks,

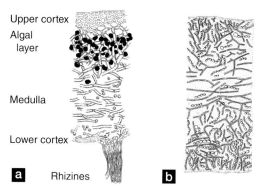

Upper cortex
Algal layer
Medulla
Lower cortex
a   Rhizines   **b**

Fig. 3.13 *Organization of the lichenic thallus (modified from Ozenda 1963): (a) heteromeric—the upper cortex, the algal layer, and the medulla (the lower cortex is present only in foliose and squamulose species, which can attach to the substrate by means of rhizines); (b) homeomeric—the hyphae and the cells of the photobiont (which are coccal or filamentous cyanobacteria) interweave to form a homogenous and undifferentiated whole. The thalli are usually gelatinous when damp. In some genera (e.g.,* Leptogium*) a well-differentiated cortex may be present.*

the penetration into the substrate can be particularly massive (endolithic lichens), because both the mycobiont and the photobiont are able to actively dissolve the carbonate matrix, which is thus colonized to a depth of a few millimeters and, in exceptional cases, deeper than one centimeter (Pinna et al. 1998) (Plate 6).

About 45% of lichens give rise to more complex forms of thallus (fruticose and foliose lichens, or macrolichens). The foliose thallus has a two-dimensional form, with dorsoventral organization (the upper surface differing from the lower surface) and a development generally parallel to the substrate. Adhesion occurs by means of sheaths of hyphae, which originate on the lower surface of the cortex of the thalli (rhizines). The presence of a well-developed lower cortex allows the detachment of the thallus from the substrate without too much difficulty (Plate 8).

The fruticose thallus is characterized by more or less numerous ramifications (laciniate), which can give it the appearance of a small bush; generally, the thallus anchors itself to the substrate with a basal attachment disk (Plate 8).

For the most part, lichens—including many crust-forming species—have a thallus that is internally layered (with a heteromeric organization) (Fig. 3.13a). Three fundamental layers are distinguishable: the cortex, the layer of the photobiont, and the medulla. The cortex, usually formed by an agglutination of hyphae immersed in a hydrophilic mucilaginous matrix, represents a mechanical barrier against microorganisms, while also playing a fundamental role in the transmission of light, in the exchange of gases, and in the absorption of water. Immediately under the cortex are the cells of the photobiont, which therefore benefit from a very favorable position for the exchange of gases and the reception of adequate quantities of light. The medulla—located under the cells of the photobiont and quite variable in thickness—consists of a loose system of hyphae, more or less hydrorepellent, in which intercellular spaces abound. The medulla may connect with the lower cortex if it is present (in foliose lichens), or put itself into direct contact with the substrate (crustose lichens). Overall, this type of organization, in which the mycobiont is the quantitatively dominant symbiotic partner, could be compared to a sophisticated culture chamber, in which the cells of the photobiont are hosted, contained, and controlled inside a sheath formed by hyphae (Honegger 2001).

Other lichens do not have layered thalli (homeomeric organization), and the cells of the photobiont are dispersed, seemingly without order, throughout the whole thickness of the thallus (Fig. 3.13b). In this case, the spaces between the hyphae of the mycobiont and the cells of the photobiont are occupied by mucilaginous substances that swell greatly in the presence of water, conferring a characteristic gelatinous consistency to the thalli. In homeomeric lichens, the quantitatively dominant partner is generally the photobiont, to the point that it can determine the form of the thallus, as is the case with microfilamentous lichens of many Lichinaceae (Ozenda 1963).

Under appropriate conditions, the mycobiont can be cultivated in the laboratory in the absence of its photosynthesizing partner. In this case, it produces a very simple, very slow-growing vegetative body, which is generally devoid of fruiting bodies and very different from the one produced when the cosymbiont is present (Honegger 1991). The morphology of the lichenic thallus therefore depends at least in part on the photobiont, which influences its architecture and determines some of its physiological characteristics. Precisely because of this phenomenon, it is interesting to study the lichens with low photobiontic specificity, in that symbiosis with different photobionts can give rise to many different morphologies. Sometimes different photobionts are even present in different sections of the same thallus (photosymbiodemes): this situation, which is very rare and generally limited to a few genera of lichens, makes it possible to verify which morphological or physiological characteristics are controlled or determined by the photobiont (Honegger 2001).

3.4.2 Physiological and Reproductive Characteristics

In lichenic symbiosis, the mycobiont tends to secure adequate illumination for the photobiont, facilitates its gaseous exchanges, and controls its renewal, dragging its cells into the zones of active growth (Honegger 2001). It is from the photobiont that the mycobiont then takes the carbohydrates produced during photosynthesis, in a way that is similar to that of mycorrhizae and many obligate pathogenic fungi. The nutritional requirements of the mycobiont are satisfied through the transfer of acyclic polyalcohols (eritrol, ribitol, sorbitol) if the partner is a green alga, or glucose if it is a cyanobacterium. In the latter case, transfer of nitrogenous substances also takes place, because cyanobacteria retain the ability to fix atmospheric nitrogen (Ahmadjian 1993). When the cells of the photobiont are isolated from the thallus, the production and the release of these substances are drastically reduced. The cells of the photobiont are connected to the mycobiont by means of haustorial or appressorial structures produced by the hyphae, but under no circumstance does the barrier of the two cellular walls, produced by the photo- and mycobionts, ever fall (Honegger 1998). In the more evolved forms, however, the flow of

substances across this apoplastic barrier seems to be assured by the presence of a hydrophobic covering formed by proteinaceous substances and secondary phenolic metabolites. The latter, generally referred to as lichenic substances, are compounds with low water solubility, which deposit themselves on the external surfaces of the hyphae of the mycobiont (Baruffo et al. 2001). The lichenic substances seem to carry out various biological functions: from an antimicrobic activity or inhibition of the germination of spores or seeds (allelopathy); to the absorption of UV radiation in order to protect the photobiont; to having a chelating action on cations. This latter aspect plays an important role in the processes of biodeterioration of rocky surfaces and in the uptake of micronutrients that are essential for the growth of the two symbionts.

The wide distribution of lichens is assured by an efficacious reproductive strategy, either by means of spores (sexual reproduction) or vegetatively (asexual reproduction) (Ahmadjian 1993).

In symbiosis, only the mycobiont usually retains the capacity to reproduce sexually. If it is an ascomycete, the spores (ascospores) form inside structures (asci) produced by the fructifying bodies (apotheces, with an open hymenium, in the shape of a disc; peritheces, with a closed hymenium inside a structure shaped like a small flask, with a small pore at its apex) (Fig. 3.14). If instead the mycobiont is a basidiomycete, the spores (basidiospores) form on slender stalks (sterigma) from the basidium, borne by the fructifying body (basidiocarp). The fructifying bodies are in general perennial, and the release of the spores usually follows seasonal cycles, with circadian rhythms (Honegger 1991).

The formation of a new thallus following the germination of a spore is linked to the condition of finding a compatible symbiotic partner, which is potentially quite a rare event. In many lichens, the formation of a thallus is thus assured by the fragmentation of parts of the vegetative body, or through the formation and release of specialized vegetative propagules (among which the most common are the soredia and the isidia), which contain both partners (Fig. 3.15).

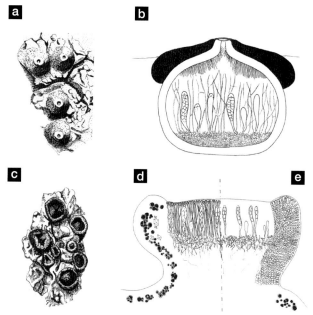

Fig. 3.14 *Main types of fructifying bodies in ascomycetes: (a–b) perithecia, in the shape of a small flask with an apical pore; (c–e) apothecia, disk-shaped and more or less flat. The apothecia may contain cells from the photobiont along the rim (lecanorine apothecia, [d]) or be devoid of them (lecideine apothecia, [e]). The fructifying bodies may contain sterile hyphae (paraphyses) as well as fertile hyphae, which form sac-like structures at the extremities (asci), where the ascospores are formed.*

Fig. 3.15 *Structures of vegetative propagation of both symbionts in lichenic symbiosis: (a) isidia; (b) phillidia; (c) schizidia; (d) blastidia; (e–g) soredia: (e) dispersed or (f–g) grouped together in soralia; (h) goniocysts. The first four types of propagula have a cortex, unlike the rest.*

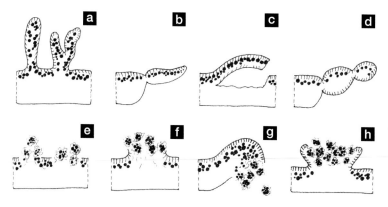

3.4.3 Ecological Characteristics

Lichens show a remarkable tolerance for environmental stresses, in particular for high and low temperature levels and (when they are desiccated) to high levels of sunlight (Rikkinen 1995). These characteristics ensure their wide longitudinal and altitudinal distribution, comparable to only a few other cryptogames, and especially make it possible for lichens to colonize environments inhospitable to other forms of life. Lichens are able to grow on the most different substrates, on the ground, on the bark of trees and bushes, and on some artificial substrates (such as glass, plastic, cement, aluminum or iron sheets, etc.). However, lichens are most frequently found on rocky substrates; in Italy, more than 50% of the species are epilithic (Nimis 2003).

Substrate variables such as physical peculiarities (structure and hardness of the constituent minerals), chemical characteristics (lithochemistry), and ecological aspects (inclination and exposure of the substrate) determine the development of highly diverse communities. These, in turn, may undergo considerable modification in their specific composition following the introduction of nitrophilic species, which develop thanks to the nitrogen compounds provided by animal excretions (birds, small mammals, etc.) or as a consequence of massive use of fertilizers. Works in stone offer much the same ecological conditions as those of natural rock efflorescences; it is therefore not surprising that they host a very similar flora (Nimis et al. 1987, 1992).

The capacity to colonize such a great variety of substrates is linked to the fact that lichens are poikilohydric organisms, i.e., able to sustain, depending on the availability of water in the surrounding environment, repeated cycles of desiccation and rehydration without suffering damage, thanks to their capacity to enter a phase of metabolic quiescence in the absence of water (cryptobiosis). In this aspect, lichens are very similar to bryophytes and, conversely, very different from higher plants, which must maintain a relatively constant water content in order to survive (homoiohydric organisms). All these factors together explain why lichens are particularly abundant in certain biomes, such as deserts and tundra, where they can represent the ecologically most important photoautotrophic component (Nash 1996).

Within the lichen category, there is a substantial physiological difference between lichens with cyanobacteria (cyanolichens or gelatinous lichens) and those with green algae (chlorolichens): while the first require liquid water in order to resume metabolic activity, the second can become active on sole atmospheric humidity. This explains, at least in part, the different distribution and ecology of these two groups of organisms, and in particular the high density of chlorolichens in areas that are characterized by masses of humid air in movement, as, for example, on coastlines and mountain peaks, where condensation phenomena often take place on exposed surfaces.

The survival capacity of lichens in difficult environmental conditions is generally accompanied by very restricted growth, which in any case is not constant, but articulated by phases with a duration that is species-specific. The annual increase of body mass is often less than 1%. The longevity of the individual seems to be greater in the crustose forms—with individuals often many centuries old—than in the foliose and fruticose forms.

The extent of radial growth of crustose thalli, which are almost circular in shape, can provide information on the age of the colonized surfaces, if correlated to the factor of time. Lichens have therefore been widely used—on the basis of geometrical methodologies—in order to date lithic surfaces (lichenometry), either as an alternative to or along with other dating techniques (C14, dendrochronology, thermoluminescence, etc.). Initially, the methodology—introduced in Italy in the northwestern section of the Alps toward the end of the 1950s—used lichens as indicators of absolute age, on the basis of the fixed ratio between the diameter of a crustose thallus and the age of the colonized rock formations. In the early 1990s, the procedure was reexamined critically, and its effective means of application were questioned, as it was recognized that many other factors could influence the growth and development of the lichenic thallus. It was thus demonstrated that it is not possible to consider the colonized surfaces and the measured thalli as coeval, even though the lichenometric technique can be useful in chronological reconstructions in areas of architectural and historic interest (Gallo and Piervittori 1993; Piervittori 2003).

Lichens are in general very sensitive to atmospheric pollution, such as sulphur dioxide and the oxides of nitrogen (Nash 1996), to the degree that they disappear in the most polluted areas (lichenic desert). Epiphytic species are widely used in studies of environmental biomonitoring, as bioindicators as well as bioaccumulators, of metals for instance (Nimis et al. 2002). The lichens that colonize rocky substrates (epilithic) are more rarely employed to this end, since the substrates on which they grow often contain carbonates or other compounds with a high buffering capacity and therefore significantly mask or reduce the effects of polluting substances. As a consequence, research is limited to determining the different densities of species that are tolerant and species that are sensitive to pollutants (Aptroot and James 2002). It might be of greater interest, instead, to emphasize that lichens may also have a bioprotective role, in addition to their biodeteriogenic action: in the presence of a high level of atmospheric pollution, the thalli do indeed provide an effective barrier against the pollutants that attack the mineral components of the substrate (Nimis et al. 1992).

3.5 BRYOPHYTES

by Sandra Ricci

Bryophytes are eukaryotic organisms; they are photoautotrophic, not vascular, connected evolutionarily to green algae, with alternating generations (gametophyte n/sporophyte 2n), in which the haploid phase is dominant. At present, the general consensus seems to be a classification scheme that subdivides bryophytes into three systematic classes: Anthocerotae, Hepaticae, and Musci; only the latter two classes play a significant role in the problems of biodeterioration of materials.

The generally small thallus of the bryophyte lacks tissues specifically designed for the transport of water and of mineral and synthesized substances. Bryophytes are, therefore, organisms that are strongly tied to water and that spread mostly in humid environments, having not only problems of reproduction, but also not possessing cell walls adapted to minimize water loss.

Mosses and liverworts are known colonizers of stone, on which they can exercise a biodegrading action. They penetrate the substrate by means of rhizoidal structures. A study carried out on *Tortula muralis*, a moss widely distributed on stones in outdoor environments, has shown that rhizoids were able to penetrate the crystalline matrix of oolithic calcareous stone up to a depth of five millimeters (Hughes 1982). Clearly, the penetration of rhizoids benefits from the destructive action of frost on the stone and, as a result, occurs more frequently on stone works in the outdoors and on substrates that absorb more water because of their structure and composition (calcareous rocks and tuffs).

Many mosses play an important role in the formation of organogenic carboniferous sedimentary rocks, with a spongy appearance and a more or less compact consistence, such as calcareous tuffs and travertine. In an aquatic environment, mosses and some algae form structures able to retain the calcium bicarbonate present in water and, subsequently, to form considerable incrustations (Charrier 1960). The deposition process is encouraged by the photosynthetic activity of organisms, which removes carbon dioxide and continuously moves the equilibrium of the reaction to the right, leading to the progressive precipitation of calcite, which coats the organisms while faithfully retaining their morphological details.

$$Ca(HCO_3)_2 \leftrightarrows CaCO_3 + H_2O + CO_2$$

The formation of organogenic calcareous deposits is more rapid in areas where the water flow is also rapid, both because of the increased evaporation and the increased potential for growth of the mosses. The lower portions of the gametophytes are progressively enrobed in the calcareous deposit, while the upper section remains free and allows the structures to grow.

In temperate zones, the mosses most involved with the formation of calcareous deposits are: *Eucladium verticillatum*, considered the most important and the most common crustose moss, *Barbula tophacea*, *Cratoneuron* spp.—a genus that includes many hydrophilic species—*Gymnostomum* spp., and *Philonotis* spp. Often associated with these are microorganisms such as cyanobacteria, green microalgae, diatoms, and also liverworts, which, as a whole, define the communities responsible for the biogenesis of travertine rocks. The mosses that form part of the deposit process take on an important role in the degradation of artistic fountains, which can at times be completely coated with thick incrustations. A striking example is that of the Fountain of the Dragons in Villa d'Este in Tivoli, where the development of crustifying mosses had led to incrustations several decimeters thick, completely obliterating the legibility of the work.

3.5.1 Structural and Morphological Characteristics

The class of Musci includes about 20,000 species, present in almost all types of environment. The dominant phase is the haploid generation—the gametophyte (producer of gametes)—usually constituted by small green cushions or felt-like structures adhering to a substrate, which are almost always perennial and able to live autonomously because of their capacity to photosynthesize. On the gametophyte, the sporophyte is formed, which represents the diploid generation; it lives as a parasite and has a short life (seasonal or annual), because it is not able to carry on photosynthetic activity (Fig. 3.16, Plate 9).

The morphology of the gametophyte is more reminiscent of higher plants than of tallophytes; it is made up of an elongated portion, called caulid, which bears flattened lateral appendices (the filloids) and basal filaments (the rhizoids), which act as an anchor. The caulid is variable in length, ranging from a few millimeters to several decimeters and can be either simple or branched. It is made up of two types of cell, hydroids and leptoids, the function of which is the conduction of water inside the plant as well as support (Giordano 1993; Ligrone 1993). The filloids are of different sizes and dimensions, are constituted by a single layer of cells, and can also exhibit a thickening of their middle section, similar to the veins found in the leaves of higher plants. The form of the cells that constitute the filloids can vary considerably and is used diagnostically to recognize the various species. The rhizoids are made up of cells devoid of chloroplasts, either colorless or pigmented a reddish-brown color; they have the double function of anchoring the moss to the growth substrate and of absorbing water and mineral salts from the environment.

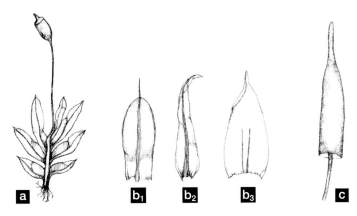

Fig. 3.16 *Morphology of Musci: (a) gametophyte with adult sporophyte; (b) different philloid morphologies; (c) sporophyte with caliptra*

The sporophyte is generally devoid of chlorophyll and lives as an epiphytic parasite on the gametophyte. It is made up of the seta, a multicellular filament, which can vary in length and has a foot at its bottom end. At the apex it has a capsule, also called urn or theca, used for the formation of spores. When maturity is reached, the capsule opens through the detachment of a kind of lid, the operculum, and from this opening the mature spores are released. The capsule can be of different shapes and sizes: it can be smooth or ornamented and can be covered by the caliptra, the presence and morphology of which can provide further diagnostic characteristics.

In nature, mosses are found in so-called growth forms, which are associations of several individuals of the same species that present constant morphological characteristics in their normal habitats (Gimingham and Birse 1957; Magdefrau 1982). Such growth forms are closely linked with the structure of the gametophyte and the position of the sporophyte on it. Mosses can be distinguished into acrocarpous and pleurocarpous mosses.

Acrocarpous mosses include species with a well-defined elongated main axis, on which the archegones are formed. They have two main growth forms: cushions, with individuals placed radially and having lateral ramifications with a hemispheric structure; and carpets, with individuals growing parallel to one another, either with a uniaxial or a branched growth habit. Acrocarpous mosses are widely distributed both in terrestrial and stony habitats because they have the advantage of an excellent reproductive strategy, characterized by a brief protonemal phase and by a high germination potential of the spores. In addition, they possess a high resistance to atmospheric pollutants and are therefore also widespread in urban environments (Proctor 1984).

Pleurocarpous mosses are characterized by a creeping growth, with highly developed lateral ramifications, on which the sporophytes are formed. They have the following growth patterns: the mats, made up of gametophytes that develop horizontally to the substrate and that have rhizoids distributed along all of the axes, and the turfs, formed by individuals gathered together in interwoven structures.

The class of Hepaticae includes some eight thousand species, belonging to one hundred and eighty genera, found particularly frequently in humid environments. In a similar way to mosses, liverworts have a dominant gametophyte generation, with a gametophyte with a highly variable morphology due to the existence of foliose and thalloid forms (Fig. 3.17, Plate 9). In the foliose cormoid forms, the tiny leaves are composed of a single cell layer; they are bilobed and are often incised so that they almost appear to be divided into two parts. The form and the distribution on the axis of the gametophyte as well as the shapes of the margins can also be used as diagnostic characteristics.

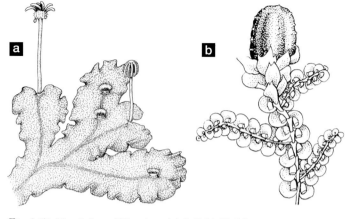

Fig. 3.17 *Morphology of Hepaticae: (a) thalloid; (b) foliose*

The thalloid forms have a laminar gametophyte, often multilayered and branched dichotomously. We can distinguish a lower layer—fixed to the substrate by means of the rhizoids and devoid of photosynthetic cells—whose function is to anchor the organism and provide reserves; and then some more superficial layers, similar to small lamina, which are intended for photosynthesis because of the presence of chloroplasts. The most external layer of cells is endowed with pneumatic chambers for protection—the contours of which are also visible to the naked eye as minute reticulations on the surface. The air present in these chambers reduces the effects of desiccation and also allows the necessary exchanges with the outside environment by means of chimney-shaped openings, which have a similar function to the stoma of higher plants.

3.5.2 Physiological and Reproductive Characteristics

Bryophytes have the same basic requirements as higher plants; but unlike plants, which have developed roots and an efficient system of conduction, bryophytes have a reduced capacity for absorption from the soil and are not able to control the loss of water from tissues (many bryophytes have no cuticle). They have therefore developed the capacity to absorb water vapor from the air and to utilize liquid water whenever it is available in the substrate, employing latent life forms as a survival strategy during unfavorable periods. As to their anatomic appearance, adaptations have been observed in the arrangement of filloids in both the hydrated state and the dehydrated state: in the latter, a drawing together of the tiny leaves along the axis takes place and, at times, also a rotation that gives the axis a helicoidal form.

As is the case with all autotrophic organisms, light—both natural and artificial—is an environmental factor that plays a fundamental role in the life of the bryophytes in that it determines photosynthetic activity.

Bryophytes have a life cycle that includes a haploid generation, in the form of the gametophyte, and a diploid generation, represented by the sporophyte. The gametophyte produces two types of multicellular gametanges, which form male and female gametes; it can be unisex in the dioic species that bear only one of the two gametanges, or hermaphrodite, in the case of monoic species in which both gametanges are present (Fig. 3.18).

Mosses can reproduce either sexually or asexually (vegetative reproduction or by means of spores). Sexual reproduction occurs through the fusion of flagellated haploid male gametes, the antherozoids, and female ones, the oospheres. The antherozoids have two flagella and are produced inside the male gametanges, the antheridia. The female gamete is the oosphere and is produced inside the female gametangium, the archegonium. Dur-

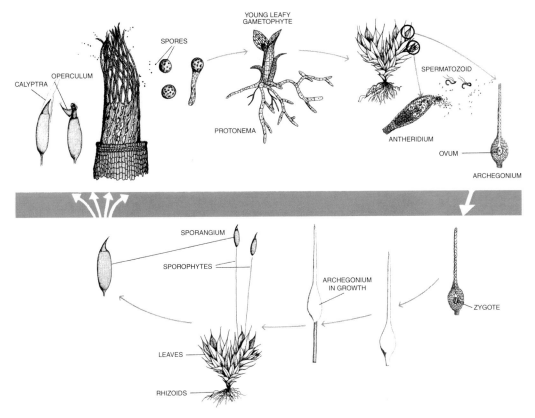

Fig. 3.18 *Reproduction and life cycle of bryophytes, in which the alternation between the haploid and diploid phases is visible (modified from Raven et al. 2002)*

ing the process of fertilization, the bryophytes are highly dependent on the presence of water, which allows the antherozoids to move and reach the upper opening of the archegonium, where fertilization occurs. The fusion of the two gametes forms a diploid zygote, which divides itself until it forms an embryo from which the sporophyte will originate.

Reproduction through sporogony occurs inside the capsule of the sporophyte where the haploid meiospores originate by meiosis. Once they fall into suitable environments, the meiospores germinate forming the protonema, which consists of a single line of cells containing chloroplasts, from which the gametophyte will then be formed.

Vegetative reproduction occurs through the formation and detachment of buds, which are groups of cells capable of reproducing the gametophyte. These structures develop in different parts of the gametophyte: in the axil of the fil-

loids, at the top of the caulidia, on the rhizoids, or on the protonema. Growth of protonemal buds has been noted in poorly lit environments, and their formation represents a selective advantage in that it allows the colonization of new environments (Whiterhouse 1980).

In liverworts, the reproductive structures are formed on the gametophyte: in the foliose species they are found at the extremities of the axes, while in the thalloid forms they establish themselves on specialized structures produced by the thallus. In the Marchantiaceae, a group of thalloid liverworts, thin umbrella-like pedunculate structures are formed, which bear both the male and female reproductive organs.

After fertilization, the zygote grows to form the sporophyte, which remains inside the archegonium for a long time. The spores are liberated when they reach maturity, either through the degeneration of the wall of the capsule, or through the formation of a circular fissure in it, or when the teeth open, as occurs in mosses. The germination of the spores produces a protonema that will eventually lead to the formation of the gametophyte. For liverworts, too, the life cycle is haplo-diplo-biontic, with clear dominance of the gametophytic generation.

3.5.3 Ecological Characteristics

Many of the species of mosses can be considered to be pioneer organisms, in the same way as algae and lichens, because they can colonize exposed rocky surfaces that are devoid of humus and, in some cases, are able to survive high levels of temperature and radiation from the sun as well as long periods of drought. Because of their capacity to survive in a latent form and then to return to normal metabolic activity with the restoration of favorable conditions of humidity, bryophytes are also considered to be poikilohydric organisms. There have been verified cases where their capacity for survival has remained unchanged after several months of dehydration, in which the water content of the tissues had fallen to as low as 10% or less of the dry weight.

It has been observed that the different growth forms have different capacities for absorption. The cushion or very dense carpet forms seem to be the most efficient, in that they allow an accumulation of water equivalent to 200–400% of the dry weight and have a strong capacity to reduce the rate of evaporation. The liverworts, more so than mosses, are dependent on habitats that are sufficiently damp or wet and are only found on walls that are rich in water or on paving rich in humus.

In general, bryophytes are equipped with adaptations that allow them to carry on good photosynthetic activity, even at very low light intensity: in comparison to a pine or a larch, which requires the equivalent of 1/9 of daylight, for many mosses the equivalent of 1/2,000 is sufficient. Studies carried out in caves have found mosses growing at light levels equivalent to 1/2,875 of those found outside (Cortini Pedrotti 1978). Species that live in shady conditions, such as caves or artificially illuminated hypogean habitats, show physiological and morphological adaptations that are linked to the scarcity of light. Both mosses and liverworts modify their reproductive cycle and are often without sporophytes (Tomaselli 1955). The thallus undergoes a number of modifications because it adapts to the dark and to the low temperatures (Proctor 1984). Because of the scarcity of light, the gametophytes become very slender and ribbon-like, the green coloring becomes paler, the internodes become elongated and the filloids smaller and thinner, and the vein network—when present—becomes less evident or disappears completely. In order to optimize the capture of available light, the filloids bend and orient themselves with the lamina towards the sources of light (Tosco 1970).

As far as temperature is concerned, many mosses are able to exceed the critical values without suffering permanent damage, because of the capacity of their cytoplasm to slow down the changes in temperature between cells and the environment.

As far as the influence of the substrate is concerned, the pH value represents the most influential factor for the growth of many bryophytes. There are species (acidophiles) with a high degree of affinity for substrates with an acid pH value, while others prefer alkaline substrates (alkalophiles), and still others that can be defined as indifferent to this factor.

Being exposed to processes of chemical degradation mostly linked to the corrosive and incrusting action of water, the surfaces of rocks in an outdoor environment exhibit incrustations with mineral characteristics that can sometimes be quite different from those of the initial lithotype. This phenomenon further affects the distribution of species relative to their affinities for particular mineral compounds. Furthermore, extraneous substances (such as atmospheric particles, humus, organic residues) can be deposited in the cracks of the rock, creating a microsoil that can be colonized even by species that are not strictly epilithic.

Bryophytes, in the same way as lichens, are highly sensitive to atmospheric pollution and are therefore very useful in the study of atmospheric deposits of anthropic origin (Cenci and Dapiaggi 1998). In early studies, mosses were used in the evaluation of pollution by heavy metals in Scandinavian countries (Ruhlimg and Tyler 1970) and were subsequently used in environmental monitoring (Rasmussen and Johnsen 1976; Cenci and Palmieri 1997; Aleffi 1992; Allegrini et al. 1994).

Because mosses do not possess a true root system and lack tissue for the conduction of water, they absorb nutrients from the atmosphere and, along with these, they absorb pollutants such as heavy metals. The concentration of such elements within the filloids can therefore be directly related to atmospheric deposits, excluding processes of absorption from the substrate.

There are, however, some limitations in the use of bryophytes as environmental indicators, mainly due to the high sensitivity of these organisms to desiccation, to their poor survival rate after transplantation, and to the frequent absence in urban environments of those species that are the most suitable for the evaluation of such deposits (Cenci and Dapiaggi 1998).

3.6 VASCULAR PLANTS

by Ettore Pacini and Maria Adele Signorini

Vascular plants (tracheophytes or cormophytes) are eukaryotic photosynthesizing organisms that are adapted to life on Earth and are made up of cells that exercise different functions; cells are organized into tissues with specific activities, and tissues are in turn structured into organs. Tracheophytes are diversified by an increasing level of organization and complexity and adaptation to different environments (Fig. 3.19).

Pteridophytes are vascular plants that reproduce by means of spores and that, for physiological and reproductive reasons, are confined to humid areas. Gymnosperms and angiosperms reproduce by means of seeds and are therefore collectively termed spermatophytes (plants with seeds). Gymnosperms carry their bare ovules (which will later become seeds) on cones or other specialized structures, while angiosperms have ovules and then seeds contained within an ovary, which is located inside a complex organ with a reproductive function: the flower.

In vascular plants, the roots form a complex system, which during growth exerts a considerable pressure on the substrate. When the substrate is constituted by ancient walls, architectural structures, or monuments, or in the case of hypogean environments, the growth of roots can cause severe physico-mechanical damage. The variables in the growth form and the characteristics of the root system—which will be discussed later—form the basis for a classification, proposed by Signorini (1995, 1996), of the levels of danger that plant species present for architectural works.

Several authors (Hruska Dell'Uomo 1979; Celesti Grapow et al. 1993–1994; Celesti Grapow and Blasi 2003; Caneva et al. 1992b, 1995a, 2003; Poli Marchese et al. 1997, 1998; Di Benedetto et al. 2000; Lisci et al. 2003) have compiled floristic lists of the plants most frequently encountered in Italian archaeological sites and monuments of historical interest; depending on the extent and complexity of the site studied, the compilations list several dozens of species and can reach several hundred in the case of large-scale monuments. In certain cases, this type of analysis has been extended to whole communities of plants (Caneva et al. 1995a), showing the most frequently encountered coenoses in ruderal environments present in archaeological contexts.

3.6.1 Structural and Morphological Characteristics

The morphological characteristics of vascular plants vary considerably according to their taxonomic classification.

In our latitudes, pteridophytes (which include Psilophytes, Lycopodiophytes, Equisetophytes, and Polypodiophytes) are generally perennial grasses, often endowed with underground-modified stems of the rhizome type. Thanks to the rhizomes, some pteridophytes are able to spread widely over an area; this is the case with certain equisetes (*Equisetum* spp.) in damp regions. The leaves differ according to the various species: in the equisetes (Equisetophyta) they are very small and replaced in their functions by green ramifications disposed as verticilla, while in the ferns (Polypodiophyta)

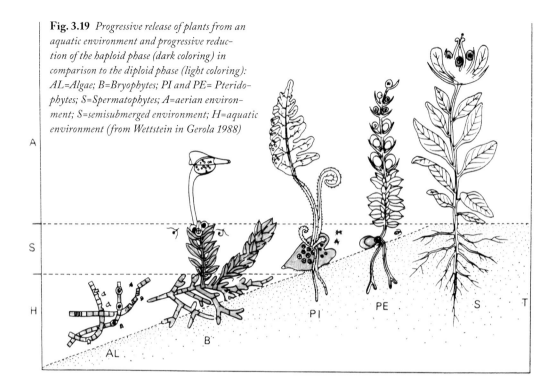

Fig. 3.19 *Progressive release of plants from an aquatic environment and progressive reduction of the haploid phase (dark coloring) in comparison to the diploid phase (light coloring): AL=Algae; B=Bryophytes; PI and PE= Pteridophytes; S=Spermatophytes; A=aerian environment; S=semisubmerged environment; H=aquatic environment (from Wettstein in Gerola 1988)*

they are well developed and generally composite, i.e., with the lamina divided into sections. On the underside of the leaves of many ferns, there are brown agglomerates formed by the sporangi, the organs in which the spores are produced (Fig. 3.20).

The gymnosperms or Pinophyta (coniferous and similar plants) are all woody plants, for the most part trees, but sometimes shrubs, as is the case of junipers (for example, *Juniperus communis*) (Fig. 3.21). The leaves are usually needle-shaped, as in pines and spruce, or else scale-shaped, as in cypresses. Unlike many angiosperms, only a limited number of gymnosperms have the capacity to regenerate secondary shoots or suckers from the stump after felling. In most gymnosperm species, the ovules are carried in specialized structures (called cones or strobili, commonly known as pine cones), formed by scales arranged around a central axis. When the ovules have matured into seeds, the scales of the cone open or fall in order to allow dispersion.

Angiosperms or Magnoliophytes are plants that bear flowers. The flowers may be large (for

example, *Magnolia*) or hardly visible (*Quercus*), depending on the mechanism of pollen dispersion: species that make use of insects as carriers generally have large showy flowers, with shapes, colors, or smells that attract insects; species that entrust their pollen transport to the wind generally have small flowers, without smell and often without a corolla. Angiosperms are characterized by a great variety of forms: they range from gigantic trees to annual grasses, shrubs, lianas, epiphytic plants, succulents, etc. The organs of the plants may also take on very different forms (Plate 10), thus making it possible for the main vegetative and reproductive functions to be carried out in the most varied environments.

In a cormophyte, we generally recognize three vegetative organs, each with different functions and morphologies: the root, the stem, and the leaf. To these are associated the reproductive structures of flowers and seeds.

The root fulfills several functions: by means of the root hairs located near its apex, it absorbs water and salts from the ground (as long as there are no mycorrhyzic symbioses, i.e., with

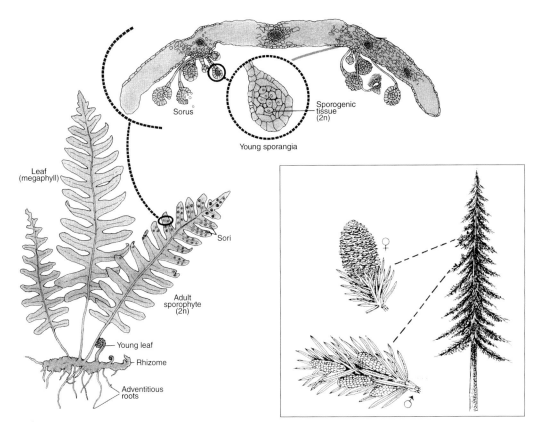

Fig. 3.20 *General morphology and structure of Pterido-phytes with detail of the reproductive structures (modified from Raven et al. 2002)*

Fig. 3.21 *General morphology and structure of gymno-sperms, with male and female reproductive structures*

specialized fungi); it stabilizes the plant on the soil; and it can accumulate reserve substances. The root system of a plant can be of the tap-root type, with a main root and with or without secondary roots; or of the fasciculate type, with roots that are all more or less of the same size (Fig. 3.22). Taproots tend to grow in more depth, which can prove dangerous for certain types of architectural works. Fasciculate roots develop more along the surface: this means that they can be less damaging if they belong to grasses, but not if they are those of a tree. Adventitious roots originate from the stem or from other organs rather than from the roots, as in the case of a branch of ivy that attaches itself to a wall. Sometimes roots act as storage organs for nutritional substances and thus become enlarged and modified through the accumulation of reserves, as is the case of the

taproots of the *Ferula communis* and *Smyr-num olusatrum*, both of which are commonly found in archaeological contexts (Fig. 3.22). The dimensions of the root system of a plant are very variable, and the roots of a tree can expand to cover tens of meters, both on the surface and in depth (Cutler and Richardson 1981; Kutschera and Lichtenegger 2002). The root system of plants growing in arid environments is usually more extensive than that of plants growing in a damp environment.

The surface of plant root tips is usually acid; G. Caneva and A. Altieri (1988) report that for certain ruderal species the values are between pH 5 and 6, while for other species even lower values have been reported in the literature. Various acid substances that are on the surface of the roots can be the cause of biocorrosion (Chapter 1, section 4).

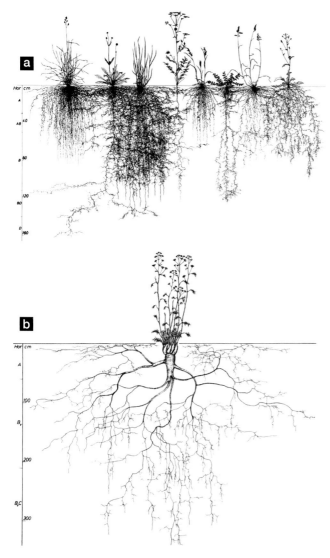

Fig. 3.22 *Different typologies of root systems: (a) grasses in a central European grassland; (b)* Ferula communis *with hypogean organs with a reserve function (from Kutschera and Lichtenegger 2002)*

enlargement of the cells. The stems of grasses and of woody plants are very different from each other. Normally they grow vertically, but in some plants also (and in some cases exclusively) horizontally: if they grow along the ground these horizontal stems are called stolons; if they grow below ground, they are called rhizomatose stems or rhizomes. The rhizomes are articulated into nodes and internodes, and the leaves are replaced with scales (for instance, *Iris*). Other underground stems are tubers—which are rounded in shape and have buds on the surface (like the potato)—or bulbs, with their apices protected by fleshy leaf sheaths, such as the onion. Underground stems are organs for vegetative reproduction and for the accumulation of reserve substances. Other types of modified stems can be found in arid environments, as in the case of the Indian fig (*Opuntia ficus-indica*) or the Knee Holly (*Ruscus aculeatus*)—which have aerial stems that look like leaves (cladodes), while the true leaves are reduced to scales or thorns in order to limit transpiration—or the case of succulent plants, with modified stems that function as water reserves, while the leaves are often reduced to thorns.

Leaves are quite varied in their morphology, but usually they possess an expanded part (the lamina)—which can have very different characteristics in its base, apex, and edges—and a petiole, which sustains the lamina and connects the vascular tissues of the leaves (the veins) with those of the stem, but which is not present in sessile leaves. The leaves are covered by the epidermis, a protective layer covered with a waxy surface, the cutin, and endowed with stomata, the opening of which can be regulated and which allow gaseous exchanges to occur. Inside the leaves are the parenchyma, which are the tissues where photosynthesis and transpiration take place. Leaves are very plastic structures that

The stem bears the photosynthesizing organs (leaves) as well as the reproductive ones and ensures the connection between these and the roots. Located in the stem are the conductive tissues: xylem, or wood, and phloem, or liber. The point where the leaves attach to the stem is called the node, and the stretch between the nodes is the internode. The stem grows in length and branches out due to the action of the meristems at the apex and the subsequent

can be modified considerably to meet different ends, even reproductive ones, such as in flowers. Natural selection has favored the appearance of many different kinds of leaves with the capacity to carry out photosynthesis and transpiration in very different environments. Plants in sunny environments have small leaves, with a thick layer of cutin; in shady and humid environments, the leaves are thin and broad instead. Even on the same plant it is possible to find thin shade leaves on one branch and thicker sun leaves on another (Heterophyllia). Heterophyllia can be even more extreme: in ivy, the leaves that creep along the ground in the shade have a different form from those exposed to light, which also bear flowers.

Woody plants can be either evergreen or deciduous: in the first, the leaves last for two or more years, and each year a few leaves fall

and are replaced by new ones; in the second, the leaves appear in spring and fall during autumn when unfavorable conditions begin. A few Mediterranean species (also common in archaeological sites) such as the broom (*Spartium junceum*) lose their leaves at the beginning of summer as a safeguard against aridity.

The Biological Forms define the physiognomy and the general structure and specify the habitus and the life cycle of the plant (Fig. 3.23); the different forms are distinguished by the different degree of protection of the buds. This is closely correlated to the climatic conditions of the site; as a result, the percentage of the various biological forms (biological spectrum) in a given area is commonly used as a bioclimatic indicator.

In annual grasses (T = therophytes), the life cycle comes to an end after a single season's growth, in which the seed germinates, vegetative growth occurs, reproduction (that is the formation of flowers and fruit) takes place, and finally the plant dies. Usually, annual grasses are small, but some, such as maize, can grow to considerable proportions. *Conyza canadensis*, which can grow along roads and on buildings, can exceed a meter in height and has a deep taproot that can cause severe damage to works.

In their first year of existence, biennial grasses (Hbienn = biennial hemicryptophytes) develop a basal rosette of leaves and a taproot, in which they accumulate a reserve of nutrient substances; in their second year, they produce flowers and fruit, and then die.

Perennial grasses have a life cycle of many years and can have buds at ground level (H = perennial hemicryptophytes), as in the case of the dandelion (*Taraxacum officinalis*); or else, the aerial part can disappear during the adverse season, but the plant survives by means of buds beneath the surface and perennial underground organs such as rhyzomes, bulbs, or tubers (G = geophytes). Dog's tooth grass (*Cynodon dactylon*) and the stinging nettle (*Urtica dioica*) are examples of rhizomatose geophytes that also grow in ruderal

Fig. 3.23 *Schematic representation of the main biological forms: (1) phanerophytes; (2) chamaephytes; (3) hemicryptophytes; (4) geophytes; (5) therophytes (modified from Raunkier, in Pignatti 1995)*

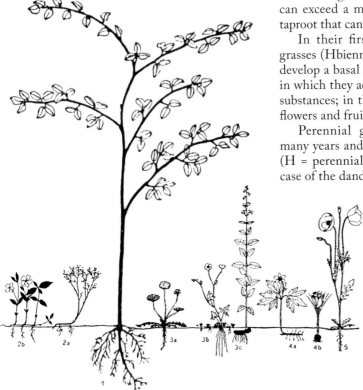

Table 3.4 Classification of species according to their level of danger for architectural works

Biological Form	Invasiveness and Vigor	Root System
0 Annual plants	0.0 Non-creeping with normal development	0.0.0 No taproot 0.0.1 Weak taproot 0.0.2 Strong taproot
	0.1 Creeping, with normal development	0.1.0 No taproot 0.1.1 Weak taproot 0.1.2 Strong taproot
	0.2 With very vigorous development	0.2.0 No taproot 0.2.1 Weak taproot 0.2.2 Strong taproot
1 Biennial plants	1.0 Creeping and non-creeping	1.0.0 No taproot 1.0.1 Weak taproot 1.0.2 Strong taproot
2 Perennial grasses	2.0 Mosses and lichens	2.0.0 No taproot
	2.1 Non-invasive grasses, or with weak development	2.1.0 No taproot 2.1.1 Weak taproot 2.1.2 Strong taproot
	2.2 Invasive grasses, or with very strong development	2.2.0 No taproot 2.2.1 Weak taproot 2.2.2 Strong taproot
3–4 Shrubs	3.0 Suffruticous plants	3.0.0 Minimal invasiveness 3.0.1 Average invasiveness 3.0.2 Great invasiveness
	4.0 Not sucker-forming shrubs, or of small size	4.0.0 Minimal invasiveness 4.0.1 Average invasiveness 4.0.2 Great invasiveness
	4.1 Shrubs with suckers	4.1.0 Minimal invasiveness 4.1.1 Average invasiveness 4.1.2 Great invasiveness
	4.2 Shrubs with root suckers	4.2.0 Minimal invasiveness 4.2.1 Average invasiveness 4.2.2 Great invasiveness
5 Lianas	5.0 Not sucker-forming	5.0.0 Minimal invasiveness 5.0.1 Average invasiveness 5.0.2 Great invasiveness
	5.1 Sucker-forming	5.1.0 Minimal invasiveness 5.1.1 Average invasiveness 5.1.2 Great invasiveness
6 Trees	6.0 Not sucker-forming	6.0.0 Minimal invasiveness 6.0.1 Average invasiveness 6.0.2 Great invasiveness
	6.1 Forming stump suckers	6.1.0 Minimal invasiveness 6.1.1 Average invasiveness 6.1.2 Great invasiveness
	6.2 With root suckers as well	6.2.0 Minimal invasiveness 6.2.1 Average invasiveness 6.2.2 Great invasiveness

The danger level of each parameter increases from top to bottom.
Signorini's Danger Index (DI, or IP, Indice di Pericolosità, see p. 143) is obtained by adding the numbers in the last column; it ranges from 0 to 10.

DI up to 3: not very dangerous species
DI 4–6: averagely dangerous species
DI 7 and above: highly dangerous species

environments; garlics, both wild and cultivated (*Allium* spp.), are bulbous geophytes.

Suffrutices (Ch = chamaephytes) such as the lesser calamint (*Calamintha nepeta*) are plants with a small perennial woody part and an abundant grassy section, which bears the flowers and dies at the end of the vegetative season.

Shrubs (Pcaesp = Caespitose phanerophytes), such as myrtle (*Myrtus communis*), are woody plants, branched at ground level; in the bramble (*Rubus ulmifolius*), the branches can root at the apex, permitting a rapid diffusion of the plant.

Lianas (Plian = Lianose phanerophytes) are perennial plants that climb on supports, for example, ivy (*Hedera helix*) and the wild clematis, or old man's beard (*Clematis vitalba*).

Trees (Pscap = Scapose phanerophytes) are woody plants branched at a certain distance from the ground. In general, they are of a considerable size and have an extensive root system. With the exception of most gymnosperms, trees will issue suckers from the stump after the tree has been felled. Particularly invasive trees produce suckers even from the roots, the false acacia (*Robinia pseudoacacia*), for example, and the tree of heaven (*Ailanthus altissima*).

On the basis of the biological forms, Signorini (1995, 1996) has proposed a classification of the levels of danger presented by plant species to architectural works (Table 3.4).

3.6.2 Physiological and Reproductive Characteristics

Aside from a few rare exceptions of parasitic species, all vascular plants carry out photosynthesis, which takes place thanks to the photolysis of water, which breaks down and donates electrons; the process takes place mostly in the leaves and the green stems. The starch accumulated during the day in the chloroplasts is broken down into simple sugars during the night; these are then either distributed to all the nonphotosynthesizing areas of the plant so that these may carry out their metabolic functions, or they are accumulated in other organelles that act as reserves (amyloplasts). Another part of the absorbed water evaporates through transpiration, thus favoring the uptake of more water and mineral salts for the metabolism of the cells.

The development of the plant occurs thanks to the multiplication of the cells of particular tissues (meristems), followed by dilation. The apical meristem is responsible for longitudinal growth. In woody plants, part of this growth activity also takes place through the vascular cambium, a meristem that forms wood on the inside and phloem on the outside, while another meristem (cork cambium) produces the protective tissues that roughly correspond to what we know as the external bark in trees.

The three parts of the corm (roots, stem, leaves) are traversed by two systems of conduction/transport: one rising and the other descending. The ascending, or woody, system carries raw lymph, i.e., water containing mineral salts in solution; the descending system, also called phloem, transports the nutrients from the leaves toward all the other parts of the plant.

Vascular plants have a haplo-diplo-biontic ontogenetic cycle with the dominant phase being the diploid one (sporophyte); the haploid phase is masked and limited to unisexual gametophytes, which develop in the structures devoted to sexual reproduction (Fig. 3.19). They have different means for reproducing, i.e., to grow in number and spread into neighboring environments, exploiting the characteristics of the environment in which they live. They can reproduce by means of sexual reproduction, i.e., give rise to a new individual through the union of two gametes (fertilization); or else by asexual reproduction, also known as vegetative or agamic propagation, which occurs through the process of regeneration. The concept of reproduction also includes that of dispersion, i.e., the diffusion into environments adjacent to those of the parent organisms. This happens with both types of reproduction, although the structures dispersed with sexual reproduction usually spread farther than those propagated vegetatively.

In pteridophytes, the reproductive structure dispersed into the surrounding environment is the spore, produced in the sporangia. The spore germinates in the soil to form a microscopic plant (prothallus), which in turn produces the male and female gametes. From the union of the gametes originates an embryo that comes into being on the prothallus and then develops into the adult plant.

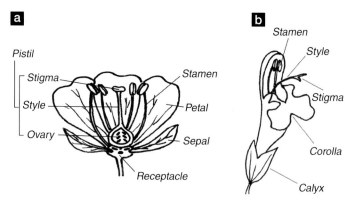

Fig. 3.24 *Schematic drawing of a hermaphroditic type of flower, composed of superimposed verticilli inserted on the receptacle of (a) an actinomorphic flower and (b) a zygomorphic flower*

Seed-producing plants (gymnosperms and angiosperms) have two reproductive structures that are dispersed into the environment: the pollen and the seed. The first is the means by which the male gamete is transported; the second contains a latent embryo, accompanied by reserves and protected by teguments. Pollen is made up of granules, within which the male gametes are formed, while the female gametes are located inside the ovules. In gymnosperms, the granules of pollen reach the ovule, which is in contact with the air, and produce a little tube, which allows the male gametes to reach and fertilize the female ones. The fertilization originates the embryo while the surrounding ovule is transformed into the seed. In angiosperms, the reproductive processes take place inside the flower, a system of highly specialized and perfected structures (Fig. 3.24).

Proceeding from the outside inward, we observe the following structures. Sepals form the calyx, the function of which is to protect the other elements of the flower in the early stages of their development. Petals form the corolla, whose main role is to attract pollinating insects onto the flower. The androecium is formed by the stamens (the male reproductive organ), and their fertile section (the anther, carried on the top of filaments) produces the pollen. The gynoecium is formed by one or more pistils (the female reproductive organs); in each pistil, it is usually possible to discern an ovary, which contains one or more ovules, as well as a style and a stigma, where the grains of pollen land.

Not all flowers have all the elements: for example, flowers that are wind pollinated often lack the corolla and sometimes even the calyx; unisexual flowers often have no stamens or pistils. The grains of pollen carried by insects or the wind are deposited on the stigma and emit a small tube through which the male gametes reach the female gametes inside the ovules. Once fertilization has occurred, the ovule transforms into a seed, and the ovary becomes the fruit.

The seed is the reproductive structure of spermatophytes and contains and carries the embryo. The size and weight of seeds can vary: from less than a millimeter to almost a decimeter, and from a few milligrams to over a kilogram. This entails a different investment from the mother plant as well as different modalities for the dispersion and growth of the new plant. The seeds are dispersed in order to avoid competition with the mother plant and its sibling seeds and also to colonize new environments. The dispersed structure can either be the seed itself, which comes out of the fruit at maturity, or the fruit when it contains only one seed.

The efficiency of the dispersal and the distances achieved during the process are influenced by the dispersion modalities, by the type of seed (or fruit), and by the period during which dispersal occurs. The different forms of dispersal can be categorized as: autonomous, active, passive, and induced (Pacini 1991). Dispersion is autonomous when the explosive opening of the fruit ejects the seeds; the distance achieved is rarely more than a few meters. In active dispersal, the seeds (or fruits) are provided with structures such as wings, hairs, or setula; in favorable conditions they can cover distances of more than one hundred meters. Plants with this method of dispersal—such as the dandelion, but also trees such as elms—normally live in open spaces, as such a method would not be very effective in a forest. Passive dispersal takes place with small seeds (or fruit) transported by wind or water; the distances covered easily exceed a hundred meters when the carrier is the wind. In induced dispersion, the seeds and fruit structures induce some animals to gather (actively or passively) and subsequently disperse them. Fleshy fruits are ingested by birds or animals, which then disperse the seeds through their feces—the structure of the seeds ensures that they are not damaged by gastric juices. Seeds carried actively or passively by birds or mammals can be released and subsequently grow on buildings (Lisci et al. 2003). Seeds with appendices with high lipid content attract ants, which gather and transport them (Lisci et al. 1993). When birds are the carriers, the seeds may even be scattered over several kilometers, while ants seldom carry them for more than a hundred meters.

Vegetative propagation is based on a unique characteristic of plants: regeneration, i.e., the capacity to regenerate missing parts. Unlike animal cells, plant cells retain the capacity to divide and form new types of cells.

Spontaneous vegetative regeneration is widely found in cormophytes. Brambles spread with branches that root at their tips; many perennial grasses such as dog's tooth grass (*Cynodon dactylon*) propagate through the fragmentation of stems modified to become stolons, rhizomes, bulbs, etc. Woody plants such as the tree of heaven (*Ailanthus altissima*) spread through suckers and secondary roots from the stump after disturbance.

In general, plants that propagate vegetatively also maintain the capacity to reproduce sexually. Vegetative propagation allows a rapid diffusion in uniform environments, while sexual reproduction allows the possibility of the development of more favorable characteristics, in case environmental conditions change.

3.6.3 Ecological Characteristics

Plants that grow on monuments have specific anatomical and ecophysiological characteristics that allow them to survive in such an inhospitable environment. Indeed, the soil available for growth is limited on a wall, unless it backs onto an embankment, as is the availability of water and the possibility of retaining it (Plate 10).

The morphology of cormophytes exhibits considerable variations in size, appearance of the various parts, and duration of the life cycle. We could say that the characteristics of each species are a response to the environment in which it lives and to the competition that is established between the various organisms. In each environment, the types of plants and the number of individuals of each species that are present are the result of a series of interactions leading to equilibrium among all the components.

The three parts of the corm can change by shrinking, growing, or specializing themselves according to the characteristics of the environment; this is what produces the differences in the structure of cormophytes. The availability of water limited to certain periods of the year has induced certain plants to: reduce their leaf mass; store water in certain organs; make growth periods correspond with those of availability of water and nutrients; encourage growth of roots and subterranean stems toward sites where there is greater availability of water; or slow down the metabolic rate until the arrival of optimal conditions. Modifications in plants can also be of a physiological order, i.e., take place through biochemical mechanisms that enable water to be retained and employed most efficiently. Succulents such as *Sedum* are endowed with specialized tissues that can retain water even in extreme conditions and can exhibit modifications in the metabolic processes adapted to an arid environment; some commonly found ferns such as *Ceterach officinarum* and *Polypodium vulgare* slow down their life processes in periods in which there is limited availability of water.

There is a succession in the colonization of walls (Lisci et al. 2003). In the beginning, it is generally mosses that collect atmospheric particulates, thus increasing the substrate; on this substrate will develop plants that are not very demanding, usually small and annual; and finally, larger plants take root. Sometimes, but rarely, vascular plants may also be pioneers.

The settlement occurs mostly with wind-dispersed seeds (for instance, *Parietaria* sp.) or through animals, which eat fleshy fruits and release the seeds in their feces. Diffusion on the architectural structure itself occurs mostly though vegetative reproduction, more rarely through seeds. The most common plants on walls are perennial grasses; annuals are more sporadic or else are limited to the initial stages of the colonization. Weed control intervention by chemical treatments or by mowing—performed in archaeological sites with the intent of stopping the development of vegetation toward more mature stages—can result in an increase in annual species (Celesti Grapow et al. 1993–1994).

For plants living on walls, M. Lisci and E. Pacini (1993) have distinguished eleven microsites that differ in position and ecological characteristics (Fig. 3.25).

Man-made pollution disrupts—to different degrees depending on the kind of pollutant—the equilibrium reached between all the organisms in a given environment. As a consequence, it is possible that conditions favorable to the growth of certain plants may be reduced, thus causing the plants to be damaged or disappear altogether. Indeed, one of the most common responses to pollution is the reduction in the number of species per unit of surface. Only the most resistant species, i.e., those able to survive the altered equilibrium, manage to stay. After this, starting with these species, it is slowly possible to reach a new equilibrium, determined by the new environmental conditions.

As we mentioned, atmospheric dust can favor the formation of a substrate for the growth of plants. However, other kinds of pollutants can act on plants, for instance, sulphur dioxide, which forms sulphuric acid in the presence of humidity in the air; although the plants most commonly found on walls do not seem particularly sensitive to pollutants.

Fig. 3.25 *Main kinds of microsites for the establishment of plants on buildings. Inside cavities: (1) at ground level; (2) on an inclined surface; (3) at the interface between two different construction materials; (4) on a homogenous vertical surface; (5) on a top horizontal surface; (6) at the juncture of a horizontal and a vertical surface; (7) between two vertical surfaces. On the formed substrate: (8) on a porous horizontal surface; (9) on ruins composed of a variety of materials; (10) in the space between a wall and a slab of stone or marble; (11) from an underground stem originating from a plant rooted in the ground, which has grown through the wall and exited on the opposite side (adapted from Lisci and Pacini 1993)*

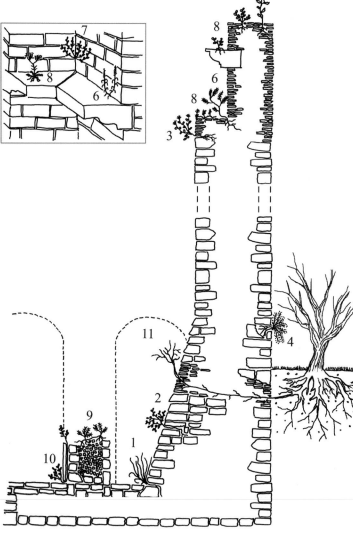

CHAPTER 4

BIODETERIORATION PROCESSES IN RELATION TO CULTURAL HERITAGE MATERIALS

Biodeterioration phenomena exhibit specific characteristics in relation to their constituent materials; in this chapter we will analyze the fundamental mechanisms and the taxonomic categories involved in the processes.

In order to give an objective evaluation of the role played by the different microorganisms in the degradation of materials—especially of those whose function as nutritional source for the groups involved is not evident—it is necessary to establish common, standardized laboratory procedures, so that the results obtained by the various researchers can be compared. The methodologies used for the identification and the quantitative assessment of microorganisms can greatly influence the results. In addition, molecular identification techniques should be used alongside the more traditional methods; only in this way is it possible to arrive at a deeper understanding of the microbial ecology of deterioration (Chapter 9).

The subject of this chapter is closely related to previous chapters as far as the general processes of biological alteration, the influence of environmental parameters, and recurring examples of biodeteriogens are concerned; these will therefore not be covered again. At the same time, the information given here is a necessary introduction to better understand which are the most important processes of deterioration in relation to the various environmental typologies where cultural heritage can be preserved (Chapter 5).

4.1 MATERIALS OF PLANT ORIGIN

4.1a General Characteristics of Materials of Plant Origin

by Corrado Fanelli, Oriana Maggi, Anna Maria Persiani, and Paola Valenti

In this section, we will describe the fundamental characteristics of those substances of plant origin that perform a structural action and that are found in the constituent materials of works of cultural heritage. Other substances—those that have a reserve function (starch, lipids, some proteins), or are catalysts (enzymatic proteins), or else act as depositories of genetic information—will be omitted because they play a negligible role in the plant tissues under consideration here.

Cellulose is the most important polymer present in the walls of plant cells and also the main component of plant-based materials (paper, wood, cotton, linen, etc.). From a chemical point of view, it is a polysaccharide composed of molecules of D-glucose (the monomer), held together by β-1,4 glucosidic bonds between the hydroxyl group in position 1 of one molecule and the one in position 4 of the adjacent molecule, forming a linear polymeric chain of more than 10,000 glucose residues. Each glucose molecule is rotated by 180° in relation to the adjacent one, and each chain is bound to another by means of hydrogen bonds and Van der Waals forces (Fig. 4.1).

The cellulose chains are placed parallel to one another and come together in bundles, which are called microfibrils. In some areas (crystalline zones), the chains of cellulose are positioned in a precisely defined crystalline network, i.e., they are rigorously parallel to one

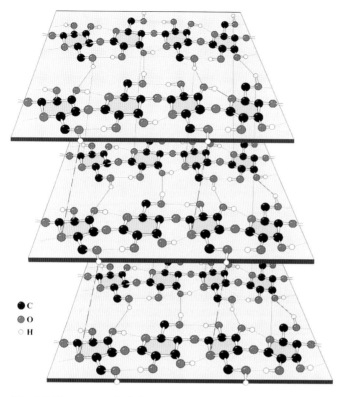

● C
◐ O
○ H

Fig. 4.1 *The structure of cellulose*

called fibrils, and it is these that, together, make up the fibers (Figs. 4.2 and 4.3). The chemical formula of cellulose is $(C_6H_{10}O_5)_n$, where n represents the degree of polymerization (DP), i.e., the number of times the unit (or monomer) is repeated to make up the individual molecule; this value defines its mechanical resistance.

In many plants, the cellulose fibers are closely associated with other substances, especially hemicellulose and lignin, as well as tannins, minerals, resins, and pectic substances.

Hemicelluloses are located between the cellulose fibers and lignin and, depending on the kind of wood, they make up about 20–30% of the plant biomass. Hemicelluloses are heteropolymers, linear or branched, of D-xylose, L-arabinose, D-mannose, D-glucose, D-galactose, and D-glucoronic acid; many hemicelluloses contain between two and six of

another and at fixed intervals. This arrangement makes it difficult for outside agents to attack. In other areas (amorphous zones), the chains are positioned in a disorderly and less compact fashion, and these areas are more liable to attack by microbial agents. The microfibrils are in their turn grouped together to form larger filaments, these sugars, among which the most important are the xylans and the glucomannans. Their structure is similar to that of cellulose, with the difference that the average degree of polymerization of hemicelluloses is generally between 100 and 200, and the molecules are often branched.

Fig. 4.2 *From left to right: (a) plant cells; (b) macrofibrils; (c) microfibrils; (d) cellulose chain; (e) glucose molecules (modified from Raven et al. 2002)*

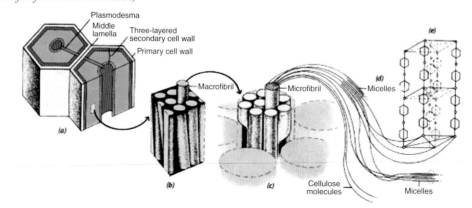

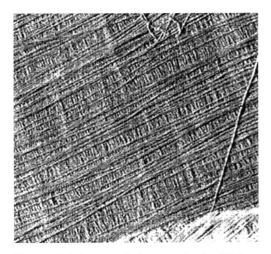

Fig. 4.3 *Cellulose microfibrils examined under SEM*

Lignin is the most abundant aromatic compound on Earth, and it is second only to cellulose in its contribution to the Earth's biomass. Lignin is a general term that covers a wide group of aromatic polymers, insoluble in water, with a highly complex three-dimensional structure formed by

Fig. 4.4 *Part of a lignin macromolecule*

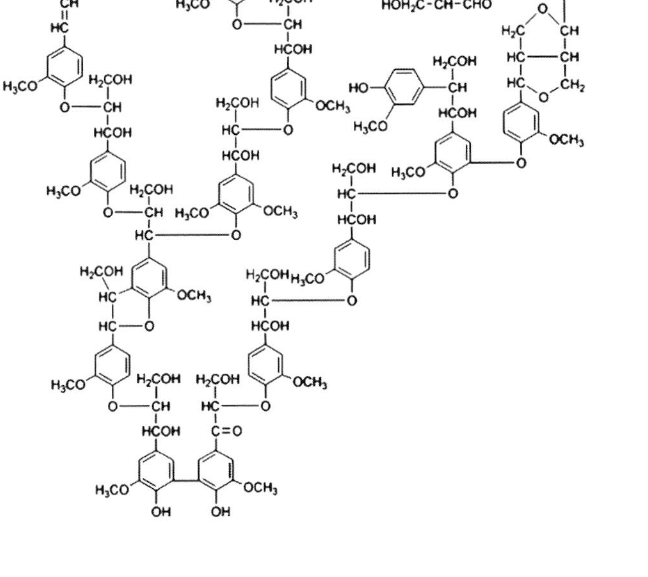

the repetition of units that have the basic structure phenylpropane (essentially made up of three aromatic alcoholic monomers: cumaric, coniferyl, and sinapyl alcohols, in varying percentages depending on the kind of wood) (Sjöström 1993) (Fig. 4.4). Lignin has an amorphous structure, low viscosity, and is extremely resistant to concentrated sulphuric acid. Its molecular weight in wood is very high and not easily measured.

4.1b General Processes of Biodeterioration of Materials of Plant Origin

by Giovanna Pasquariello, Oriana Maggi, and Anna Maria Persiani

Cellulose degradation is carried out by cellulolytic microorganisms able to produce a system of enzymes (C_1, C_x, β–glucosidase) acting in succession, known as cellulase complex; these hydrolyze cellulose, producing water-soluble sugars. The enzyme C_1 (exo-β-glucanase) hydrolyzes the β-glucosidic bonds that are near the endings of the cellulose molecules, releasing fragments of varying lengths. Subsequently, the enzyme C_x (endo-β-glucanase) hydrolyzes the β-1,4 glucosodic bonds between two adjacent glucose molecules inside the cellulose molecule. A third enzyme, a β-glucosidase, completes the degradation by hydrolyzing the water-soluble cellobiose and cellodextrins to glucose.

Although in the depolymerization of cellulose the three main classes of enzymes (endo- and exocellulase and β-glucosidase) are always involved, it is known that many organisms produce a variety of isozymes and cellulases that attack cellulose with a different stereochemical modality; sometimes, an additional enzyme is produced: glucohydrolase. Many deuteromycetes of the soft rot in wood, for example, are not able to produce exocellulase, and recent studies have shown that a system of oxidizing enzymes acts alongside endocellulase in the degradation of cellulose (Eaton and Hale 1993) (Fig. 4.5).

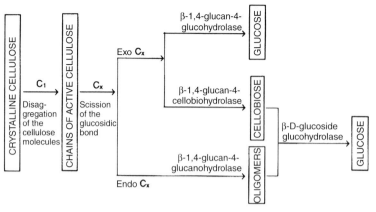

Fig. 4.5 *Schematic representation of the hydrolysis of cellulose by enzymes of the cellulase complex*

The enzymes of the cellulase complex are inducible: their synthesis is induced by the presence of cellulose and inhibited by the presence of sufficient quantities of glucose for metabolism to take place (Eriksson et al. 1990). Many cellulolytic microorganisms (bacteria, fungi) are not able to synthesize enzymes with an exoglucanasic activity (C_1) and therefore will only act on amorphous cellulose (Allsopp and Seal 1986). Some microorganisms—belonging to the ascomycetes (*Chaetomium*) and the deuteromycetes (*Trichoderma*)—are able to break down cellulose in its native form (for instance, cotton fibers) because they are able to produce both C_1 and C_x enzymes and are therefore considered to be real cellulolytics.

Hemicelluloses are hydrolyzed by most bacteria and fungi through the production of hemicellulase, both glucoside hydrolase and carbohydrate esterase (Vasella et al. 2002); these enzymes can be constitutive—produced independently of the growth substrate—or inductive—produced only in the presence of hemicellulase in the substrate.

Lignin is particularly difficult to biodegrade and it also reduces the bioavailability of the other constituents of the cell wall, thus functioning as a physical barrier. Some microorganisms, in particular white-rot fungi (basidiomycetes), have the necessary enzymatic equipment required for its degradation. The three main fungal enzymes are lignin-peroxydase (LiP), manganese-peroxidase (MnP), and laccase (Lac); they are not always produced by the same fungus and are responsible

for the initial depolymerization of lignin (Hatakka 1994). These enzymes are oxidizing, extracellular, and inducible. Depending on the genus of the fungus, different combinations of lignolytic enzymes are present: some secrete LiP and MnP (and not Lac), others produce MnP and Lac (but not LiP). Both Lac and MnP can be found in several isoenzymatic forms. Since lignin consists of carbon-carbon units and ether bonds, the enzymes are oxidizing rather than hydrolizing. They act in a nonspecific way on other molecules (for instance, on hydrogen peroxide, veratryl alcohol), generating free radicals that are highly reactive; unstable; that give rise to a series of oxidation reactions in the presence of oxygen; and that lead to the breakdown of lignin (Blanchette 1995). In spite of recent discoveries, not all the mechanisms involved in the process of degradation are fully understood yet. Because lignolysis is a nonselective process with low specificity, lignolytic enzymes are also able to oxidize—thanks to their high powers of oxidation—a large range of organic aromatic pollutants (dioxins, pentachlorophenol, polycyclic aromatic hydrocarbons).

In nature, lignin is probably broken down by a whole range of microorganisms: fungi (basidiomycetes, some ascomycetes and deuteromycetes) and aerobic bacteria, among which are the actinomycetes, as well as some anaerobic bacteria and fungi (Durrant 1996).

4.1.1 Wood

by Stefano Berti, Corrado Fanelli, Sabrina Palanti, and Flavia Pinzari

4.1.1a *Structure and Composition*
Wood, or xylem, consists of a complex system of fundamental tissues with the function to conduct and to act as a support; they are produced by a ring of meristematic cells, called the vascu-

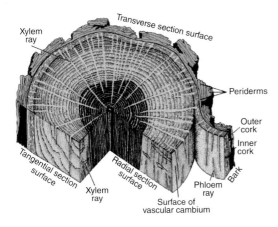

Fig. 4.6 *Radial section of a tree trunk showing the relationship between bark and xylem, and distinguishing the duramen and the alburnum on the inside (modified from Raven et al. 2002)*

lar cambium, which becomes active during the second year of life in shrubs and trees and determines secondary growth and girth. In arboreal species, the greater part of the biomass—in particular that of the trunk, roots, and branches—is formed by wood, while all the tissues external to the cambium form the bark (Fig. 4.6).

With only a few exceptions, Gymnosperms are by and large all woody plants: among these, conifers or needle-forming species (such as those belonging to the Cupressaceae, Pinaceae, and Taxidiaceae) take on particular relevance as they include many arboreal species that produce highly prized woods. Numerous angiosperms (broad-leafed) are also woody plants: they include many arboreal species (in the dicotyledons or Magnoliopsida) that are important for the production of timber (for instance, in the Fagaceae, Juglandaceae, Salicaceae, Tiliaceae, Aceraceae). In monocotyledon angiosperms (Liliopsida), there is no true formation of wood, although some species at times take on an arboreal habit (for instance, palms). At an anatomic level, the various cells present can give rise to a more or less homogenous structure (the homoxylous wood found in Gymnosperms or softwood), or to a heterogeneous one (the heteroxylous wood of angiosperms or hardwood) (Fig. 4.7). Once differentiation has been completed, two types of cells can be distinguished: dead cells with thick walls and a cellular lumen of more or less ample proportions, which fulfill the functions of conduction (tracheas and tracheids, respectively, in broad-leafed and coniferous species) or support (fibers), and live cells that function as reserves (parenchyma cells) or as secreting cells (for example, resiniferous canals in coniferous species and laticifers and tanniniferous canals in broad-leafed ones) (Fig. 4.8).

The structure of wood is therefore composed of clearly distinct cells with their own cell walls, which confer upon them a high degree of morphological, functional, and technological individuality; but, at the same time, they are bound together through the presence of a common wall—the middle lamella—that contributes important properties to the tissue as a whole.

Fig. 4.7 *Transverse section of: (a) homoxylous wood; (b) heteroxylous wood (AR = annual rings; RC = resin canals; TR = tracheids; V = vessels; P = pits; R = parenchyma rays; F = fibers)*

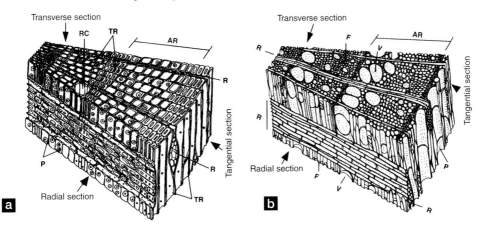

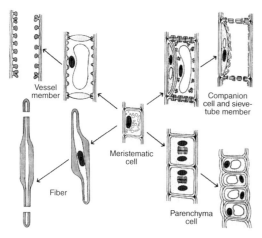

Fig. 4.8 *Differentiation of the various cell types in wood, starting with the meristematic cells in the cambium*

The cell wall is composed of different layers of varying thicknesses, deposited one over the other during the development of the cell; we can distinguish a primary wall, which is mostly composed of pectic substances, of lignin, and of hemicellulose; a secondary wall, which is the thickest and which consists almost exclusively of cellulose and lignin; and a tertiary wall, which is closest to the lumen of the cell and which contains a considerable portion of pectic substances as well as cellulose and hemicelluloses (Fig. 4.2a).

Wood exhibits varying characteristics, depending not only on the significant differences among the individual species, but also on the genetic differences within a single species, as well as on the inevitable influences that environmental conditions have on the growth of the plant from which the wood is obtained.

To better understand the structure of wood, we must remember that the activity of the cambium produces successive layers of xylematic cells toward the interior (xylem) and phloematic cells toward the exterior, the latter forming the inner bark. In spring, the new cells are larger so as to be able to handle vegetative reawakening, and they form the so-called spring wood or earlywood; at the end of the summer and in the autumn, on the other hand, autumn wood or latewood is produced, in which the cells are smaller but the cell walls thicker, conferring (on a macroscopic scale) a darker tonality to the wood. The color contrast produced with the cyclical alternating of the seasons is visible with the annual growth rings, in those environments where the seasons are clearly defined from the climatic point of view (Fig. 4.7).

The interior portion of the stem, which therefore consists of older wood, is called duramen or heartwood and, in some species, the substances that accumulate within it over the years give it a dark coloring as well as greater durability, while the relatively younger sections—the alburnum or sapwood—are lighter in color and less durable (Fig. 4.6). Whether or not it is possible to distinguish these two zones with the naked eye depends on the characteristics inherent to each species of wood; in some woods the alburnum and the duramen are undifferentiated (beech wood, for instance), while in other woods the two are clearly distinguishable (for instance, oak).

With the passing of years, certain substances (tannins, for example) are deposited inside the older cells; this gives the wood greater durability and better mechanical resistance. Durability indicates the capacity of a particular wood to resist the attacks of agents of degradation and must not be confused with hardness, which refers the wood's resistance to penetration by an external body and which therefore belongs to the category of mechanical properties.

As mentioned, the main components of wood are of a polysaccharidic nature (cellulose and hemicellulose) and of a phenolic nature

Table 4.1 Percentage chemical composition (based on dry weight) of coniferous and broad-leafed woods (from Higuchi and Chang 1980)

Wood	Cellulose %	Hemicellulose %	Lignin %	Trace substances %
Conifers (softwood)	40–45	20	25–35	<10
Broad-leafed species (hardwood)	40–45	15–35	17–25	<10

(lignin) and vary qualitatively and quantitatively according to the cell wall layer, the species of origin (i.e., coniferous or broad-leafed wood) (Table 4.1), the age of the tree, and the conditions prevailing during its development.

4.1.1b *Biodeterioration of Wood*

In addition to insects—which, along with other animals, will not be discussed in this context—the main organisms responsible for wood degradation are certain specialized fungi and, to a lesser extent, the bacteria that are able to use the main constituents of the woody cell wall as nutrients.

In addition to the natural durability of the wood species, the factor that most determines attacks by microorganisms is the moisture content. There is, indeed, a very clear correlation between the conditions of the environment where the wooden artifact is located and the attacks by destructive biological agents.

The Italian (UNI) and European (EN) standardization bodies have drawn up a classification of the durability of wood in relation to its susceptibility to biological attack, published by the European Committee for Standardization (CEN) (Norm UNI EN 335-1, 1992). Five different classes of degradation risk are identified in relation to the levels of humidity to which the wood is exposed in the various conditions of use (Table 4.2).

More specifically, the sapwood of all species of wood is not durable (Table 4.3), while the heartwood can have different resistance levels to biodeteriogens depending on the species (Table 4.4).

Fungi will attack wood directly on the living plant (parasitic fungi) and also once the tree has been felled, gathered, stored, or used (saprophytic fungi). Those responsible for the most serious damage to processed wood used in artifacts or for timber are the basidiomycetes. Fungi can be responsible for legions of large proportions (the different kinds of rot) or, in the case of chromogenic fungi, for what are essentially chromatic alterations. In the field of cultural heritage, while it is true that large lesions occurring on wooden structures and works of art are extremely serious examples of damage, alterations in color can also constitute a serious problem that requires prevention and care.

Depending on which component tends to get degraded (Fig. 4.9) and the consequent appearance of this degradation, the phenomena of wood alteration are called white rot, brown rot, and soft rot; the latter is the kind most frequently encountered in the context of cultural heritage.

The fungi that cause white rot (Plate 11) break down both lignin and cellulose, thus involving the entire structure of the wood. Wood that has suffered this kind of damage appears whitened—because of the breakdown of lignin

Table 4.2 Class of risk for wood in relation to its susceptibility to biological attack (from UNI EN 335-1, 1992)

Class	1	2	3	4	5
General situation	Not in the soil, sheltered	Not in contact with the soil, sheltered	Not in contact with the soil and not sheltered	In contact with soil or fresh water	In salt water
Level of humidity	None	Occasional	Frequent	Permanent	Permanent
Deteriogenic organisms	Coleoptera, termites	Basidiomycetes, chromogenic fungi, coleoptera, termites	Basidiomycetes, chromogenic fungi, coleoptera, termites	Basidiomycetes, soft rot, chromogenic fungi, coleoptera, termites	Basidiomycetes, soft rot, chromogenic fungi, coleoptera, termites, marine organisms

Table 4.3 Durability of the various wood species against attack by fungi and by xylophagous insects (from Giordano 1971; Gambetta and Orlandi 1982)

Wood Species	Durability against Attacks by Fungi	Resistance to Attacks by Xylophagous Insects
Conifers		
Mediterranean cypress (*Cupressus sempervirens*)	ED	VR
Yew (*Taxus baccata*)	ED	VR
Scots pine (*Pinus sylvestris*)	D (d) SD (a)	NR
Norway spruce (*Picea excelsa*)	SD	NR
Aleppo pine (*Pinus halepensis*)	D (d) ND (a)	R (d) NR (a)
Cedars (*Cedrus libanotica, C. atlantica, C. deodara*)	ED	VR
Douglas fir (*Pseudotsuga menziesii*)	D	R
Redwood (*Sequoia sempervirens*)	ED	R
Broad-leafed Species		
Oaks (except the Turkey oak) (*Quercus* sp. pl.)	ED (d) SD (a)	VR (d) NR (a)
Black locust (*Robinia pseudo-acacia*)	ED (d) SD (a)	R (d) NR (a)
Chestnut (*Castanea vesca*)	VD (d) SD (a)	R (d) NR (a)
Olive (*Olea europaea*)	VD (d) SD (a)	R (d) NR (a)
Eucalyptus (*Eucalyptus globulus, E. camaldulensis*)	D	R
Turkey oak (*Quercus cerris*)	D (d) SD (a)	R (d) NR (a)
Walnut (*Juglans regia*)	D (d) SD (a)	R (d) NR (a)
Ash (*Fraxinus* sp.)	SD (d) ND (a)	R (d) NR (a)
Beech (*Fagus sylvatica*)	ND	NR
Mahogany (*Swietenia* sp.pl.)	VD	VR
Black walnut (*Juglans nigra*)	VD	R
Sweetgum (*Liquidambar styraciflua*)	D	NR
Red oak (*Quercus nigra*)	SD (d)	NR (d)
Teak (*Tectona grandis*)	ED	VR
Brazilian rosewood (*Dalbergia nigra*)	D	R

Key: d = duramen; a = alburnum; ND = not durable; SD = slightly durable; D = durable; VD = very durable; ED = extremely durable; NR = not resistant; R = resistant; VR = very resistant

Table 4.4 Durability of the duramen of various species of wood in contact with the soil (from Highley 1995)

Less than 5 Years	5–10 Years	10–15 Years	15–25 Years
Ash	Elm	Douglas Fir	Oak
Beech	Fir	Larch	Sequoia
Alder	Pine	Thuja	Oriental Red Cedar

(brown)—as well as lighter and more fibrous; the weightbearing capacity of the wood and its resistance to traction are lost, with severe consequences in cases where wood is employed in a building in a load-bearing capacity. White rot can be divided into simultaneous rot—in which the breakdown of lignin and cellulose occur at the same time—and white rot proper, in which there is an initial breakdown of lignin, followed by the degradation of the other components of the wood. Among the species responsible for white rot are *Trametes versicolor, Phanerochaete chrysosporium,* and *Pleurotus ostreatus.* In addition to these basidiomycetes, certain ascomycetes can also cause this kind of degradation in woods, for instance, *Daldinia, Hypoxylon, Ustulina* and *Xylaria,* although some of these genera (*Hypoxylon* and *Xylaria*) are mentioned in the literature as also being responsible for soft rot (Eaton and Hale 1993). These fungi are mostly responsible for attacks on live plants, although they have often been found on timber, especially when employed architecturally (ceiling beams, coffered ceilings, etc.) or, in any case, when it is in contact with damp structures (as happens in churches and hypogean environments), and also on archaeological woods in contact with the soil. They are also frequently found on wood pulp used in the paper industry.

The fungi of brown rot (Plate 11) only break down cellulose and hemicelluloses, giving the wood a darker coloring due to the presence of lignin; when the attacked wood dries out, fissures appear both longitudinally and transversally (cubic rot). The enzymes involved in the degradation are those of the cellulase complex and the hemicellulase. The fungi that are most frequently cited as being responsible for this type of rot are basidiomycetes such as: *Coniophora puteana, Antrodia vaillantii, Poria placenta, Gloeophyllum trabeum, Serpula lacrymans,* etc. The latter represents the most dangerous species for wooden artifacts or timber, because it can develop even when the humidity of the wood is at relatively low levels (around 22%) and at ambient temperature (7–26°C); for this reason, the damage caused to the wood is called dry rot (Fig. 4.10). This fungus also grows on walls, mortars, and stucco and can spread for several meters from the original

Fig. 4.9 *Cell walls of wood with fungal attack seen in TEM: (a) brown rot (H = hypha, S = secondary wall, ml = median lamella); (b) white rot; (c) soft rot (from Blanchette 2000)*

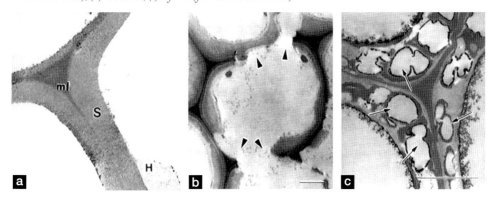

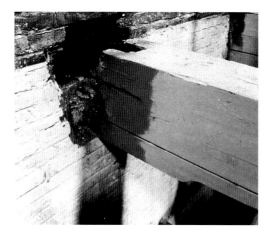

Fig. 4.10 *Attack by* Serpula lacrymans *on a working beam* – Photo S. Berti

source of nutrients, thanks to its ability to produce specialized strands of mycelium (rhizomorphs) that enable it to discover new sources of healthy untouched wood (Singh 1993).

Soft rot (Plate 11) is caused by ascomycetes and certain deuteromycetes and attacks wood exposed to conditions of high humidity or to conditions that are not favorable for the development of basidiomycetes, such as, for instance, when the water content is high and the ventilation is low, pH values are high, or when there is low availability of oxygen, or else in the presence of extractives and substances that inhibit colonization. Furthermore, the attack is favored by high concentrations of soluble nitrogen compounds as well as by high temperatures. It is therefore mostly encountered in water-logged and archaeological wood, in wood that is in contact with the soil, and in industrial cooling towers. The fungi responsible for soft rot are mostly ascomycetes—such as *Chaetomium globosum* and numerous species of the genus *Xylaria*—and deuteromycetes such as *Lecythophora hoffmannii* in land environments, terrestrial *Monodictys putredinis* in freshwater environments, and *Humicola allopallonella* in seawater. These fungi have a very low capacity for penetration; consequently, the damage is usually limited to the outer layers and is located in the S_2 layer of the cell wall. The attacked wood usually exhibits a dark, spongy surface that is often eroded. When it dries out, it exhibits a reticulate structure due to the presence of both vertical and horizontal fissures.

The chromatic alterations of wood can have different origins (they may, for example, be the result of the felling techniques used and the metal tools employed), but those that are most relevant here—because they are the most common in works of artistic and historical interest—are the ones resulting from the action of chromogenic fungi. Certain species of wood are particularly susceptible to the action of light and will alter in color due to exposure to sunlight. The chromogenic action of fungi can be attributed to the colored metabolic substances produced by the hyphae between the wood fibers, although a considerable proportion of known alterations are in fact due to the color of the hyphae themselves (melanins), or else to the optical diffraction effects generated by the particular nature of the fungal mycelium. Among the chromatic alterations that occur on live wood or immediately after sawing, we note:

— pinkish alterations (*rosature*), found in some coniferous woods, which can be observed in the initial stages of attack by basidiomycetes;
— bluish alterations (*azzurrature*), typical of certain species of *Pinus* (Scotch pine, larch, and black pine), which are mainly due to ascomycetes of the type *Ceratocystis*;
— grayish alterations (*grigiature*), typical of beech wood, mainly due to ascomycetes.

In general, however, a broad distinction can be made between chromogenic fungi that are able to penetrate wood and cause very obvious chromatic changes and fungi that develop only on the surface. The latter mostly belong to species of deuteromycetes or ascomycetes such as *Trichoderma lignorum*, which attacks mostly conifers, or various species of *Aspergillus* and *Penicillium*, which utilize reserve substances such as starch and simple sugars, propagating along the alburnum and without attacking the duramen, which is much poorer in nutrients. The most commonly encountered chromatic alteration in structural timber is a bluish one (*azzurratura*), but it is primarily due to the action of *Aureobasidium pullulans* (Unger et al. 2001). These species do not alter the mechanical characteristics of wood, but they can cause an increase in hygroscopic-

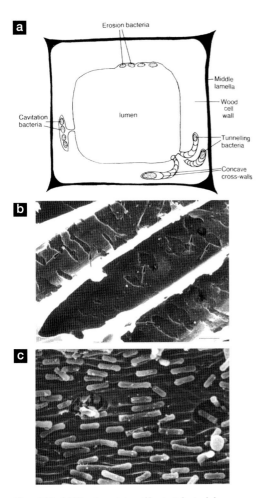

Fig. 4.11 *(a) The three types of bacteriological damage; (b and c) SEM images of walls altered by bacteria (from Eaton and Hale 1993; Blanchette 2000)*

Recent studies have shown that there are various groups of bacteria in existence able to attack and degrade the cellulosic components of wood, and they also show that the damage inflicted is specific to each group of microorganisms (Singh and Butcher 1991; Daniel and Nilsson 1998). Thanks largely to the examination of thin sections and fragments of ancient woods under TEM (Transmission Electron Microscope) and SEM (Scanning Electron Microscope), three distinct types of bacteriological deterioration have so far been identified: erosion, cavitation, and tunnelling (Fig. 4.11). The process of biodeterioration carried out by bacteria seems to be very slow, especially in comparison to fungal attacks. Many different bacteria have been identified on degraded wood, and these include the genera *Alcaligenes, Bacillus, Brevibacterium, Cellulomonas, Cellvibrio, Clostridium, Cytophaga, Flavobacterium, Pseudomonas,* and *Sporocytophaga*. However, we do not as yet have sufficient evidence of their roles in the various processes of deterioration, although two genera, *Bacillus* and *Pseudomonas*, have been identified more frequently (Eaton and Hale 1993; Blanchette 2000).

The bacteria responsible for erosion break down the layers of the secondary walls, utilizing the cellulose and the hemicellulose in the wood. As a result, the residual material in the cell wall takes on a porous appearance. The bacteria responsible for cavitation create irregular and sometimes diamond-shaped cavities within the secondary cell wall, which are perpendicular to the longitudinal direction of the wood fibers (Blanchette 2000). Tunnelling bacteria, on the other hand, create narrow channels within the secondary cell wall; these channels can sometimes also reach the primary structure and even the middle lamella, which is strongly lignified, forming along the bordered pits in softwoods.

All the bacteria responsible for the deterioration of wood in an underwater environment (and therefore reducing bacteria) are thought to be anaerobic, or at least able to metabolize at very low levels of oxygen, although there is no evidence to date of wood degradation taking place in conditions of complete anoxia (Blanchette et al. 1990; Daniel and Nilsson 1998).

Understanding bacteriological biodeterioration mechanisms can yield important information in archaeological contexts. To some extent,

ity, thus rendering the wood more susceptible to new microbial attack.

While in most land environments the biodeterioration of wood is mostly due to the activity of insects and fungi, wood that is either under water or buried in water-logged soil is primarily attacked by bacteria (Blanchette 1995, 2000; Bjordal et al. 1999). In river, lake, and marine environments, even algae can play an indirect role in the degradation of submerged materials (Chapter 5, section 4), while in subaerian environments their presence has only been noted in particular cases, especially in cold-temperate zones where wood still plays an important role as a building material.

it can reveal the conditions of preservation of the finds and, in some instances, their previous history, as in the case of underwater archaeological sites or in the study of ancient ships, a particular focus in Scandinavian countries (Björdal et al. 1999). The traces left by bacteria on submerged wood are such that they can still be observed after hundreds of years, and the micromorphological characteristics of the alterations make it possible to trace the typology of the microbial attacks, whether ancient or recent.

Finally, the bacteria that take part in the process of degradation of wood seem to play a subsidiary role in wood rot. Singh et al. (1990) have put forward the hypothesis of the existence of a biocoenosis, in which the fungi of soft rot and the bacteria cooperate—especially deep inside the wood—in the breakdown of the various constituents, proceeding in the degradation until only the empty lignin skeleton is left as a residue.

4.1.2 Paper

by Giovanna Pasquariello, Paola Valenti, Oriana Maggi, and Anna Maria Persiani

4.1.2a *Structure and Composition*

Paper—one of the earliest man-made materials —is composed of cellulose and other substances of varying types and quantity, depending on the nature of the raw materials used as well as on the processes employed in its manufacture. Essentially, paper consists of equimolar quantities of water and cellulose; water is not a constituent part of the cellulose matrix and is present rather in the pores, which are thought to be about 1.5 nanometers in size. These pores are completely surrounded by amorphous cellulose, and the crystalline cellulose is well spaced out; the water cannot be removed without causing the total collapse of the paper structure. Cellulose has a DP (degree of polymerization) varying from a few hundred to a few thousand units; the variation is due to the original plant species and also to the effects of the chemical treatments during the extraction process involved in paper production. The native cellulose has a high DP and is highly resistant; this index is reduced when the fibers are transformed into paper, because of the different treatments and substances used in its manu-

facture or because of external factors. Paper is a hygroscopic material and therefore sensitive to changes in relative humidity and temperature of the environment; such changes can determine different levels of absorption or release of moisture, resulting in potential biodeterioration phenomena and/or degradation phenomena of a chemical nature (Mantovani 2002).

In order to understand the different composition of the various kinds of paper that have been produced over the centuries and employed in manuscripts, documents, books, engravings, photographic prints, etc., we can divide paper production into two broad historical periods: the one before the 1800s—defined as the period in which paper was made by hand ("ancient paper")—and the one after the 1800s—that of machine-made paper ("modern paper").

The main sources of cellulose for paper manufacture are: textile fibers, wood, and cereal straws. Up to the nineteenth century, paper was manufactured from rag pulp coming from selected textiles: cotton, linen, and hemp rags, all raw materials containing almost pure cellulose. This kind of paper is a product of the highest quality, not having been excessively refined and being sized with glues of a plant or animal origin (starch and gelatine, respectively), neither of which inherently affects the acidity of the paper. Its main characteristics are therefore a high DP and a good resistance to attacks by external agents, including those of a biological nature. Toward the end of the seventeenth century, the increasing demand for paper, along with the mechanization of the production processes, brought about the first decline in the quality of paper with the coming into use of the refining machine called "the Hollander," which caused a diminishing of the mechanical paper properties of paper due to the shortening of the cellulose fibers. Also, paper manufacturers started adding rock alum to the gelatine used for sizing, which increased acidity levels, thus accelerating the natural aging process of the product. With the discovery of the bleaching properties of chlorine in the middle of the eighteenth century, the process of bleaching the fibrous rag paste represented a further aggression on the cellulose polymer.

From the 1800s onward, sizing with starch and gelatine was replaced by sizing with tree resins, such as colophony (from the trunks of conifers) added to alum, which made the paper acid and compromised its stability over time. Subsequently, aluminum sulphate was employed (the so-called papermakers' alum), which caused an even worse degradation of the paper. With the industrial revolution, the so-called modern period of paper production began, marked by the use of wood—still practiced today—from conifers (fir, pine, cypress) and broad-leafed trees (birch, beech, poplar, chestnut) as well as cereal straws (mostly wheat). Straw is almost never used nowadays, because of the polluting chemical substances used for extracting the cellulose. The raw material for current paper production is therefore wood pulp, the prime constituents of which are cellulose (45–55%) and other substances such as lignin (20–30%) and hemicellulose (15–25%), which are called incrusting substances. Industrial processes remove the various incrusting substances from the cellulose fibers with a variety of treatments. The various extraction processes produce paper pulps that can be classified into:

—mechanical or wood pulp, obtained by grinding down the wood by means of simple mechanical action on the wood, from which only the bark has been removed and without the use of chemical reagents. This type of paper pulp contains high percentages of lignin, which remains associated to the cellulose, and it is used for papers that do not require any particular degree of stability or mechanical resistance (newspaper, for instance). While the paper is easily degradable from a chemical point of view as well as physically (yellowing), it is less susceptible to biodegradation thanks to the presence of lignin, a substance that is not easily attacked by biological agents;
—semichemical pulp is obtained through a mechanical process and a light chemical treatment that only partially eliminates the incrusting substances, so that the fibers are for the most part separate but still encased within a layer of lignin. This is an intermediate product between mechanical and chemical pulps; the resulting paper is of medium quality;
—chemical pulp is obtained with thermochemical treatments; the most

commonly used processes are those employing soda, sulphate, calcium bisulphite, sodium sulphite, and chlorosoda. Chemical pulp does not contain lignin, nor any other incrusting substance. Paper manufactured with chemical cellulose (first in world consumption) is stable in terms of aging properties and is considered of good quality.

The vast array of papers we find today preserved in museums, libraries, and archives is therefore the result of a complex manufacturing process (Grandis 1985a; Koch 1993; Baldi 2000; Impagliazzo and Ruggiero 2002), which depends on the type of cellulose used, the processes and treatments employed, as well as all the other substances added to the fibrous pulp, which we list below:

—nonfibrous substances called mineral fillers (aluminum silicate, i.e., kaolin; magnesium silicate, i.e., talc; the carbonates of calcium and magnesium; titanium dioxide, called satin white; the oxides and sulphides of zinc; fossil flour; etc.) Being bases, these substances provide alkaline reserves that act as buffers to the acidity of external agents; moreover, they give the paper different levels of opacity, smoothness, and whiteness;
—natural adhesives (gelatine, starch, colophony) and synthetic adhesives (acrylic and anhydride resins), which make the paper suitable for writing and printing;
—bleaching agents (optical correctors), water-soluble substances that increase the whiteness of the paper and the effects of which are due to fluorescence;
—colorants, which provide the paper with a stable color.

4.1.2b *Biodeterioration of Paper*

When assessing the potential long-term durability of paper, i.e., its capacity to maintain its physico-chemical characteristics, the important criteria are:

—its stability, mainly exhibited by its resistance to environmental agents (such

as oxygen, ozone, humidity, heat, light, and microbial or insect attacks);
—its durability, mainly a result of resistance to changes linked to the inherent characteristics of the material (the kinds of fibrous pulp used, the nature of the substances added during the manufacture of the sheet of paper, the degree of acidity, etc.); to the stresses it is exposed to; to how it is used; and to the effects of time.

But in any case, it is chemical changes that are mainly responsible for the weakening and the physical disintegration of the materials (Feller 1994).

The factors causing degradation can act either directly or indirectly, and at times in synergy: for example, high levels of temperature and relative humidity promote the development of biodeteriogens, and celluloses depolymerized through chemical or physical agents are more easily attacked by microorganisms.

Only when the relative humidity exceeds 65%, and the water content of paper nears the 10% boundary, are conditions established for the development of microorganisms. In addition to these factors, a determining role in biodeterioration is played by water activity (a_w) (Chapter 2). The different characteristics of the various paper materials result in different susceptibilities to fungal growth, precisely because of the different percentages of water content and values of a_w (Florian 1994, 1997, 2000; Gallo et al. 1999b; Pasquariello 2001).

Dust, which also contains airborne particles of various nature—among which are fungal spores—represents a nutritional resource for fungi and can form a microenvironment on the surfaces on which it deposits, thus allowing the survival of the spores themselves (Gallo 1993; Florian 1997; Maggi et al. 2000). Dust contains many volatile and semivolatile organic components some of which—called MVOC (microbial volatile organic compounds)—are products of the secondary metabolism of the microorganisms themselves (Filer et al. 2001; Menetrez and Foarde 2002), while others are released by a variety of construction materials, or else are introduced from the outside air. These substances are also

important for humans because they can cause allergic reactions and respiratory irritations.

The biodeterioration of paper is largely due to the activity of microorganisms (fungi and bacteria) that utilize its constituents (cellulose, hemicellulose, lignin, etc.), transforming them into simpler molecules that can then be assimilated as nutrients.

Microfungi are the most commonly found agents of degradation—they can be distinguished, according to their capacity to degrade the various constituents of paper, into fungi with cellulolytic activity and those able to degrade adhesives and/or additive compounds.

The list of species of microfungi biodeteriogenic to paper-based materials is rather long (Table 4.5) (Gallo 1992, 1993; Nyuksha 1994; Zyska 1997; Pasquariello 2001); in the case of bacteria, only a few species of heterotrophic bacteria belonging to the genera *Cytophaga*, *Cellfalcicula*, and *Cellvibrio* as well as some actinomycetes (Nocardiforms, Actinobacteria, streptomycetes) have been identified.

Degradation due to microbial development essentially takes on the following forms:

—chromatic alterations in the form of stains that exhibit considerable variety in color (purple, yellow, brown, black, red, etc.), in shape, and in size; they are caused by the presence of pigmented mycelium or fungal spores or else by the exopigments released by bacteria or fungi;
—structural alterations of the main components, caused by enzymes (cellulase, protease, etc.) produced by various kinds of microorganisms; these changes show in the paper's fragility or even in its partial destruction;
—alterations of the essential added components, which become evident with the loss of some of the characteristics inherent to the materials and are caused by an attack by the microorganisms on those substances that, although they are part of the material, do not constitute its main elements (glues, plastifiers, antioxidants, etc.) (Plate 12).

The study of the chromatic changes found on paper (but also on textiles, photographs, and

Table 4.5 Species of fungi and types of degradation activity on paper and its constituent materials (from Nyuksha 1994)

Species	A	B	C
Acremonium charticola		B	
Alternaria alternata	A		
Alternaria tenuissima		B	
Aspergillus candidus		B	
Aspergillus flavus		B	
Aspergillus fumigatus	A		
Aspergillus sydowii			C
Aspergillus terreus	A		
Aspergillus versicolor		B	
Aspergillus wentii		B	
Aureobasidium pullulans			C
Botryotrichum piluliferum			C
Chaetomium bostrycodes		B	
Chaetomium chartarum	A		
Chaetomium elatum		B	
Chaetomium globosum	A		
Chaetomium indicum	A		
Chrysonilia sitophila		B	
Cladosporium herbarum	A		
Doratomyces stemonitis	A		
Epicoccum purpurascens			C
Eurotium herbariorum			C
Fusarium culmorum	A		
Fusarium solani	A		
Geomyces pannorum			C
Mariannea elegans		B	
Myrothecium verrucaria			C
Myxotrichum chartarum	A		
Myxotrichum deflexum	A		
Oidiodendron cerealis	A		
Penicillium brevicompactum			C
Penicillium canescens		B	
Penicillium citrinum			C
Penicillium decumbens		B	
Penicillium funiculosum	A		
Penicillium luteum	A		
Penicillium miczynskii		B	
Penicillium oxalicum			C
Penicillium puberulum			C
Penicillium purpurogenum	A		
Phialophora fastigiata			C
Preussia fleischhakii			C
Rhodotorula glutinis			C
Scopulariopsis brevicaulis	A		
Sordaria fimicola		B	
Stachybotrys chartarum	A		
Trichoderma koningii	A		
Trichoderma viride	A		
Trichothecium roseum		B	
Ulocladium chartarum		B	
Verticillium tenerum	A		

Key:
A: Always found on paper; they penetrate the fibers and cause disintegration of the substrate.
B: Often found in paper; they cause certain types of damages to its texture.
C: They only assimilate certain specific components of paper (paraffin, synthetic polymers, waxes, gums, etc.).

parchment), commonly known as foxing or fox spots (Plate 12), is particularly relevant. Foxing (so called because of the similarity of its coloring to that of the fur of the red fox) is a complex phenomenon that takes the form of small, isolated rust-colored, brown, or yellow spots of various shapes. Examined in ultraviolet light, these spots of foxing exhibit fluorescence (Fig. 4.12).

How these spots originate has been a matter of debate for decades in scientific circles, and up to this day an unequivocal answer as to the causes of the phenomenon has not been found. Research has developed primarily in two directions: investigations and microbiological experiments to determine the biological origins of the spots (Fig. 4.12c) and to deliver their characterization; and chemical investigations to establish possible correlations between the presence of metal elements (for instance, iron) within the structure of the paper and their oxidation as well to attempt a definition of the different types of foxing (Gallo and Pasquariello 1989; Florian 2000; Arai 2000; Szczepanowska and Cavaliere 2000; Rebrikova and Manturovskaya 2000; Bicchieri et al. 2001; Pasquariello et al. 2003; Montemartini Corte et al. 2003).

The most recent and most exhaustive studies on the aspects linked to the biological hypothesis have been carried out by Florian (2000) and Arai (2000). Florian suggests a nomenclature of the fox spots based on their superficial appearance and on the color and shape of the chromatic alterations found on the paper. For the most commonly encountered alterations, she proposes the terms irregular fungal fox spots and corroded iron spots, thus distinguishing the spots of fungal origin from those caused by the corrosion of iron. The first are irregularly shaped, of a uniform rusty red color, and migrate onto adjacent pages, while the second are circular, appear only on one side of the paper, and do not migrate onto adjacent pages.

Fig. 4.12 *Examples of foxing as seen with (a) natural light and (b) ultraviolet light* – Photo ICPL; *(c) SEM image of fungal colonies from foxing stains (from Gallo et al. 1989)*

These observations were mostly made with the use of the stereomicroscope.

In the numerous studies he conducted on fox spots, Arai has identified the presence of both obligated and facultative xerophilic fungi. By analyzing the organic acids, the sugars, and the amino acids in the rust-colored areas in which the fungi had been identified, he also showed the presence of:

—organic acids (fumaric, malic, lactic, acetic);

—sugars (glucose, cellobiose, cellotriose, cellotetrose, cellopentose, and cellohexose);

—various amino acids, of which the most frequently found was γ-aminobutyric acid.

Compared to facultative xerophilic fungi, obligated xerophilic fungi produce a greater quantity of malic acid and of glucose. Based on these results, according to Arai the rust coloration is due to the presence of melanoidins, which are the product of the reaction between the amino acids and the sugars produced by the fungi; this reaction is commonly known as the "Maillard reaction."

The studies of Arai and Florian have therefore unequivocally shown the link between fox spots and fungal development, but they certainly have not resolved the question of whether the phenomenon of foxing is always exclusively caused by fungi, because other factors—both physical (environmental) and chemical—concur, and all of these are aspects that must be taken into account in terms of the overall systemic effect.

Among the Italian examples that are of particular interest—because of the nature of the materials involved and the severeness of resulting damage or because of the identification of the potential causes of damage—we refer to the study by Montemartini Corte et al. (2003) on fungi as the biological agents of the foxing that was found on geographical maps belonging to the collection of the city of Genoa and that date from the seventeenth to the twentieth century. Fluorescence techniques were used to detect biological action, and cultures in the laboratory were used to isolate microfungi from fragments of paper that had exhibited positive fluorescence. A variety of species of fungi (Chapter 5, section 1) were identified by the authors with a certain frequency.

4.1.3 Textile Fibers (Cotton, Linen, and Other Fibers)

by Maria Pia Nugari

4.1.3a *Structure and Composition*

Plant-based textiles are made of fibers found in nature; because of their chemical structure, their morphology, and properties such as elasticity, resistance, and flexibility, they lend themselves to being spun and then woven.

Such fibers are obtained from certain types of plants and, depending on the part from which they are extracted, they can be divided into those that originate from the seed, as is the case with cotton, or from the stems, as with linen, hemp, and jute.

The chemical composition of the fibers therefore varies greatly depending on the plant of origin and the part of the plant utilized; for instance, the percentage of cellulose present in unrefined yarns is 94% in cotton, 77% in hemp, 75% in flax, and 63% in jute.

Below, we will discuss the characteristics of cotton and linen (we will also include a brief mention of hemp), since they are the most widely encountered textiles of plant origin in the field of cultural heritage.

Historically speaking, it is still not known exactly when man began to make use of plant and animal fibers in order to create man-made materials. The most ancient textile fiber seems to have been flax, which was used in Egypt as early as the fifth millennium b.c.e., and it was employed to make the linen bandages in which mummies were enveloped as well as for clothing, sails, and ropes, as shown by many of the archaeological finds in existence. In addition to the Egyptians, it was also cultivated by the Phoenicians, the Babylonians, and other peoples of the Middle East who then disseminated its use to the Greeks and Romans. From the Middle Ages onward, the cultivation of flax spread through Europe, and Flanders became one of the main centers of production. Only later was flax supplemented by hemp.

Among the most ancient examples of cotton to reach us are fragments of woven cloth and pieces of rope from the Indus Valley (today's Pakistan), and they date back to 3000 b.c.e. Other finds in Egypt can be dated to 4000–3000 b.c.e. The introduction of cotton to Europe can be traced back to the Arabs who brought it to Greece around 350 b.c.e.; from here, its use spread to the whole continent around 1300 c.e. Yet, the production of linen cloth remained prevalent until the beginning of the nineteenth century, when cotton went from being a highly prized product destined for the high nobility to a product for the masses as a result of the introduction of industrial technology for spinning and weaving.

The flax fiber is obtained from the stem or, to be more specific, from the phloem fibers of

Linum usitatissimum (Linaceae), the seeds of which are used to extract an oil that is widely used in many artistic techniques. Flax is harvested during the period between the flowering and the ripening of the fruit. If it is harvested when the stem is still green and the fruit has only just formed, the resulting fiber is very thin but not very resistant; it is called *lino azzurro* (blue linen) and is used mostly in the manufacture of lace and embroideries. If it is harvested when the fruit is yellowish-green and the stem is yellow, a more resistant but less flexible fiber is obtained (*lino bianco*, or white linen). Finally, if the harvest takes place when the fruit is brown and the stem is dark yellow, the resulting fiber is extremely resistant but very coarse.

After harvesting, the stems are laid out to dry, before being macerated in water or in chemical substances, and then beaten in order to separate the fibers from the woody residues of the stems. The fibers are then combed in a specialized machine and subsequently spun. Every individual fiber has a cylindrical form with a thin central lumen and with transverse striations along its length, giving it a bamboo-like appearance.

Chemically, flax is made up of 70–80% cellulose, the remainder consisting of lignin, waxes, fats, and water.

Cotton is obtained from soft, hair-like fibers that grow around the seeds of the genus *Gossypium* (Malvaceae), in particular *G. hirsutum* and, to a lesser degree, *G. barbadense*, but thirty-nine or so species are known and used that have the same designation. The seeds, thickly covered with unicellular hairs that are between 25–55 mm long, silken and generally white, are produced inside a capsule that is subdivided into three or five sections. On reaching maturity, this capsule explodes, exposing the seeds covered in their cotton-wool, which is their anemophilous mechanism. The cotton fiber is ribbon-like, with spiral windings that are frequent and regular in high-quality fibers (Fig. 4.13).

In cross-section, four different parts are visible: cuticle, primary wall, secondary wall, and lumen. The cuticle is an external, extremely thin membrane that is not cellulosic, while both the primary wall and secondary walls are formed of cellulose fibrils that intersect in the primary wall and are wound spirally in the secondary wall. The lumen is the central part of the fiber and is also not cellulosic. Each microfibril is composed of approximately two thousand chains of cellulose that are parallel to one another, thus forming highly crystalline zones.

Chemically, unrefined cotton is made up of cellulose (87–90%), small quantities of nitrogenous substances (0.2–0.4%), fats and waxes (0.6%), mineral substances (1%), and water (8–12%).

Hemp is also a textile fiber originating from the phloematic fibers of a plant, mostly from *Cannabis sativa* (Cannabaceae): the stem from which the fiber originates has two parts: an external section rich in fibers and an internal section that is mostly woody. The fibers are obtained from the stems, which are harvested, dried, and macerated; then the internal section is separated from the external one, and the latter is used for the making of textiles, ropes, etc.

4.1.3b *Biodeterioration of Textiles of Plant Origin*

The chemical nature of woven fibers of plant origin influences their biodeterioration, which is mostly carried out by cellulolytic microbial species. Susceptibility to degradation changes in relation to the differing percentages in the chemical composition of the fibers; the variations are linked not only to the origin of the

Fig. 4.13 *Cotton fiber: (a) transverse section; (b) telescopic image*

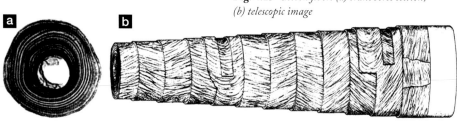

fibers, but also to the production processes of the yarns and the manufacture of the objects.

Microbial attack is made more difficult by the presence of high proportions of lignin and natural waxes, but it is favored by the presence of more degradable organic substances such as pentosanes, pectins, and starches. In relation to the presence of lignin, resistance to biological attack is greater in jute, followed by hemp, cotton, and linen, although hemp is the fiber most susceptible to fungal attack (Vigo 1977; Kowalik 1980b). However, it should be taken into consideration that the manufacturing processes for the purification of the fibers reduce the presence of lignin, thus reducing their natural resistance to biological attack. For example, jute becomes more biodegradable than cotton if the lignin is removed from it, because of the presence of other noncellulosic components, such as hemicellulose and mineral salts. The sizing of textiles with starches and dextrins also renders them more liable to attack by biodeteriogens, because these molecules are water soluble and therefore easily metabolized.

Biodegradability is further influenced by other characteristics of the cellulose fibers, such as the degree of polymerization (DP) and therefore the length of the chains, their degree of crystallinity, and the orientation. As a consequence, damages of a mechanical, chemical, or photochemical nature can increase susceptibility to biodeterioration by modifying the structural characteristics of the fibers; the state of conservation of the objects can therefore become a highly influential factor (Vigo 1980; Sagar 1988; Caneva et al. 1994a). Even the type of weave can influence susceptibility to biodeterioration: textiles with an open weave are less resistant than tightly woven fabrics, as they gather more dust and biological contaminants between their fibers, thus creating conditions of higher risk.

In general, we can say that the susceptibility to biological deterioration of plant textile fibers depends on the following factors:

—the chemical composition of the fibers (percentage ratios of cellulose, lignin, and other compounds and their degree of polymerization), which varies in relation to the origin of the fiber (type and part of plant utilized), the manufacturing treatments (removal or addition of substances), and aging (oxidation or photochemical reactions, mechanical damage, etc.);
—the structural characteristics of the artifact, such as: the type of weave (thread count, etc.), the form of the artifact (pleated, or with reliefs, gathers, or embroideries, etc.), the presence of constituent parts of a different nature (paper, leather, metal, etc.).

Certain conservation interventions that include linings and reconnecting fragments of ancient textiles by means of adhesives can make these works more prone to biodegradation; such objects must therefore be conserved in controlled environments after restoration. Even the custom of preserving fragments of prized cloth sandwiched between two sheets of glass can encourage biological damage by promoting condensation phenomena.

The microorganisms most frequently encountered in the degradation of textiles are fungi and bacteria, and among the latter particularly the actinomycetes; fungi and bacteria can act individually or in synergy. They often find ideal conditions for their development in the environments where textiles are kept, especially sacristies, museum storage areas, or generally in damp and poorly ventilated environments. Bacteria require, as is known, higher moisture levels than fungi (Chapter 3) and are therefore more frequently found on the damp archaeological textiles from excavations; actinomycetes are more dangerous because of their high level of cellulolytic activity. Because of their greater adaptability to environmental conditions, fungi are found more frequently than bacteria, although a sufficient water content in the textiles is required for their development; if the a_w is lower than 0.61, fungal development does not take place (Dekker 2001). The more critical genera belong to the deuteromycetes, such as *Alternaria*, *Aspergillus*, *Fusarium*, *Memnoniella*, *Myrothecium*, *Neurospora*, *Penicillium*, *Scopulariopsis*, *Stachybotrys*, *Stemphylium*, and *Trichoderma*. Frequently found and particularly damaging because of their high cellulolytic capacity

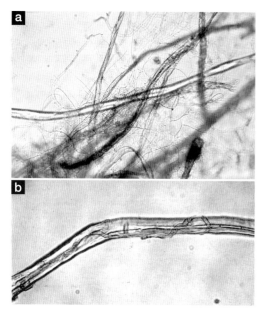

Fig. 4.14 *Fungal hyphae in the interior of fibers*
– Photo ICR

are the ascomycetes of the genus *Chaetomium*. Also recurring and responsible for the rapid colonization of textiles are zygomycetes such as *Mucor* and *Rhizopus*. The most frequently cited species in the literature are *Trichoderma viride*, *Chaetomium globosum*, *Myrotechium verrucaria*, and *Aspergillus niger*, although the latter does not exhibit particularly pronounced cellulolytic abilities (Mahomed 1971; Kowalik 1980a; Vigo 1980; Montegut et al. 1991; Giuliani and Nugari 1993).

The specificity of action of the fungal species can be such that in textiles of a mixed composition, for example cotton and silk, cellulolytic fungi will only attack the cellulose fibers through enzymatic hydrolysis, damaging fibers of a proteinaceous nature only indirectly (Plate 13) (Giuliani and Nugari 1993).

As with other cellulosic materials, the process of degradation of cellulose occurs through the activity of the enzymatic complex of extracellular cellulase, which acts directly on the substrate, and also through the action of intracellular enzymes.

Fungi can develop on the surface of textiles or else penetrate deeply into the fibers, reaching the internal structure of the fibers themselves; in cotton, for example, fungal hyphae are often observed inside the lumen (Fig. 4.14).

At times, the development of fungal hyphae is not very evident, but more frequently microbial colonization causes alterations in color as well as the formation of patches of different colors: reddish-purple, brown, blackish, or green. The blackish-brown spots are mostly linked to the presence of fungal species with cell walls pigmented with melanin, while the reddish-purple ones are the product of the release of indelible exopigments (Plate 13); dangerous producers of exopigments are certain deuteromycetes such as *Fusarium*, *Aspergillus*, etc., and ascomycetes such as *Chaetomium* and *Myxotrichum* and, among the bacteria, streptomycetes in particular. In addition to the aesthetic damage caused by the formation of more or less obvious stains, the microbial colonization of textiles is always accompanied by the degradation of the structure of the textile and of its fibers. The action of the hydrolytic enzymes and of the organic acids, which are emitted as metabolites, leads to a reduction in the elasticity and resistance to traction of the yarns and the cloths; if protracted over time, this degradation activity will inevitably lead to the partial or total destruction of the objects.

4.2 MATERIALS OF ANIMAL ORIGIN

by Maria Pia Nugari

4.2a General Characteristics of Materials of Animal Origin

Many are the objects of historical and artistic interest that can be partially or totally composed of materials of animal origin; to cite but a few examples, parchment (which is the support used for so many priceless documents, illuminated and painted codices, and for the bindings of ancient books), artifacts in skin or leather (used as accessories to garments), silk textiles and embroideries, and the great variety of ethnographic materials that make use of feathers and quills. Even human bodies, which may have been conserved through specific techniques such as mummification or because

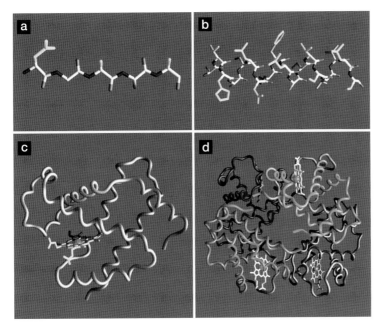

Fig. 4.15 *(a) Primary, (b) secondary, (c) tertiary, (d) quaternary structure of proteins (elaboration by F. Polticelli)*

of accidental events (ice), constitute precious testaments to our past, requiring preservation.

Proteins are the chemical components that form the basis of these materials; they are the fundamental substances that characterize living matter and represent 50% of the dry weight of animal tissues. Proteins are complex molecules that are essentially composed of four elements: carbon, hydrogen, oxygen, and nitrogen, which, for this reason, are called quaternary substances. Proteins often also contain other chemical elements such as sulphur, iron, magnesium, etc. The basic units of proteins are the amino acids (twenty-two known), which are linked together by peptide links (-CO-NH-); the sequences of amino acids give rise to chains of varying lengths, known as peptides.

The succession of amino acids constitutes the primary structure, while the secondary structure refers to the way in which the polypeptide chain is wound or extended, for example, the structure of β-keratin in silk or the helical structure of α-keratin in hair. Tertiary structure, on the other hand, refers to the three-dimensional structure and is typical of globular proteins. And finally, there is also a quaternary structure, which relates to the relationship between different polypeptide chains within a protein (Fig. 4.15).

Proteins carry out a variety of functions that depend on their structure. Proteins that are elongated in shape or fibril-like and are insoluble in water have structural and support functions (structural proteins): for instance, keratin in skin and hair, collagen in connecting tissue, and elastin in elastic connective tissue. Those that are spherical or globular in shape and are generally water soluble play an important role within cellular membranes, in cytoplasm, and in the liquids of organs (functional proteins). To this group belong plasmatic proteins, immunoglobulins, peptidic hormones, enzymes, and proteins with carrier functions, such as haemoglobin or myoglobin for instance, which carry oxygen.

There are often fatty fractions present in the composition of proteinaceous materials, linked to the raw material used or to the manufacturing processes of the object, such as polishing, waterproofing, etc. Fats and lipids are substances made up of carbon, hydrogen, and oxygen. They can be of plant or animal origin and have either structural or reserve functions. In general, lipids are insoluble in water but dissolve in organic solvents; they include triglycerides (which constitute 77–80% of all lipids found in skin), phospholipids, waxes, and free fatty acids.

4.2b General Processes Involved in the Biodeterioration of Materials of Animal Origin

Proteins can provide a nutritional substrate (source of carbon and organic nitrogen) for many heterotrophic organisms. The biodeterioration of objects of a proteinaceous nature is nevertheless linked primarily to the action of some microorganisms with the ability to

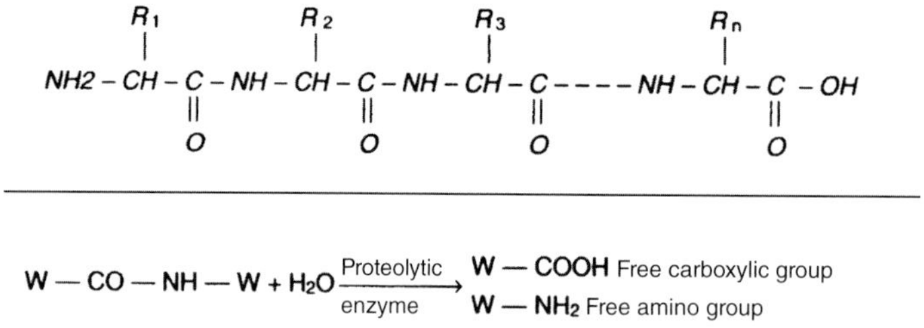

$$NH_2 - CH - \overset{R_1}{\underset{\|}{C}} - NH - CH - \overset{R_2}{\underset{\|}{C}} - NH - CH - \overset{R_3}{\underset{\|}{C}} - - - - NH - CH - \overset{R_n}{\underset{\|}{C}} - OH$$

$$W - CO - NH - W + H_2O \xrightarrow[\text{enzyme}]{\text{Proteolytic}} \begin{array}{l} W - COOH \text{ Free carboxylic group} \\ W - NH_2 \text{ Free amino group} \end{array}$$

Fig. 4.16 *Peptide linkage and lysis (R = typical radical of every amino acid; W = portion of the molecule not involved in the hydrolysis)*

metabolize proteins, thanks to the presence of extracellular proteolytic enzymes, which sever the peptide linkages by means of enzymatic hydrolysis (Fig. 4.16).

Proteins, which are not normally able to diffuse into cells, are therefore converted into soluble peptides that are able to penetrate through the cell wall; these are subsequently broken down by peptidase into their constituent amino acids. Microorganisms that have these proteolytic capacities are bacteria (among which the actinomycetes are particularly active) and certain fungi, ascomycetes and deuteromycetes. The species involved in the degradation can vary greatly in relation to the chemical characteristics of the proteins in the work; these can be very diverse depending on the animal from which they originate (mammal, insect, bird), the part that was used, and the manufacturing processes. As a general rule, proteins will be all the more resistant to biological deterioration the greater their molecular weight, the lesser their solubility, the greater the molecular organization (cristallinity), and if strong intermolecular bonds are present.

The presence of fats confers a certain degree of hydrorepellence to the materials, which generally slows down the process of biodeterioration. Fatty substances are somewhat resistant to biological degradation; however, they become more susceptible if they are in a water-based emulsion or if additional nutritional sources are present that are easy to metabolize, such as simple sugars. Some species of bacteria and of fungi are able to produce lipase, which hydrolyzes the ester bonds present in both animal and vegetable fats. The most active species belong to the deuteromycetes and zygomycetes fungi (especially phycomycetes) as well as to actinomycetes. Among the best-known producers of esterase, and often identified

on cultural heritage materials, we would like to mention the following deuteromycetes: *Aspergillus flavus, A. niger, A. oryzae, A. versicolor,* some species of *Penicillium,* and the streptomycete *Nocardia paraffinae.*

In the following paragraphs we will be discussing the structure, the chemical composition, and the biodeterioration phenomena of some materials of animal origin that are often elements of artifacts of historical-artistic or archaeological interest, such as parchment, leather, and the textile fibers silk and wool.

4.2.1 Parchment and Leather

4.2.1a *Structure and Composition*
The main chemical components of parchment and leather are collagen, keratin, and elastin, in addition to smaller quantities of albumin and globulin, and lipid substances.

Collagen is a structural protein and is the essential unit in the intercellular substance of the connective tissue. It is a fibrous protein, composed primarily of three amino acids: glycine, proline, and hydroxyproline. The basic structural unit is tropocollagen, composed of three polypeptide chains of equal length, wound around each other as a triple helix that is stabilized by numerous intermolecular bonds, both covalent and hydrogen. The longitudinal, parallel bundling together of the tropocollagen forms fibrils, which in their turn come together to form fibers. The collagen fibers are bound together with weak linkages, such as hydrogen bonds (Fig. 4.17). The impenetrable alignment and ordered structure of the fibrils into fibers gives the molecule an almost crystalline structure, although amorphous areas are also present where the structure is more disordered and open and therefore more susceptible to degradation. The mechanical properties of leather and parchment are due precisely to this three-dimensional structure of the protein fibers. Any alteration in this structure causes changes in the mechanical properties and in the resistance to degradation of the material.

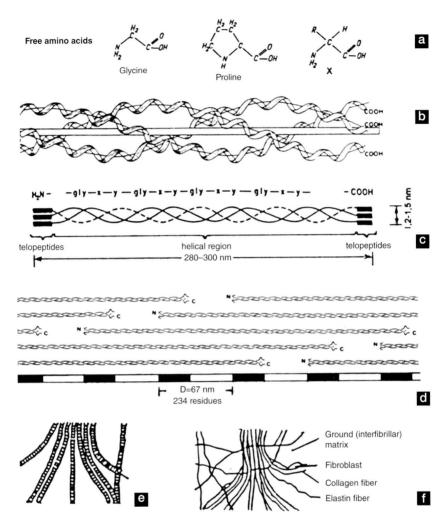

Fig. 4.17 *Molecular structure of collagen: (a) free amino acids; (b) fragment with triple helix; (c) molecule of tropocollagen; (d) organization of the tropocollagen into collagen fibrils; (e) typical striations found on collagen; (f) collagen fibers made up of numerous chains of fibrils (as can be observed under the optical microscope) (modified from Karbowska-Berent and Strzelczyk 2000)*

Other chemical components are: keratin, a structural protein rich in sulphur (sclero-protein), and elastin, also a scleroprotein and a fundamental component of elastic connective tissue; it is chemically similar to collagen but more resistant to hydrolyzing agents and characterized by considerable elastic properties which, however, diminish with time.

Parchment and leather are materials originating from the skins of animals, usually sheep or goats, but the embryos of lambs and calves have also been used for parchment, and bovine

skins for leather. The processes involved in the preparation of the skins vary considerably according to period and geographical provenance; the production processes, mostly artisanal, are determined by cultural and technological knowledge as well as by the intended use of the material. As a general guideline, the main stages of production were maceration, scraping, stretching and drying of the skins, and—for leather—the subsequent curing with plant or mineral substances to avoid putrefaction caused by microbial attack. Depending on the use of the end product, further treatments might be carried out to improve some of the properties, such as strength, resistance, or impermeability for leather, and transparency and gloss through the use of greases or glycerin for parchment. Dyeing was an additional possible treatment.

The use of prepared skins originates in the depths of time, as it was one of the earliest materials used by man to protect himself from the rigors of climate; there is much evidence of the use of skins—more or less treated and sometimes with primitive tanning procedures—that goes back to prehistoric times. For example, the mummified man found in the Tyrolean mountains at Similaun and thought to date from 3350–3100 B.C.E. was wearing shoes with leather soles, trousers, and an overgarment of tanned goat skin.

In the context of decorative arts, prepared skins were used at various times and in various ways. Leather used in conjunction with wood, for example, was used in the manufacture of furniture from the 1300s onward among the Arabs and the Spaniards. In later centuries, the use of leather became more of an art form—leather was decorated with carvings, reliefs, intaglios, and frequently painted. From the fifteenth through the eighteenth century, the use of artistic leather became very popular, both in the production of objects for liturgical use (reliquaries, altar frontals, etc.) and in the manufacture of small luxury objects as well those of more common usage, such as cases and bags. The use of leather for the binding of books is instead considered as an art form in itself, which begins with the first rudimentary bindings in monasteries (twelfth century).

The first written documents using skins prepared as parchment are from Egypt and date from the Fourth Dynasty (2575–2467 B.C.E.). Initially, parchment was used in the form of scrolls made from a succession of individual sheets glued together, and it was only later that these were substituted by codices more similar to present-day books. In Europe, the use of parchment as a writing material became prevalent during the Late Roman Empire (fourth century C.E.) and in the Middle Ages and continued until the fifteenth century when it was slowly replaced by paper.

4.2.1b *Biodeterioration of Leather and Parchment*

The factors influencing the susceptibility to deterioration of leather and parchment can be both internal and external to the material. The internal factors relate to the animal of origin (health, age, gender, diet, etc.) and to the phases, type, and duration of the manufacturing processes (skinning, liming, tanning, drying, etc.). For example, it has been observed that the skins of calves are more prone to biodeterioration than those of an adult bovine; or, that tanning treatments using chemical substances such as chrome impart a greater resistance to biodeterioration in comparison to treatments with substances of plant origin (Strzelczyk et al. 1987, 1989). External factors, on the other hand, are those relating to the environment in which the artifacts are kept, such as humidity, temperature, light, and atmospheric pollution (Tanasi 2002). Susceptibility to biodegradation depends therefore on a number of factors connected to provenance, manufacturing processes, substances employed during the processing and finishing of the objects, the presence or absence of painted decorations, and also and above all, to the state of conservation of the object itself and the environmental conditions in which it is kept.

In the biodeterioration of precious objects made of leather and parchment, the most serious damages are those resulting from the action of microorganisms possessing specific proteolytic enzymes (collagenase, keratinase, etc.). Collagen is considered to be one of the most resistant enzymes to enzymatic breakdown and few microorganisms are able to hydrolyze it. Until the 1960s, it was thought that native collagen was only hydrolyzed by the anaerobic bacterium *Clostridium histolyticum*, but subsequently it was discovered that other anaerobic species belonging to *Clostridium* had analogous capacities. Further studies have shown that collagenase is also produced by aerobic species of bacteria belonging to the genera *Vibrio*, *Bacteroides*, *Pseudomonas*, *Bacillus* and *Cytophaga*, as well as various species of actinomycetes of the genus *Streptomyces*. Among the fungi, *Trichophyton schoenleinii* (dermatophyte), *Enthomophtora coronata*, *Aspergillus oryzae*, and *A. niger* are all endowed with collagenolytic enzymes (Karbowska-Berent and Strzelczyk 2000).

The environmental conditions for the preservation of leather and parchment are such that they rarely encourage the growth of bacteria. The most serious damage linked to the activity of these microorganisms is therefore on objects that have been buried (archaeological finds, clothing accessories in burial grounds, etc.), where there is contact with the soil, a high level of moisture and, at times, absence of oxygen. What is more likely and frequent, instead, is the colonization by actinomycetes and fungi.

Yet, in line with other complex molecules, the highly organized structure of collagen can undergo chemical changes induced by abiotic factors, which make it more susceptible to biodeterioration; these include: the loss of the helical structure due to the rupture of the hydrogen bonds; changes in the organization of the individual fibrils and fibers through the loss of the forces of attraction; loss of crystallinity; fragmentation of the collagen fibers due to hydrolysis or chemical oxidation reactions with consequent rupture of peptidic linkages, of the -C-C- and -N-C- bonds, and of lateral chains (Tanasi 2002). Over time, less than favorable conditions of preservation (strong light, air pollution, thermohygrometric fluctuations, etc.) can modify the chemical organization of collagen and of the other proteinaceous components of leather and parchment, favoring the development of microorganisms with nonspecific enzymatic action (protease, peptidase). As a result, the biodeterioration of leather and parchment initially manifests itself with a nonspecific biological attack on the most easily degradable organic components (proteins, fats, mineral substances, impurities, etc.) that are either linked to their manufacture or that form part of the objects' component materials; only subsequently, with the continuation of microenvironmental conditions that are favorable for biological development, will collagen be attacked.

Referring more specifically to ancient objects made of leather, the research relating to biological degradation has primarily been concerned with book bindings, hangings, or objects of daily usage, such as footwear, bags, etc., both recent and dating back to archaeological periods. With regard to parchment, research has focused on its use as a support for writing and painting, in particular pastel, but also in the production of drums and other musical instruments, for sealing windows, and for inclusion within garments and vestments (miters, for instance), etc. (Baynes-Cope 1971; Orlita 1977; Voronina et al. 1981; Kuroczin and Krumbein 1987; Strzelczyk et al. 1987; Gallo 1992; Strzelczyk et al. 1997; Matè 2002). Many studies have been carried out in tropical countries where biodeterioration problems are particularly serious (Sharma and Sharma 1980; Zainal et al. 1983; Razdan and Chatpar 1991; Sharma 1991).

The main colonizers of leather objects are fungi, but colonies of actinomycetes have also at times been identified in the early stages of the attack (Plate 14). In these cases, it is the fats and water-soluble substances that support their development. Because of the rapidity of their growth and their capacity to utilize fats as a source of energy, the first fungi to develop are the phycomycetes (Chapter 3); however, they are unable to hydrolyze molecules with a high molecular weight such as collagen and, because of this, their attack is short-lived and is rapidly succeeded by that of other fungal species belonging to the genera *Chaetomium*, *Scopulariopsis*, *Trichoderma*, *Penicillium*, and *Aspergillus*. Consequently, the most frequently isolated species are lipolytic ones such as *Paecilomyces varioti* and *Aspergillus niger*; the latter can also develop a collagenolytic action after the source of lipids is exhausted. The fungi responsible for the breakdown of noncollagenic materials present in skin are therefore predominant in the biodeterioration of leather, and their development is favored by the characteristics of an acid pH in the substrate (Table 4.6 and Fig. 4.18).

The action of actinomycetes, on the other hand, is initially mostly impaired by the presence of fungi and the acidity of the skin, but once the attack is established, actinomycetes are particularly damaging because they are able to break down collagen fibers (Strzelczyk et al. 1987). Fortunately, their development is inhibited by the predominance of fungi, but if the acidity of the substrate is reduced, or if the source of nitrogen is increased (which can occur as a result of some treatments and conservation of the leather), the material's susceptibility to attack by these microorganisms is considerably increased.

Table 4.6 Most frequently isolated fungi on parchment and leather artifacts (Gallo 1992; Caneva et al. 1994a; Karbowska-Berent and Strzelczyk 2000)

Alternaria sp.	P	L
Aspergillus sp.	P	L
A. flavus	P	
A. niger		L
A. versicolor	P	
Aureobasidium sp.		L
Botryotrichum pilluliferum	P	
Cephalosporium sp.	P	
Chaetomium sp.	P	L
Cladosporium sp.	P	L
C. cladosporioides	P	
Epicoccum sp.	P	
Fusarium sp.	P	L
Monilia sp.	P	
Mucor sp.	P	L
Mucor scinrinus	P	
Ophiostoma sp.	P	
Paecillomyces sp.		L
Penicillium sp.	P	L
P. chrysogenum	P	
P. commune	P	
P. meleagrinum	P	
P. notatum	P	
P. variabile	P	
Rhizopus sp.	P	L
Rhodotorula sp.		L
Scopulariopsis sp.	P	L
Sepedonium sp.	P	
Stemphylium sp.	P	L
Stysanus sp.	P	
Trichoderma sp.	P	L
Trichotecium sp.	P	
Verticillium sp.	P	

Key: P = parchment; L = leather

However, the element that most determines the biodeterioration of leather is the type of tanning employed in its preparation. Tanning with minerals (and chromium in particular) makes the objects more resistant than those tanned with plant-based products. The plant extracts used in the tanning process contain various glucosidic substances that can sustain microbial growth. On the other hand, treatment with chromium-based compounds does confer upon the skins a certain degree of fungistatic and bacteriostatic properties, but it does not render them completely resistant to biological attack. Indeed, although the skins treated this way are on the whole more durable, they can nevertheless sustain the development of certain species of fungi if the environmental conditions of preservation are particularly humid. Different species of fungi have different levels of tolerance to this metal; for example, it has been observed that some species of *Aspergillus* and *Penicillium* can grow in the presence of 10% concentrations of chromium (even though their development slows down gradually), while other species such as *Curvularia*, *Alternaria*, *Cunninghamella*, and *Cephalosporium* can only tolerate concentrations of about 0.5% (Von Endt and Jessup 1986; Strzelczyk et al. 1987; Sharma 1991; Caneva et al. 1994a).

In the case of ancient objects made of parchment, the state of conservation and the resistance to biodeterioration are predominantly related to the factor of humidity, and in particular to the water content of the objects themselves. The hygroscopic capacity of parchment, which is greater than that of paper, is such that its water content may allow the development of numerous microorganisms even at relatively low levels of RH (RH > 55%).

Which microbial group becomes dominant in the colonization of parchment depends on the moisture content of the latter and on the duration of exposure to damp conditions. In the case of objects that are saturated with water or particularly damp, the most active biodeteriogens are bacteria, the prevalent genera being *Streptococcus*, *Micrococcus*, *Bacillus*, and *Bacterium* (Table 4.7). During the drying-out phase of archaeological finds, it is fungi or actinomycetes that can take the upper hand. These microorganisms also prevail when the

Fig. 4.18 *Fungal attack on parchment (from Karbowska-Berent and Strzelczyk 2000)*

environment in which the objects are kept has an RH > 65%; the most recurrent fungi are ascomycetes and deuteromycetes with lipolytic and proteolytic capacities (Table 4.6).

Normally, degradation begins on the side of the parchment that corresponds to the internal surface of the skin of the animal. In the more advanced stages of deterioration, the external surface in which the collagen and elastin fibers form a dense closely woven net is also attacked (Gallo and Strzelczyk 1971; Karbowska-Berent and Strzelczyk 2000).

In the case of the parchment used in books, the microorganisms initially develop on the edges where there are greater exchanges of oxygen and moisture. The attacked edges become thinner and thinner and then begin to shrink, so that the oxygen is able to reach the inner portions of the book, thus making them prone to attack from microorganisms. The sheets that have been colonized appear worn, darker in hue, and covered with stains (Fig. 4.19, Plate 14). Sometimes the degraded areas and those adjacent to them seem saturated with a sticky substance, which is due to the presence of glycopolysaccharides secreted by the microorganisms. The final result is the complete disintegration of the material. On the flat surfaces of parchment documents—in contrast to the deterioration occurring in books—

the microbial colonies are evenly distributed.

When leather or parchment is the support of a painting, biodeterioration is influenced by the type of pigment and binder used: easily metabolized organic substances (such as sugars and proteins with low molecular weight, water-soluble substances, etc.) that are present in the binders of some of the pigments (pastels and temperas) increase the biodegradability of the object, while the presence of heavy metals, such as lead or zinc present in other pigments, tend to increase their resistance (Strzelczyk 1981; Caneva et al. 1994a).

Among the different kinds of alterations induced by microbial development, the most frequent are stains, varying in size, shape and color. These are easy to remove when they are due to the pigmentation of the cells of the

Table 4.7 Most frequently isolated bacteria on parchment and leather artifacts (Gallo 1992; Caneva et al. 1994a; Karbowska-Berent and Strzelczyk 2000)

Bacillus sp.	P	
B. licheniformis	P	
B. megaterium	P	
B. pumilus	P	
B. subtilis	P	L
Bacterium sp.	P	
Micrococcus sp.	P	
Pseudomonas sp.	P	
Streptococcus sp.	P	
Streptomyces spp.	P	L
S. fimicarius	P	
Nocardia sp.	P	

Key: P = parchment; L = leather

Fig. 4.19 *Page from a parchment book attacked by streptomycetes; a loss in the material as well as numerous stains and deformations are visible (from Karbowska-Berent and Strzelczyk 2000)*

4.2.2 Textile Fibers (Silk and Wool)

4.2.2a *Structure and Composition*

Essentially, there are two types of textile fibers of animal origin: wools and silks. Wools are produced from the hairs of the fleeces of a variety of mammals; among those most frequently employed are those of the common and the merino sheep, the angora and cashmere (originating from Kashmir) goats, the alpaca, the vicuña, the llama, the camel, and the angora rabbit. Silks, on the other hand, are made from the filaments produced by certain serictery animals, such as certain kinds of insects (from which silk proper is produced) and mollusks, from which byssus is obtained.

Chemically, wool is composed of keratin (85%) (Fig. 4.20); lipid substances, especially lanolin; and mineral salts. Keratin is a highly insoluble protein that contains sulphur and is highly resistant to enzyme attack; its amino acid chains have a spatial structure of an α-helix and are therefore highly amorphous; this confers good elasticity to the fiber. The α-helix structure is stabilized by hydrogen bonds between the strands of the helix. When the wool is put under tension, the α-structure has the capacity to transform itself into the β-structure (similar to a folded piece of paper), but its elasticity is reduced. Wool is highly hygroscopic and is considered one of the most hygroscopic textile fibers; indeed, it is able to retain up to 30% humidity without appearing wet (Table 4.8).

To obtain wool, the hairs of the animal are cut, an operation termed shearing; this is followed by spinning, which, through twisting, joins several fibers together to make a yarn.

Wool fiber is a corneous filament; viewed in cross section it is composed of three concentric layers: the cuticle, the cortex, and the medulla (Fig. 4.21). The cuticle, the most external part of the fiber, is composed of scales

colonizing species and more difficult to remove when they are linked to the release of colored metabolites (exopigments) (Plate 14). The development of chromogenic fungi or actinomycetes, which produce pigments as secondary metabolites, is frequent (Chapter 1, section 4). Among the microorganisms responsible for colored stains, especially red and pink-purple in color, are actinomycetes of the family Streptomycetaceae (Table 1.1); more specifically, *Streptomyces fimicarius* has been isolated on illuminated codices (Karbowska-Berent and Strzelczyk 2000).

Structurally speaking, the most serious damage to objects made of leather and parchment is due to the emission of organic acids, which have corrosive effects on the surface, or of extracellular enzymes responsible for the hydrolysis of the chemical components of the objects. The degradation manifests itself with modifications of the original color—sometimes in the form of bleached patches in the area in which the agent of degradation is developing—and with loss of the physico-chemical and structural characteristics of the material. The production of organic acids is a negative influence on materials also because it changes the pH on the surface and therefore modifies the object's susceptibility to degradation; as a consequence, the establishment of alternating colonizations becomes possible, with the upper hand taken by acidophilic species.

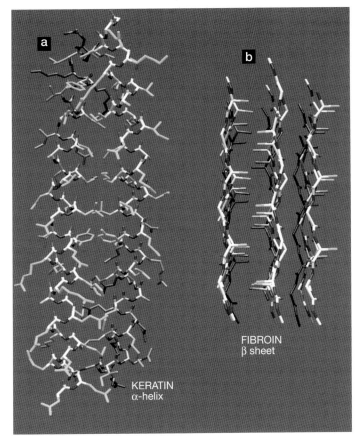

FIBROIN
β sheet

KERATIN
α-helix

Fig. 4.20 *Structure of keratin and fibroin (elaboration by F. Polticelli)*

are numerous pores through which the air circulates.

Instead, silk fiber is composed chemically of the proteins fibroin (60–80%) and sericin (20–30%) (Fig. 4.20), in addition to waxy materials and mineral salts. In contrast to wool, the proteins of silk do not contain sulphur. Each fiber is composed of two fibrils of fibroin joined together with sericin. Fibroin is a scleroprotein belonging to the β-keratin group, with the polypeptide chains folded over in a zigzag, as though they were arranged on a surface with multiple folds; the amino acids on the side chains are linked with hydrogen bonds on the same plane and with bonds between the lateral chains on parallel planes. Because of its spatial structure, fibroin is a highly crystalline and insoluble molecule, and the fibers derived from it are resistant but not very elastic (Table 4.8). Sericin, in contrast, is an albuminoid water-soluble protein.

arranged like tiles, and each scale consists of one cell. The cortex is the middle section and represents 90% of the weight of the fiber; it is formed by millions of spindle-shaped cells, completely keratinized, and composed of bundles of microfibrils grouped together into larger macrofibrils. The medulla is the most internal part and can be either continuous, i.e., along the whole length of the fiber, or discontinuous and located only in the thickest sections of the fiber. Inside the medulla

Silk fibers are among the longest found in nature: they are produced as a single filament that can reach 800–1,500 meters in length. In the most widely accepted meaning of the term, silk is the solidified filament of a viscid fluid produced by the lepidopterus *Bombyx mori* in two sericery glands located on the sides of the intestines; it is used to build the cocoon prior to the insect transforming itself into a chrysalis and then a butterfly. The *Bombyx mori*, or silkworm, has been raised in China since

Table 4.8 Main chemical, physical, and mechanical characteristics of animal fibers

Fiber	Polymer	Structure	Strength	Elasticity	Hygroscopicity
Wool	Keratin	Highly amorphous	Weak	High	Very high
Silk	Fibroin	Highly crystalline	Strong	Low	High

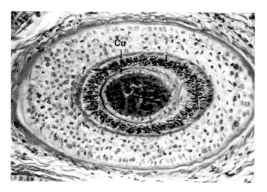

Fig. 4.21 *Section of a hair follicle bulb from sheep's vellum showing three concentric layers (cuticle [Cu], cortex, and medulla)*

3000 B.C.E.; today, this practice has spread to all regions where the mulberry tree—which provides the nutrients for the larvae of the silkworm—can be grown. Other insects, for instance the *Antheraea mylitta* in India, the *A. perni* in China, and the *A. yama-mai* in Japan, are able to produce a silk called *toussah*, which, however, is inferior in its characteristics to that produced by the *Bombyx*.

In the production of the silk yarn, the insect is killed at high temperatures before it can exit the cocoon and break the fibers. The cocoons are then immersed in hot water at 80°C in order to free the end of the silk filament. The spinning brings together several strands of the filaments from a number of cocoons that, on cooling, bind together because of the hardening of the sericin. Other operations linked to the spinning of silk are the degumming and the weighting.

The function of the process of degumming is to remove some or all of the sericin that surrounds the filaments and that makes them opaque and rough. The process involves the immersion of the yarn into a highly concentrated solution of water and neutral soap, heated to 90°C. The silk thus obtained is called degummed or boiled silk and is soft and glossy. Weighting, on the other hand, is a treatment that involves the impregnation of the thread with mineral salts (the salts of tin, zinc, iron, or tannic acids) that are then absorbed by the fibroin. A moderate weighting will enhance the qualities of the yarn, while one that is too heavy will make it more fragile.

Wool and silk fibers are commonly found in the context of cultural heritage, where they are components of objects of daily use as well as of prized objects of artistic and historical value; as such, they represent precious archaeological, ethnographic, or artistic records. Silk in particular, because it is considered a prized fiber, is widely found in tapestries, sacred vestments and hangings, as well as being used as a support for paintings and prints. It is also often found in objects made of multiple materials, either as a yarn (for instance in embroideries) or as a cloth, in altar frontals, banners, book covers, etc.

Wool, on the other hand, is found less frequently, although carpets, tapestries, and elements of clothing have all been made out of wool from the earliest times. There are numerous references to woolen cloth dating back to Roman times, often colored with purple dye or bleached with sulphates.

Because of the delicate nature of these materials, textile archaeological finds are relatively recent, and more ancient examples have only reached us because of the particular environmental conditions in which they were preserved over time, such as for instance the woolen clothes worn by the Simulaun man found in the ice or the Tibetan Tankas on silk (richly woven, embroidered and painted cloths representing religious scenes), which originate from dry, cold climates.

Among other sericeous textiles, *byssus* should be mentioned: it is produced by a type of bivalve mollusk that emits secretions to anchor itself to rocks. These fibers reach a maximum length of twenty centimeters, are either brown or green, and appear smooth and flattened. Extremely difficult to work with, byssus was greatly valued in the past and was therefore employed in the creation of precious royal robes. Nowadays, it is no longer produced, but the term is still used to denote certain types of cotton or hemp cloth that imitate the appearance of this particular type of cloth by creating a loosely woven and thin texture.

4.2.2b *Biodeterioration of Fibers of Animal Origin*

As with all organic materials, the susceptibility of textile fibers of animal origin to microbial attack is strongly linked to their chemical nature and to the treatment and manufacture processes the threads are subjected to. Among the chemical characteristics, the degree of crystallinity, the hygroscopicity, and the purity of the polymeric constituents are the main factors determining the degree of resistance to microbial attack: the higher the degree of crystallinity and the lower the presence of impurities and the affinity for water, the greater the resistance to degradation. The methods of preparation modify the characteristics and the chemical composition of the yarns; for instance, when silk fibers are degummed, sericin (which is a water-soluble protein and therefore more subject to microbial attack) is removed, thereby increasing resistance to biodeterioration.

In general, we can say that proteinaceous fibers are less prone to deteriorate than cellulosic fibers, although they can be damaged both by bacteria and fungi, particularly when there is a high content of organic impurities in the fibers, and the artifacts are kept in a warm and humid environment (Hueck 1972; Vigo 1977; Kowalik 1980b; Caneva et al. 1994a; Bearle et al. 1998). However, the most serious types of damage to wool or silk artifacts are usually linked to the action of insects. The microbial degradation of proteinaceous fibers takes place either as a result of the direct action of proteolytic microorganisms endowed with specific enzymes, or indirectly through the metabolic activity of nonproteolytic microbial strains; the latter use for their growth the various substances added to the fibers or to the textiles during the processes of manufacture—whether of the cloth itself or of the artifact—or other elements that may be occasionally present.

The creation of artifacts made of multiple materials—which includes the use of a variety of materials with a higher susceptibility to biodegradation—can also increase the object's overall susceptibility to degradation. For instance, painted silks are often found glued onto cardboard or paper supports, and this makes them more susceptible to attack by microorganisms thanks to the presence of cellulosic materials and the use of starch-based plant glues.

Embroideries in relief as well as pleats and gathers that are elements in the creation of banners, altar frontals, garments, and costumes can encourage dust deposit and accumulation, thus providing a vehicle for biological pollution and favoring the establishment of microbial colonizing microorganisms. Even certain types of weave that include the creation of raised designs or areas of contrast between high and low gloss—such as found in damask and brocade—can result in greater fragility. The following, for example, have been identified as being responsible for a grayish-brown staining: *Cladosporium herbarum* and *Penicillium chrysogenum* on a silk damask cloth, and *Chaetomium* sp. on a medieval banner in embroidered silk. Because of their reduced or specific proteolytic capacities—but not for those molecules of which silk is composed (keratinolytic)—these fungi can be linked with the dirt and dust deposited and retained on the surfaces (Nugari 1992; Domsch et al. 1993; Giuliani and Nugari 1993).

As with proteinaceous materials in general, for this type of textiles also, susceptibility to biodegradation is increased if the constitutive molecules are structurally altered due to aging phenomena of a physical, chemical, or mechanical type (through exposure to high intensities of light, chemical air pollution, wear, etc.). This means that ancient textiles are more prone to biological attack than modern ones, even though they are often of better quality, which generally offers a greater resistance to decay.

When evaluated in and of themselves, and not as part of the objects they form, silk fibers are considered among the most resistant to biodeterioration, but their high sensitivity to photooxidation phenomena and to physico-chemical stresses often entails an increased fragility and therefore an increased susceptibility to biological deterioration (Berardi and Giuliani 2002).

In theory, wool fibers are more resistant than silk fibers because they are composed of keratin, which as a sulphur-containing protein is less prone to hydrolysis by bacteria and fungi. Wool degradation is more commonly linked to attacks by proteolytic bacteria rather than fungi, as long as the water content is sufficiently high. This condition is more easily met in wool than in other textiles because of its high hygroscopicity; however, a temperature of 25°C would still require an RH of 90%. The most dangerous bacteria are keratinolytic (able to hydrolyze keratin), and among these are a number of streptomycetes (*Streptomyces albus*; *S. fradiae*). Microfungi may also attack wool, but this is less frequently the case; the most damaging species belong to the genera *Fusarium*, *Aspergillus*, *Chaetomium*, *Trichoderma*, as well as *Trichophyton* and *Microsporum*, which, being dermatophytes, can also cause mycosis in man (Hueck 1972; Kowalik 1980b; Domsch et al. 1993; Caneva et al. 1994a). The isolation and systematic identification of the microorganisms responsible for decay in wool, especially in the case of archaeological finds, is particularly difficult. Often, biological alterations are observed but no additional elements are present that may be useful for identification, as is the case, for example, in the studies carried out on woolen Coptic textiles (Bearle et al. 1998).

In contrast, the microbial strains that degrade silk include species that attack proteinaceous materials in general, while microorganisms with specific action are not well known. Among the genus *Aspergillus*, the species of the group *A. flavus* play an important role because of their proteolytic and in particular sericinolytic action (Plate 13); the enzymes produced by these fungi were employed industrially for the degumming of silk (Raper and Fennel 1965; Giuliani and Nugari 1993). Numerous bacteria, and in particular streptomycetes, also hydrolyze sericin. Degummed silk, i.e., purified fibroin, does not seem to be colonized by fungi, but remains susceptible to attack by certain bacteria—bacteria have indeed been isolated on degummed silks of different origins and epochs. The development of bacteria remains, however, a somewhat rare phenomenon, and the colonization process is slow; this explains why this material, which is considered fragile, can still exhibit a relatively good state of preservation even in very ancient objects. Akai (1997) makes the point that good-quality kimonos are passed on from generation to generation without exhibiting any of the damage associated with microbial deterioration, even in humid climates like that of Japan. Recently, laboratory studies have shown that certain soil bacteria such as *Micrococcus luteus*, *Bacillus megaterium*, *B. licheniformis*, and *B. subtilis* have the capacity to utilize fibroin as their sole source of carbon and nitrogen; more specifically, it seems that the species *Variovax paradoxus*, previously classified as *Pseudomonas* sp. or *Alcaligenes paradoxus*, is able to produce an acid protease, probably specific to fibroin (Seves et al. 1998; Forlani et al. 2000; De Rossi and Ciferri 2003).

Overall, the microorganisms that are most frequently mentioned in relation to the degradation of wool or silk artifacts are the bacteria of the genera *Bacillus* (*B. mesenthericus* and *B. subtilis*), *Proteus* (*P. vulgaris*), *Pseudomonas*, *Serratia*, and certain actinomycetes (*Streptomyces albus* and *S. fradiae*) (Seves at al. 1998; Nugari et al. 2003). The bacterium *Pseudomonas aeruginosa* is known to be responsible for red and green stains observed on silks; the color of the stains varies according to the pH of the substrate (Table 1.1). Among the microfungi we should mention the genera *Aspergillus*, *Fusarium*, and *Trichoderma* (Vigo 1977; Kowalik 1980b; Caneva et al. 1994a).

Microbial colonization on silk and wool fibers and textiles produces different kinds of damage depending on the kind of object and the nature of all its components; on the whole, we can say that the most frequent alterations consist in stains of various shapes, size, and color; in modifications of the original color; and in changes in the resistance to traction with various levels of increased fragility, up to complete hydrolysis of the fibers.

4.3 STONE AND RELATED MATERIALS

by Daniela Pinna and Ornella Salvadori

According to the definition given by the NORMAL Commission (NORMAL 1/80) from its very inception, the term "stone materials" refers to "marbles and rocks as well as stuccos, plasters, *intonaco*, and ceramic products used in architecture (bricks, tiles, and terracotta)." All of these materials, whether they be natural (rocks) or artificial (obtained from the transformation of rocks by different processes), exhibit similarities both in structure and composition. The mechanisms of biodeterioration to which they are susceptible are also by and large similar; however, in our overview we prefer to separate natural from artificial stones and highlight the characteristics of biodeterioration for each. As far as artificial materials are concerned, mural paintings will be discussed in some depth (below in this chapter, section 3.2). Indeed, these works have been the subject of countless studies for a variety of reasons: their artistic and historical importance, their widespread geographical presence, and the ease with which they are biodeteriorated, due to either their physico-chemical characteristics or the environmental conditions in which they are often located.

Natural and artificial stones are the most studied materials in the field of biodeterioration of works of art because of their wide distribution—they are the main element of the archaeological and monumental heritage—and because they are more subject to biodeterioration as they are mostly located outdoors; in addition, the processes of deterioration affecting them are the same that play an essential role in pedogenesis.

It is a widely held opinion that physical and chemical agents (abiotic ones) play a more important role in the deterioration of stone than biological agents. This is partly based on the fact that often the role played by organisms is taken into account only when their presence is macroscopically verifiable (plants, mosses, and variously colored patinas), and partly due to the evident increase in the degradation phenomena since the last century, an acceleration caused by atmospheric pollutants. But in reality, even though they may often not be visible, microorganisms already begin to settle on stone during extraction, forming a complex ecosystem similar to that of soil (Orial 2002).

The susceptibility to deterioration of stones is influenced by their chemical nature, their physical structure, and their geological origin, in addition to other environmental factors (Chapter 2). Important is also the amount of degradable minerals present (for instance, feldspars as well as clay and iron minerals): when it represents more than 5% of the weight/volume it passes a critical mark because it can then satisfy the organisms' nutritional requirements for mineral salts (Warscheid and Krumbein 1996). Beyond their chemical composition, the porosity and roughness of the stone surfaces are also particularly relevant for the establishment and growth of the organisms.

4.3.1 Natural Stone Materials

4.3.1a *Structure and Composition*

Rocks are aggregates of one (monomineralic rocks) or more minerals (polymineralic rocks). They can be classified, according to their genesis, as igneous or magmatic, sedimentary, and metamorphic (Casati 1985).

Igneous or magmatic rocks are formed by the cooling and consolidation of magma, which is a silicatic liquid-gaseous mixture that formed at extremely high temperature. Depending on the chemical makeup, there are two distinct forms of magma: granitic magma—acid and saturated with a high proportion of SiO_2 (> 65%)—and basaltic magma—basic and poor in SiO_2 (< 52%)—which give rise, respectively, to acidic (or persilicic) rocks and hyposilicic (maphic and ultramaphic) rocks. Those with an intermediate content of SiO_2 are called mesosilicic rocks. The magma mass can cool slowly in the deep underlayers of the Earth's crust, giving rise to intrusive magmatic rocks, or it can reach the Earth's surface as lava and then cool rapidly, producing effusive magmatic rocks. Intrusive magmatic rocks are entirely crystalline and have a granular or crystalline structure (holocrystalline) (for example, granites, gabbros, diorites). In contrast, effusive magmatic rocks do not have a crystalline structure: they are amorphous or glassy, or else exhibit a porphyritic structure with any crystals embedded in a glassy or microcrystalline base paste (for example, trachites, obsidians, basalts, andesites).

Sedimentary rocks usually form layered deposits in superimposed beds; they include clastic, organogenic, and chemical rocks. Clastic sedimentary rocks derive from the physical and chemical degradation of previously existing rocks into clasts; these are transported by water (rivers, lakes, and seas), by winds, and by glaciers, and finally give rise to the formation of new kinds of rocks through sedimentation and diagenesis. Organogenic and chemical sedimentary rocks, on the other hand, do not derive or derive only partially from preexisting rocks.

The sediments forming clastic rocks are classified according to their granulometry into rudites (> 2 mm), arenites (between 2 and 1/16 mm), lutites (between 1/16 and 1/256 mm), and argillites (< 1/256 mm). The rudites give rise to incoherent rocks, such as gravel and pebbles, or to coherent rocks called conglomerates (breccias, puddingstones); arenites give rise to coherent (sandstones, quartz sandstones, microbreccias, greywacke, arkose) or incoherent rocks (sands). Pyroclastic rocks derive from the accumulation of fragments projected by volcanoes and subsequently further shaped to varying degrees by waters (tuffs, volcanic breccias, cinders, cinerites, lapilla).

Biogenic and/or physico-chemical rocks include carbonate rocks—essentially made up of calcite, aragonite, and dolomite—and nondetritic siliceous rocks of various origins. Organogenic rocks derive from the accumulation of skeletons and fragments of aquatic animal organisms (lamellibranchia, gasteropods, brachiopods, foraminifera, etc.) and vegetable ones (for example, algae) which sometimes are preserved within the rocks as fossils. Among these, we would like to mention the following limestones: nummolitic, red ammonitic, coralliferous and madreporitic, oolitic and pisolitic, sometimes soft, but usually quite compact and well diagenized. To these we would like to add dolomitic rocks, which were formed as a result of a diagenetic metasomatic process (transformation with an exchange of elements).

Sedimentary chemical rocks are instead formed through the deposit of salts separated out of solution by water evaporation or by oversaturation. They include the evaporites (gypsums, rock salts, anhydrites), the carbonates or carbonate rocks (limestones and dolomites), and the siliceous rocks (flints and jaspers). Travertine is an evaporitic rock formed by the precipitation of calcium carbonate from the fluvial waters that flow through calcareous waters (Fig. 4.22). Calcium carbonate precipitates in the water if there is a lowering of the level of CO_2 dissolved in it. The following organisms can contribute to the precipitation of $CaCO_3$ and to the formation of various kinds of rocks (therefore called biochemical): bacteria (oolitic limestones), incrusting algae (stromatolites and oncolites), cyanobacteria, algae, and mosses (travertines).

Metamorphic rocks derive from transformations of preexisting rocks caused by changes in pressure and/or heat (metamorphism) and are characterized by the complete crystallinity of their mineralogical components. Among

these we would like to mention gneiss, mica schists, quartzites, phyllites, sericitic schists, serpentine schists, talc schists, chlorite schists, and crystalline limestones (saccharoid marbles) (Fig. 4.22).

Each rock is characterized by its own mineralogical composition, petrographic structure, and chemical composition. The physical characteristics that can be seen under the microscope are structure and texture. The structure is defined by the form of the individual minerals, by their dimensions, and by their reciprocal relationships (for example, holocrystalline, microcrystalline, porphyric, glassy, etc.). The texture indicates the spatial relationships among the minerals, also in relation to the forces that shaped it (for example, schistose, fluidal, foliated, etc). The porosity of a rock is constituted by open pores, communicating with the air, as well as closed pores; in addition to the overall porosity, it is important to know the shape and size of the pores as they influence the processes of water absorption

Fig. 4.22 *Crystalline structure of two lithotypes: (a) Carrara marble; (b) travertine*

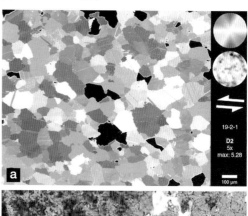

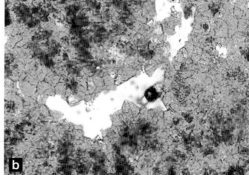

and the associated processes of degradation (for example, salt crystallization, freeze-thaw, biodeterioration).

The employment of rocks as construction material, or for the realization of sculptures, bas-reliefs, etc., requires a good knowledge of their technical properties, such as their compactness, hardness, resistance to various factors (compression, traction, flexure, shock, wear, atmospheric agents), divisibility, imbibition capacity, permeability, workability, and color. Sedimentary rocks are the most commonly used as construction materials (in particular limestones and sandstones), while a huge array of different rocks—magmatic, sedimentary, and metamorphic—have been used for decorative finishes in architecture and for sculpture.

4.3.1b *Biodeterioration of Natural Stone*

In a way, a stone exposed outdoors is not a substrate favorable to biological colonization, because its surface and, to some degree, its interior represent a very difficult environment, both in terms of physical parameters and in terms of the scarcity of nutrients. The stone surfaces exposed to direct sunlight can reach extremely high temperatures (> 50°C) (Garty 1990) and, depending on the climate, they can be subjected to freeze-thaw cycles. Vertical or subvertical walls in particular do not favor the retention of water by the substrate and can often be considered at the same level as desert conditions, where a high degree of aridity is dominant, with intervals of sometimes even violent rainfall that may cause localized water streaming. Nevertheless, biological colonization occurs also in these circumstances, which may even favor the establishment of not only photoautotrophic pioneer organisms such as cyanobacteria, algae, and lichens, but also of other oligotrophic or poikilotrophic microbial groups such as certain groups of fungi with good resistance to environmental stresses (Gorbushina and Krumbein 1999). To these organisms we should add chemolithotrophic and chemoorganotrophic bacteria, whose biodeteriogenic role is well known. On the other hand, the endolithic environment—beneath the surface of the stone—has many advantages (water retention is greater; radiation from the sun is lower and not limiting for phototrophs; and, to a certain degree, the action of wind is also limited) and thus represents an environmental niche for many microorganisms therefore called endolithic (cyanobacteria, green algae, fungi, and lichens) (Pohl and Schneider 2002).

According to the different types of environment in which the stone works are located (indoors, outdoors, semiconfined areas), specific biodeterioration problems arise; these will be analyzed in Chapter 5.

In the initial stages (Pochon and Coppier 1950; Pochon and Jaton 1971), and for many years subsequently, research on stone degradation focused on the role played by chemolithotrophic bacteria that are able to utilize, as a source of energy, the oxidation of inorganic compounds such as ammonia, nitrites, hydrogen sulphide, and elemental sulphur. To this group belong the bacteria operating in the sulphur and nitrogen cycles, which have been studied in more depth since they are autotrophic. What motivated the research in this direction was the fact that stone is an inorganic material—and therefore autotrophs were the first that required investigation—as well as the fact that the organisms that, in theory, could be the most dangerous were those able to draw energy from the chemical processes that were typical of their energetic metabolism (therefore, chemoautotrophs even more than photoautotrophs). In addition, with the sense of urgency created by the acceleration of the processes of stone decay due to the widespread problems associated with sulphation and nitration, the hypothesis of a biological cause (at least as a contributory cause) was clearly plausible. The autotrophs' capacity to deteriorate stone has, in fact, been demonstrated experimentally in the laboratory, even though the true dynamics related to the different typologies and environmental conditions have not been established.

The sulphates, frequently found on stone, may be caused by the sulphur-oxidizing bacteria (among which is the genus *Thiobacillus*, resistant to acids and able to resist up to values of pH 1), which, by utilizing various reduced compounds of sulphur or elemental sulphur, produce sulphuric acid; this, in turn, reacts with the calcium carbonate of the stone and gives rise to gypsum ($CaSO_4 \cdot 2H_2O$) (Fig. 4.23).

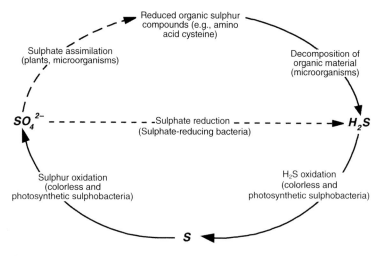

Fig. 4.23 *The sulphur cycle*

A substantial problem is that only very rarely are sulphur or reduced sulphur compounds present in stone materials; therefore, in order to explain a biological hypothesis for the phenomena of sulphation in stone, we must presume the utilization of either atmospheric SO_2 or of other sources of organic S (as it is possible in complex communities), or else the association of sulphate- and sulphur-reducing bacteria, which are able to transform the sulphates into sulphides in conditions of anaerobiosis. These bacteria (e.g., *Desulfovibrio desulfuricans*)

Fig. 4.24 *The nitrogen cycle*

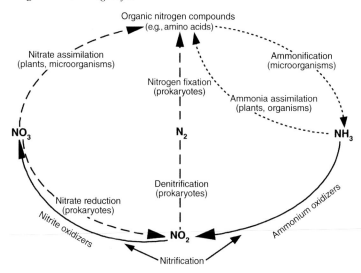

are abundant in the soil, and the reduced compounds they generate can move up the walls along with water rising by capillary action; this can cause a true cycle, which, however, could only be sustainable in the lower parts of walls. It is of interest to note that nowadays, because of their metabolic characteristics, these microorganisms can be used positively during the cleaning phase of a conservation operation to remove the black incrustations particularly rich in sulphates, even though their anaerobiosis poses application problems not easily resolved (Atlas et al. 1988; Ranalli et al. 1997) (Chapter 8, section 5.1). Nitrifying bacteria cause the oxidation of ammonia (deriving from the degradation of organic material and, for instance, from agricultural fertilizers) or of other nitrogen-based compounds (deriving from atmospheric pollution), thus producing nitrous and nitric acids (Fig. 4.24). These bacteria consist of two groups: ammonia-oxidizing bacteria (genera *Nitrosomonas*, *Nitrosovibrio*, *Nitrosococcus*, *Nitrosospira*, *Nitrosoglobus*)—which transform ammonia into nitrous acid—and nitrite-oxidizing bacteria (genera *Nitrobacter*, *Nitrococcus*, *Nitrospira*), which transform nitrous acid into the highly corrosive nitric acid.

Analogously, recent laboratory experiments testing the utilization of denitrifying bacteria (*Pseudomonas stutzeri*) for the removal of nitrates from stone have yielded positive results (Ranalli et al. 1996, 2000).

Large quantities of sulphur-oxidizing bacteria

have frequently been found by Italian and French researchers on marble and limestone, as opposed to nitrifying bacteria, which have only rarely been identified and usually in smaller numbers (Jaton 1972; Barcellona et al. 1973; Tiano et al. 1975; Urzì and Realini 1998; Bartolini and Monte 2000). Still, the problem of quantifying the role of sulphur-oxidizing bacteria, at least in Mediterranean climates, has not been sufficiently investigated and merits further in-depth research.

Studies carried out in the last few years on historical German monuments, mostly composed of sandstone, have shown that the microorganisms found in the highest number are the chemoorganotrophs, followed by fungi and nitrifying bacteria, while sulphur bacteria are either not found at all or else in insignificant numbers (Wilimzig et al. 1992; Mansch and Bock 1998).

Many chemoorganotrophs are able to remove iron (Fe^{2+}) and manganese (Mn^{2+}) cations from the stone by oxidation, inducing color changes through the deposition of oxides. Recent studies have emphasized their importance, along with photoautotrophic microorganisms, as primary colonizers of stone (Warscheid and Braams 2000). Chemoorganotrophic and chemolithotrophic bacteria have also been found in significant numbers on deteriorated works in granite (Leite Magalhaes and Sequeira Braga 2000).

The chemoorganotrophic bacteria most frequently encountered on stone are grampositive and belong to the genera *Bacillus*, *Micrococcus*, *Streptomyces*, *Geodermatophilus*, *Micromonospora*, *Arthrobacter*, *Clavibacter*, *Aureobacterium*, *Rhodococcus*, and *Brevibacterium*. Although they are commonly present in the soil, gram-negative bacteria are only infrequently found on exposed stone surfaces, as they are more sensitive than gram-positive bacteria (Warscheid and Braams 2000). Chemoorganotrophic bacteria produce many compounds with a chelating action (e.g., the following acids: 2-ketogluconic, lactic, glycolic, citric, succinic, gluconic, galacturonic), which form complexes with various elements such as Ca, Cu, Ni, Mn, and other metals. In addition to calcite, dolomite, and aragonite, many other minerals can be attacked, e.g., haematite, goethite, limonite, and pyrolusite.

The salts present within the rock interact with the microorganisms, increasing the density of extracellular polymeric substances (EPS) and influencing their hydric properties. Laboratory experiments have shown that the combination of physical agents (salts) and microbiological agents (heterotrophic, sulphur-oxidizing, halotolerant, and moderately halophilic bacteria) of deterioration results in a noticeable increase in deterioration as compared to that obtained with any one of the agents alone (Papida et al. 2000). The importance of the role played by the extremely halophilic and alkalophilic Eubacteria and Archaea in the deterioration of mural paintings (Rölleke et al. 1998) has recently been discovered using molecular techniques; this result could also be relevant for the decay of stone. Saiz-Jimenez and Laiz (2000) have demonstrated how the addition of salts (sodium chloride or magnesium sulphate) to culture media allows for the isolation of a greater number of halotolerant or halophilic heterotrophic bacteria, in particular those belonging to the genus *Bacillus*, which have a particular osmotic adaptation to high levels of salinity.

It is well known that pollutants (SO_2, NO_x, aliphatic hydrocarbons, and aromatic polycyclic hydrocarbons) can have different effects on the microflora found on stone (Chapter 2, section 2.5). Mitchell and Gu (2000) compared the biofilms of polluted and nonpolluted tombstones in two American localities and showed that the presence of pollutants reduced both the diversity and the density of the microorganisms. In general, pollution reduced the number of the groups of microorganisms under consideration (fungi, heterotrophic and chemolithotrophic bacteria); within the chemolithotrophs, however, a sharp rise was noted in the number of sulphur-oxidizing bacteria. Saiz-Jimenez (1997) has shown how certain heterotrophic bacteria (genera *Pseudomonas*, *Bacillus*, *Nocardia*) are able to mineralize phenanthrene and probably also other anthropogenic organic compounds present in the black crusts of stone monuments. According to Pitzurra et al. (2003), the biological action of some fungi (*Cladosporium cladosporioides*, *Cryptococcus*

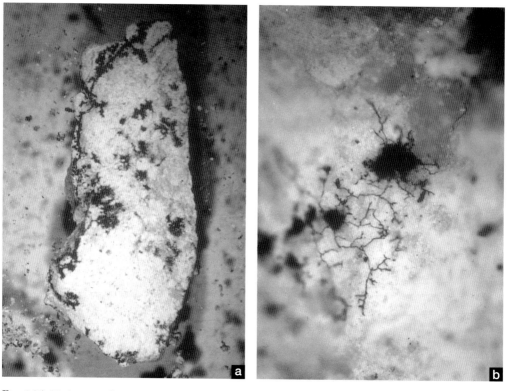

Fig. 4.25 *Meristematic fungi on a stone substrate* – Photo D. Pinna

inuguttulatus, Rhodotorula glutinis) in the deterioration of limestone in urban environments is greater than the one exercised by pollutants over the same period of time (one year).

Black fungi, or Dematiaceae, are considered by some authors the most damaging in the deterioration of stone (Krumbein and Diakumaku 1996; Urzì et al. 2000a). Many of the fungi found on stone, especially on surfaces exposed to direct sunlight, do indeed exhibit a dark pigmentation due to the presence of melanin, which provides them with protection against physical agents (e.g., UV radiation, x- and γ-rays) and cellular lysis. These are the so-called meristematic black fungi or microcolonial fungi (MCF) and black yeasts (Fig. 4.25; Chapter 3, section 2.4). Although already known because of their naturally occurring outcrops, these fungi have been the object of many studies in the past decade and have been identified on monuments and in archaeological sites (Gorbushina et al. 1995; Wollenzien et al. 1995). The fungi identified so far have been

attributed to the genera *Acrodictys, Aureobasidium, Capnobotryella, Coniosporium, Exophiala, Hormonema, Hortaea, Lichenothelia, Monodictys, Phaeococcus, Phaeococcomyces, Phaeosclera, Sarcinomyces,* and *Trimmatostroma* (Gorbushina et al. 1997; Urzì et al. 2000a). Meristematic fungi seem to produce organic acids very rarely; according to some authors, the decay they cause is therefore of a purely physical nature (Gorbushina et al. 1995; Dornieden and Gorbushina 2000).

Many deuteromycetes belonging to the families of hyphomycetes and coelomycetes that are common both in the soil and in the air and that can be producers of melanin or not (such as *Alternaria, Cladosporium, Aureobasidium, Ulocladium, Epicoccum, Phoma, Penicillium, Aspergillus, Fusarium*) can also develop on stone, but only in more favorable conditions (greater availability of organic substances or nonextreme thermohygrometric conditions); they tend to be involved more frequently in the degradation of wall paintings and restored stone artifacts (below in this chapter, section 3.2b and section 6). Their capacity for producing organic acids and dissolving carbonates has been demonstrated in numerous laboratory experiments and depends essentially on the conditions of growth and on the amount of available carbon sources. Among organic acids are the following: tartaric, gluconic, glyoxylic, oxalic, citric, fumaric, pyruvic, succinic, malic, propionic, etc., which can mobilize cations with their chelating action (Hirsch et al. 1995a). Some of these acids, for instance citric and oxalic, act as chelating agents; citric acid has a greater dissolving action on stone than oxalic acid (McNamara et al. 2003). The excretion of oxalic acid results in the extensive corrosion of the primary minerals and the complete decomposition of the ferruginous clay minerals; in addition, iron oxides, amorphous gels, and oxalates are precipitated.

Limestones, silicate minerals (especially mica and orthoclase), minerals containing iron and magnesium (for example, biotite, olivine, pyroxene), and different phosphates are also attacked. Basic igneous rocks are more susceptible to fungal attack than granitic rocks, which seem highly resistant.

Many fungi identified on deteriorated stone have the capacity to oxidize iron and manganese either directly (through enzymes) or indirectly, by forming extracellular incrustations on the hyphae (Warscheid and Braams 2000); these fungi could play a role in the formation of some patinas (rock varnish) in desert climates (Chapter 6, section 2.1).

In dolomitic and calcitic stones, the hyphae penetrate the calcite crystals not only along the planes of the crystals, but also across transverse lines, and the surface of the stone appears clearly corroded when observed under SEM (Koestler et al. 1985). The filamentous structure of the hyphae facilitates their penetration into the substrate; some fungi are endolithic and produce pitting phenomena (see below).

Fungal growth can also be found on stone that has been restored, on which resins have been applied either to consolidate the materials or as water repellents (below in this chapter, section 6).

As far as photoautotrophic microorganisms are concerned, it should be noted that,

on the basis of various studies carried out on European monuments, the biodiversity of photosynthetic microorganisms living on outdoor stone is rather broad. Indeed, the specialized literature describes around two hundred different taxa of Cyanobacteria, Chlorophyceae, and Bacillariophyceae (Tomaselli et al. 2000a). Cyanobacteria are in the majority (thirty-five genera and ninety species, e.g., *Phormidium*, *Nostoc*, *Microcoleus*, *Plectonema*, *Gloeocapsa*, *Myxosarcina*, *Chroococcus*, *Stichococcus*), followed by the Chlorophyceae (twenty genera and thirty-five species, e.g., *Klebsormidium*, *Chlorella*, *Muriella*, *Apatococcus*), while the Bacillariophyceae (for example, *Navicula*, *Nitzschia*) are less common and have been isolated on works in contact with water (for instance, fountains) or located in coastal areas. The Rhodophyceae (red algae) have been isolated very rarely, on monument fountains or in very humid underground environments (Chapter 5).

There seems to be no evident correlation between the different species and the mineralogical composition of the substrate, although among the cyanobacteria it has been observed that species of the genus *Nostoc* seem to have a preference for artificial substrates, while unicellular cyanobacteria favor calcareous substrates, and filamentous species without heterocysts prefer siliceous substrates (Tomaselli et al. 2000a, 2000b, 2000c). The most commonly identified Chlorophyceae are cosmopolitan and are found on every sort of substrate.

A research carried out on painted buildings in five countries of Latin America (Argentina, Bolivia, Brazil, Mexico, and Peru), has led to the identification of thirty-four genera of cyanobacteria and forty-six genera of algae (Chlorophyceae, Bacillariophyceae, and Xantophyceae) (Gaylarde and Gaylarde 2000). According to these authors, green algae are the initial colonizers of these works, while cyanobacteria become predominant only in more stable populations, for instance, when the stone surface has not undergone any treatments or maintenance. Analogously, an earlier study conducted in Singapore showed that the photosynthetic organisms found most frequently on buildings belonged to the green filamentous algae of the genus *Trentepohlia* (Wee and Lee 1980).

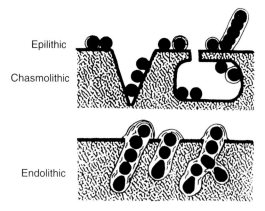

Epilithic

Chasmolithic

Endolithic

Fig. 4.26 *Schematic representation of the growth of algae and cyanobacteria in relation to the substrate (modified from Golubic et al. 1975)*

The damages caused by photosynthetic microorganisms to natural stone materials, although still at times a matter for controversy, are generally linked to their growth forms (Fig. 4.26).

Aside from the aesthetic disturbance caused by the variously colored patinas formed by these microorganisms, the main consequences of their development are: water retention with consequent damage linked to freeze-thaw cycles; increased ease of establishment for other organisms (for instance, fungi and macroorganisms); and, depending on the growth forms, the mobilization of calcium ions, the shared responsibility for the detachment of scales of material (casmolithic and cryptoendolithic algae), or the dissolution of the substrate (endolithic algae) (Chapter 3).

Chasmoendolithic and cryptoendolithic algae develop in the stone's interior; the first

Fig. 4.27 *Schematic representation of the detachment of scales by chasmoendolithic organisms*

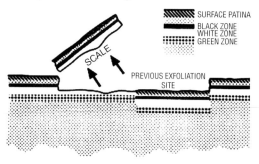

| | SURFACE PATINA |
| BLACK ZONE |
| WHITE ZONE |
| GREEN ZONE |

SCALE

PREVIOUS EXFOLIATION SITE

choose cavities and fissures that are preexistent, while the second generate them themselves. In both cases, the increase in the biomass inside the stone builds up a pressure that, in the end, causes a fracture in correspondence of the colonized area, with the consequent detachment of a scale of the material. This process can occur repeatedly, involving areas deeper and deeper within the structure (in relation to the surface), as illustrated in Figs. 4.27 and 4.28. Euendolithic species are the most damaging in that, by dissolving the very substance of the substrate, they develop within it, forming microcavities of varying morphology according to the species present (Plate 15). The growth of these microorganisms is limited by the need for light: penetrations of a few millimeters have been recorded.

A study conducted on a Scottish castle built of sandstone—colonized mainly by *Trentepohlia aurea*—has shown that this organism can contribute to the mechanical disaggregation of the stone as well as to its chemical degradation through the production of lac-

Fig. 4.28 *(a) Scales of stone detached through the action of chasmoendolithic organisms and (b) internal growth* – Photo G. Caneva

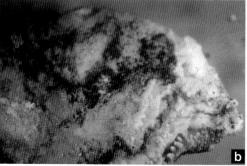

tic acid, which acts as a chelating agent and mobilizes calcium ions (Jones and Wakefield 2000). The dissolution of stone accompanied by the mobilization of calcium is also carried out by calcifying cyanobacteria (for example, *Scytonema julianum*, *Loriella osteophila*), which reprecipitate calcium as calcium carbonate in the polysaccharide sheaths. These, in turn, act as centers of nucleation for the neoformation of minerals (Ariño et al. 1997a; Hernandez-Marine et al. 1999; Albertano 2002).

In general, the colonization of stone surfaces by microflora, whether it is bacterial, fungal, and/or algal, produces different kinds of alterations, such as patinas, stains, and pitting.

Many authors have emphasized the importance of the microbial origin of patinas occurring on stone materials located outdoors, whether from natural outcrops or monuments (Krumbein 1992; Urzì et al. 1992; Dornieden et al. 2000; Gorbushina and Krumbein 2000; Warscheid and Braams 2000). The patinas are of various colors, from black to beige, to green, brown, orange, etc., and have different microstructures and thicknesses. The nature of the patinas can also vary, ranging from those that are essentially a biofilm (for instance, algal patinas) to others with a high mineralogical content and a crusty consistency; the main component in the latter patinas is generally calcite with—depending on the cases—calcium oxalates, quartz, gypsum, etc. Often, a biological genesis is recognized for these patinas as well.

An in-depth examination of the so-called algal patinas—thus named because of the predominance of photosynthetic microorganisms—shows that they are actually made up of biofilms containing a variety of microorganisms such as algae, cyanobacteria, fungi, actinomycetes, and other bacterial groups, as well as protozoa that feed on all of the above (Flores et al. 1997; Gaylarde and Gaylarde 2000). The varying colors of these patinas (green, black, yellow, orange, purple, red, etc.) depend on the type of biocoenosis, on the stage of their development, and on the growth phase of the prevailing alga. It is known, for example, that patinas made up largely of the green alga *Haematococcus pluvialis* can take on a red color in conditions that are stressful for this microorganism (Ricci et al. 1985).

Gram-positive bacteria, cyanobacteria, algae, and fungi give rise to patinas of different colors depending on the microorganisms involved and on the pigments produced by them. Krumbein et al. (1993) have identified gram-negative bacteria able to form orange colorings on marble samples in the laboratory.

Indeed, various attempts have been made to classify and characterize these patinas. Warscheid and Braams (2000) distinguish three different kinds of patina—indicating for each which biodeterioration processes are involved—according to their morphologies (film, corroded surface, crust); the type of rock; the petrographic characteristics (size of the grains, levels of porosity and size of the pores, water content); and the distribution and type of microflora.

Garcia-Valles et al. (2000) have recently characterized—both from a petrographic and a biological standpoint—the patinas differing in color and thickness present on a Turkish quarry marble. They distinguished them into five types: microstromatolithic crusts, microlaminate patinas, regular and irregular single-layer patinas, biopenetrated dark surfaces (from black to gray). The different patinas are formed according to the inclination of the stone surface, to the degree of exposure to rainfall, and to the different microorganisms involved (bacteria dissolving calcium carbonate, pigmented and nonpigmented bacteria, algae, cyanobacteria, fungi). According to these authors, the processes of biomineralization and biodeterioration occur simultaneously on the surface, and the equilibrium can shift in favor of one or the other depending on certain environmental conditions such as sunlight and availability of water and nutrients.

Another alteration attributed by some authors to the development of microorganisms is the formation of pinkish-orangy-red stains, which have been found on several Italian marble works. It was possible to isolate some heterotrophic pigmented bacteria from these alterations (*Micrococcus roseus*, *Micrococcus* sp., *Flavobacterium* sp.) as well as a yeast, *Rhodotorula minuta* (Tiano and Tomaselli 1989; Sorlini et al. 1994), while the search for chemolithotrophic and photosynthetic microorganisms proved fruitless. Some studies have correlated the stains with the presence

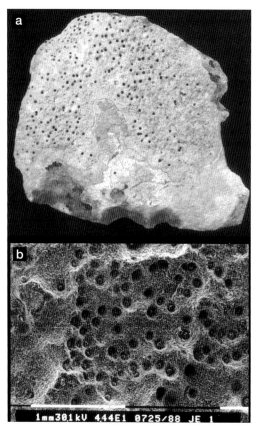

Fig. 4.29 *Biopitting: (a) endolithic lichens observed under the stereomicroscope; (b) microphotograph under SEM of* Lichenothelia *sp.* – Photo O. Salvadori

may also have a biological origin (Figs. 2.14 and 4.29). Micropitting is induced by the action of endolithic microorganisms (cyanobacteria, green algae, fungi). This type of growth is generally widespread on unprotected surfaces—especially if they have a certain degree of inclination that favors water flowing—and can cover large areas of the works and sometimes the whole surface (for instance, tombstones) (Salvadori 2000; De Leo and Urzì 2003). The microscopic alteration appears simply as an overall variation of the stone's natural color, which takes on varying shades of gray, from light to dark and up to black (Plate 15). For this reason, the phenomenon is largely underestimated as it is often confused with abiotic alterations such as deposits of atmospheric particles. Endolithic microorganisms are not found on the surface of the stone, but develop within it; the patterns vary according to the species present. Observing the stone surface under magnification, it is possible to see these microorganisms lodged in the microcanals and cavities that they themselves formed by dissolving the stone and that open onto the surface, creating the micropitting. Because of this, their identification is quite difficult, and it becomes impossible if the sample is taken only from the surface of the stone instead of through removal of a scale of material (Fig. 4.30).

The endolithic biodeteriogens known at present are photosynthetic (cyanobacteria, green algae, lichens), to which bacteria and chemoorganotrophs (fungi, especially meristematic ones) can sometimes be associated. The density of the colonization, the maximum

of minium (Pb_3O_4), and some of the isolated bacteria (*Bacillus* sp. and *Staphylococcus* sp.) showed resistance to varying concentrations of lead nitrate (500–1,500 ppm) (Zanardini et al. 1997; Giamello et al. 2004). However, the role of these microorganisms in the production of stains has not been objectively established yet (Zanardini et al. 2003).

Pitting is a type of deterioration characterized by the presence of numerous blind holes varying in size. Depending on the diameter of the hole, it can be distinguished into micropitting (from 0.5 to 20 μm), mesopitting (from 20 μm to 1 mm), and macropitting (from 1 mm to 2 cm) (Gehrmann et al. 1992). Since micro- and mesopitting are produced by organisms, the term biopitting is also used (Fig. 4.29), although it is not excluded that macropitting

Fig. 4.30 *Photograph of endolithic microorganisms in cross-section and under SEM* – Photo O. Salvadori

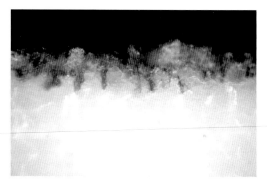

penetration, and the biomass per surface unit vary considerably according to the endolithic species present. The depth of penetration generally varies from a few dozen to a few hundred of microns. Cyanobacteria and green algae (both free and as photobionts) generally reach depths of 100–300 µm, and some filamentous species can penetrate up to a depth of a few millimeters (as in the case of the Fontana del Nettuno, Trento) (Salvadori 2000). As mycobionts, fungi can penetrate farther, reaching depths of several millimeters into the stone.

In addition to dissolving the stone through the penetration of the thallus, endolithic lichens also cause grave damage to the substrate through the development of ascocarps (apothecia or perithecia). The perithecia are produced inside the stone, under the photobiont, and protrude toward the exterior to allow the release of the spores. When they die, the surface of the stone is left with the holes (mesopitting), the diameter of which varies according to the species (Gehrmann and Krumbein 1994; Wessels and Wessels 1995). These holes, which have a high density per square centimeter, can then be enlarged by the action of water and merge to form larger holes (Fig. 4.29).

As mentioned, the presence of macropitting—with holes measuring a few centimeters in diameter (Fig. 2.14)—is not uncommon on many monuments (for example, the south front of Trajan's Column or the lower section of the south face of the Cestia Pyramid). Such macropitting represents a very favorable microhabitat for microorganisms, which almost always colonize them instead of the surrounding areas. In many instances, it is difficult to establish whether or not the microorganisms are the cause of the formation of these holes, but it is certain that once bacteria, algae, and fungi are established in these holes, they contribute to their enlargement acting in synergy with other abiotic factors (e.g., water) (Caneva et al. 1992c).

The process of biological succession on outdoor stone surfaces is frequently initiated by lichens, in addition to bacteria, cyanobacteria, and algae. Lichens therefore behave as a pioneer species able to extensively colonize rocks. The most frequently encountered growth form on rocks is crustose. The thallus can develop

above the stone surface (epilithic lichens) or inside the stone, either completely (endolithic lichens) or partially (hemiendolithic lichens) (Chapter 3, section 4). Endolithic lichens have recently been the object of particular focus in the conservation field (Pinna et al. 1998; Pinna and Salvadori 2000; Tretiach et al. 2003). Despite the fact that they frequently colonize limestone monuments, they received little attention for a long time. Such organisms can seriously weaken the substrate as they are immersed in it, but they are sometimes difficult to spot by nonspecialists because their thallus is of the same color as the stone.

Most of the species found on Italian monuments have a very wide ecological and geographical distribution, while the endemic species, which are linked to more specific areas, are somewhat rare. Lichen flora found in archaeological sites is often quite rich, both for the variety of substrates and for the characteristics of the works of art, which provide a greater variety of specific growth environments (niches, vertical surfaces with different exposures, etc.) than natural outcrops of rocks (Nimis et al. 1987). Nitrophilic species are also very frequent, preferring eutrophized surfaces on which nitrogenous substances of various origins accumulate; such environments often exist in locations with human presence. Lichen species also differ according to the substrate, i.e., calcicolous and silicicolous lichens, which prefer, respectively, calcareous and siliceous rocks (Nimis et al. 1992).

Lichens, in particular foliose and crustose species, damage the stone mechanically because of the contractions and expansions of the thallus in relation to the absorption of water and the penetration of hyphae, which enlarge the pores and cause the loss of stone cohesion (Fig. 4.31). The hyphae of the mycobiont are even able to penetrate minerals such as calcite or quartz (Modenesi and Lajolo 1988; Pinna et al. 1998; Salvadori and Tretiach 2002).

The depth of penetration into the substrate reached by epilithic crustose lichens is linked not only to the species, but mainly to porosity, texture, and physico-chemical deterioration of the stone, and, to a lesser degree, to its mineralogical composition (Salvadori and Tretiach 2002) (Plate 16). The thickness of rock

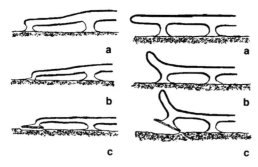

Fig. 4.31 *Schematic representation of the detachment of rhizines—left,* terminal rhizine in crustose lichen, *right,* subterminal rhizine in foliose lichen—*as a result of the degree of hydration: (a) hydrated to (c) dehydrated (from Mellor 1923)*

deteriorated by lichens varies from a few microns to 16 mm (Ascaso et al. 1982, 2002; Chen et al. 2000; Bjelland et al. 2002). On the other hand, the capacity for penetration of endolithic lichens seems to be linked to individual species, which can penetrate thicknesses from 380 μm to 3 mm (Pinna et al. 1998; Pinna and Salvadori 2000) (Plate 16). Although found more frequently on limestone, endolithic lichens can also colonize other substrates. *Lecidea* aff. *sarcogynoides*, for instance, has an average penetration in sandstone of 2.94 mm, but is able to reach depths of 5.80 mm (Wessels and Wessels 1995). These authors have also shown that lichens grow yearly 4 mm horizontally and 0.1 mm vertically. Such a speed of growth, unusually high if compared to the values proposed by Hale (1973)—from 0.5 to 3 mm every hundred years—can be explained by the particularly favorable microclimatic conditions of the examined site.

The continuing growth of lichens detaches from the substrates mineral fragments such as muscovite, quartz, feldspar, albite, calcite, micas, etc., which are then incorporated in the thalli (Salvadori and Lazzarini 1991; Bjelland et al. 2002; Salvadori and Tretiach 2002) (Fig. 4.32). Other fragments brought by the wind can also be added to these.

An interesting aspect, which refers in particular to epilithic crustose lichens, is represented by the presence of microorganisms in the lichen-substrate area of contact. Some algae and/or cyanobacteria can be charac-

terized by endolithic growth and can live in intimate contact with the mycobiont without, however, being part of the symbiosis (Ascaso et al. 1998).

The chemical processes of deterioration attributed to lichens are the production of carbonic acid, organic acids, and substances with chelating properties that extract cations from the substrate (Chapter 1). The lichen-substrate interface is therefore a zone rich in chemical activity (Jones and Wilson 1985).

Carbonic acid, formed during the reaction between carbon dioxide and the water retained by the thalli, reacts with stone and removes basic cations (K^+, Na^+, Mg^{2+}, Ca^{2+}) and silicates.

Lichens produce a variety of organic compounds, generally called lichenic substances, among which are a number of characteristic organic acids. The main groups of lichenic substances are: depsides, depsidones, depsones, benzylesters, dibenzofurans, usnic acids, chromones, xanthones, anthraquinones, higher aliphatic acids, terpenoids, and derivatives of pulvinic acid. These compounds are usually produced in particular areas of the thallus such as the cortex, the medulla, the apothecia, or the soralia. Bjelland et al. (2002) identified these substances inside the degraded rock as well, under the thalli of some crustose species on sandstone, thus proving their direct involvement in the deterioration of the substrate.

Although the studies carried out on the chemical degradation mostly concern crustose lichens, some of the results seem to indicate that

Fig. 4.32 *Thin section of* Lecanora campestris *with incorporated minerals: calcite, quartz, and neoformation products (oxalates)* – Photo O. Salvadori

foliose species may have degrading properties as well. According to Adamo et al. (1993), the capacity for deterioration varies more according to the physiology of the individual species than to the morphology of growth.

Among the substances produced by lichens, oxalic acid is the most active in the alteration of stone, because of its chelating properties. It is an intermediate product of cellular metabolism and may have oxaloacetate as its precursor, which is itself an intermediate compound in the Krebs cycle. The oxidation of glucose leads to pyruvate and water, which is followed by the carboxilation of the pyruvate to oxaloacetate and subsequently by the hydrolysis to form oxalic acid and acetic acid.

Oxalic acid is also produced through another metabolic path: the glyoxylate cycle, which is a deviation of the Krebs cycle. In it, the isocitrate undergoes enzymatic scission into glyoxylic acid and succinic acid; the glyoxylic acid is then oxidized by glyoxylic-dehydrogenase into oxalic acid.

Oxalic acid can also be produced through a third metabolic path, which has ascorbic acid as its precursor (Chapter 1, section 4.2).

The action of oxalic acid on the substrate is easy to recognize: the crystals of calcium oxalate—an almost completely insoluble salt formed by the reaction between oxalic acid and calcium carbonate—precipitate into the extracellular spaces on the lichen surface, inside the thalli, at the lichen-substrate interface, or even inside the rock in the areas where the hyphae have penetrated (Gorgoni et al. 1992) (Fig. 4.33). Lichens are also able to form other kinds of oxalates when they grow on substrates that contain other minerals. Magnesium oxalate dihydrate (glushinskite), manganese oxalate dihydrate, anhydrous ferric oxalate, and hydrous oxalate of copper (moolite) have all been found in lichens that were present on serpentinites and on minerals containing magnesium and copper (Wilson et al. 1981; Ascaso et al. 1982; Ascaso 1984; Purvis 1984). The capacity to produce oxalic acid (and therefore oxalates) varies according to the lichen species; it generally increases with the age of the lichen; and is higher in calcicolous lichens than in silicicolous ones. In addition, the oxalate content can vary between thalli of the same species depending

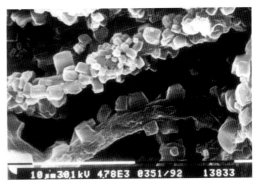

Fig 4.33 *Oxalate crystals on the external surface of the hyphae of the mycobiont* – Photo O. Salvadori

on the nature of the substrate (Salvadori and Lazzarini 1991; Bjelland et al. 2002).

Calcium oxalate is the most widespread of the neoformation products in lichens and exists in two forms of hydration: whewellite and weddellite, characterized by monocline crystals of tabular habit (whewellite) or tetragonal crystals (weddellite). The factors influencing the formation of one mineral form or the other are still under examination. Some authors suggest that it is a biologically controlled process of mineralization, in which the production of the two mineral forms is a taxonomic character that is constant at the species level. Others propose instead that the mineralization is a biologically induced phenomenon. In this hypothesis, the influence of the organism is minimal and the interactions with the organic matrices of the thallus are scarce, while the decisive factors are external (Giordani et al. 2003). On the other hand, studies carried out on some endolithic lichens have shown the complete absence of oxalates (Pinna et al. 1998; Pinna and Salvadori 2000; Tretiach et al. 2003).

The capacity of lichens to produce calcium oxalate has induced some authors to hypothesize that the films or patinas found on many monuments and prevalently composed of calcium oxalate were originally produced by lichens, which are no longer present today because of atmospheric pollution (Del Monte and Sabbioni 1987; Del Monte and Ferrari 1989). Other authors affirm, on the contrary, that these layers are the result of the transformation (probably operated by microorganisms) of organic substances applied on the stone for protective and/or aesthetic purposes (AA.VV. 1989; Lazzarini and Salvadori 1989; Realini and Toniolo 1996). Only in a very limited number of cases, which cannot be generalized, could the patinas be traces of past lichenic growth.

The presence of lichens on monuments has at times been considered beneficial for the preservation of the stone surface. Some authors indeed suggest that the lichen covering at times exercises a protective function on the stone rather than being detrimental to it (Lallemant and Deruelle 1978; Ariño et al. 1995; Wendler and Prasartet 1999; Carballal et al. 2001; Chiari and Cossio 2001). However, the experimental proofs presented to confirm this hypothesis are somewhat limited. According to Wendler and Prasartet (1999), lichens can have a protective effect against the natural decay of porous stones by reducing the intensity of water exchanges between the substrate and the environment. More widespread exfoliation, saline efflorescences, flaking, powdering, and honeycombing of other, noncolonized surfaces would seem to support this hypothesis. Still, it should be taken into account that lichens are not able to establish and grow on highly unstable surfaces, such as the areas subject to the crystallization of salts, because the removal and loss of substrate is more rapid than the establishment and growth of lichens.

In the case of highly porous stone, lichens can exercise a certain degree of protective action, mostly against atmospheric agents of deterioration such as water, wind, marine aerosol, and pollution (Ariño et al. 1995). Endolithic lichens seem to have a water-repelling effect on stone surfaces, but when they die, the stone loses these hydrophobic properties (Modenesi, unpublished). Nonetheless, we should emphasize that, although the protective effect of lichens deserves further research, the phenomenon cannot be generalized and must be evaluated on a case-by-case basis. The literature concerning the degradation of stone by lichens in relatively short periods of time is very copious, well documented, and with such solid experimental proof that any confutation must provide equally solid results to be considered valid (Piervittori et al. 1994, 1996, 1998, 2004).

In the past, the presence of lichens on stone was appreciated and even encouraged through the use of various substances, such as cow dung and water, urine, skimmed milk, as well as bulbs of *Scilla maritima* in Mediterranean coastal regions (Capponi and Meucci 1987).

Bryophytes (mosses and liverworts) and vascular plants grow abundantly on buildings and archaeological areas when the substrates and the environmental conditions are favorable. Bryophytes are more widespread in humid areas, while vascular plants exhibit more ecological diversification. Both, however, require sufficient water content, adequate illumination for photosynthetic activity, and a suitable porosity to allow the retention of moisture as well as the penetration of rhizoids and roots.

Colonization of bryophytes depends on the chemical nature of the substrate (Altieri and Ricci 1994, 1997). A study of the mosses occurring in various Roman archaeological sites has shown that the most frequent species are xerophilic (*Grimmia pulvinata*, *Tortula muralis*, *Barbula convoluta*, *Dydimodon fallax*, and *D. vinealis*) and mesophilic (*Bryum donianum*, *Brachythecium velutinum*, *Rhynchostegiella tenella* and *Cephalozia bicuspidata*) (Altieri and Ricci 1994). Bryophytes are known for their pedogenic action, determined also by the capacity to accumulate particles from the atmosphere and to accelerate the transformation processes of rock into soil.

With regard to the deterioration caused by mosses, research results are not sufficient to be able to affirm that mosses damage the substrate mechanically, although a study on *Tortula muralis*, one of the most common mosses found on stone, shows that rhizoids penetrate oolithic limestones to a depth of 5 mm (Hughes 1982). In terms of chemical deterioration, mosses extract calcium ions from the substrate for their metabolic requirements (Altieri and Ricci 1994).

The colonization of stone by vascular plants usually occurs after the substrate has been attacked by pioneer organisms (cyanobacteria, algae, lichens, mosses) and proceeds to structurally more complex and evolved stages. In pioneer conditions, i.e., during the early stages of colonization or when a further evolution is not possible, the species most commonly found on walls in Europe belong to the phytosociological class of Parietarietea judaicae. The most common are *Capparis spinosa*, *Parietaria diffusa*,

Cymbalaria muralis, Anthyrrinum majus, A. tortuosum, and *Centranthus ruber,* characteristically found in a variety of associations (Caneva et al. 1995a). *Ficus carica, Hedera helix,* and *Ailanthus altissima* are among the most damaging and most frequent species found in Italian archaeological sites (Caneva and Roccardi 1989). Other woody species (*Pinus halepensis, Cupressus* sp., *Robinia pseudoacacia, Ulmus minor, Celtis australis, Quercus ilex, Pistacia terebinthus, Rubus ulmifolius,* and *Spartium junceum*) belonging to forest plant communities are present in dynamically more evolved situations (Plate 22).

From an ecological point of view, pioneering woody species such as figs, brambles, and ivies behave like calciophilic ones and tend to prefer more hygrophilic conditions; elms and nettle trees seem to prefer slightly more xerophilic conditions; robinias and especially trees of heaven exhibit remarkable ecological flexibility and an enormous capacity for regeneration; in the absence of human intervention, they rapidly form structurally more complex populations and copses with a tendency to monospecificity. Other species, such as broom, buckthorn, coronillas, holm oaks, etc., seem linked to more advanced stages, starting from pioneering wall coenoses, which develop when the state of abandonment of the walls allows their further evolution. As mentioned, to these should be added the shrub and grass forms typically found on walls and therefore pioneers from the earliest stages, such as capers, pellitories, snapdragons, henbane, ivy-leafed toadflax, red-spur valerians, or stocks (Chapter 3, section 6). Their capacity to degrade the substrate varies considerably among these species, and their aesthetic appearance is, in some cases, even quite appreciated.

The deterioration caused by plants is both mechanical and chemical. The roots grow mainly in the mortars between the bricks or stones or else beneath the plaster layers covering the walls, that is in the areas of least resistance. But the more compact areas can also be colonized when a reduction in the cohesion of the materials occurs as a result of other physical and chemical factors of deterioration. The pressure exercised by the growth and radial thickening of roots (which can be up to fifteen atmospheres) can cause serious damage to the substrate.

In addition to the production of carbonic acid through respiration processes, chemical action is due to the acidity of the root tips and the acidity and chelating properties of the exudates (Chapter 1). The level of acidity can vary from a pH of about 5–6 for some ruderal plants to a pH of 3–4 for some cultivated plants (Keller and Frederickson 1952; Caneva and Altieri 1988). Certain plants, for instance ivy, can also cause a change in stone color as a result of the release of organic compounds, which is clearly visible if the stone is light in color (Lewin and Charola 1981).

In archaeological zones, trees can cause severe problems due to the expansion of the root system, which can develop for several meters, both laterally and in depth. This is very dangerous for hypogean environments (Chapter 5, section 1.4) and for wall structures when the trees are growing in their vicinity (Caneva 1997). The problem is even greater in the case of clay soils: even when the roots are at a distance, they cause contractions in the soil volume due to water absorption, triggering the subsidence of the foundations (Cutler and Richardson 1981).

The habitus of the species—exhibited by their growth form and associated to the general characteristics of the root system—provides useful elements in evaluating the potential damage to walls; on this basis, Signorini (1996) proposed a Danger Index (DI, or IP, Indice di Pericolosità, see p. 92) that expresses the level of danger to the stone object for each plant species present in the area under examination (Table 3.4). This index has been applied in the study of various archaeological and monumental sites, such as Ostia Antica, Syracuse, Noto, etc. (Poli et al. 1997; Lattanzi and Tilia 2004). These indications are very useful when choosing conservation interventions. The knowledge of the deterioration caused by each species in an archaeological site is necessary in order to plan the most suitable interventions for the preservation of cultural heritage, but it is also useful to protect the floristic and vegetative aspects that are valuable from a naturalistic point of view if they are compatible with the archaeological structures (Caneva 1994, 1997, 2004).

Some of the positive aspects of the development of flora are related to contributing to useful changes in the characteristics of the microclimate, such as the reduction of possible phenomena due to wind erosion; the lowering of the water table in relation to water pumping; and the reduction of pollutants thanks to the adsorption phenomena on the leaves. On the other hand, negative effects due to the presence of plants, such as increase in RH and water stagnation and reduced exposure to sunlight, can favor the growth of microflora (Chapter 7, section 1.2c).

4.3.2 Artificial Stone

by Maria Pia Nugari, Daniela Pinna, and Ornella Salvadori

4.3.2a *Structure and Composition*

Stuccos, mortars, concretes, intonacos, frescoes, and the ceramic products used in architecture (bricks and terracottas) are all considered artificial stone materials, in that they are man-made and contain materials of geological origin. These materials may have a chemical composition that is not entirely inorganic, for instance when the mixture is enriched with pigments and binders which may influence their susceptibility to biodeterioration.

Clay is the raw material from which ceramic materials are manufactured; it is a loose sedimentary rock with little cohesion (Fig. 4.34), which is impasted with water and corrected with appropriate substances that have a variety of functions, i.e., natural degreasers (siliceous sand) and fluxes (feldspars and feldspathoids), and man-made ones (*coccio pesto*— ground terracotta). Depending on the level of water absorption, a distinction is made between porous ceramic products (bricks, faenza, maiolicas, earthenware), which absorb more than 5%, and more compact ones (grès, porcelain), which absorb less than 5%. Firing transforms the paste into a solid mass that varies in porosity and hardness depending on the original mineralogical composition and the temperatures employed (e.g., 800–1,200°C for bricks, 1,200–1,500°C for porcelains).

The use of brick, a ceramic product with a porous matrix, became established primarily to compensate for the scarcity of construction materials (wood and stone) in plains or areas near lakes and rivers, which instead are rich in clay deposits. Clay is a sedimentary rock, loosely bound with little cohesion; it is formed by one or more clay minerals (kaolinite, chlorite, halloysite, montmorillonite, illite), a sand-based frame-structure (quartz, mica, feldspar, calcite), and various impurities (fossil remains, iron oxides—which give clay its characteristic red color—iron sulphides, gypsum, organic substances) (Menicali 1992).

Bricks are made of a basic paste or matrix, resulting from the firing (800–1,200°C) and the destruction of the clay, of a sandy frame-structure of natural or artificial origin, and of pores; as such, they are therefore a highly porous construction material. It is not known when brick manufacture started.

The earliest bricks were raw, or unfired, and were simply placed in the sun to dry (adobe); to the clay paste were added chopped straw or fragments of rocks to serve as a frame-

Fig. 4.34 *General structure of clay materials*

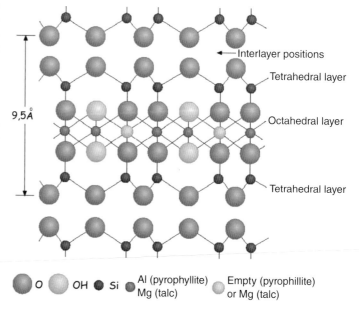

9,5Å

Interlayer positions
Tetrahedral layer
Octahedral layer
Tetrahedral layer

O   OH   Si   Al (pyrophyllite) Mg (talc)   Empty (pyrophillite) or Mg (talc)

structure as well as dung, etc. In desert or subdesert climate regions (e.g., Yemen or North Africa), in tropical regions (Latin America, Sri Lanka), and in poorer areas in general, this same type of unfired brick is still very much in use today—suffice to think that one third of the world's population lives in houses made of raw, unfired earth.

Mortars, cements, plasters, and stuccos are obtained by mixing a binder (lime, gypsum, clay, or cement) with a natural mineral aggregate (river sand, volcanic lapilla such as pozzolane) and an artificial one (fragments of stone, brick powder or brick fragments, metal slag). The binder is responsible for the hardening of the paste, which can occur in the open air (air-hardening binders) or in contact with water (hydraulic binders). Sometimes organic additives, such as vegetable fibers, animal hairs (for example, horse hair), coal, etc., are also present.

The quality of artificial stone materials depends both on the materials used and the methods of preparation as well as firing (when it is part of the process).

Stucco refers to a fine plaster made of slaked lime mixed with fine marble dust, impasted with water until it reaches a malleable and plastic consistency. A variety of organic materials can be added to improve the adhesive qualities (e.g., animal and vegetable glues). Once hardened, it is generally smoothed until it achieves the appearance of marble. Mineral pigments are at times also added to the paste, or else the stucco is painted after hardening.

The term wall paintings includes all the techinques employed to decorate walls. The support, i.e., the bearing structure, is more often than not a brick wall, but can also be natural stone or wood. The painting is executed either directly onto the support or after the application of an intonaco layer. In general, the following layers are to be found on a wall painting (Fig. 4.35):

—the *intonaco*, which is constituted by one or more successive layers (depending on the period) made of lime or gypsum and sand in varying proportions mixed with water, at times with the addition of ground terracotta and marble dust. Usually, there are at least two distinct

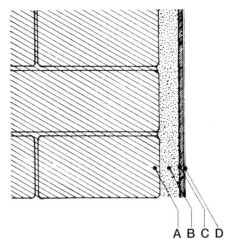

A B C D

Fig. 4.35 *Layers of a fresco painting. A = supporting wall; B = arriccio; C = intonaco; D = paint layer*

layers: 1. the *arriccio*, the inner layer, which is thicker and rougher compared to 2. the *intonaco* proper, constituted of finer particles and therefore smoother, on which the paint is applied. In Asia, the intonaco for many mural paintings is traditionally prepared with clay mixed with plant fibers (for example, chopped straw, sawdust, linen, jute, or hemp fibers) in the place of sand, and the surface is then smoothed with a thin final application of clay, gypsum, or lime.

—the paint layer, executed using a variety of techniques. In paintings executed in *buon fresco*, i.e., true fresco technique (frescoes), the colors are applied onto a damp lime plaster, and the pigment particles are englobed by the calcium carbonate that forms during the drying process of the intonaco; in this case the constituents are by and large inorganic. In wall paintings executed with the *a secco* technique, the pigments are applied onto dry plaster. Adhesives or binders are used (for example, egg, casein, vegetable gums, oils, and animal glues) to help the colors adhere to the surface.

The materials constituting wall paintings are highly porous and represent a physico-chemical system that is open to exchanges with the surrounding environment.

4.3.2b *Biodeterioration of Artificial Stone Materials*

Mortars, cements, plasters, and stuccos are very susceptible to colonization by microorganisms and organisms because of their high porosity as well as their mineralogical structure; they therefore have a high degree of bioreceptivity. This is easy to see when looking at brick walls or mosaics, where it is evident that biological attack begins on mortars and then expands onto the bricks and the tesserae of the mosaic. Among fungi, the most frequently isolated genera are *Cladosporium* and *Alternaria* (De Leo and Urzì 2003).

A study by Ariño et al. (1997a) showed that the extent of deterioration carried out by epi- and endolithic lichens on various mortars was related to the mineral composition and the quantity of the aggregates: the mortar with the highest percentage of calcite was the one that was most altered chemically and that was penetrated most deeply by hyphae because of dissolution phenomena.

As a rule, wall paintings are frequently affected by biodeterioration, both because of their high porosity and because the indoor environments where the majority of frescoes are situated are where conditions favoring biological growth are more common (high RH, phenomena of capillary rise, and/or condensation of water). Interaction with the atmosphere can bring about the deposit of gaseous particles and pollutants; in addition, it is not uncommon that the repeated conservation and/or maintenance interventions these works of art are subjected to throughout their history—and that often include the use of organic substances—contribute to changes in the surface environment, inducing massive growths of microorganisms.

In the case of *a secco* technique, the surface often contains a variety of organic substances and should, in theory, be more susceptible than true fresco to biological attack by heterotrophs. However, some authors have found no correlation between the presence of areas painted in tempera and increases in biological deterioration (Saiz-Jimenez and Samson 1981a), and according to others, the presence and development of organisms is mostly dependent on the water content of the surfaces (Petersen et al. 1995; Caneva et al. 2002). Very high values of RH often nearing saturation, condensation,

and poor air circulation are frequently associated with biodeterioration of wall paintings. In addition, the negative effects associated with unsuitable heating systems should not be underestimated, nor indeed should the microclimatic changes induced by the presence of visitors, especially in small indoor spaces.

The earliest discoveries of microorganisms on wall paintings date from the 1940s, when Augusti (1944a, 1944b, 1948) first described some alterations caused by fungi appearing as white surface films and small black stains.

Many studies have been carried out in order to identify the microorganisms responsible for visible alterations, such as black, brown, or green stains, purple patinas, whitish efflorescences, etc.; other authors have instead investigated the presence of microorganisms on deteriorated frescoes exhibiting no evident signs of microbial colonization. In both cases, the most frequently isolated microorganisms are fungi (Fig. 4.36), bacteria (important roles being played by actinomycetes, halophiles, and cyanobacteria), and green algae (Plates 17 and 20). These biodeteriogens are responsible not only for a variety of stains, but also for the detachment of fragments of the paint film, induced especially by fungi, the hyphae of which penetrate to a depth of 10 mm or more inside the intonaco (Eckhardt 1985).

Nitrifying bacteria have also been frequently isolated, and some authors consider them among the earliest colonizers of these works (Saiz-Jimenez and Samson 1981b; Karpovich-Tate and Rebrikova 1991). The death and lysis of these bacteria could favor the development of het-

Fig. 4.36 *Colonies of fungi on a fresco* – Photo ICR

erotrophic bacteria that utilize the bacterial structures and the extracellular polysaccharides as nutrients. Among the genera most frequently found on wall paintings are *Anthrobacter*, *Bacillus* (*B. subtilis*, *B. polymyxa*, *B. megaterium*), *Pseudomonas*, *Micrococcus* (*M. roseus*, *M. luteus*), *Kocuria* (Karbowska-Berent 2003).

Biodeterioration often coexists with other kinds of alterations, for example saline efflorescences, as both are caused by high levels of humidity. The presence of salts can provide a favorable environment for many fungi and bacteria. Many of the species of bacteria found on frescoes do indeed exhibit different degrees of halotolerance or even of halophilism; these are

Fig. 4.37 *Whitish patina caused by actinomycetes on wall paintings (Dambulla, Sri Lanka)* – Photo O. Salvadori

generally gram-positive heterotrophic bacteria (Schostak and Krumbein 1992; Heyrman et al. 1999). Recently, thanks to the application of molecular techniques, it has been possible to isolate *Clostridium*, *Frankia*, and *Halomonas* on wall pantings with high saline levels; *Halomonas* can grow on substrates that contain up to 32% of salts (Rölleke et al. 1996). The same research methods applied to *Archaea*, to which belong certain extremely halophilic bacteria, have also yielded positive results (Rölleke et al. 1998; Piñar et al. 1999). It is interesting to note how *Penicillium* and *Aspergillus* species are among the most resistant to high concentration of salts and are also among those most frequently found on deteriorated frescoes (Saiz-Jimenez and Samson 1981a).

Actinomycetes, and in particular *Streptomyces*, *Micromonospora*, and *Nocardia*, have been identified in great numbers on wall paintings in caves, crypts, and tombs (Giacobini et al. 1988) (Plate 20 and Chapter 5), but also in other environments, although more infrequently (Saiz-Jimenez and Samson 1981b; Hadjiulcheva and Gesheva 1982; Agarossi et al. 1986; Karpovich-Tate and Rebrikova 1991; Petersen and Hammer 1992). Their development can give rise to extensive, thin, and pulverulent whitish patinas or to thicker and more localized forms, which closely resemble saline efflorescences (Fig. 4.37); because of their appearance, their genesis is indeed often misinterpreted when they

are observed with the naked eye. At times, bacteria and fungi will develop successfully in conditions of microaerophilia and of high alkalinity (pH 10.5) beneath the lime wash (even if present in several layers) that was sometimes applied in the past to cover the frescoes (Appolonia et al. 1989; Petersen et al. 1995).

An examination of the literature clearly shows that the most frequently isolated fungi on mural paintings are the deuteromycetes (Dematiaceae family), and in particular those belonging to the genera *Cladosporium*, *Penicillium*, *Aspergillus*, *Trichoderma*, *Acremonium*, and *Alternaria*. Even *Serpula lacrymans*—a brown-rot fungus of wood that is also fairly common on stone—is capable of chelating calcium, iron, and silica ions, as well as other metal cations, from walls and plasters thanks to its high production of oxalic acid. This would explain the considerable affinity of this fungus for buildings; stones, bricks, and plasters would favor its degrading capacity for wood, which is not in itself particularly rich in minerals, thus ensuring mineral supply (Low et al. 2000) (Plate 17).

Cyanobacteria and algae form patinas of various colors or pulverulent formations (usually pinkish in color, but at other times yellow, gray, or white) depending on the biocoenosis, the phase of development, and the nature of the predominant alga (Tomaselli et al. 1979; Ricci et al. 1985). Very often the colonization by photosynthesizing microorganisms is favored by the presence of artificial lighting systems (Plate 2).

The degradation of wall paintings is usually greater near windows and doors and close to the ground. In these areas, microorganisms are found in greater amounts, evidence of the importance of air flow, in terms of the deposit of spores made easier by the air movement; the higher degree of humidity due to a lower temperature and, for the lower wall areas, due to the capillary rise of water, are also factors, as is light in the areas near windows. Zanotti Censoni and Mandrioli (1979) carried out an aerobiological investigation of the conditions inside and outside the Scrovegni Chapel in Padua. Their findings showed no significant differences in the composition of the microflora between the inside and the outside of the chapel, but the concentration of fungal spores inside the chapel was half of the outside levels, and on the outside there was a higher concentration of bacteria. Subsequent to scientific investigations or conservation operations, an increase in microbial concentration was found on the wall paintings in the tomb of Nefertari (Arai 1987) and in the Sistine Chapel (Montacutelli et al. 1992), although in these cases the measurements were taken from the affected surfaces rather than from the air, as in the Scrovegni Chapel. The most frequently identified spores on indoor wall paintings are, in decreasing order: *Cladosporium*, *Penicillium*, *Aspergillus*, *Alternaria* and *Fusarium* (Nugari et al. 1993c).

Some authors have conducted laboratory investigations in order to judge whether the presence of pigments, binders, or products used in conservation can favor or inhibit microbial growth (Boustead 1963; Strzelczyk 1981; Sorlini et al. 1987; Sampò and Luppi Mosca 1989; Dhawan et al. 1992). According to Boustead (1963), the presence of most pigments plays no part whatsoever in the development of fungi; some pigments could be considered inhibitors (for example, zinc oxide, titanium white, lead white); and still others could be used as micronutrients. Earth pigments are among the most susceptible to biological attack (Strzelczyk 1981).

It is well known that pigments containing metals can change color as a result of oxidation, reduction, or transfer of metal ions. Azurite (blue), for example, can change to malachite (green), lead white can change from white to brown (lead dioxide). Petushkova and Lyalikova (1986) have shown that some bacteria (*Anthrobacter* and mycoplasms) are able to transform lead white ($2PbCO_3 \cdot Pb(OH)_2$), massicot (PbO), and minium (Pb_3O_4) into lead oxide (PbO_2). Fungi, on the other hand, are not able to oxidize bivalent lead and are inhibited by lead white. Other bacteria can transform lead-containing pigments and lead acetate into black lead sulphide, thanks to the hydrogen sulphide they produce. The "blackening" of frescoes, due to the transformation of lead white into lead sulphide, is a well-known phenomenon although generally it is not attributed to biological causes.

Binders can also have varying degrees of susceptibility to fungal attack depending on their chemical composition: casein, egg tempera, tempera emulsions, and linseed oil are among the most vulnerable. As always, all substances in the substrate being equal, the triggering factors are the environmental conditions and the RH in particular. Synthetic products, such as polyvinyl alcohols and polyvinyl acetates, or acrylic emulsions and resins, which are often employed in conservation, can induce instantaneous microbial growths in conditions of high moisture content in the substrate, but can be used without problems in suitable environmental conditions (below in this chapter, section 6).

The growth of lichens on wall paintings depends essentially on the presence of light and a sufficiently high level of humidity. Lichens usually grow on outdoor wall paintings and on surfaces exposed to rain, although some species that do not require to be wetted by rain can also grow in semienclosed spaces. One of the most studied cases is that of the frescoes in Palazzo Farnese (Caprarola) (Fig. 5.16). Large portions of the surface were covered by *Dirina massiliensis* f. *sorediata*, which had formed incrustations of various thicknesses (0.5–2 mm). The lichens seemed to favor growth in areas of yellow and brown ochre, while red pigments appeared to have an inhibitory effect on the growth, probably due to the presence of lead or mercury compounds (Seaward and Giacobini 1988). The deterioration induced by this lichen—studied through Raman spectroscopy—consists of the incorporation of particles of the substrate (gypsum and calcium carbonate) into the thalli and of the formation of calcium oxalates (whewellite and weddellite) (Seaward et al. 1989; Edwards et al. 1991, 1992, 1994). The mechanical removal of the thalli revealed the absence of color in the underlying areas, due to the physico-chemical action of the lichen, the hyphae of which could easily penetrate the porous substrate.

Plants can also play an important role in the deterioration of frescoes, especially those located in subterranean environments, as the depth and distance reached by the roots can be considerable (sometimes dozens of meters). Studies carried out in Roman hypogean environments (*Domus Aurea, Domus Tiberiana, Dolocoenum* on the street Via San Domenico) have shown that roots grow down following a rising water gradient (Caneva 1985); thus, the roots exert a mechanical action on the walls of these environments, insinuating themselves into the areas of least resistance. They also tend to grow beneath the plasters or the wall paintings, gradually lifting them up and eventually causing their detachment (Fig. 5.8). The trees of the genera *Pinus, Quercus, Celtis,* but especially *Ficus,* are particularly dangerous because of the great spread of their root system (sometimes reaching as far as 50 m from their point of origin).

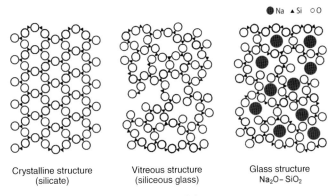

Na ▲ Si ○ O

| Crystalline structure (silicate) | Vitreous structure (siliceous glass) | Glass structure Na₂O– SiO₂ |

Fig. 4.38 *The structure of vitreous materials*

4.4 METALLIC AND VITREOUS MATERIALS

4.4.1 Glass

by Giulia Caneva and Simona Ceschin

4.4.1a *Structure and Composition*
Glass is a ceramic material (an inorganic, nonmetallic solid) composed of mixtures of silicates, obtained through fusion and characterized by a particular state of aggregation called vitreous, in which the molecules have the disordered and random spatial arrangement characteristic of liquids. Indeed, glass is generally considered a super-cooled liquid or, better yet, an amorphous and inorganic solid produced through fusion and then cooled without crystallization.

The amorphous organization of glass is achieved through the process of vitrification of the raw materials—all inorganic and of a crystalline nature—combined in precise proportions according to their specific functions. In order for the vitreous mass components to lose their individual specificity and achieve an amorphous condition, the raw materials are heated and fused in a process that requires a great input of energy.

The main constituent of glass is amorphous silica (about 70%), which is used as the vitrifying substance; its structure can be represented as a disordered and aperiodic succession of tetrahedral molecules that have an atom of silica at their center and four atoms of oxygen at their vertices (Fig. 4.38).

The precise origins of the invention of this material remain unknown, but there are Mesopotamian finds that date back to 1700 B.C.E. and small pieces of jewelry of even more ancient origin, both from China and Egypt. Over time, the technology of glass evolved considerably and, beginning in the first century B.C.E., the technique of blown glass was developed. Among the most important centers of production up to the thirteenth century was Constantinople, and then Venice, which achieved technological superiority and maintained it until the last century.

The basic constituents used were siliceous sands or powdered quartz (often ground river pebbles), to which various kinds of substances were added, depending on the processes of fabrication and the kind of glass to be produced. The master glassmakers maintained the utmost secrecy around their techniques and ingredients, safeguarding the art and technology and keeping it linked to specific territories, which relied on them for a good portion of their economy (e.g., the island of Murano in the Venetian Republic; Moretti et al. 2004).

In general, the manufacturing process involved an initial fusion to form a "frit" that was worked on in the furnaces over a period of days, during which various other substances would be added as agents to the siliceous material:

—fluxes (for example, potassium, calcium, or sodium carbonate, to about 15%), especially in the form of soda ash of brackish plants such as *Salsola soda* or *S. kali*, in order to form the *allume catina*, which lowers the fusion temperature, thus facilitating its workmanship;
—stabilizers (for instance, carbonates and oxides of magnesium, barium, zinc, and aluminum, to about 10%), in order to make glass—composed only of silica and a few alkalis (sodium and potassium)— insoluble in boiling water. These can be substituted with lead oxides, thus obtaining glass with a high index of refraction (lead glass, called "crystal");
—decolorizing agents to remove the naturally greenish tinge of glass and to increase its transparency;
—refiners to make glass more homogenous, facilitating the emission of dissolved gases;
—colorants, generally metal oxides.

Because of the different substances added to the siliceous structure of glass, varying quantities of atoms of sodium, manganese, calcium, barium, and potassium are found in the interstices between one tetrahedron and the next; this strongly influences the general characteristics of the vitreous mass and, of course, its stability and resistance to deteriogenic agents.

As a general rule, glass exhibits low levels of elasticity and, consequently, considerable fragility; it tends to break during elastic deformation, giving rise (as is the case with all ceramic materials) to the so-called fragile fracture (a fracture that is characterized by the absence of plastic deformation).

Despite its fragility, glass possesses considerable mechanical resistance: in particular, its resistance to compression often has values of 150–200 kg/mm^2. Glass is also the material with the lowest thermal expansion coefficient and can be endlessly recycled through fusion and cooling. The fundamental aspect of its thermal resistance (which, typically, is inversely proportional to its thickness) is its ability to tolerate sudden changes in temperature.

The transparency of glass is determined by its amorphous structure and by the chemical composition of its constituents, while its capacity for transmitting light is limited in part by reflexion phenomena on its surface, by absorption by the silica of extreme UV radiation (150 nm) and of middle-of-the-range (6,000 nm) infrared wavelengths, and—in the visible spectrum—by the presence of inorganic constituents, added for decorative purposes or present as contaminants of the raw materials.

4.4.1b *Biodeterioration of Glass*

Although theoretically it possesses a considerable degree of chemical inertness (only fluoric acid and highly concentrated alkaline solutions can degrade it to any significant degree), vitreous material can nevertheless be subject to negative interactions with the substances with which it comes in contact.

In particular, the sodium present in the amorphous reticulate of glass is not strongly bound to the structure and can be released if the glass is subject to the so-called acid attack; this corresponds to an ionic exchange between hydrogen and alkali ions (Na^+, K^+).

Much more dangerous is the alkali attack, because it results in the destruction of the reticulate structure and, therefore, the release of the silicates, a process that is accelerated by temperature. It should also be noted that when glass remains exposed to the air for a long time it releases alkalis through the action of humidity and tends to become opaque—although the process is gentler and more gradual.

Biological attack on vitreous materials, although already observed as early as the nineteenth century and described in its broad outlines in the case of lichens during the 1920s (Mellor 1923; Fry 1924; Mattirolo 1928), has become the object of more focused studies only relatively recently.

In the case of microflora, it has been shown that certain groups of bacteria and fungi are able to attack vitreous surfaces exposed to the outdoors or located in a marine environment, where algal species can also play a certain role (Krumbein et al. 1991, 1993; Weissman and Drewello 1996; Drewello 1997; Gorbushina and Palinska 1999; Rölleke et al. 1999; Schaberaiter-Gurtner et al. 2001; Danilov and Ekelund 2001; Gallien et al. 2001; Müller et al. 2001). The described alteration phenomena range from opacifications to blackenings, microfractures, surface erosions, incrustations, and pitting.

In marine environments in particular, it has been observed how iron bacteria and sulphur bacteria can induce blackenings of the surface as a result of iron sulphide deposits.

Laboratory experiments were carried out in order to prove the action of fungi and cyanobacteria and demonstrate the reproducibility of processes encountered on glass of historical/artistic interest in outdoor environments. A dense growth of biofilm was obtained on glass surfaces, which resulted in the formation of micropitting and fissures, with morphologies that were completely in keeping with those found on the ancient stained-glass windows of cathedrals (Gorbushina and Palinska 1999). In particular, the synergetic role played by airborne biological populations in the degradation process of these materials was clearly demonstrated, underlining how in laboratory conditions—which are much more favorable than those found in natural environments—there is a distinct acceleration of those processes of deterioration that have a simple physical or chemical origin, with the production of fissures 10–15 times wider than those found in the controls. This has been ascribed to a variety of causes, including the heterogeneity of the surface caused by the presence of any biofilm, with subsequent formation of localized microniches, and the differential release of exopolymeric substances, resulting in differences in aeration and oxygenation of the surfaces; this is similar to what occurs on metals (Ford and Mitchell 1991).

It has not always been made clear if the biological processes of attack on glass are the result of the emission of aggressive chemical substances or if they should rather be imputed to processes of a physical nature. It is highly probable that the processes of microbial corrosion are the result of the combined action of biogeophysical and biogeochemical agents. The first occur in the presence of filamentous forms (especially fungal ones) that exercise a mechanical action with the capacity to generate and enlarge surface microfractures. Biogeochemical processes are the result of the emission of organic and inorganic acids, enzymes, and exopolymeric substances. Microorganisms in particular have the capacity to induce a chemical attack on glass thanks to the development of a biofilm that allows the substitution of ions englobed in the siliceous matrix and the simultaneous diffusion of H^+ ions. Once formed, this layer may exercise a protective function, but its removal triggers a new round of corrosion processes. A further mechanism of chemical attack could be due to the interaction between the typical products of biomineralization—such as oxalates and carbon, sulphur, and phosphate derivatives—with the metal cations drawn from the substrate (Weissman and Drewello 1996; Müller et al. 2001).

Analyses carried out on German historical stained-glass windows revealed the frequent occurrence of biological patinas, the diversity of which is surely underestimated considering how difficult it is to cultivate these species in the laboratory, and given the need to adopt biomolecular techniques of gene amplification in the investigations (Drewello 1997; Rölleke et al. 1999; Müller et al. 2001). In particular, observations made with the confocal microscope have demonstrated the wide distribution of filamentous fungi, even though there did appear to be a certain heterogeneity among the samples; this was attributed largely to the different composition of the glass under examination, thus hypothesizing a precise correlation between the type of vitreous material and the type of microflora able to colonize it, in addition to the different durability of the material itself and the presence of additional organic material on the surfaces (Müller et al. 2001). In contrast, the presence of lead was found to exercise an inhibitory action.

The fundamental role played by fungi—which has long been emphasized ever since the presence of microetchings even on the surface of polished lenses was first observed (if they are exposed to favorable environmental conditions)—has received further confirmation through microbiological studies carried out with molecular techniques, in which the deteriogenic potential of a wide range of fungal species was tested: *Aspergillus, Aureobasidium, Coniosporium, Capnobotryella, Engyodontium, Geomyces, Kirschsteiniothelia, Leptosphaeria, Rhodotorula, Stanjemonium, Ustilago,* and *Verticillium* (Schaberaiter-Gurtner et al. 2001). On the other hand, among the species of bacteria recently isolated, again, from ancient stained-glass windows of German cathedrals, we note: *Flexibacter* (Cytophagales), *Nitrospira* (Proteobacteria), *Micrococcus, Streptomyces* (actinomycetes), *Arthrobacter, Frankia, Geodermatophilus* (Rölleke et al. 1999).

As mentioned, lichenic attack on glass was the first biodeterioration phenomenon to be identified, while the first systematic identification of the lichens present on stained-glass windows was carried out almost a century later on French cathedrals dating between the twelfth and the sixteenth century, as a result of the dismantling of the windows during World War I (Brightman and Seaward 1978). Thus, in the 1920s, Mellor (1923) and Fry (1924) studied, respectively, the biochemical and biophysical deterioration induced by lichens. The chemical action can be imputed to several factors, such as the emission of chelating substances and weak acids, among which is the carbonic acid emitted during respiratory processes. The physical action is due to both the penetration of the fungal hyphae into the material and the expansion and contraction of the thallus and, consequently, of the organs of attack, in relation to their different degrees of absorption, thus giving rise to a sort of "peeling" effect (Fig. 4.31c).

The lichenic flora growing on windows can be specific (glass-growing lichens), but often it seems to be composed of species that also grow on stone, often with a crustose type of thallus. Among the most frequently encountered species are *Diploicia canescens, Parmelia scabrosa, Pertusaria leucosoria,* and *Lepraria flava.* Even though the action is slow, the damage that a lichenic attack can produce on glass can be very serious in that opacization, iridescence, and pitting to a depth of 5–6 mm can be induced. Lichens are also active on the lead frames joining the panes of glass, where sometimes the degradation processes are even accelerated, perhaps because of the greater retention of water in these areas. As evidence of the great powers of adaptation of lichens to extreme conditions (as long as they are in environments that are not affected by pollution), we would like to mention that traces of lichenic colonization were identified on fragments of glass found in Antartica (Fig. 4.39) (Schroeter and Sancho 1996).

Fig. 4.39 *Lichenic attack on vitreous materials (from Schroeter and Sancho 1996)*

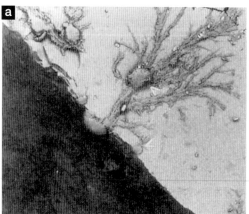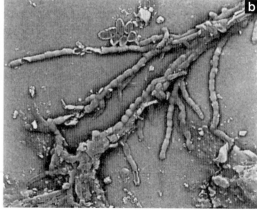

4.4.2 **Metals**

*by Elisabetta Zanardini, Francesca Cappitelli,
Giancarlo Ranalli, and Claudia Sorlini*

4.4.2a *Structure and Composition*

The nature of metals is a very complex subject
from the physico-chemical point of view; the
presence of their specific properties implies the
existence of a particular type of interatomic
bond, called indeed metal bonding, in which
the electrons of the bond are not directional, but
are delocalized within the entire volume of the
material. Metals can have either cubic or hexag-
onal crystalline structures (which represent the
most compact solution for filling a volume with
spheres of an equal size); because of the nature
of the bonds, they have high electrical and thermal conductivity (Fig. 4.40). Other characteristics of metals are their particular sheen (termed
metallic luster) and the mechanical properties of
ductability and malleability, which allow them
to undergo deformations without their crystal
structure suffering any great alterations.

Metal materials employed for artistic purposes are seldom constituted by a single metal;
more often than not they are alloys that derive
from the fusion of one metal with at least one
other element, either metal or metalloid (e.g.,

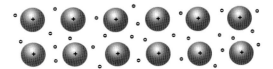

Fig. 4.40 *General structure of a metal*

carbon). Binary alloys are constituted by two
elements, ternary and quaternary ones by three
and four elements, respectively, and so on. The
chemical composition of alloys is expressed
as a percentage composition (either by weight
or mass); from the relative percentages will
depend the various characteristics and, hence,
the different applications. However, it should
be taken into account that, in archaeological
materials especially, there can be significant
compositional inhomogeneities, even within a
single artifact.

The most commonly found metals until
the end of the Middle Ages were copper, iron,
gold, silver, in addition to lead and tin, which
were mainly used in the pewter and bronze
alloys (Table 4.9); only rarely would the metals
be employed "pure." All metals exhibit a highly
ordered crystalline structure that is formed
during solidification.

Table 4.9 Metals and alloys (from the late Neolithic to the third century c.e.) (modified
from Pedelì and Pulga 2002)

"Pure" Metal	Alloy
Silver	+ Gold = Electrum (natural alloy)
Gold	+ Silver = Electrum (natural alloy)
Iron	+ Carbon = Steel (archaeological iron)
Copper	+ Arsenic + Tin = Bronze + Tin + Lead = Bronze + Lead = Bronze + Zinc = Brass + Zinc + Lead = Brass + Silver + Gold
Lead	+ Tin = Pewter
Tin	+ Lead = Pewter + Copper = Bronze

The first man-made alloy to be used on such a large scale as to give its name to an era was bronze (the Bronze Age), which is composed of copper and tin. In antiquity, bronze contained a constant percentage of tin (around 10%), while subsequently—and still today—the percentage would vary according to the applications to which the bronze was destined; in bells, for instance, it could reach 21% (Leoni 1984). Bronzes may contain other elements as well (lead, zinc, nickel, silver, arsenic, antimony, iron, bismuth, cobalt), added deliberately or else already present in the minerals.

With the exception of gold, all metal artifacts are chemically unstable, i.e., they tend to return to a more stable state, similar to that of the raw minerals from which they were extracted.

Because of this, they are subject to chemical corrosion—reacting with compounds present in the air or the soil (O_2, CO_2, SO_2, salts, etc.)—and electrochemical corrosion in the presence of water and humidity (Fig. 4.41a).

Chemical corrosion consists in the reaction between a metal and a liquid or a metal and a gas, as for instance the reaction between iron and atmospheric oxygen resulting in the production of oxides: the iron yields its electrons to the oxygen with the formation of positively charged metallic ions and negatively charged oxygen ions, which bond together forming a layer of corrosion products.

When metals are immersed in an electrolyte, electrochemical corrosion occurs through two main reactions:

—at the anode; oxidation reactions occur transforming the metal, which will either dissolve as ions or return to its compound oxide state;
—at the cathode; reduction reactions take place, with the consumption of electrons produced by the anodic reaction and conducted through the metal. On the surface of the metal, differences in potential are formed that give rise to the passage of a current through the liquid. The metal thus goes into solution at the anodic end, and hydrogen ions are formed at the cathode.

In an oxygenated aqueous environment, the following equations can take place:

$$2\,Fe \rightarrow 2\,Fe^{2+} + 4e \qquad \text{anodic reaction}$$

$$O_2 + 2\,H_2O + 4e \rightarrow 4\,OH \qquad \text{cathodic reaction}$$

while in the absence of oxygen, at the cathode we have either the reduction of hydrogen ions or of water, as shown in the following equations (Allsopp and Seal 1986):

$$4\,H^+ + 4e \rightarrow 2\,H_2$$

$$2\,H_2O + 4e \rightarrow H_2 + OH^-$$

4.4.2b *Biodeterioration of Metals*

Microbial corrosion can occur when microorganisms use these metals directly, or else when they metabolize one of the products of the electrochemical process, for example the hydrogen released at the cathode (Fig. 4.41b).

The establishment of a microbial colony on the surface can lead to corrosion of the metal through the products of metabolism: carbon dioxide, elemental sulphur, hydrogen sulphide, ammonia, the production of cellular exopolymers, and the

Fig. 4.41 *(a) Metal corrosion linked to the percolation of water – Photo G. Caneva; (b) schematic representation of the corrosion mechanism produced by a colony of microorganisms*

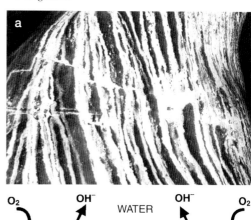

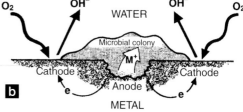

release of organic acids such as citric, oxalic, succinic, fumaric, gluconic, ketoglutaric, etc.

The most frequently involved microbial groups in the biocorrosion of metals belong both to aerobic and anaerobic species of bacteria.

Of particular importance are the bacteria of the sulphur cycle represented by sulphur-oxidizing bacteria (chemolithotrophs)—which, in the presence of oxygen, are able to oxidize the reduced inorganic compounds of sulphur to sulphuric acid—and the sulphur-reducing bacteria, which instead utilize organic compounds as electron donors and reduce not oxygen but sulphate to hydrogen sulphide.

The sulphur-oxidizing bacteria belonging to the genus *Thiobacillus* dissolve heavy metals in aerobiosis, causing the corrosion of the metal and biopitting. The attack on metals through the production of metal suphides (FeS_2, MoS_2, PbS, ZnS, etc.) by *Thiobacillus ferrooxidans*, *Thiobacillus thiooxidans*, *Leptospirillum ferrooxidans*, *Metallogenium* spp., *Acidianus* spp., and *Sulfobulus* spp. seems to depend on the action of Fe^{3+} ions and/or protons, with the additional involvement of cellular exopolymers excreted by the bacteria and containing Fe^{3+} ions that form complexes with the residues of glucuronic acid (Bondonno et al. 1989; Sand et al. 2001).

In conditions of anaerobiosis, it has been reported that certain microorganisms are able to utilize the Fe^{3+} ion as an electron acceptor, transforming it into Fe^{2+}; the sulphur-reducing bacteria, belonging to the genus *Desulfovibrio*, produce H_2S, which induces precipitation of the metallic sulphides, as can be seen in the following reactions:

$$4\, Fe \rightarrow 4\, Fe^{2+} + 8e^- \qquad \text{anodic reaction}$$

$$8\, H^+ + 8e^- \rightarrow 4\, H_2 \qquad \text{cathodic reaction}$$

$$8\, H_2O \rightarrow 8\, OH^- + 8\, H^+$$

$$SO_4^{2-} + 4\, H_2 \rightarrow S^{2-} + 4\, H_2O$$

$$Fe^{2+} + S^{2-} \rightarrow FeS \qquad \text{corrosion product}$$

$$3\, Fe^{2+} + 6\, OH^- \rightarrow \qquad \text{corrosion product}$$
$$3\, Fe(OH)_2$$

where the complete reaction would be:

$$4\, Fe + SO_4^{2-} + 4\, H_2O \rightarrow$$
$$FeS + 3\, Fe(OH)_2 + 2\, OH^-$$

Figure 4.42 illustrates the role played by sulphur-reducing bacteria in the corrosion of ferrous metals. Hydrogen sulphide has a corrosive action on metals, the extent of which can be correlated to the speed of growth of the sulphur-reducing bacteria.

As far as the biodeterioration of artistic metal materials is concerned, it is well known that the fundamental factor is the formation, on the attacked surfaces, of biofilms that result from the attachment of bacteria on solid substrates; this depends mostly on the physico-chemical characteristics of both the material and the cell surface. Determining factors in the formation of biofilm are the degree of hydrophobicity of the surface and the electrostatic charge (Van der Waals forces); from these concepts originates the theory of colloidal stability

Fig. 4.42 *Role of sulphur-reducing bacteria in the corrosion of ferrous metals (modified from Allsop and Seal 1986)*

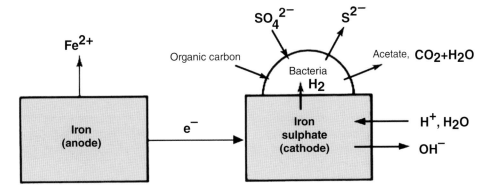

Derjaguin-Landau-Verwey-Overbeek (DLVO), which takes into account the electrostatic forces present in the interaction at the interface between two surfaces (Little et al. 1998).

The possibility of specific bonds being formed between bacteria and metals depends on the chemical composition of the surface of the metal as well as of the most external layer of the bacteria's cell wall: for example, gram-negative bacteria have an external extracellular polysaccharidic layer composed of phosphate groups and carboxylic acids. *Pseudomonas fluo-rescens* (a bacterium typically present in soil and in water) is able to attack materials containing copper, zinc, and brass based on the presence of surface oxides, which appears to be a key factor for the processes of adhesion of bacteria (Valcarce et al. 2002).

Over the years, the different mechanisms involved in the biocorrosion of metals by microorganisms have been discussed and investigated by various authors, both in general and specifically in the field of the biodeterioration of cultural heritage (Iverson 1968; Calderon et al. 1968; Hummer et al. 1968; Mara and Williams 1972; Brown and Masters 1980; Cragnolino and Tuovinen 1984; Beloyannis 1985; Guidotti 1985; Dhawan 1987; Krumbein et al. 1994; Little et al. 1998).

The microbial attack of metals can assume a particular significance in the field of biodeterioration of works of art, both in aquatic and terrestrial environments.

In marine environments, the role played by associations of anaerobic bacteria is particularly important in the corrosion of metals, because of the formation of biofilm on the surfaces of these materials (Mara and Williams 1972; Gaylarde and Vileda 1987; Little et al. 1998).

In the case of metal works of art exposed to the outdoors, the literature reports attacks on objects made of gold by two anaerobic bacteria belonging to the genus *Clostridium*, which bring about a light corrosive action due to the production of organic acids (Pares 1965).

The presence of black stains on a variety of bronze objects in European museums was already noted in 1977; the role of atmospheric pollutants is of particular importance in such processes of corrosion. However, fungi belonging to the genus *Cladosporium* were also isolated

from these stains, and they seem to participate in the attack on metals alongside the H_2S present from pollution. It is therefore sometimes difficult to discriminate between microbial and abiotic deterioration in metals (Guidotti 1985; Gilbert 1987; McNeil et al. 1991).

The corrosion, accompanied by the formation of sulphides, that was found on works of art made of copper/lead alloy and located on the banks of the river Thames in London is attributed to the metabolic activity of sulphur-reducing bacteria (Duncan and Ganiaris 1987).

Even in tropical hot/humid environments, the processes of biocorrosion of copper, and metals in general, have been attributed to the action of bacteria such as *Thiobacillus* and *Ferrobacillus* because they are producers of sulphuric acid; to bacteria that are corrosive through the production of organic acids; and to sulphur-reducing bacteria (Hummer et al. 1968; Dhawan 1987).

The condition and the corrosion products of the copper of the figure of Atlas situated under the gold Sphere of the Fortuna in Venice have been the object of investigations utilizing a variety of analytical techniques: among the corrosion products, copper oxalate was also investigated as a possible product of the activity of fungi such as *Aspergillus* and *Penicillium* (Staffeldt and Kohler 1973).

Still in Venice, the deterioration of the Horses of St. Mark's has also been investigated, and a variety of conclusions have been drawn; some have considered the presence of oxalic acid on the horses as a product of biological activity, while others have correlated its presence to the oxidation of hydrocarbons in atmospheric smog. The studies on the condition of the horses and the procedures required for their preservation and conservation also date from that time (Vittori and Mestitz 1976; Paleni et al. 1978).

4.5 COMPOSITE MATERIALS

Artifacts of historical/artistic interest are rarely composed of one single material, but are mostly made up of a variety of organic and inorganic materials. It is almost impossible to list all the types of work that are composed of different materials and even more difficult to indicate their susceptibility to biological deterioration.

In general, we can say that the risk of biological deterioration for these artifacts is linked to the most vulnerable element and is the result of the various effects that favor or inhibit biological development of the different components of these objects. Consequently, the appearance of the damage and the degradation mechanisms will be those of the most susceptible of the main components.

In the field of cultural heritage, many objects and artifacts are composed of a variety of materials; suffice to think of paintings, ethnographic collections, and some archaeological finds that are made up of such diverse plant and animal materials as fibers, woven materials, wood, feathers, bone, ivory, etc.; or also natural history collections, which involve the preservation (with different methods) of animals and plants.

A vast and varied number of constituent materials are also present in paintings, which include a range of chemical substances depending on the support, on the technique of execution, as well as on the period in which they were painted and the artist who created them. The variety of artistic materials of the past was further expanded in modern and contemporary art, which has seen the creation of works with the widest possible diversity of materials. A striking example is the production of "pop art," which envisages the provocative use of the most varied objects of daily life for the purpose of removing from the work of art its character of unique and subjective experience. In addition to paintings, recent centuries brought the development of photographic materials, which have radically changed in their constituent elements and materials over time. These two types of artifacts—paintings and photographs—will therefore be examined in greater detail.

4.5.1 Easel Paintings

by Maria Pia Nugari

4.5.1a *Structure and Composition*

The term easel painting includes all painted works with the exclusion of those painted on walls, i.e., all the paintings that may be moved from the site in which they are located. Moveable painted works, or easel paintings, are usually classified according to the type of support and the technique of execution; for instance, a distinction is made between paintings executed in oil on canvas or on panel, in tempera on paper, etc.

The support of a painting can be made of several different materials such as: canvas, wood, paper and pasteboard, metal plates, glass sheets, etc. Among the various painting techniques, the traditional ones include oil, tempera, pastel, and watercolor, to which must be added the infinite variations in use in modern and contemporary works, which include mixed media, acrylics, enamel, etc.

The most common structure for a painting is multilayered: the support, the preparatory layer, which varies according to the type of support and the painting technique adopted by the artist, the paint layer, and in some cases also the varnish layer (a homogenous and optically uniform layer that adds sheen and protects the paint layer).

Oil paintings are the most widespread; the technique is thought to date from the beginning of the fifteenth century. To give an idea of the complex compositional structure of an oil painting on canvas we list below the different layers and their respective constituent materials:

—the support, usually a woven fabric of a cellulosic nature, mostly linen or hemp, more rarely cotton or jute;

—the preparatory layer, composed of glues and inert materials; its function is to make the support suitable for paint and it usually involves the application of several layers. The glues are usually of animal origin and can be distinguished into strong glues, which are rich in impurities, and gelatins, which are pure glues consisting almost exclusively of collagen. These are therefore essentially proteinaceous substances containing small quantities of other substances, both organic and inorganic; among the amino acids present in the proteinaceous products of the hydrolysis of the glues, the prevalent one is glycine and, in lower quantities, hydroxyproline (Matteini and Moles 1989). The most commonly used inert materials were flour, gypsum, and also pigments such as lead white, lead-tin yellow, verdigris, and the earths. Drying oils, honey, varnishes, resins, and wax could also be used as additives. The choice of the inert material and of the binder as well as their relative proportions were dictated by structural requirements—such as flexibility and porosity—as well as by aesthetic considerations—surface texture, color of the preparatory layer—depending on the pictorial exigencies of the artist;

—the paint layer, composed of pigments and their binders. Pigments can be distinguished into organic (lakes) and inorganic (of mineral origin such as oxides, carbonates, sulphides, and sulphates of the various metals). In oil paintings, the binder is an oil that, thanks to its cohesive and adhesive properties, binds the pigment together and attaches it to the support. These oils—called drying or siccative oils—polymerize in the presence of air through a process of oxidation and thus change from a liquid to a solid state. They are essentially vegetable oils, such as linseed, poppy, or walnut oil, and are composed of mixtures of unsaturated triglycerides (esters of glycerine with long-chain unsaturated fatty acids such as oleic, linoleic, and linolenic acids) as well as mixtures of saturated triglycerides (esters of saturated fatty acids such as palmitic and stearic acids);

—the final varnish layer; this usually consists of a mixture of natural resins, mostly of plant origin (for example, mastic, dammar, colophony), which is brought to a fluid state and then brushed on to the paint surface. Oleoresinous varnishes are composed of one or more natural resins and an oil, generally heat-treated linseed oil.

4.5.1b *Biodeterioration of Easel Paintings*

The degradation of easel paintings is strongly determined by the complex stratification as well as the chemical nature of the materials of which they are composed. Consequently, problems of biodeterioration can be specific to the supports (canvas, wood panel, paper, etc.)—already described in previous paragraphs—while the various materials and substances added for the execution of the works can have an effect on the speed of development as well as the type and modality of the biological degradation.

In general, when organic materials such as vegetable glues, egg, casein, and others have been used, susceptibility to biological attack is more marked; such substances are indeed a potential source of nutrients for a vast group of heterotrophic microorganisms. Among these substances we note starch, which can be present for instance as size in the textile materials used as supports or as a glue, for example in the flour paste or glue used in the lining of deteriorated canvas supports. The fact that starch is so susceptible to biological attack means that, given appropriate environmental conditions, the paintings that have undergone conservation with this type of glue are the most prone to attack by microfungi.

The substances of which paint layers are composed tend to be less susceptible to biodeterioration, which is influenced by the nature of the pigments and of their binders; the pigments made of the oxides and salts of heavy metals such as zinc, copper, lead (lead white, zinc oxide, etc.) can, for instance, block the infection locally, while other pigments such as earth colors (siennas and umbers) do not influence their development, or else favor it if they are rich in microelements (Ionita 1971; Strzelczyk 1981; Berovic 2003; Szczepanowska and Cavaliere 2003). Among the binders, the oils are more resistant than egg- or casein-based ones; indeed, although various groups of bacteria and fungi have the ability to attack fatty acids and triglycerides through lipase, the level of polymerization of the oils binding the pigments makes them resistant to microbial attack (Lazar and Schroeder 1992). Finally, finishing varnishes may also play a protective role by virtue of their hydrorepellence.

Because of the environmental conditions in which these works are generally kept (museums, churches, villas, and palaces), the biodeterioration phenomena encountered most frequently are those linked primarily to attacks by microfungi; indeed, the humidity values are generally too low to allow colonization by heterotrophic bacteria.

The site of the fungal attack varies primarily in relation to the location of the work, the nature of the support material, and the painting technique. Paintings on canvas are predominantly attacked on the reverse, because the canvas is used as the nutritional substrate. Microclimate conditions that are favorable to fungal development—such as low ventilation and higher humidity through contact with wall surfaces that are often colder and damper (Plate 27)—can also be created between the painting and the wall; in these cases, a prolonged attack, in addition to damaging the mechanical properties of the support, can also cause small fractures and detachment of the paint layer due to the penetration of hyphae from inside the support into the painted surface (Plate 18).

Paintings on paper and parchment are also mostly attacked by fungi and only rarely by bacteria (mostly actinomycetes); in these cases, the colonizations take place on the side of the painting that is the most exposed to contact with airborne microbial pollution (earlier in this chapter, section 2.1b). The painting techniques used on these supports (pastel, watercolor, tempera) do not inhibit colonization from the painted side (contrary to what happens with works on canvas, usually oil paintings treated with varnishes) (Fig. 4.43). In these cases, it is difficult to establish if the fungal development is connected with the support, the pigments, the binders, or with the presence of dirt or dust on the surface of the work. Panel paintings, on the other hand, are rarely affected by microbial deterioration, thanks to the high natural durability of the woods usually employed as supports and also because long periods of high relative humidity are necessary for biodeteriogenic fungi to develop on wood. These types of works are instead often damaged by attacks by xylophagous insects.

In the case of framed and glazed works, the surfaces that are particularly at risk are those in contact with the glass: the latter is a cold surface and therefore favors condensation phenomena, creating an area of stagnant humidity for the materials that are in contact with it.

The fungi responsible for damage on paintings belong mostly to the genera *Alternaria*, *Aspergillus*, *Aureobasidium*, *Cladosporium*, *Fusarium*, *Nigrospora*, *Penicillium*, and *Trichoderma*. The most frequently recurring species are ubiquitous and are able to utilize the most varied organic substrates; this makes it difficult to establish which substance has most supported the attack. Overall, the deuteromycetes (*Aspergillus*, *Botrytis*, *Penicillium*, *Trichotecium*, *Verticillium*) and the ascomycetes (*Chaetomium*) are those most frequently isolated on the reverse of canvas paintings; however, in the presence of glue, it is the zygomycetes (*Mucor* and *Rhizopus*) that prevail. On paint layers—both oil and casein tempera—species of *Alternaria*, *Cladosporium*, *Stemphylium*, and *Geothricum* have been found; the latter will also develop on finishing varnishes. The genus *Aureobasidium* has been associated primarily with oil binders. On paint-

Fig. 4.43 *Formation of stains on a pastel on parchment caused by the development of* Penicillium *sp. (seen under the optical microscope in the magnified portion of the image)* – Photo ICR

ings on a paper support—which hardly ever have a preparatory layer separating the pigments from the substrate—the dominant species are often cellulolytic ones. Recent studies on pastels on paper have shown the presence of deuteromycetes of the genera *Alternaria, Aspergillus, Aureobasidium, Cladosporium, Fusarium, Nigrospora, Trichoderma*, the zygomycetes *Rhizopus nigricans* and *Mucor michei*, the ascomycetes *Chaetomium*, and the yeasts *Rhodotorula* and *Saccaromyces*. The genus *Thielavia* has recently been identified with molecular techniques of systematic identification (Bonaventura et al. 2003).

Among the species most cited in the degradation of paintings are *Aspergillus fumigatus, A. nidulans, A. niger, A. oryzae, A. terreus, A. versicolor, A. flavus, A. parasiticus, Cladosporium oxysporum, Fusarium oxysporum, Penicillium chrysogenum*, and *P. roqueforti* (Ionita 1971; Strzelczyk 1981; Dhawan and Agrawal 1986; Inoue and Koyano 1991; Berovic 2003; Bonaventura et al. 2003; Szczepanowska and Cavaliere 2003).

As far as bacteria are concerned, few studies exist that examine their role in the degradation of easel paintings, most likely because the conditions in which paintings are housed rarely attain the levels of humidity required for bacterial development. The species involved seem to be almost exclusively gram-positive and are favored by prior fungal colonization (Seves et al. 1996).

The damages resulting from microbial colonization of the support have morphologies that are typical of the material constituting the support (woven fabrics, wood, paper, etc.), while on the paint layers they more often than not appear as chromatic alterations: bleaching of the paint layer or stains of various forms and colors depending on the species present and the nature of the substrate. The grayish-brown stains associated with fungi belonging to the group Dematiaceae, which have melanin-rich cell walls, are found frequently. The formation of stains is also often caused by the production of exopigments or the release of acid metabolites onto the paint surface; these can then react with the pigments of the painting and alter them chemically. In the case of pastels, for example, the various organic acids emitted (especially citric, oxalic, malic, and fumaric acids) can induce reactions with the pigments giving rise to the formation of their relative salts with consequent alterations

in the color and in the optical properties of the pigment (Berovic 2003). Varnishes are instead rendered opaque in the areas corresponding to the emission of metabolites.

The development of fungi takes on different morphologies as the attack is protracted in time and as the vital stage of the strain changes; consequently, mycelia that were initially white and fluffy can become gray and powdery. A strong sporulation of the colonizing species may also often occur; in this context, it is worth pointing out that many species identified on canvas paintings—for instance, *Aspergillus niger* and *A. fumigatus*—can cause human respiratory illnesses or allergies.

4.5.2 Photographic Materials

by Donatella Matè, Giovanna Pasquariello, and Maria Carla Sclocchi

4.5.2a *Structure and Composition*

Photographic materials are composite, layered artifacts, consting of inorganic and organic matters. The structure and composition of a photograph may be sketched as follows:

— a primary support, varying in nature, which represents the base for the photographic image;
— the elements forming the "image layer," which are the photosensitive substance and the binders.

Alongside these basic components, historical photographs also include secondary supports of various origins, which have a decorative role or a supporting function. The complexity of the subject requires some chronological information regarding the main types of photographic material and the various processes employed over time in the history of photography (Table 4.10) (Crawford 1981; Reilly 1980, 1986; Cartier Bresson 1987; Lavédrine 1990a, 1990b; Stroebel et al. 1993; Berselli and Gasparini 2000; Residori 2002a, 2002b).

The photographic processes employed over time can be classified into silver-based and non-silver based. The first use halides of silver as their photosensitive substances (silver chloride, silver bromide, and silver iodide), while

Table 4.10 Classification of the main types of photographic materials

Daguerreotype (from 1839 to ca. 1865)	Support: silvered copper plate, polished, made sensitive with iodine vapor and developed with mercury vapor; no binding medium. The image is unique and can neither be printed nor duplicated. The image is in reverse; to be viewed, the daguerreotype must be placed against a dark surface.
Callotype (negative) (from 1840 to ca. 1851)	Support: paper, no binding medium. Negative image from which positives can be made.
Salt paper print (positive) (from 1840 to 1860)	Support: paper, no binding medium. The first printing material to be blackened directly.
Albumen glass plates (from 1847 to 1855)	Support: glass plate; binding medium: albumen. For the first time, a medium is added as binding agent for the particles that compose the image in silver.
Albumen prints (from 1850 to 1920)	Support: salt paper; binding medium: albumen. Many papers are tinted with aniline dyes (blue, pink, purple) before being printed.
Collodion plates (from 1851 to ca. 1880)	Support: glass plate; binding medium: collodion. The collodion is almost always covered with a protective layer of varnish.
Ambrotype (from 1854 to ca. 1870)	Support: glass; binding medium: collodion. Unique image, which can neither be printed nor duplicated. The positive image is produced directly and requires a black background to be viewed; black varnishes or dark glass is employed for this purpose.
Tintype or *ferrotype* (from 1855 to the end of the nineteenth century)	Support: iron plate, paper; binding medium: collodion. Unique image, which can neither be printed nor duplicated. The positive image appears directly; protective varnishes are used.
Gelatin plate (from 1871 to the present day)	Support: glass plate; binding medium: gelatin.
Papers that blacken directly (from 1880 to ca. 1910)	Support: paper; binding medium: collodion or gelatin. Presence of a layer of baryte between the paper and the emulsion to cover the fibers of the paper and to improve the adhesion of the binding medium to the support.
Emulsion papers requiring development (from 1885 to the present day)	Support: paper; binding medium: gelatin. A layer of baryte is present. RC (resin-coated) papers are also worth mentioning: these have a polyethylene-coated paper support.
Black-and-white film (from 1880 to the present day)	Support: cellulose nitrate, acetates, polyester, etc.; binding medium: gelatin. At this point in time, plastic materials enter the history of photography.
Color photography (from 1903 to the present day)	Support: glass, plastic, paper, etc.; binding medium: gelatin. The first color images are the "autochromes"; glass plates made sensitive with the use of silver salts and colored potato starch.

the second employ other substances (e.g., iron salts and alkaline bichromates) (Crawford 1981; Reilly 1986; Cartier Bresson 1987; Mina and Modica 1987; Lavédrine 1990a, 1990b; Jacob 1991; Scaramella 1999; Berselli and Gasparini 2000).

In addition, there are photomechanical processes that—by means of photographic techniques—create initial images (technically, these are not true photographs) that are then transferred onto a matrix and printed on paper with a typographic process (*héliogravure*, phototype, photogravure) (Crawford 1981; Cartier Bresson 1987; Scaramella 1999).

In general, the photographic process is based on a series of complex chemical reactions which, starting with the action of light on the photosensitive substances, allows the photographic image to arise, either as a positive or as a negative. The image is negative when the blacks, resulting from the photochemical reduction of the silver salts, correspond to the light areas of the subject. The light tones, on the other hand, correspond to the silver halides that are not affected by light exposure and by the subsequent treatments. These are the methods used to obtain black-and-white photographs.

The earliest photographic process, such as the daguerreotype and the callotype, did not make use of any kind of binder, and the image was formed directly onto the primary support. With the evolution of photographic processes, the role of the binder became essential for the definition of the image, so that it is now employed in all photographic techniques.

The mechanics of image formation are the following: the process of direct blackening (the image is formed through prolonged exposure to light) and the development process (after a short exposure to light, a latent image arises that is not visible to the naked eye and requires a development treatment to be revealed) (Reilly 1980, 1986).

In historical photographs, various chemical treatments are often used for decorative and conservation purposes, for instance the practice of toning (using gold, platinum, copper, selenium, etc.), or the use of aniline to prepare the paper support, and also the frequent use of dyes and pigments (both natural and synthetic) for the hand coloring of images (Craw-

ford 1981; Reilly 1986; Mina and Modica 1987; Residori 2002b). As far as color photography is concerned, the same basic elements are still in use as for black-and-white photography, with the addition of different kinds of dyes that undergo chemical reactions until the image is fully formed (Lavédrine 1990b).

Primary supports and binders are the basic constituents of photographic materials.

Among primary supports, the most important are paper, glass, metals (copper, iron, and silver), and plastic materials; other kinds of materials should also be kept in mind such as porcelain, textiles, and wood. The types of paper used vary from high-quality linen and cotton mixtures (albumen prints) to relatively coarse ones (callotypes). The plastic materials employed as photographic support are: celluloid, cellulose nitrate, cellulose acetate, cellulose diacetate, cellulose triacetate, polyesters, and polyethylene terephthalate (PET).

The binders used in photography are: albumen, collodion, and gelatine. Albumen is a water-soluble globular protein and derives from the egg white. Collodion is a solution of nitrocellulose (also called collodion cotton) in ethyl alcohol and ether. Its characteristic of not being very water soluble—which is a limit in photographic procedures—has been partially obviated by adding hygroscopic and hydrophilic materials (such as sugars, syrups, beer, and gelatine) that augment the sensitivity of the photographic emulsion. Gelatine is a water-soluble protein, obtained from the bones and skins of animals that are transformed into collagen after being boiled in water. It is a compound consisting of solid particles (the colloid) dispersed in water (the solvent). It is usually composed of several different gelatines as well as of impurities; these secondary products, as well as the gelatine itself, are of great importance because they affect the general characteristics of the photographic emulsion.

Table 4.11 lists the main constituents of the secondary supports and of the adhesives most commonly used in the processes of mounting and preservation of photographic materials (Cartier Bresson 1987; Mina and Modica 1987; Lavédrine et al. 2000; Matè et al. 2002; Matè and Residori 2002; Sclocchi et al. 2003).

Table 4.11 Main types of secondary supports and adhesives

Secondary supports	Paper, cardboard, leather, metals, textiles, glass	Paper and cardboard varying both in quality and color are used as secondary supports for various types of prints. This type of mount is used to reinforce the photograph and also for aesthetic enhancement. Leather, glass, metals, and textiles are used to make the cases in which daguerreotypes and ambrotypes are kept.
Adhesives	Products of vegetable origin (starch, etc.) and animal origin (glue from animal bones, fish glue, albumen, casein, etc.)	These substances, which are able to hold together two surfaces, are classified as contact, thermoplastic, and thermosetting adhesives. They are used to: attach the photograph to the secondary support (e.g., albumen prints); in the assembly of the protective covers (e.g., for daguerreotypes and tintypes); and during preventive conservation procedures (e.g., facing, filling, etc.).
	Natural resins (colophony, Arabic gum, gum tragacanth, etc.), synthetic resins	These materials become an integral part of cases, albums, and frames and are also used during various conservation operations.

4.5.2b ***Biodeterioration of Photographic Materials***
The main biological agents causing the degradation of the different kinds of photographic materials are microfungi and bacteria, in addition to certain species of insects (Kowalik 1980a, 1980b; Flieder 1985; Flieder and Lavédrine 1987; Gallo 1992; Pasquariello 1992, 1993; Pasquariello et al. 1996; Florian 1997; Matè 2002; Matè et al. 2002; Sclocchi et al. 2003).

The organic and inorganic constituents of the supports, of the emulsions, and of the binders and the adhesives are potential sources of nutrients for microorganisms, and the damage can be incurred by one or more components of the photograph. The attack, which only occurs in favorable environmental conditions, can sometimes begin in the secondary support and then extend to the photographic image, thus either affecting a very small part of the document or its entirety (Fig. 4.44).

The biodeterioration of photographic materials is a complex aggregate of degradation processes, which can involve one or more species of microorganisms belonging to different genera that frequently act in succession. Studies of the mechanisms of biodeterioration in photographs are rare compared to studies on other types of works of cultural heritage (Flieder 1985; Flieder and Lavédrine 1987; Pasquariello 1992; Pasquariello et al. 1996; Pileggi et al.

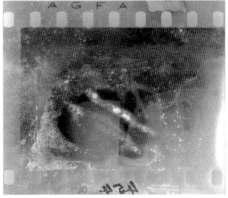

Fig. 4.44 *Black-and-white negatives showing microfungal deterioration (E.U.R. Fund, A.C.S. concession, digital reproduction by M. C. Sclocchi)*

1996; Flieder and Capderou 1999; Matè et al. 2002; Sclocchi et al. 2002).

The mechanisms of biodeterioration are both of a physical and chemical nature (Allsopp and Seal 1986; Pitt and Hocking 1985). Those of a physical nature result in the more or less marked structural and mechanical alteration of the support, caused by the development of biological species that can be either bacterial or fungal. The penetration of the hyphae of microfungi produces such a high degree of pressure as to induce the fragmentation of the photographic support. When the attack is invasive, it can also result in the detachment of the photographic emulsion; in the case of printing papers, these can exhibit a felt-like appearance, with an increase in fragility and a tendency to crumble. The extent of the damage is therefore linked to the capacity of microorganisms to exert mechanical actions, to the degree of competition between the various microbial species involved, and to the different types of photographic materials.

The chemical mechanisms of biodeterioration are much more complex and widespread and are linked to the transformation of the actual chemical nature of the photographic supports; they may be ascribed to two distinct processes (Chapter 1): assimilation—i.e., using as nutrients the different constituent components of photographic materials by means of enzymatic activity—and emission of metabolic substances (such as acids, alkalis, and pigments). Cellulolytic microfungi and bacteria have the ability to degrade in particular photographic material on a paper support. The selective degradation of gelatine and albumen, achieved through the action of certain species of microfungi and bacteria, occurs as a result of the action of specific extracellular and intracellular enzymes called proteinase and peptidase. Gelatine can become fragile and soft, taking on a viscous appearance and at times even liquefying. In addition to the swelling of the gelatine, the damage caused by the hydrolysis process can determine the complete detachment of the image from the support.

Various degrees of chromatic alterations can occur on photographs; they can be caused by degradation and oxidation processes, or else by the production of organic acids (citric, lactic, succinic, etc.) and colored metabolites (red, brown, orange, or yellow anthraquinones) (Plate 19) (Fanelli 2001). Another very common type of alteration is the appearance of small foxing stains, which can be found on the emulsioned paper support, on certain photographic images, and on the secondary supports employed in their mounting (Plate 19) (earlier in this chapter, section 1.2). The presence of certain kinds of stains found on the various supports can be ascribed to the actual pigmentation of the hyphae of some species of fungi (e.g., *Alternaria, Stemphilium, Epicoccum*, etc.)

Although glass is attacked with difficulty by microbial agents, some damage has been observed on glass supports and can be ascribed to the action of certain species of xerophilic fungi and nonspecific autotrophic bacteria; the damage consists in the formation of micropores (pits) that corrode the glass giving rise to furrows, opacification, and dark stains (Onions et al. 1981; Allsopp and Seal 1986; Matè 2002).

Photographic materials on plastic supports can also become sources of nutrients for certain microbial species because they contain plastifiers. In fact, it is the additives that are the first to be degraded, with a loss in the flexibility of the material (Matè 2002).

Microbial attack of the metal supports of photographic materials (daguerreotypes and ferrotypes) can only occur in conditions of high humidity and is not easily distinguishable from an attack of purely electrochemical origin (Plate 19). A surface corrosion takes place in which the microorganisms involved (mainly sulphur bacteria and iron bacteria) can attack the substrate producing acid metabolites (earlier in this chapter, section 4.2). The degradation is therefore related to the nature of the metal as well as to the presence of possible fissures and extraneous substances on the surface (Allsopp and Seal 1986; Florenzano 1986; Caneva 1994).

The groups of biodeteriogens that give rise to alterations on photographs as well as the damages they produce on supports and binders are listed in Table 4.12.

Table 4.12 *Main groups of biodeteriogens and description of the kinds of damage*

Supports and Binders	Groups of Biodeteriogens	Kinds of Damage
Paper	Heterotrophic bacteria: schizomycetes (*Cellvibrio, Cellfalcicula*), mixobacteria (*Cytophaga, Sporocitophaga*), actinomycetes (*Streptomyces, Serratia*)	Erosions, stains, mechanical alterations
	Microfungi: ascomycetes (*Chaetomium*), deuteromycetes (*Alternaria, Aspergillus, Cladosporium, Fusarium, Penicillium, Stachybotrys, Scopulariopsis, Stemphylium, Trichoderma*), zygomycetes (*Mucor, Rhizopus*)	Stains, pigmentations, and discolorations, structural and mechanical alterations
Glass	Heterotrophic bacteria: (*Bacillus*)	Erosions, opacifications, and pits
	Microfungi: deuteromycetes (*Aspergillus glaucus* group, *Aspergillus restrictus* group)	Superficial or deep erosions, stains, opacifications
Metals	Chemoautotrophic bacteria: ferrobacteria (*Sphaerothilus, Crenothrix, Leptothrix*) and sulphur bacteria (*Thiobacillus*)	Erosions and pits
Plastic materials (cellulose nitrate, cellulose acetate, cellulose triacetate, polyester, etc.)	Heterotrophic bacteria: schizomycetes (*Cellvibrio, Cellfalcicula*), actinomycetes (*Streptomyces, Serratia*)	Variations of the physico-chemical characteristics
	Microfungi: ascomycetes (*Chaetomium*), deuteromycetes (*Alternaria, Aspergillus, Cladosporium, Fusarium, Penicillium, Stachybotrys, Stemphylium, Trichoderma*)	Pigmentations, stains, variations of the physico-chemical characteristics
Proteinaceous materials (albumen, gelatin)	Heterotrophic bacteria: schizomycetes (*Cellvibrio, Cellfalcicula*), mixobacteria (*Cytophaga, Sporocitophaga*), actinomycetes (*Streptomyces, Serratia*)	Loss of physico-chemical characteristics, liquefaction and detachment from the support
	Microfungi: deuteromycetes (*Alternaria, Aspergillus, Cladosporium, Fusarium, Penicillium, Stachybotrys, Stemphylium, Trichoderma*), zygomycetes (*Mucor, Rhizopus*)	Loss of physico-chemical characteristics, liquefaction and detachment from the support, formation of stains

4.6 PRODUCTS EMPLOYED IN CONSERVATION

by Maria Pia Nugari and Ornella Salvadori

For the conservation of a work of art it is critical to know not only the constituent materials but also the previous interventions the work has undergone as well as the nature of the products employed to this end. A restored artwork—in a fragile condition to begin with since it requires an intervention in the first place—can at times be made more susceptible to biodeterioration precisely as a result of the nature of the substances used in the intervention.

The products used in conservation may:

—support microbial growth;
—induce microbial colonization;
—prevent and inhibit biodeterioration phenomena.

In the first case, it is usually organic materials that provide a nutritional substrate for numerous microbial groups: organic glues of vegetable and, less frequently, of animal origin are examples of this.

The substances that can induce microbial colonization are those that, because of their methods of application or the type of solvents needed, favor the growth of microorganisms by modifying the characteristics of the substrate and acting directly as a stimulus for spore germination. At times, therefore, biological growth occurs during or immediately after the application of conservation products, even when no microbial colonization was present before; this happens particularly in environments with high humidity and high concentrations of biological contaminants in the air (Bartolini et al. 2000a, 2000b). It has been observed, for instance, that numerous solvents (alcohols, acetone, etc.) for the products used in conservation can provide the stimulus for the germination of fungal spores in the areas around the intervention area. A study carried out by Florian (1993) showed that the activation of fungal spores is stimulated by the contact with numerous substances used in the restoration (e.g., alcohols, ketones, some detergents) as long as the concentrations attain certain values.

An additional element favoring microbial colonization is the stickiness of many of the employed substances—especially during their polymerization—which fosters the capture of particles in the air and, therefore, the subsequent contamination and microbial development. Examples of this are the waxy substances frequently used in the past and still used experimentally today in synthetic formulations (microcrystalline waxes, for instance). Because of their hydrorepellence, these substances were used on frescoes and on outdoor stone and metals sculptures; although they do indeed exhibit a considerable degree of resistance to biological attack, their stickiness encourages the attachment of biological and organic particles that support the superficial colonization by heterotrophic microorganisms. A research aimed to identify protective materials for metal sculptures exposed outdoors demonstrated that both natural and synthetic waxes are subject to microbial attack, with a slight difference in favor of microcrystalline waxes (Nugari and Bartolini 1997).

Other materials used in conservation might instead prevent or reduce biodeterioration, because of their capacity to modify certain characteristics of the substrate on which they are applied or because toxic biocides are added to them. For example, the application of natural and synthetic resins with consolidating and protective functions reduces the porosity and increases the hydrorepellence of materials, thus containing their susceptibility to biodeterioration—certain synthetic resins employed in the conservation of stones, or polyethyleneglycol (PEG), which is used in the consolidation of archaeological wood, fulfill these functions. Still, the inhibiting effects depend on the conditions of preservation.

The treatment of a mortar with silicone resins is effective in the prevention of colonization by algae and lichens, but it does not inhibit the growth of chemoorganotrophic fungi and bacteria (Mansch et al. 1999). A precocious and considerable development of meristematic fungi can also be observed on stucco fillings made of marble powder and acrylic resins (Pinna and Salvadori 1999). The application of protective substances on the sculpted parts of fountains seems, on the other hand, completely useless in the inhibition of the development of algae because of the high values of water present (Tiano et al. 1995b; Nugari and Pietrini 1997).

Sometimes the products used in conservation are mixed with biocides with the aim of preventing biological growth and extending the life to the intervention, but often these mixtures are empirically determined and only seldom based on scientific experiment (Nugari and Salvadori 2003b). In the field of works in stone, studies have recently been carried out on plasters and synthetic polymers that include substances with biocidal action (Quaresima et al. 1997; Corain et al. 1998; Ferone et al. 2000).

Among the organisms that attack conservation products, the most frequently encountered are fungi: because of their great capacities for metabolic adaptation, they have the ability to colonize and utilize as a nutritional substrate even synthetic polymers. Usually this necessitates the production of adaptive enzymes that require long periods of time and the presence of small organic fractions that are easy to metabolize, in order to set the metabolic functions in motion and accumulate the energy required in order to activate the process. In the case of certain polymers in which the organic fraction is practically nonexistent, the development of fungi is linked to the frequent presence of organic additives in commercial formulations, as happens in the biodeterioration of silicone-based resins used as protective and consolidant agents on stone.

Among the factors that most influence biological colonization of natural and synthetic substances used in conservation, the following should be noted:

— the chemical nature of the product and its coformulations (for instance, additives, catalysts, etc.);
— its affinity for water;
— the chemical and physical characteristics of the materials on which they are applied;
— the environmental conditions in which they are employed (T and RH);
— the length of time they are kept in humid conditions;
— the levels of environmental contamination by organic substances (dust, pollutants, etc.) and biological ones (spores, vegetative propagules);

— the duration of their exposure to contaminants.

When considering the suitability of a product for conservation, very frequently its susceptibility to biological attack is not taken into consideration. Instead, it is critical that the products employed in conservation not be more susceptible to biodeterioration than the materials or artifacts on which they are applied.

Organic materials of plant and animal origin, which are easily susceptible to microbial colonization, are especially used in the case in interventions carried out with traditional techniques. These interventions, developed on the basis of practical experience in conservation workshops, often limit themselves to verifying the immediate results and do not take into account the long-term effect on the work. The examination of the main Italian manuals on conservation of paintings (Secco Suardo 1866; Forni 1866) offers an overview of a vast range of materials and recipes used in the past, many of which are still consulted or used today. The substances used during the various interventions are often of organic origin and of very diverse typology. Table 4.13 lists some of the most curious substances employed, along with their function (Cecchini et al. 1995).

At present, in the great majority of cases, an effort is made not to use substances known to be particularly susceptible to biological degradation, but some conservation techniques have not yet found valid substitutes. In the conservation of paintings on canvas, for instance, it is sometimes necessary to carry out a lining or a relining of the support; in this process, one or more new canvases are attached to the reverse of the original canvas, which has lost its supportive power, using natural adhesives of vegetable or animal origin (for example, glue-paste or wax-resin, the latter no longer in common use) or synthetic adhesives (for example, Plextol, BEVA, etc.). Among the adhesives of vegetable origin, glue-paste, or starch glue, is undoubtedly the one most frequently used because of its elasticity and reversibility. But this operation enriches the structure of the painting with an organic, water-soluble material and thus provides an ideal nutritional substrate for many microorganisms; after the

Table 4.13 Some substances used in the past as components in conservation recipes (from Cecchini et al. 1995)

Substance	Function	Process	Painting
Limewater	Solvent	Cleaning	Canvas and panel
Garlic	Reduces surface tension	Fixing and transport of the paint film	Canvas and panel
Beer	Protective	Cleaning	Canvas and panel
Ashes	Hygroscopic	Treatment of panels	Canvas and panel
Lupin flour	Solvent	Cleaning	Canvas and panel
Ox gall	Solvent and emollient	Cleaning	Canvas and panel
Glycerin	Emollient	Consolidation and transfer	Canvas and panel
Milk	Protective	Protection of the paint layer	Wall
Honey	Emollient	Cleaning and preparation of glues	Canvas and panel
Bread (soft part)	Abrasive	Cleaning	Canvas, panel, and wall
Linseed, walnut, and poppy oils	Emollient	Cleaning	Canvas and panel
The root of *Saponaria officinalis*	Solvent	Cleaning	Canvas and panel
Saliva	Solvent	Cleaning	Canvas and panel
Spirits of wine	Solvent	Cleaning	Wall

intervention, if the painting is placed in a humid or otherwise unsuitable environment, it will inevitably be attacked by fungi, bacteria, and sometimes insects. Because this risk is well known, the composition of the glue-paste usually includes an antifermentation product or a biocide: for example, today potassium sorbate or an ortho-phenyl-phenol is used in place of the phenol (no longer used because of toxicity) and of the rock alum, which was widely used in the past. However, these do not manage to definitively ward off a biological attack. Moreover, the addition of a biocide to the glue can sometimes considerably reduce its powers of adhesion (Tiano et al. 1997).

Materials based on starch also find wide employment (as glues, strengtheners, etc.) in the conservation of library and documentary materials, because they do not interfere either chemically or aesthetically with the substrate; ease of application and reversibility of the process are additional reasons for use.

The awareness of the risks of biodeterioration linked to the use of starch-based products has stimulated research on possible alternatives such as synthetic adhesives (methyl-hydroxyethyl-cellulose, i.e., Tylose, polyvinyl alcohol, and polyvinylpyrrolidone) or starches modified through etherification. Comparative tests carried out on these substances have shown that synthetic adhesives offer a greater resistance to biological attack while starches—both natural and modified—undergo microbial colonizations with equal rapidity and degree of intensity. In addition, the results are strongly influenced by the materials on which the product is applied (e.g., paper, parchment, silk, or leather) and in particular by their affinity with water; after

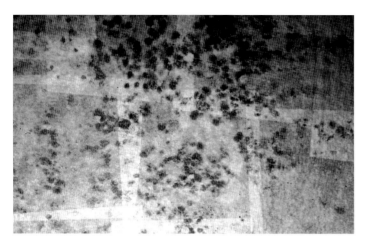

Fig. 4.45 *Fresco with a PVAC facing, showing the development of fungal colonies* – Photo O. Salvadori

treatment with the various glues, paper, parchment, and silk were all found to be more susceptible than the leather treated with tannins (Pierantonelli et al. 1984–85; Gallo et al. 1989; Plossi Zappalà 1999; Gallo and Valenti 1999)

As a preliminary procedure to the conservation of canvas or panel paintings, and also frescoes, it is at times necessary to apply a facing to the surface in order to protect the paint layer from the detachment of paint fragments during the operation. Sheets of facing Japanese paper or tissue are applied to the surface using natural adhesives (colletta, a weak glue solution, or wax-resin) or synthetic adhesives (acrylic resins or polyvinylacetates). Given favorable environmental conditions, these routine operations can also induce the development of microorganisms, even when synthetic adhesives are used, which in general are less susceptible to biological attack (Fig. 4.45).

The introduction of synthetic materials in conservation (employed as adhesives, consolidants, or water repellents) led to the erroneous assumption that the problems associated with the risks of biodeterioration could be disregarded. The synthetic resins used in conservation belong to various categories such as: acrylics, epoxies, silicons, perfluoropolyethers, polyvinyls, polyvinylacetates, polymethylmethacrylates, polyesters, cyanoacrylates, and polyurethanes. On these synthetic materials fungi, bacteria, actinomycetes and algae can all develop, although generally it is the micro-

scopic fungi that play the most important role (Favali et al. 1978; Laurenti Tabasso et al. 1991; Gu 2003; Lugauskas et al. 2003). Only on rare occasions has lichen growth been found on stone treated with acrylic resins (Pinna and Salvadori 1999) or polyester resins (Ariño and Saiz-Jimenez 1996b).

Synthetic polymers represent potential substrates for heterotrophic microorganisms that can utilize them as a source of carbon or else partially degrade them; in particular, a number of bacteria and fungi that produce exoenzymes are able to hydrolize the bonds present in certain synthetic polymers (for instance, esterase in the biodeterioration of polyurethanes) (El-Sayed et al. 1996; Horward and Blake 1998). In addition, heterotrophic microorganisms secrete a vast range of organic acids that can contribute to the degradation both of the polymers and of the materials on which they are applied.

The biodegradability of a polymer depends on numerous factors including molecular weight and crystallinity; in general, the greater the molecular weight and the degree of crystallization, the lesser the biodegradability of the polymer. Formulations that contain additives are in general more easily colonized.

The smoothness and the integrity of the surfaces treated with polymers also considerably influence biological colonization; dust, spores, and substances of varying natures are more easily deposited on a rough surface or on one with microfractures, and these deposits are the initial source of nutrients for microorganisms that land and take root on the surface. Usually, the colonization of materials treated with polymers increases with longer exposure, because, over time, the resins undergo processes of physicochemical degradation that alter their performance characteristics (such as hydrorepellence, for instance), usually lowering them. A study carried out on samples of sandstone treated with a variety of consolidants and water repellents and then exposed to the open air for twenty years did

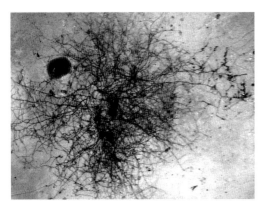

Fig. 4.46 *Resin figure colonized by fungi*
– Photo M. P. Nugari

not provide unequivocal answers to what the relationship might be between microbial growth and the variations in water absorbency of the treated surfaces (Von Plewe-Leisen et al. 1996).

The most frequently found biodeteriogens of synthetic materials (Fig. 4.46) employed in conservation are fungi, usually cosmopolitan and airborne; among these are the genera *Penicillium, Aspergillus, Alternaria, Cladosporium, Acremonium, Trichoderma, Chaetomium* (Lugauskas et al. 2003).

Many experiments have been carried out on samples of stone treated with consolidants and/or water repellents, inoculating them with microorganisms (nitrifying bacteria, heterotrophic bacteria, or fungi) found on deteriorated stone, with and without the addition of culture media (Domaslowski and Strzelczyk 1986; Koestler and Santoro 1988; Salvadori and Nugari 1988; Petushkova and Grishkova 1990; Leznicka et al. 1991; Tiano et al. 2000). The different classes of polymers respond in a non-uniform fashion to fungal growth: within the same family of products (for example, natural resins and acrylic resins), we can therefore encounter substances that are highly biodegradable (such as dammar, a natural resin of animal origin, and Acryloid F-10, which is an acrylic polymer) alongside others that exhibit a moderate resistance (such as gum lac, a natural resin of vegetable origin, and Acryloid B-72), or highly resistant ones (Rhoplex AC234). We should add that certain additives and catalysts contained in artificial resins (for instance, dibutyltin dilaurate, formaldehyde, acetic acid) may

act as biocides (Koestler 2000). In many cases, the presence of such substances can only be presumed, because manufacturers are not obligated to list them and are covered by industrial secrecy. A study was conducted on some linen textiles of Egyptian origin treated with twelve different resins and polymers and inoculated with fungi: an acrylic polymer (Acryloid F-10) and a polyvinyl acetate (Mowilith)—both widely used in the conservation of archaeological textiles—were found to be the most susceptible to biodeterioration, while BEVA 371, Plexisol P-550, Paraloid B-72, and carbamate starch were found to be the most resistant (Abdel-Kareem 2000).

The development of microorganisms—especially of fungi even in the stage of sporulation—on resins used in conservation has been confirmed in the majority of studies. What is more rarely determined is whether these growths induce modifications on the performance characteristics of the resins. In many cases, tests conducted with different methods, for example, measurement of the contact angle and the absorption of water by capillarity (FTIR, Fourier Transform Infrared Spectroscopy), have not shown transformations in the resins under examination, or at least, modifications that are below the measurement thresholds of the instruments in question (Sorlini et al. 1991; Koestler 2000). On the other hand, some authors found an alteration in the IR (infrared ray) absorption spectra (Petushkova and Grishkova 1990). Leznicka et al. (1991) also found a significant reduction in the water-repellent properties of two silicon resins following fungal growth.

Recently, tests have been carried out to evaluate the resistance to biological deterioration of certain synthetic polymers (ethers of cellulose, polymers of acrylic acid, wax emulsions), which are used as carriers for solvents in the cleaning of easel paintings, in order to limit the diffusion of solvents inside the paint layer and in the air. The polymers of acrylic acid (Carbopol 940 and Ultrez) were found to be susceptible to biological colonization; as a result, it is very important to make sure that all residues of these products are carefully removed from the paint surface after employment (Talarico et al. 2001).

CHAPTER 5
PROBLEMS OF BIODETERIORATION IN RELATION TO PARTICULAR TYPES OF ENVIRONMENTS

This chapter highlights the conservation problems of the previously discussed materials as specifically related to the type of environment in which they are preserved.

As mentioned, the different environments of the various contexts in which cultural heritage is preserved play a considerable role in the physico-chemical and biological deterioration processes of the works.

We will therefore illustrate the characteristics of these environments in relation to problems of biodeterioration, beginning with a broad distinction between enclosed, semienclosed, and outdoor environments: these categories are discriminating factors because of the influence of the surrounding climate on the works in question.

Within each of these categories, we will highlight those contexts that are most significant in our discussion, such as conservation and exhibition areas, and we will also touch on some specific but recurring habitats, such as marine and edaphic ones.

5.1 ENCLOSED ENVIRONMENTS

Enclosed environments are those characterized by reduced exchanges with the outdoor environment. Different types of problems may arise depending on their specific function, particularly relating to the nature of the materials and/or collections that are held there, but also to the nature of the building in which they are housed as well as its site and geographical location.

5.1.1 Libraries and Archives

by Giovanna Pasquariello, Paola Valenti, Oriana Maggi, and Anna Maria Persiani

5.1.1a *Characteristics of the Environment and of the Materials*

Libraries and archives can be defined as "ordered collections of books, documents, and materials of various types, placed in premises suitable to this end and put at the disposal of the public for reading, study, and information purposes" (Vigini 1985). The durability of this immense cultural heritage—represented by manuscripts, printed books, documents, graphic works (prints and drawings), and photographs—is determined, over time, by the intrinsic characteristics of both the various components and the environment in which they are preserved.

These components may be of plant origin (cellulose, starch, vegetable gums, resins, etc.) and of animal origin (collagen, bone and fish glues, casein, albumen, gelatin, etc.); in addition, there is a vast range of plastic materials deriving both from natural and synthetic substances (cellulose acetates, polyesters, etc.). Given the complexity of these cultural goods, it is rather difficult to identify and to unequivocally define the correct environmental parameters suitable for the conservation of all the different categories of materials involved.

Such a problem is particularly acute in rare books or "historical" libraries and in the special collections departments, where books are kept alongside documentary, graphic, and photographic heritage; a similar situation is found in archives.

This state of affairs is often due to variations, transformations, and extensions that occur over time in these types of premises, which may have been adapted for various uses, such as storage, conservation, consultation, and often also for temporary and permanent exhibition. Ongoing monitoring of environmental conditions such as temperature, relative humidity, light, ventilation, and atmospheric pollutants is therefore essential to slow down the processes of physico-chemical and biological deterioration (Appelbaum 1991; Lull 1995; Ogden 1999).

As in the case of museums, the buildings that house libraries and archives exhibit different characteristics that influence the microclimate conditions of the spaces in which works are kept. An initial differentiation may be established on the basis of:

—geographic location: in Italy, for example, there is a bioclimatic transition that separates the North and the Apennine regions—characterized by a temperate central-European climate—from the central and southern coastal regions and the islands, which are characterized by a Mediterranean climate (Chapter 6);

Fig. 5.1 *Example of a library in a historic building: the Vallicelliana Library (Rome)* – Photo ICPL

—architectural characteristics: e.g., ancient buildings or monuments—often characterized by rooms with extensive walls and windows (Fig. 5.1)—or modern buildings or buildings constructed for a distinct purpose, with specific areas with assigned use, etc.;
—type of location: e.g., urban and extra-urban areas; proximity to parks or green areas, or to factories or other installations emitting pollutants; marine or freshwater areas, etc.

Other characteristic elements inside buildings are the absence or the presence (in which case, also the type) of:

—climate control and air-conditioning systems (including dehumidification);
—seasonal heating systems;
—air-filtering systems;
—thermoventilators and other ventilation systems.

The various types of (artificial and/or natural) lighting systems must also be taken into account, in terms of general lighting and the illumination inside the showcases, as well as the absence or presence of screens on the windows; these are all factors that can influence the environmental microclimate conditions.

In Italy, libraries and archives with climate control systems that are also equipped with air filtering are still rare. The various rooms and spaces inside libraries and archives have different intended purposes (UNI 10586, 1997; ISO/DIS 11799, 2003); the possible conservation problems vary accordingly, depending also on the location of the particular space within the building (underground, half-underground, in basements, mezzanines, or attics). Specifically, we can distinguish between:

—warehouses or storage areas, in which the works are stored and preserved for considerable periods of time;

—consultation and study areas, designed for the reading and the study of texts as well as the consultation of documents;
—exhibition areas, in which works are exhibited, either temporarily or permanently;
—transit and service areas, in which the works are in transit.

Fig. 5.2 *Different types of bookcases: (a) wooden; (b) compact shelving*
– Photo ICPL

In all of these areas, the bibliographic, documentary, and graphic heritage is stored with different types of holding or shelving systems (cabinets, bookcases, shelving, and containers).

Generally, the cabinets and bookcases that are most frequently found in libraries and archives are made of wood or metal. In historical libraries and archives, the shelving is almost always made out of wood, while in modern buildings the furniture is more often metallic (Fig. 5.2). The different materials out of which cabinets and bookcases are made can determine different microclimate conditions, which, in turn, influence the microenvironment of the cultural goods and therefore also biodeterioration.

Wooden furniture has the advantage of being hygroscopic, which means that it constantly absorbs humidity from the air and releases it, maintaining a precise equilibrium; this prevents possible condensation phenomena, which can, on the other hand, occur with metal shelvings when relative humidity is high. However, wood can be attacked by xylophagous insects, and, in addition, various species are particularly damaging to some organic materials in library, archival, and graphic heritage.

As far as bookcases are concerned, what needs to be taken into account is whether they are open or enclosed, and, if the latter, whether they are enclosed with metallic netting (with more or less open mesh) or completely enclosed with full doors.

Open shelving encourages better air circulation, but fully enclosed shelving has the advantage of protecting books from dust and from everything that is contained within it, and also limits any thermohygrometric fluctuations.

The most recent generation of shelving is the compact shelving unit, which is an enclosed metallic shelving cabinet that ensures optimal book preservation. The use of such units entails an ongoing monitoring of the thermohygrometric parameters (temperature and relative humidity of the air) and of the moisture content of the materials stored within, in order to avoid the development of stagnant damp air, which may trigger biological attacks (Fig. 5.2).

This kind of shelving can be considered the best solution only in environments provided with regularly monitored climate control and filtering systems. However, we should add that, overall, there is no ideal type of shelving; it is therefore necessary to evaluate each case individually and choose the most suitable solution based on the nature of the materials and on their surrounding environment.

If the thermohygrometric parameters are set correctly, the risk of biodeterioration of materials (with the exclusion of damage caused by insects) in libraries provided with climate control systems that include air filtering is practically nonexistent, unless exceptional circumstances occur that would require emergency interventions (Chapter 7, section 1.1a).

Where climate control is not available, the following should be kept in mind:

—the greater the mass of books and/or documents and their compactness, the longer it will take for them to enter into equilibrium with the environment;
—the hygroscopicity of the individual components of the artifacts varies;
—when there is a capillary water rise in the walls, the situation can become particularly critical in the spring/summer when the temperature rises above 20°C and a warm-humid microclimate can be established;
—the scanty aeration can favor condensation;
—the microclimates within the bookcases are influenced by the environment to varying degrees, in relation to their structural characteristics and the materials of which they are made.

5.1.1b *Problems of Biodeterioration*

The different types of materials and containers used to preserve or protect collections, series, or groups of individual items (folders, mats, paper or plastic envelopes) or of items bound in volumes can determine a variation in susceptibility to biodeterioration; consultation and exhibition, which involve human contact, can also contribute to an increase of such phenomena.

Numerous microbiological investigations carried out in national and international archives and libraries have found the presence of more than two hundred species of microorganisms, mostly microfungi and, to a lesser degree, bacteria (Gallo et al. 1999a).

Among the most interesting cases is the study of Montemartini Corte et al. (2003) on the foxing (Chapter 4, section 1.2) on geographical maps dating from the seventeenth to the twentieth century and belonging to the collection of the city of Genoa. First, fluorescence techniques were used to determine biological activity, followed by culture methods to isolate the microfungi from the fragments of paper that fluoresced positively. The species most frequently identified in relation to foxing stains were: *Chaetomium globosum*, *Ulocladium botrytis*, *Cladosporium sphaerospermum*, *Eurotium pseudoglaucum*, *Penicillium chrysogenum*, *P. canescens*, *P. citrinum*, *P. variabile*, *Peziza ostracoderma*, and many colonies of nonidentified yeasts and bacteria.

Microbiological air pollution and the presence of possible sources of infection are the main potential risk factors of biodeterioration; but the role of dust should not be overlooked. Dust—or, more precisely, the particles deposited on the surface of objects—varies considerably in composition and is influenced by environmental, geographic, climate factors, etc. Studies carried out by Florian (1997) on samples of dust from books have shown that dust is often a contributory factor of biodeterioration, not only because its characteristics favor the retention of humidity, spores, and other depositing particles, but also because it provides a source of nutrients for the development of microorganisms and of microfungi in particular. This study also pointed out the organic component of dust as mostly fragments of textile fibers, human and animal epithelial cells, hairs, soot, fragments of dead insects, etc.

For a good maintenance of these environments it is therefore particularly important to remove dust—and consequently any contaminating microorganisms—from library and archival material as well as from the furniture.

As far as public libraries are concerned, dusting is at present an activity regulated by law (G.U.–D.P.R. n. 417-7/1995) (Chapter 7, section 1). Indications on equipment to be used and procedures to be followed for the removal of dust in libraries and archives are specified in the general norms; these are provided by an expert team of the *Istituto Centrale per la Patologia del Libro* (Central Institute for Book Pathology) and of the *Centro di Fotoriproduzione degli Archivi di Stato* (Center for Photo Reproduction of the State Archives) (Aruzzolo et al. 1997). On the other hand, the verification procedures of a correct implementation of the dusting are still in the process of being drawn up.

Recent investigations of the microbial composition of dust were carried out on works on paper (documents, prints, drawings, and books) in three different types of conservation environments: the warehouse of the

Archivi di Stato di Roma (the State Archives in Rome), the consultation rooms in the *Istituto Nazionale per la Grafica* (National Graphic Arts Institute), and the *Sala Borromini* of the *Biblioteca Vallicelliana* (the Borromini Hall of the Vallicelliana Library in Rome) (Maggi et al. 2000; Gallo et al. 2002; Sclocchi et al. 2002). These investigations have shown the potential risk represented by dust in relation to the airborne fungal spores deposited in it; living spores were indeed found in the dust, albeit with variations according to the type of environment and conditions of preservation. Some of the species identified were those that are often isolated from paper with varying degrees of biodeterioration.

Among fungal species isolated from the dust in storage and preservation environments, and cited in the literature as being frequently isolated on paper, are: *Alternaria alternata, Aspergillus flavus, A. fumigatus, A. versicolor, Chaetomium globosum, Cladosporium herbarum, Epicoccum purpurascens* (syn. *E. nigrum*), *Penicillium brevicompactum, P. citrinum, P. miczynskii, P. oxalicum, P. puberulum.*

An additional point to be kept in mind relates to the type of space designated for storage and preservation, given the extensive amount of cultural property (books, archival materials, and works on paper) that must be preserved over time. Because of the volume of materials, the selected spaces are often warehouses that are sometimes located in damp environments, with poor ventilation, and partially or completely underground; all of these are conditions that facilitate processes of biodeterioration.

This state of things is often confirmed during on-site visits to these types of premises. One such instance involved the storage areas of the *Biblioteca Centrale Giuridica del Ministero di Grazia e Giustizia* (Central Law Library of the Justice Department) (Gallo and Valenti 1999). The transfer operations of the library holdings to a new site revealed that a good part of the collection—placed on metal shelving and in unsuitable conditions—had undergone microbial attack, especially by microfungi. The magnitude of the phenomenon was determined by the thermohygrometric conditions, aggravated by an infrequent monitoring of both the prem-

ises and the state of preservation of the books. It therefore proved necessary to completely disinfect the mass of volumes in those areas in which the infection had occurred.

The problem of biodeterioration results from a complex series of interactions between biotic and abiotic factors that occur in the storage and preservation environments. For this reason, a good maintenance requires a careful assessment of the state of conservation of the materials, making prevention the only real method of preservation of cultural property for future generations (Chapter 7).

5.1.2 Museums

by Maria Pia Nugari, Oriana Maggi, and Anna Maria Persiani

5.1.2a *Characteristics of the Environment and of the Materials*

Museums are depositories for objects of historical, artistic, or scientific value, which are collected, preserved, and displayed for purposes of study, education, and enjoyment by the public. The *Testo Unico* (the legislative applicable text) of regulations concerning cultural and environmental heritage (D.L. 490/99 art. 1) classifies a museum as "a structure (however it may be denominated) organized for the preservation, the use and fruition, and the public enjoyment of collections of cultural heritage" (G.U. n. 229/L. 27 December 1999).

For historical reasons, most Italian museums are located in historic buildings, public structures or palaces built by the nobility, villas and palatial residences, royal palaces, ecclesiastical complexes or convents, all of which are strongly entrenched within both the state and the religious "cultural heritage systems" and present throughout major and minor centers around the country. Only rarely, therefore, are exhibition spaces and buildings suitably designed for conservation, and the factors necessary for a correct conservation of the works of art are not always maintained under strict control.

The microclimate conditions inside museums—similarly to libraries and archives—are strongly influenced by the location and geography of the site and by the architectural characteristics of the building. The microclimatic parameters and the quality of the air also vary greatly, often even within the same building, in relation to the presence and the type of heating systems, climate-control or air-conditioning systems, ventilation, and air-filtration systems, to artificial lighting installations, and also to the specific location within the building of the areas under consideration (top floors, cellars, attics, etc.).

Of critical importance when determining the level of quality of a museum is also its management; all the operations that occur within it can indeed interact with the internal environmental conditions. The opening hours and the conditions of public access, the activities of cleaning and maintenance of the objects, the installation of temporary exhibitions, cultural events, etc., can all induce fluctuations in thermohygrometric parameters and in the levels of chemical and biological pollutants (De Guichen 1980; Thomson 1986; Stolow 1987; Brimblecombe 1990; Camuffo 1990; Mukerji et al. 1995; Artioli et al. 2000).

It is therefore fairly common to find, especially in historic buildings, at-risk conditions for degradation (also biological) of the collections. Indeed, condensation phenomena as well as infiltration and capillary water rise are frequently encountered, with a consequent increase in relative humidity; also, ventilation is often insufficient or inadequate, especially in areas used for storage. Dangerous thermal fluctuations are often registered, resulting from the particular characteristics of exposure of the building, or the presence of skylights or large glass surfaces in walls or windows, or because of nonefficient window frames, etc. The choice as to what particular use is assigned to each space—i.e., exhibition rather than storage or warehouse, or use for a temporary exhibition or a cultural event, etc.—is often made without taking into consideration how suitable it is in terms of the conservation requirements of the objects to be exhibited.

Sometimes, particularly risky situations are created for the very works that are the most sensitive and would therefore require greater care and protection; this happens, for instance, when objects are exhibited in showcases for reasons of preservation and security. The inadequate design of these showcases—sometimes finalized to the achievement of only one specific objective—leads to particularly dangerous situations. For instance, an incorrect choice of lighting system and an ineffective monitoring of the thermohygrometric conditions within the showcase can have negative effects on the preservation of the objects and increase their susceptibility to biodeterioration. It should also be noted that most existing showcases do not always provide the objects with sufficient protection from dust and often have a purely aesthetic function.

The complexity and the variety of objects that are in museum collections make it very difficult to identify and define in absolute terms the limits and ranges of the environmental parameters—at both the critical and the optimal levels of the scale—that are necessary for good preservation; as a result, the objects often find themselves in conditions that are not specifically determined for the characteristics of each category (Thomson 1986; Cassar 1995; Brown et al. 2002). The definition of categories of artifacts with different "conservation affinities" (as cited in DL 112/98 art. 150, G.U. n. 244, 2001) is very difficult; indeed, the definition of different classes of objects has to consider not only the constituent materials, but also the state of preservation, the type of collection, its value, etc. (Chapter 7). A further complication in the management of conservation is the vast range of materials that can be found—in various combinations—in the composition of some objects, as for instance in the case of ethnographic collections.

Moreover, it is not a rare occurrence to find objects with very different conservation requirements housed together in the same space due to particular historical or cultural requirements. It is sometimes possible to find ancient panel paintings exhibited next to canvas paintings, which are less susceptible to variations in temperature and humidity; and stone works, which are resistant to biodeterioration in indoor environments, can be found next to archaeological finds of an organic nature, which are highly susceptible to biological degradation (Fig. 5.3).

Fig. 5.3 *Example of a museum with a mixed collection*
— Photo M. Ramolaccio

5.1.2b ***Problems of Biodeterioration***

The complex interrelation of situations encountered in museums may, at times, result in environmental parameters reaching values that allow the development of microorganisms on the objects. Biological deterioration of museum objects is mainly linked to the development of heterotrophic microorganisms, especially bacteria and fungi, and organic artifacts are most susceptible to degradation as they offer a more accessible source of nutrients. The environmental conditions generally found in exhibition rooms and in showcases in museums are such that the selected species are mostly mesophilic and xerophilic, since it is unusual to find temperature extremes and RH values higher than 65% in such environments. An exception to this are storage areas, where the humidity levels are sometimes high and the ventilation poor—both conditions that facilitate the development of a wider variety of species with broader ecological requirements.

In addition to increases in relative humidity, other factors that significantly influence microbial development are: exchanges in moisture between the environment and the collections; fluctuations in temperature; the levels of natural and/or artificial illumination; the physical characteristics of the surface of the objects; air movements, etc. Indeed, it is known that all of these parameters interact with the capacity of a material to absorb and release moisture, the element that is at the root of the real risk of biodeterioration; it is thus possible to have conditions with rather high values of RH without the artifacts suffering any biodegradation phenomena. For instance, the hydrorepellence of the paint layer, induced by the presence of the finishing varnish on the oil painting, makes it more resistant to biological degradation even in relatively damp conditions, but often to the detriment of the support or lining canvas where, indeed, the attack usually begins. The possibility of an object being subjected to biological colonization is linked not only to the availability of water, but also to the chemical nature of the available nutrients, the pH of the materials, the dust deposits, and the gaseous composition of the air (the concentrations of carbon dioxide or other gases, as occurs at times in showcases).

As a general rule, we can say that if the environmental conditions are suitable for the development of microorganisms, damage to artifacts occurs differentially: cellulosic materials first (such as books, textiles, paintings, furniture, wooden sculptures), being the most susceptible to biodeterioration; proteinaceous materials second (parchment, leather, mummies, etc.); and last, synthetic and other materials (stone, ceramics, glass, etc.).

In the case of artifacts of an organic nature, the microorganisms that are mainly responsible for problems of biodeterioration are fungi and bacteria. Of these, microfungi—and in particular deuteromycetes—are certainly the most common, because many species have a great capacity for adaptation, even to conditions of very low RH. The genera and species involved in degradation phenomena can be specific to the type of main material component of the object (Mukerji et al. 1995; Kramer et al. 1998; Sbaraglia et al. 1999). Tables 5.1 and 5.2 list, respectively, the fungi and the bacteria most frequently isolated on the artifacts and in the air in museums.

Bacteria are found less frequently than fungi because of their water requirements, which usually tie them to damper conditions than those found within museums. Among the various groups of bacteria, the actinomycetes deserve particular attention, both because of their biodeteriogenic potential for materials of a cellulosic and proteinaceous nature and also because of their greater capacity for environmental adaptation.

Table 5.1 Most frequently identified fungal species from the artifacts and from the air in museums (from Marcone et al. 2001; Valentin 2003)

Alternaria tenuis	C	P	A
A. solani	C		A
Aspergillus flavus	C	P	A
A. fumigatus	C	P	A
A. glaucus	C	P	A
A. nidulans	C	P	A
A. niger	C	P	A
A. tamarii	C		A
Botrytis cinerea			A
Cephalosporium sp.	C		
Cladosporium herbarum	C	P	A
C. cladosporioides	C	P	A
Chaetomium globosum	C		A
Curvularia lunata	C		A
Epicoccum nigrum			A
Fusarium roseum	C	P	A
F. oxysporum	C	P	A
Geotrichum sp.	C	P	
Gliocladium catenulatum	C	P	
Humicola grisea	C		
Memnoniella sp.	C		
Microsporum sp.	C		
Myrothecium verrucaria	C	P	A
Mucor racemosus	C	P	A
Neurospora sp.	C		
Ophiostoma sp.		P	
Paecilomyces variotii	C		A
Penicillium brevicompactum	C	P	A
P. commune	C	P	A
P. frequentans	C	P	A
P. notatum	C	P	A
Pestalotia oxyanthi	C	P	A
Phoma betae	C	P	A
P. pigmentovora	C		
Rhizopus nigricans	C	P	A
Scopulariopsis acremonium	C		A
S. brevicaulis			A
Sporotricum pulverulentum	C		A
Stachybotrys atra	C	P	A
Stemphylium botryosum	C		A
S. vesicarium	C	P	A
Trichoderma viride	C		A
Ulocladium consortiale	C		A
Verticillium nigrescens	C		A

Key: C = cellulosic materials;
    P = proteinaceous materials; A = air

Table 5.2 Most frequently identified bacterial species from the artifacts and from the air in museums (from Valentin 2003)

Acinetobacter sp.	C	P	A
Aeromonas hydrophila	C	P	A
Bacillus subtilis	C	P	A
B. cereus	C	P	A
B. circulans	C		A
B. mesenthericus		P	
Cellfalcicula sp.	C		
Cellvibrio sp.	C		A
Clostridium sp.	C	P	
Corynebacterium sp.	C	P	
Microbacterium sp.	C		
Micrococcus luteus	C	P	A
M. roseus	C	P	A
Nocardia sp.			A
Proteus vulgaris		P	
Pseudomonas aeruginosa	C	P	A
Sarcina sp.		P	
Serratia marcescens	C	P	A
Sporocytophaga myxococcoides	C		
Streptococcus viridans	C	P	A
Streptomyces albus	C	P	A
S. fradiae		P	A

Key: C = cellulosic materials;
    P = proteinaceous materials; A = air

The conservation history of a work of art (conservation interventions, changes of location, conditions of exhibition, etc.) can also modify its susceptibility to biodeterioration, inducing physical and chemical alterations in the constituent materials. As an example, because the visual enjoyment of objects in a museum is heavily dependent on how they are lit, it is possible to find situations of excessive or unsuitable illumination; this can encourage the development of photooxidation phenomena of the materials, especially in artifacts of an organic nature. By modifying the chemical structure of the molecules through the partial depolymerization of compounds such as cellulose, proteins, etc., these phenomena increase their susceptibility to microbial attack, enabling noncellulolytic and nonproteolytic species to also colonize the object. Another example is the lining process undergone by canvas paintings with the use of a starch-based glue, a molecule that is easy to dissolve and to metabolize, thus increasing the range of microorganisms with the capacity to damage the support of the painting.

The deteriogenic potential of fungi and bacteria varies in relation to the species and to the artifact under consideration and its state of conservation, in addition to microclimate conditions. Recently, it has been suggested that a database of the most dangerous species be compiled, which could provide a useful consultation system for defining the level of risk of biodeterioration present for an object at an early stage of contamination. This, in turn, would be most helpful in determining future procedures, such as the suitability of a particular disinfection, the type and extent of the treatment, whether or not the object under attack should be isolated, etc. (Valentin 2003).

When studying the museum environment, we should remember that here the main vehicle for spreading biodeteriogenic microorganisms, and in particular fungi, is the air; aerobiological investigations therefore take on particular importance when determining how infections propagate and how to prevent biodeterioration (Chapter 7, section 3) (Pasquariello and Maggi 1998, 2003; Sbaraglia et al. 1999; Artioli et al. 2000).

5.1.3 Churches and Crypts

by Anna Maria Pietrini, Sandra Ricci, and Maria Pia Nugari

5.1.3a *Characteristics of the Environment and of the Materials*

The main function of churches is to be places for worship, but they are also often places of cultural interest, because of the architectural, historical, and artistic merit of the buildings themselves and also because of the works of art that are housed within them.

Because of their functions and their different structural and architectural characteristics these buildings exhibit complex microenvironmental situations (Nugari et al. 1998). Generally, churches have reduced exchanges of air with the exterior, mostly limited to the entrances used by visitors and churchgoers;

this situation leads to a certain stability in the internal microclimate conditions, which are not much influenced by the daily and seasonal variations in external temperature and humidity. Usually, the primary function of the windows, generally limited in number, is to provide light or for decoration, and the glazing is usually fixed. From a thermal point of view, the height of the ceilings, often great in churches, allows air stratification with temperature gradients that can change over time in relation to external fluctuations (Camuffo 1998) (Fig. 5.4). Churches with heating installations are rare, although in colder regions the practice of placing heating systems under the benches or beneath the floors is increasingly frequent.

Churches and crypts are frequented by churchgoers participating in liturgical rites and by visitors; access to visitors is usually controlled, and limited to particular hours of the day. The presence of a large number of visitors can substantially modify internal microclimate conditions, in addition to supplying not only biological contaminants (transport of bacteria and fungi) but also chemical ones (increase in the levels of carbon dioxide and water vapor).

Variations in the microclimate can also be induced by communication with other adjacent indoor environments (sacristies, crypts, underground spaces, offices, etc.) and outdoor environments (cloisters, courtyards). The existence of older, underlying structures (archaeological sites, catacombs, tombs, etc.) can be an additional factor influencing the thermohygrometric conditions of the church; indeed, in comparison to the upper spaces, underground spaces exhibit greater thermal inertia and higher levels of relative humidity (Chapter 5, section 1.4). These microclimatic characteristics make them, when compared to the upper church, cooler in the summer and warmer in the winter: this causes alternating air flows because of the phenomenon of thermal inversion between the two levels. Such a condition becomes particularly dangerous when the church is cooler and the currents of warm-humid air rise from the underlying spaces. The formation of air currents is facilitated by the presence of communicating structures with

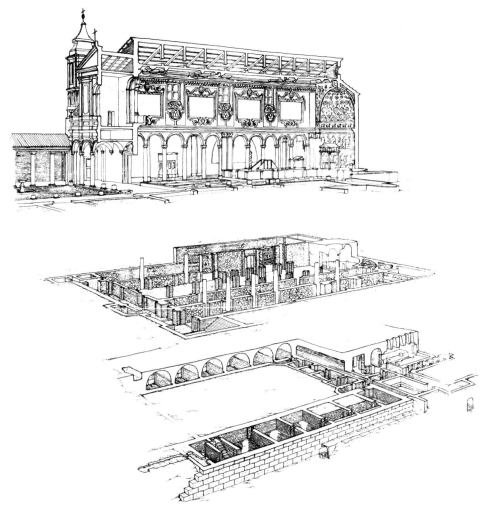

Fig. 5.4 *Example of a church with complex microenvironmental problems (Basilica of San Clemente, Rome) (drawing by V. Cosentino, from Guidobaldi and Lawlor 1990)*

these underground spaces, such as stairwells, openings, etc. These air currents constitute a risk factor in conservation because, in addition to inducing microclimate fluctuations, they can transport biological contaminants present on the floor, as was shown in a study carried out in the crypt of the Duomo of Anagni (Nugari et al. 2003). With regard to the nature of the different materials found in these environments, artifacts in stone are predominant, in addition to plasters and frescoes, but organic materials are not rare, such as canvas and panel paintings as well as wooden ceilings, which influence not only the microclimate but obviously also the biodeterioration phenomena.

A particular type of artifact that is frequently encountered in churches is the sarcophagus, in which the internal microclimate conditions can vary according to the constituent materials and the shape as well as the placement and are influenced to varying degrees by the surrounding environment. For example, the microclimate measurements carried out in the red porphyry sarcophagus of Frederick II confirmed that the conditions within the sarcophagus represented a system that was fairly isolated from the surrounding environment, with a lower RH than that found in the Cathedral of Palermo where it is located (Cacace 2002).

5.1.3b *Problems of Biodeterioration*

For this type of indoor environment, too, fluctuations in humidity are the microenvironmental factors that most influence the development of biodeterioration phenomena. The relative humidity of the air varies according to the height of the ceiling, the ventilation, and the phenomena of capillary water rise and condensation, and organisms are selected according to their water requirements.

Within the category of heterotrophic microflora, the group most frequently found on damp walls, and in particular in crypts, is that of the actinomycetes; these often deteriorate wall paintings and, because of their halophilic characteristics (preference for high salt levels), are recurrent in areas of walls where capillary water rise along with efflorescence of salts are present. They are also linked with the presence of damp earth in contact with the surfaces, which contributes salts and organic substances; in addition, because they are chemoheterotrophs, they do not require the presence of light (Bettini et al. 1982; Agarossi et al. 1986; Nugari et al. 1993c). Equally widespread on wall paintings are the damages linked to colonization by microfungi, especially by Dematiaceae; these usually manifest themselves by the presence of brown stains (Fig. 5.5), which were the earliest forms of biological deterioration to be identified in these contexts (Giacobini and Lacerna 1965). The most frequently found species are those that have adapted to temperatures that are not high and to substrates lacking in organic nutrients.

When church structures are also exposed to a sufficient amount of natural and/or artificial light, the appearance of forms of deterioration linked to the development of photosynthesizing microflora—constituted primarily by cyanobacteria and microalgae —is facilitated (Tomaselli et al. 1979; D'Urbano et al. 1998). The levels of natural light in a church are generally rather low, and artificial lighting is linked to the presence of churchgoers or visitors. A situation of particular biological risk arises when the church houses works of great artistic interest that are often exposed to high levels of illumination—although the duration of the lighting is often timed in these cases. Nevertheless, certain taxa of cyanobacteria have been found to be capable of developing with very low levels of light and, sometimes, to survive even long periods of darkness. In the Lower Basilica of San Clemente in Rome, for example, cyanobacteria proliferated on damp wall paintings with levels of illumination of 6–9 lux, given out from the lamps that had been installed for the visual appreciation of the site (Ricci and Pietrini 1994b).

In certain specific conditions—such as those found in the rock churches near Matera (Italy) (below in this chapter, section 3.2)—the high levels of humidity are compounded by the presence of solar irradiation; this condition gives rise to a strong presence of photoautotrophic microflora distributed over the painted surfaces, in strict correlation not only to the levels of humidity, but also to the parameters of intensity and length of exposure to light (Pietrini and Ricci 1993; Caneva et al. 2002).

The lighting of candles—which now has only devotional ends but which in the past also

Fig. 5.5 *Development of fungi on a fresco by Giotto caused by* Cladosporium herbarum *(Lower Basilica of San Francesco in Assisi)* – Photo ICR

had a lighting function—is a further element to be considered; a large number of lit candles can lead to a rise in temperature in adjacent areas and create conservation problems linked to the deposit of soot on surfaces. The latter can enrich the substrate with nutrients, leading to the formation of fungal colonies even on inorganic works such as wall paintings; the resulting small brown spots visible on their surfaces are mostly due to the presence of the family Dematiaceae, especially of the genus *Cladosporium* (Fig. 5.5). It is also for these reasons that traditional candles are increasingly being replaced with electrical systems that simulate the look of real candles.

A strong influence on microclimate within churches is exercised by the materials employed in the construction of ceilings, laying of floors, and creation of church furnishings; it seems obvious that there is a difference between a stone church that is bare and a church with a coffered ceiling, wooden choir stalls, altarpieces, canvas and panel paintings, etc. Stone walls provide cold surfaces on which condensation occurs, while ceilings and wooden furniture can have a "buffering" effect against increases in environmental humidity. In the latter case, however, organic artifacts may reach such levels of water content that they will instead encourage the biodeterioration of materials linked to the development of heterotrophic microorganisms, fungi, and actinomycetes.

In the degradation of wooden, woven, and paper artifacts kept in churches, fungi—and in particular deuteromycetes—are the most commonly found because of their great adaptability, even to conditions of low availability of water—this group does indeed include a considerable number of xerophilic species, such as those belonging to the genus *Aspergillus*. In conditions of high RH and contact with the soil, various cellulolytic and ligninolytic species, such as *Serpula lacrymans* and *Caniophora puteana*, can develop; because of the presence of wooden structures (for instance, beams) either behind or above them, these fungi can diffuse onto wall paintings or stuccos by forming a vegetative mycelium and rhizomorphs (Plate 17) (Blanchette et al. 1999).

Cases of fungal attack on canvas paintings are those recurring most frequently. Paintings placed above altars and in chapels are often inserted within stucco, wood, or stone frames, in niches constructed especially for them; this means that behind the works there is little ventilation accompanied, at times, by high levels of humidity linked to the water content of the wall on which they are placed. These conditions, which are ideal for the development of fungi, can give rise to dangerous degradation phenomena of the canvas support, which—in the initial phases of development—are not visible on the paint surface (Plate 18).

Although this topic is beyond the scope of this volume, it is important to remember that the various organic materials employed in the furnishings, in the wooden architectural structures, in the works of art, or in the objects of worship can also be degraded by animal organisms such as arthropods (thysanura, coleoptera, lepidoptera, hymenoptera, isoptera) and rodents.

Materials made of metal can also be present in a church in the form of frames, ostensoria, showcases. These are less prone to biological attack, although in certain situations, deterioration can take place that is linked to particular groups of bacteria (sulphur bacteria) able to produce oxidation of metals.

Aside from specific cases of very damp churches—such as those built in hypogean contexts—it is important to add that it is nonetheless rare to find situations of extensive microbial development inside churches, partly because the level of care of the works conserved within them is usually high; this means that there is good control of the instances of biodeterioration, which, on the whole, are either prevented or identified in the early stages of their development. Usually, the phenomena of biodeterioration are linked to accidental events (such as damage to gutter and water drainage systems, etc.) or environmental disasters (earthquakes, floods), which cannot be anticipated either as an event or in the gravity of their effects. However, the case of churches closed for worship in which no preventive measures are taken is becoming ever more frequent.

As far as sarcophagi are concerned, the microbial communities found within them are very diverse in their composition, although the absence of light excludes the presence of phototrophic microorganisms such as cyanobacteria and algae. The nature of materials preserved inside the sarcophagi and the composition of the indoor air can strongly influence the species present; these can be heterotrophic fungi and bacteria, both proteolytic and cellulolytic. Among the bacteria, the actinomycetes are predominant, as their growth is facilitated by the presence of calcium salts linked to the bones within the sarcophagi. The microbial species inside sarcophagi are by and large mesophilic, and the absence of oxygen—characteristic of the deeper areas if not of the entire environment in the sarcophagus—can encourage the development of anaerobic microorganisms (Arai 1984, 1987; Abdulla et al. 1999; Blanchette et al. 1999).

The state of conservation of the materials can therefore be relatively good in some cases, while in others the materials can be completely degraded. In the sarcophagus of Frederick II (Palermo Cathedral), the microbial flora was relatively limited, composed mostly by a few fungi belonging to the genus *Penicillium*, and by gram-positive bacteria, consisting mostly of actinomycetes. The materials seemed only partially degraded, but it is likely that openings of the sarcophagus in past centuries played a considerable role in the state of conservation. It should in any case be understood that the opening of sarcophagi is at all times an extremely critical moment and presents high risks (Tarsitani et al. 1998; Nugari et al. 1998, 2002a, 2002b).

The complex of the Basilica of San Clemente in Rome is a good example of the different and interrelated problems that can be encountered in churches and crypts: it includes the superimposition of buildings belonging to different eras that communicate with one another in such a way as to create a complex environmental situation, which has facilitated the development of various groups of microorganisms on the walls and on the frescoes. The building complex has three levels: the Upper Basilica, which is medieval and is situated at street level; the paleochristian Lower Basilica, which is situated at a depth of about six meters; and a third, even deeper, hypogean level, consisting of structures dating from Roman times. Because of its underground location, the latter

level shares characteristics with the types of environments that will be described in the next section. Several campaigns of investigation, aimed toward a better understanding of the biodeterioration of the complex, have shown the presence of a vast array of microorganisms (Table 5.3); they also demonstrated how the microflora was most abundant in the deepest areas, becoming progressively less abundant moving up, as a result of the lower availability of moisture in the materials and in the air (Agarossi et al. 1986; Ricci and Pietrini 1994b; Pietrini et al. 1999).

5.1.4 Tombs, Catacombs, and Other Hypogean Environments

by Patrizia Albertano, Clara Urzì, and Giulia Caneva

5.1.4a *Characteristics of the Environment and of the Materials*

The hypogean environments under consideration here include all those archaeological sites that are situated—either naturally or for historical reasons—below ground level.

In addition to tombs and catacombs (hypogean funeral chambers), we will also be considering monumental complexes (Roman in particular), mithraea, and places of esoteric and pagan worship. These subterranean sites are characterized by a widespread presence of stone artifacts and wall structures that circumscribe the archaeological sites themselves and the works contained within them; these include, for example, marbles, stuccos, mosaics, wall paintings used as decorations for the sites, statues, sarcophagi, funerary objects, and other furnishings. The substrates are often materials of high porosity that are easier to excavate and to work with, such as tuffs and calcareous arenites.

The size of the hypogean space is an important variable from the point of view of the stability of microclimate conditions; indeed, the situations that characterize single isolated tombs—even when they are part of burial complexes such as Etruscan necropolises, colombaria, mithraea, and nymphaea—are different from those found in extensive hypogea. This is

Table 5.3 List of autotrophic and heterotrophic microorganisms found in the course of several investigations in the Lower Basilica of San Clemente in Rome (from Agarossi et al. 1986; Ricci and Pietrini 1994b; Pietrini et al. 1999; internal note, Istituto Centrale del Restauro)

Heterotrophic Bacteria	Fungi	Cyanobacteria	Algae
Streptomyces albus	Acremonium camptosporum	Aphanocapsa grevillei	Chlorococcum sp.
S. aurecolorigens	Acremonium murorum	Chlorogloea sp.	Gloeocystis sp.
S. aureomonopodiales	Acremonium sp.	Chroococcus minor	Navicula gallica
S. babili	Alternaria sp.	Leptolyngbya sp.	Stichococcus bacillaris
S. cinereoruber	Aspergillus ruber	Oscillatoria sp.	
S. cyanocolor	Aspergillus versicolor	Phormidium sp.	
S. fulvoviridis	Cladosporium macrocarpum	Plectonema sp.	
S. gougeroti	Cladosporium sp.	Scytonema julianum	
S. griseoalbus	Penicillium chrysogenum	Spelaeopogon lucifugus	
S. minoensis	Penicillium lilacinum	Synechocystis sp.	
S. naganishii	Penicillium sp.		
S. olivaceus	Pullularia sp.		
S. parvulus	Verticillium lateritium		
S. pilosus	Verticillium sp.		
S. pyridomyceticus			
S. resistomycificus			
S. spiroverticillatus			
S. tendae			
S. umbrosus			

the case with some imperial Roman complexes that are still underground, with some prehistoric caves, and with catacombs, i.e., those subterranean cemeteries dug out by Christians or by Jews between the first and fourth century, which are characterized by a complex and articulated system of tunnels and mounds used for burial.

In hypogean environments, the characteristics of the terrain above (agricultural or urban), the degree of pollution in the atmosphere, and the depth at which the hypogeum

lies, all influence the conditions of humidity, temperature, and quality and circulation of the air, to which the works are exposed within it, thus defining its microclimate and the levels of internal pollution. In the case of certain volcanic rocks, the emission of toxic gases has been noted (for instance, carbon monoxide and dioxide) as well as of radioactive ones (for instance, radon), which are linked to volcanic activity.

The high and constant level of humidity (RH > 70% and often greater than 90%) is one of the characteristics of subterranean environments, and it is due in part to the actual moisture in the ground diffusing by capillarity from the interior, in part to water infiltrations from above, and, to a lesser degree, to the presence of visitors (Groth and Saiz-Jimenez 1999). The higher the humidity, the easier it is for water vapor to condense, even with as little as 1–2°C difference between the temperature of the surfaces and that of the air. In contrast, the evaporation of water due to the rise in temperature of the air, caused either by marked seasonal variations of the external climate or else by the presence of visitors (which also induces rises in the concentration of carbon dioxide), results in widespread carbonatation, which may be compounded by the soot from the oil lamps or the candles that were once used and, more generally, by any particles present in the air.

Normally, the temperature is relatively constant, between 10 and 18°C, but when influenced by the influx of visitors it can rise to above 20°C in some cases (Hoyos and Soler 1993; Pantazidou et al. 1997; Ariño et al. 1997b). Environmental conditions are thus established in which light, humidity, temperature, input of nutrients, and nature of the substrate all become facilitating and determinant factors for microbial colonization (Agarossi et al. 1985; Pantazidou et al. 1997). Direct sunlight is normally absent, but a certain amount of radiation, although weak, enables the development of phototrophic organisms in areas adjacent to entrances and exits or, deeper inside, in areas where there are openings to the surface above, such as wells, skylights, air vents, and other outlets. On the other hand, the colonization of surfaces by photoautotrophs inside these hypogean environments is strictly limited to areas near artificial sources of light, which produce sufficient levels of light for photosynthesis to take place (Photosynthetic Available Radiation, PAR) (Fig. 5.6, Plate 2). The quantity of light is almost always extremely low (<2 µmoll photons m^{-2} s^{-1})—which is approximately three orders of magnitude less than full sunlight—while the quality of the light (i.e., its spectral composition) plays a considerable role in determining the development of particular groups of photoautotrophs: emissions peaking in the blue and red areas of the spectrum encourage the development of eukaryotic microalgae, green algae, and diatoms, while emissions in the green and orange-red part of the spectrum favor cyanobacteria (Bruno et al. 2001). In the study of communities dominated by photoautotrophs, it is therefore crucial to select appropriate methods in order to assess the ability of the different kinds of biofilms to use light (Albertano and Bruno 2003). The development of certain microbial species such as fungi and actinomycetes may occur even in the absence of light if nutrients are available (Albertano and Urzì 1999) (Plate 20).

The concentration of CO$_2$—already quite high because of the slow air circulation in subterranean environments—rises even more because of the presence of visitors, reaching levels that are sometimes ten times higher than those normally found in the atmosphere, thus favoring the development of photoautotrophs.

Among physical factors, serious damage to the stability of these environments can be caused by seismic activity, vehicular traffic in the area, and most importantly, construction work taking place above the hypogea.

Fig. 5.6 *General schematic representation of the distribution of phototrophic biofilms in the form of variably colored patinas, from intense green (a), to blue-green (c and d), to brown (b), within a radius of about one meter from the lamp, inside an arcosolium in Roman catacombs*

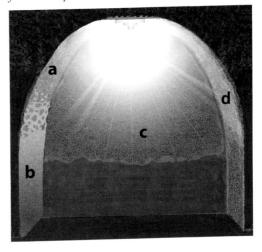

5.1.4b *Problems of Biodeterioration*

The humidity of the air, the condensation of water on the walls, and the porous nature of the substrates are the primary causes of the development and proliferation of bacteria and fungi and—if light is present—of cyanobacteria and algae, all of which become visible as colored stains and patinas of white, black, gray, and green hues. These biological colonizations are called biofilm; they may be characterized either by a prevalence of photoautotrophic microorganisms such as cyanobacteria, joined by populations of heterotrophic bacteria, or else by a composition where there is a predominance of gram-positive bacteria and, in particular, of actinomycetes. Also frequently present, only a few centimeters away from artificial sources of light, are "crowns" of mosses, mostly protonemata and gametophytes. On the other hand, the presence of lichens is very rare, limited not only by the absence of light, but also by the absence of ventilation and therefore mostly present in areas that are in direct communication with the outside environment.

Several authors have done in-depth studies on the biodeterioration caused by the growth of cyanobacteria and algae in damp sites that are lit artificially, such as caves, hypogean churches, tombs, and catacombs (Lefevre 1974; Tomaselli et al. 1979; Albertano 1991, 1998; Albertano et al. 1994, 1995; Ariño et al. 1997b; Bellezza and Albertano 2003; Hernández-Marine et al. 2003a). Of the approximately 350 taxa (i.e., distinct species or strains) listed for these environments with very low levels of light, at least two hundred belong to the cyanobacteria, followed by green algae and diatoms, and by a few other eukaryotic species of algae. Among these taxa, some authors distinguish between troglobites (i.e., obligated cave organisms) and troglophiles, which, although they grow in this habitat, develop optimally in other environments (Hoffmann 2002). In Etruscan tombs and in some Roman hypogea, species of *Nostoc* have frequently been identified, while in the catacombs the most common species of coccal cyanobacteria belong to the genera *Gloeothece* and *Eucapsis*, and the prevalent filamentous species are represented by *Leptolyngbya* spp., *Scytonema julianum*, *Fischerella* sp., and *Loriella osteophila*. All these cyanobacteria are highly sciaphilic, i.e., by their adaptation to conditions of extremely low light (Albertano and Bruno 2003).

The photosynthetic capacity of species characteristically found in hypogean environments is in any case relatively high, especially in terms of efficiency; the recording of oxygen evolution from biofilms and from culture strains showed that during the light period an increase in pH values took place that was sufficient to induce a change in the equilibrium of bicarbonates, which in turn was sufficient to favor the precipitation of carbonates (Albertano et al. 2003). Indeed, some heterocystous cyanobacteria, such as *Scytonema julianum*, are called "calcifiers" because they are characterized by the presence of calcium carbonate crystals on the surface of the polysaccharidic sheaths that surround the trichomes. These species can be particularly biodeteriogenic because they are able to mobilize Ca^{2+} from the colonized substrates. Another important aspect for the understanding of the biodeterioration produced by biofilms is the study of the biopolymers that are excreted by microorganisms and that form the capsules, the sheaths, and the mucilaginous matrix of the biofilm (Aboal et al. 1994; Albertano et al. 2000; Albertano and Bellezza 2001; Roldàn et al. 2003).

When considering colonization by algae and cyanobacteria, it is important to remember that once colonization phenomena are established, simply reducing the lighting levels is not always a sufficient control measure, since many examples have shown that these phototrophs will adapt to conditions of heterotrophic growth. This was the case in the caves of Lascaux (France), where the Paleolithic paintings were colonized by the green alga *Bractaecoccus minor*; here, although both lighting and visitor access were stopped, biocide treatments still proved necessary (Lefevre 1974). Analogous phenomena were also observed in the *Domus Aurea* (Nero's imperial villa), which is now below ground level: the widespread presence of algal patinas on the frescoes was still present many years after the removal of light sources (Albertano and Grilli Caiola 1989; Caneva, unpublished).

In the context of studies on the biodeterioration caused by autotrophs, the role of chemoautotrophic bacteria—macroscopically not very visible—has not been researched much. An exception is the case of the subterranean vaults of the Neopythagorean Basilica of Porta Maggiore in Rome, in which Agarossi et al. (1985) found that gray patinas had formed in the absence of light; these were linked to the presence of sulphur-oxidizing autotrophic bacteria and actinomycetes, which various authors found are the most frequently encountered heterotrophs in hypogean environments (Agarossi et al. 1985, 1988; Agarossi 1994; Groth and Saiz-Jimenez 1999; Laiz et al. 2000a). In general, the gravity of the attacks by these microorganisms can be linked to their capacity to grow extensively over surfaces, to produce extracellular acids and pigments, and to favor the precipitation of mineral phases. In the majority of cases studied biodeterioration was associated with the presence of microbial species of the genera *Streptomyces* and *Nocardia*. Bacteria belonging to these genera, as well as to the genus *Micromonospora*, were also isolated in Egyptian tombs (Abdulla et al. 1999).

The microbial communities found in caves are also by and large made up of actinomycetes, which find an ideal habitat in these environments, particularly when the caves are not deep and have wide openings toward the outside. In these cases, the exchanges of air, the alternation of humid and dry cycles, and any available irradiation by the sun lead to variations in the composition of the microbial communities, with a preponderance of gram-positive bacteria, including strains from the genera *Bacillus* and *Micrococcus*, as reported in the research conducted on the cave paintings of the Sierra de Cazola (Spain) (Laiz et al. 2000a). In the caves of Altamira and La Pasiega (Spain)—and also in the caves at Lascaux (Lefevre 1974)—strains of actinomycetes were isolated from the ground, the walls, and the roof vaults (Arroyo and Arroyo 1996). In addition, Somavilla et al. (1978) isolated several strains of the genus *Streptomyces* from the samples taken from dripping water. In the caves of Tito Bustillo and Altamira (Spain), the following strains were isolated: *Streptomyces, Amycolatopsis, Brevibacterium, Aureobacterium, Nocardia, Nocardioides, Rhodococcus*, as well as strains belonging to the family of the Micrococcaceae

(Groth and Saiz-Jimenez 1999). Still in the caves at Altamira, Cañaveras et al. (1999) found microbial communities associated with deposits of hydromagnesite and aragonite; they demonstrated that the predominant strains, which belonged to the genus *Streptomyces*, were able to precipitate calcium carbonate in laboratory cultures.

In the catacombs of Milo (Greece) (Pantazidou et al. 1997), the presence of fungi, cyanobacteria, and bacteria such as *Bacillus* spp. and *Streptomyces* spp. was detected. In studying the Roman catacombs of Domitilla, Priscilla, San Callisto, San Sebastiano, and Sant'Agnese, Albertano and Urzì (1999) found different species of cyanobacteria in association with *Streptomyces* as the dominant phototrophic and heterotrophic microflora causing the biodeterioration of the surfaces (frescoes, mortars, marbles). An understanding of the relationships that are established between phototrophic and heterotrophic components is therefore an additional factor that must be taken into consideration when studying the biofilms deteriorating the stone surfaces. Appropriate sampling techniques and methodological approaches devoted to these microbial communities are continuously being developed to allow the application of suitable methods of control and monitoring (Urzì and Albertano 2001; Albertano 2003).

Studies carried out in the caves of Perama (Greece) showed that the artificial lighting installed after the opening of the caves to the public in 1956 had resulted in an increase in photoautotrophic epilithic and endolithic microorganisms, along with a prolific growth of heterotrophic bacteria, fungi, and actinomycetes (Iliopoulou-Georgoudaki et al. 1993).

More recently, the study of the heterotrophic component of the biofilms in the catacombs of San Callisto and Priscilla confirmed the widespread colonization by actinomycetes—and especially strains of *Streptomyces*—and determined that each site hosted its own microbial population irrespective of the locality, the kind of substrate, and the associated phototrophic microflora (Urzì et al. 2002).

Fungi are relatively widespread in underground environments, but their presence is strongly conditioned not only by the presence of

Fig. 5.7 *Reconstructed cross-section of the trajectory of the root systems above the* Domus Aurea *in Rome (elaboration by G. Caneva and A. Merante)*

as plasters, stuccos, and frescoes), fungal hyphae may penetrate to a depth of more than 10 mm, producing flaking and loss of fragments; obviously, this is particularly serious if the plaster has been painted.

Wooden elements that are at times present or form part of the tombs can also favor the development of cellulolytic or lignolytic deuteromycetes, basidiomycetes, or ascomycetes, such as *Serpula lacrymans* and *Coniophora puteana*, which are widespread in conditions of permanent high humidity and contact with the soil (Plate 17). Usually, this is restricted to the development of the vegetative mycelium and of rhizomorphs (Plate 17) (Blanchette et al. 1999).

organic substances, but also by the fluctuations in microclimate parameters. Very stable conditions of relative humidity and temperature do not seem to be particularly favorable for their development, while the fluctuations of these parameters seem to activate the germination of the spores and encourage the colonization of the surfaces (Florian 1993). This was confirmed in numerous hypogean environments in Rome, where the microclimate variations resulting from the opening and closing to the public or the presence of conservation teams and operations favored rapid and highly visible fungal colonizations of the plasters and frescoes.

The strains that have been identified most frequently as being the cause of alterations belong to the family of Dematiaceae, such as *Cladosporium*, *Stachybotrys*, and *Alternaria*—this may be because of the particularly visible effects of their development, due to the dark pigmentation of their cells. Species of *Acremonium*, *Aspergillus*, *Penicillium*, *Curvularia*, and *Fusarium* have also been frequently isolated, and their presence has been linked to the incoming air from outdoors (Arai 1983; Tilak 1991; Deshpande and Gangawane 1995; Nugari et al. 1998).

The damage caused by the development of fungi is not only superficial; often, because of the high water content of the materials and the high degree of porosity of the artifacts (such

In subterranean environments, the transport of biodeteriogenic microorganisms by the air takes on particular relevance, because airborne microflora can accumulate, and the exchange of air in these environments is not sufficient to dilute it (Saarela et al. 2004). In addition, the conditions of high humidity encourage a particularly rapid colonization of the surfaces, and in the case of inviolate tombs that are then opened to the public, a new equilibrium in the microbial ecosystem is established with a predominance of biodeteriogenic species (Agarossi 1994; Nugari and Roccardi 2002). Studies carried out on the microflora that develops under these conditions have shown a rapid and considerable increase over time of airborne microorganisms, fungi and bacteria in particular (Arai 1974, 1983, 1988).

It is also possible to find in subterranean environments sciaphilic bryophytes, such as *Conocephalum conicum* and *Marchantia polimorpha*, and a few pteridophytes, especially *Adiantum capillus-veneris* (maidenhair fern), in addition to angiosperms such as *Parietaria diffusa* and *Trachelium coeruleum*. For certain weakly illuminated subterranean Roman sites (Tomba degli Scipioni and *Lapis Niger*) a new association, *Conocephalo-Adiantetum*, has been described (Caneva et al. 1992b; Altieri et al. 1993).

An additional, fairly common biodeterioration problem is the damage caused by the roots of the trees growing above the hypogea. It is well known that the development of roots—especially those of trees, which are obviously more extensive and vigorous—can damage architectural structures situated either beneath or in the vicinity of the plant.

To the wealth of literature on the subject of root damage to buildings in urban and industrial contexts, we should also add the literature on the damage encountered in archaeological contexts, as for instance in the Etruscan tombs in Tarquinia (Cesari and Rossi 1972; internal note, Istituto Centrale del Restauro) and Chiusi (Vlad Borrelli 1954), or in villas and mithraea in Rome (Caneva 1990, 1994). In the archaeological areas in the city of Rome—so complex because of the stratification of monuments that have succeeded one another over time—considerable damage is caused both by the mechanical action of the growth of the root systems of woody plants and by the chemical action, due to the release by the roots of acid metabolites and chelating substances. One of the most serious examples of damage is that to the *Domus Tiberiana* on the Palatine, separated by a soil thickness of a mere two meters from the *Horti Farnesiani*, which contain centuries-old trees (among which *Quercus ilex*, *Cercis siliquastrum*, *Cedrus atlantica*, *Cupressus sempervirens*, as well as exotic varieties such as *Brachychiton populneum* and *Sequoia sempervirens*) that have caused ongoing serious damage, including collapse, in the underlying archaeological structures (Caneva 1990). Similarly, in Nero's villa on the Colle Oppio (the *Domus Aurea*, which was already underground at the time of Trajan and was used as foundations for Trajan's Baths), it was observed that not only the roots of some of the trees had grown beyond the four meters that divide the vaults from the gardens, but also that some of the trees (such as *Pinus roxbourgii*) had penetrated to a depth of more than fifteen meters and covered more than twenty-five meters in horizontal extension (Figs. 5.7 and 5.8) (Caneva 1994).

In the Jewish catacombs of the Villa Torlonia, evident penetrations of roots both in the vaults and the walls have been noted for some time; these roots penetrate to a depth of about 8–10 meters and sometimes cover distances of more than fifty meters, as in the case of one particular *Ficus carica* (the common fig) (Caneva, unpublished).

The considerable distances covered by the roots can be explained by the "well effect" present here, which facilitates their development downward, following rising water gradients.

To find out which species is responsible for a particular damage situation it is essential to determine the anatomy of the root wood: the simple correlation between underground and aboveground locations is not sufficient, since there is considerable variation in the capacity

Fig. 5.8 *Penetration of roots into the vaults of the* Domus Aurea: *(a)* Pinus roxbourgii; *(b)* Thuja *sp.*
– Photo G. Caneva

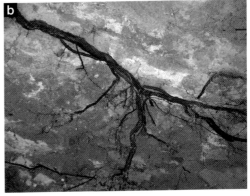

for root development between species and, therefore, the roots identified underground may not necessarily correspond to the nearest tree (Cutler and Richardson 1981; Caneva 1994; Kutschera and Lichtenegger 1997, 2002). It is possible to establish which elements are most critical and have to be kept under control by taking into consideration: the species present, the nature of the soil and the thickness of the soil layer, the climate conditions, and any irrigation systems that might be in place. Needless to say, it is to be hoped that any planning of and around archaeological sites may include a careful selection of the plant species introduced, in terms of their compatibility with conservation requirements, historical contexts, and the surrounding landscape (Caneva 1999b).

5.2 OUTDOOR ENVIRONMENTS

It is not possible to make generalizations about outdoor environmental conditions, because there is a vast range of different characteristics that vary according to the climate context under consideration (Chapter 6) and to the geographical and topographical location of the site. The outdoor environment is essentially characterized by exposure of the materials to meteorological phenomena such as rain, solar irradiation, the considerable fluctuations in temperature between night and day, and the winds, all factors that can facilitate the development of biodeteriogens and, in particular, of autotrophic organisms and microorganisms.

Nevertheless, in this context it is still useful to make a main distinction between urban and rural environments, mostly because of the significant differences in the influence of the varying levels of pollution and microclimate effects, independently of the macroclimate and the geographical location of the site on a larger scale. In addition to these, we are going to highlight some specific situations, such as those found in coastal areas—especially as related to the influence of marine aerosol—and those encountered in fountain habitats, where the water gradients are obviously very relevant —which makes them similar to submerged environments, but where the maintenance and conservation problems are completely different.

Archaeological parks are not discussed separately, because their location can be both urban and rural; they can be located at different elevations, in coastal, hilly, or mountainous areas that are characterized by microclimates with thermal and rainfall levels exhibiting their own unique characteristics (Chapter 2). As a result, their location as well as their size (which may or may not determine differing microclimates, which generally are closer to rural conditions) can give rise to different microenvironmental situations; and for these, the same considerations expressed in the introductory paragraphs are applicable.

As far as materials are concerned, it is to be noted that monuments and statues located outdoors are largely made of natural and artificial stone (rocks, bricks, ceramic materials, plasters, etc.) and only to a lesser extent of metal and wooden artifacts. Consequently, the surfaces of monuments and artifacts outdoors are frequently affected by phenomena of biological alteration that are predominantly caused by autotrophic organisms, which can produce different kinds of deterioration. Still, the role of heterotrophic organisms, such as fungi and bacteria, in the processes of alteration should not be ignored (Chapter 4).

5.2.1 Monuments and Artifacts in Urban Environments

by Giulia Caneva, Simona Ceschin, and Maria Luisa Tomaselli

5.2.1a *Characteristics of the Environment and of the Materials*

In constructing cities, man has created an entirely new type of environment, albeit artificial, that provides both the animal and the plant organisms inhabiting it with a very unique habitat. In terms of ecology, this habitat presents a structure, an evolution, and regulating mechanisms that are different from those of other ecosystems. Indeed, some species find in cities the ideal habitat and the best conditions for growth, partly because of the reduced competition with more sensitive species.

An accurate analysis of the environmental conditions unique to these contexts is very

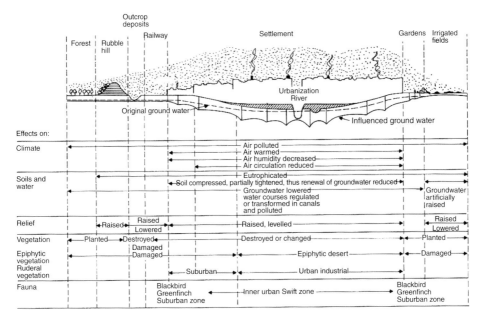

Fig. 5.9 *Characteristics of urban ecosystems (from Horbert et al. 1980)*

important, since a large proportion of works of art is located in urban environments, either exhibited directly in the open, or else preserved in museums and churches or other indoor spaces, where they are protected from the aggressive environmental factors that surround them.

The most commonly used materials in cities are stone, bricks, asphalt, and cement, all of which have a low permeability to water and a high capacity for heat absorption. The built-up areas of city centers react to climate factors in a completely different way than the surrounding areas, where the land is covered by agricultural crops or natural vegetation. When a city area reaches or exceeds a hundred square kilometers, very real changes occur in the local climate (urban meso-climate), which in general becomes warmer and drier, with direct consequences for the growth and distribution of biological species. Because of this, the urban environment has often been defined as an "island of heat," which can clearly be registered in infrared satellite images that are sensitive to heat radiation (Bullini et al. 1998).

The main characteristics that distinguish an urban from a rural and green environment are (Fig. 5.9):

—an increase in temperature, both during the day and the night; during the day the radiating flux entering the city is prevalent, and the surfaces found in urban spaces (asphalt, in particular) have a tendency to overheat because of the sun's radiation; subsequently, they themselves radiate heat out during the night hours. To this thermal energy originating from the sun is added the energy coming from combustion processes or else electricity, which in the end is dissipated in the form of heat. The urban heat island thus gives rise to a mesoclimate with "desert" characteristics, which is more limiting for the development of organisms (Bettini 1996). Even stronger than the overheating of the air is the overheating on stone surfaces, especially during the summer months in warmer cities when it can exceed 50°C, which has very limiting consequences even for the most resistant organisms;

—a decrease in relative humidity, with a tendency for the air to become ever drier as the temperature rises. This dryness is accentuated by the lack of water in the soil, because rain water that flows down roofs and onto impermeable road surfaces is rapidly diverted into the drainage system, without having the chance to establish a substantial reserve in the soil and without the plants being able to make use of it and promote the processes of evaporation and transpiration, which are so important in raising the level of microenvironmental humidity;

—a decrease in the circulation of air and winds—due to the barrier effect of buildings, characteristic of urban structures—and resulting in a reduction in ventilation of about 20–30% compared to the natural environment surrounding the urban area;

—a rise in cloud cover and rainfall as a result of the convection air currents from the countryside toward the city. During the day, the city tends to overheat, and the warm dry air rises, drawing in the damp and fresh air from the surrounding countryside. These factors affect exposure to sunlight, but not the real supply of water to organisms, because the water flows off the impermeable surfaces and drains away quickly;

—an increase in atmospheric pollution, linked to the high levels of pollutants present (noncombusted hydrocarbons, sulphur dioxide, nitrogen oxides, carbon monoxide, heavy metals, and great quantities of dust particles often saturated with pollutants), originating from the combustion processes of heating systems and vehicular traffic, as well as from the activity of industrial establishments, thermoelectric power stations, and refineries. This factor will be examined in detail, as it can have a complex and usually inhibiting effect on biological organisms.

It should be emphasized that within an urban area the intensity of the described phenomena increases progressively moving from the suburbs toward the city center, and this can be correlated to the building density, to what percentage of the land is covered by vegetation, and to the heavier traffic found in urban centers (Sukopp 1990).

5.2.1b *Problems of Biodeterioration*

As a general rule, we can say that in heavily urban environments, where the conditions of the urban mesoclimate are more pronounced and the edaphic conditions are often very limiting, the potential for growth and development of microflora and vascular flora is more limited when compared to the general conditions in the surrounding areas.

In particular, the effects of pollution on the various biological populations can be different and can change according to the nature of the pollutants and the type of organism involved. The majority of pollutants have an inhibiting effect on biological growth, especially on those organisms—such as lichens and mosses—that are not equipped with specific mechanisms of protection, which is why their presence in urban contexts is limited to the more resistant species (Chapter 3, sections 4.3 and 5.3). Acid pollutants in the atmosphere—such as NO_x, SO_2, and CO_2—indeed greatly reduce the pH of rainfall, resulting in a potentially corrosive action on the gelatinous or waxy protective surface layers of organisms; in addition, they can cause an acidification of the substrates, which inhibits the growth of neutrophilic and alkalophilic populations. Lichens, which are highly sensitive to sulphur dioxide pollution, decrease in terms of diversity of species as the presence of the toxic gas increases progressively from the suburbs toward the city center, where traffic is heavier; this leads to the so-called "lichen desert" found in the most polluted and highly urbanized areas (Nimis 1990). This characteristic forms the basis of the biomonitoring methods that employ this type of biological response to the environment to interpret the values of the environmental parameters that condition them.

In the past especially, the possibility was investigated of a biological genesis of the phenomenon of sulphation of limestone and, in particular, of the formation of "black crusts" typical of polluted areas (Ausset et al. 2000). These processes could be carried out by the sulphur-oxidizing bacteria—frequently isolated in urban contexts (Chapter 4, section 3)—which are able to catalyze the biological transformation of sulphur dioxide into gypsum. Numerous investigations have shown the recurring presence of different types of microorganisms (bacteria, algae, and fungi) in association with these black crusts, but the conclusion seems to be that it is impossible to identify a single cause for all the different forms of blackening encountered in urban environments (Chapter 2, section 2.1). In the study of several sandstone monuments in urban contexts in the Netherlands, it was experimentally shown that the processes involved in the deterioration are a result of the formation of gypsum, the deposition of particles, the microbial activity, and in particular the deposition of iron hydroxides (Nijland et al. 2004).

While mosses and lichens are particularly sensitive to pollutants, higher plants, although damaged by high levels of pollution, on the whole exhibit higher levels of tolerance. Among the plants growing on walls in urban habitats are species that are widely distributed in other environments as well (for example, *Parietaria diffusa*, *Cymbalaria muralis*, *Capparis spinosa*, *Antirrhinum majus*, *Ficus carica*, etc.); the plant selection depends mostly on microenvironmental gradients (water contents of the walls, levels of sunlight, and nutrients) (Caneva et al. 1992b).

As far as algae are concerned, there is a definite reduction in their density in urban environments; this is probably attributable not so much to the effects of pollution as to microclimate effects—cities with high levels of pollution, for example, but located in a tropical climate (e.g., Singapore) still exhibit a dense algal cover on exposed surfaces (Wee and Lee 1980). Analogous results were obtained in Rome in areas with high levels of pollution from vehicular traffic (the *muraglioni* of Lungotevere—the embankment walls along the Tiber River), where algal growth appeared to be mostly related to different water gradients (Bellinzoni et al. 2003) (Fig. 2.5).

At times, however, pollutants can have a positive effect on the development of certain biological species. This can take place indirectly—in the case of resistant organisms that find themselves in situations with no competition and form extensive populations of one single species or a few of them at most—or directly, as in the numerous occurrences of eutrophication, correlated to an enrichment of the substrate with organic nutrients, which favor the establishment of heterotrophic species (fungi and some bacteria), or else with a supply of nitrates favoring nitrophilic species, such as lichens (although the latter case is much more likely in rural areas).

Certain polluting substances—both in gaseous and particulate forms—can be used as nutrients by many microbial species, even if in low concentrations; the deposit of saturated and unsaturated hydrocarbons (deriving from the partial combustion of oil) on the surfaces of buildings and monuments encourages the growth of those heterotrophic fungi and bacteria that are able to use these substances as sources of carbon (Saiz-Jimenez and Samson 1981b; Saiz-Jimenez 1995a; Ortega-Calvo and Saiz-Jimenez 1997; Laiz et al. 2002).

However, in adverse conditions like those created in urban environments, some organisms such as cyanobacteria (e.g., *Myxosarcina spectabilis*, *M. concinna*, *Tolypothrix byssoidea*, and *Chroococcus lithophilous*) or certain species of lichens that are resistant to pollution (e.g., *Lecanora dispersa*, *Verrucaria nigrescens*) are able to survive by activating or deactivating their metabolism in response to how favorable or not favorable the environmental conditions are. Cyanobacteria can endure both long periods of desiccation, thanks to thick mucilaginous sheaths that can absorb and retain water for a long periods of time, and excessive illumination, by becoming darkly pigmented and effectively blocking the in-depth penetration of damaging radiation; the latter produces dark patinas on the surfaces of monuments, which are often confused with the dark crusty deposits from chemical pollutants (Plate 21 and Fig. 5.10).

We can generally say, therefore, that in artificial ecosystems such as those of urban environments the equilibrium between organisms and the surrounding natural environment is altered; this results in less favorable conditions to the development of the more sensitive species—which are thereby damaged or entirely eliminated—and the establishment of new conditions that favor new species.

Fig. 5.10 *Black patinas on the Cestia Pyramid (Rome), an example of a typical biological colonization in an urban environment* – Photo G. Caneva

For the reasons expounded above, which can be attributed not only to pollution but also to more selective microclimate conditions, one of the most evident effects is the decrease in the number of different organisms present per surface unit. Only the most resistant species will succeed in surviving; a decrease in the diversity of the flora and microflora takes place with varying degrees of intensity and is accompanied by a perceptible increase in synanthropic species, i.e., those linked to human activity (calciophilic, xerophilic or xerotolerant species, or those resistant to pollutants, etc.). Also, in urban environments, an increase in widely distributed species (cosmopolitan and subcosmopolitan) is registered to the detriment of those that are more specific to a particular biogeographic environment.

Precisely because of these effects, various studies have shown how in these contexts the presence of archaeological sites and urban parks is an important element in the preservation of

biodiversity; this must therefore be an additional element to be taken into account when carrying out treatments using biocides (Nimis et al. 1992; Ariño and Saiz-Jimenez 1996a; Celesti Grapow and Blasi 2003; Ceschin et al. 2003).

5.2.2 Monuments and Artifacts in Parks and Rural Environments

by Giulia Caneva, Rosanna Piervittori, Ada Roccardi, and Maria Luisa Tomaselli

5.2.2a *Characteristics of the Environment and of the Materials*

Monuments and stone works in parks and gardens in urban areas as well as in rural environments exhibit various types of biodeterioration, which are generally more extensive than in other environmental contexts since the habitat is usually more favorable to microbial development.

Their environmental characteristics are first and foremost the result of the microclimate conditions induced by forest and vegetation cover. The different effects can be summarized as follows (Horbert et al. 1980):

— a decrease in temperature, even of only a few degrees (Fig. 5.11a) in comparison to adjacent urban areas, which is variably substantial depending on the composition and the structure of the vegetation as well as other parameters such as solar radiation or wind factor; this results in a greater risk of condensation during the cold hours and a potential increase in phenomena linked to frost, in addition to a more extensive biological attack linked to greater water stagnation;
— an increase in relative humidity of around 10% more than in urban areas (Fig. 5.11b), linked to both the lower thermal values and a larger water supply, which is a result of plant respiration and lower evaporation due to larger shaded areas. Parallel to that, there is also an increase in humidity as a result of closer contact with the soil, related to effects of capillary rise, which are not negligible;

—a reduction in chemical pollution, due to the buffering effect of the foliage of the trees. The relevance of these effects can vary (up to 25% of the particles present), depending on the level of pollution, the expanse and structure of the green areas, the strength and direction of the dominant winds, and the location of the polluting sources. This has positive effects for the preservation of works, but reduces the inhibiting effects that many of the pollutants exercise on biological growth (especially on sensitive species such as lichens);

—an increase in microbial pollution, due to the widespread presence of reservoirs of spores of various origins, particularly linked to the vegetation and the various rural activities;

—a reduction in wind strength in comparison to the surrounding countryside, related to the structure and morphology of the foliage of the trees; this results in a greater localized stagnation of humid air and in a difference in the transport of airborne particles in areas of undergrowth;

—an increase in shaded areas, due to the screening provided by the foliage; this does not inhibit the growth of photoautotrophs, but it does favor the increase of sciaphilic species;

—an increase in nutrients and in the phenomena of eutrophication, related to closer contact with the soil and to the greater presence of avian fauna and fertilizers from neighboring agriculture and cultivation.

All these effects, which lead to a general increase in biodeterioration phenomena, can be somewhat counteracted by the shielding effect and the localized reduction in rainfall resulting from the presence of the tree foliage. However, if we take into account the high levels of relative humidity and the local topographic effects, what takes place on average is an increase of stagnant humidity, which favors the development of conspicuous cryptogamic patinas (algae, leprous and gelatinous lichens, bryophytes), especially in the areas with northern exposure.

Furthermore, it should be remembered that the works found in these contexts are usually made of stone and include statues, monumental staircases, and nymphaea, as well as fountains, balustrades, and other decorative elements, which will be treated separately because of their specificity. Metal or wooden sculptures are less common in these environments.

Fig. 5.11 *Reduction in T (a) and increase in RH (b) in a park in the city of Berlin (from Horbert et al. 1980)*

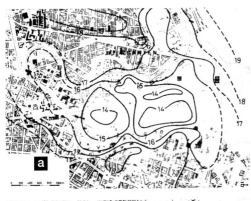

5.2.2b *Problems of Biodeterioration*

A vast range of organisms develops in these contexts, giving rise to biofilms, stains, patinas, films, incrustations, cushions of varying thicknesses and pigmentation, compromising both the appearance and the preservation of the artifacts (Fig. 5.12).

Patinas and biofilms that are green or reddish-green in color are frequently found on the exposed surfaces, generally resulting from the

Fig. 5.12 *Abundant biological colonization in rural contexts (Canopus of Hadrian's Villa, Tivoli)* – Photo A. Roccardi

development of populations of photoautotrophic microorganisms, such as algae, cyanobacteria, and at times also the crustose and foliose thalli of lichens, which have cells rich in green, reddish, and yellow-orange pigments (chlorophylls, phycobiliproteins, and carotenoids). Black patinas are usually due to the presence of cyanobacteria and fungi (Saiz-Jimenez 1995b); reddish or pinkish ones tend to be more due to the presence of certain algae (for example, *Haematococcus pluvialis*) or, if more orangey-brown, to those belonging to the genus *Trentepohlia* or to chromogenic heterotrophic bacteria (Pietrini et al. 1985; Tiano and Tomaselli 1989; Sorlini et al. 1994). The morphology of lichens is very varied, with forms that can be epilithic and others endolithic, the latter recognizable mainly through surface micropitting.

Horizontal, protruding, or moderately sloped surfaces are all particularly well suited as resting places for birds, which results in accumulation of deposits of guano. These conditions facilitate the settlement of airborne biological material (the spores of bacteria, lichenized and nonlichenized fungi, algal cells, bryophytes, lichen propagules)—especially if the deposits are substantial enough—and favor the development of nitrophilic, coprophilic, and ornitocoprophilic species. However, vertical surfaces can also be affected by these phenomena and sometimes therefore show obvious and extensive "flow marks" of a biological nature. In rural sites, the wind may favor accumulations of agricultural or gardening fertilizers on these stone surfaces, the amounts of which vary depending on the exposure. In addition, a conspicuous and extensive cryptogamic cover represents an ideal substrate for the capture of spores and seeds, with the consequent development of phanerogams (Plate 3).

Given that the development of nitrophilic organisms is generally rapid, operations aimed at reducing or eliminating the causes of eutrophication do not always have durable results, especially in rural areas (Nimis and Salvadori 1997; Piervittori and Caramiello 2001).

Just as an example of the specificity of bio-deterioration problems, we will describe a few case studies relating to historic gardens of considerable importance and complexity.

The various investigations carried out on the Boboli Gardens in Florence showed that the prevalent colonizations on the marble surfaces of the statues were by pioneers, with biofilms composed primarily of chemoautotrophic bacteria and green microalgae; in certain instances of prolonged exposure, cyanobacteria, heterotrophic bacteria, fungi, and lichens would also develop. In secondary successions—after conservation interventions had been carried out—it was possible to observe the development of a type of microalga (*Coccomyxa*) that was also present on the lichenized trunks of the surrounding trees; the latter seem to have functioned as a reservoir for microorganisms, which—carried by wind and rain—would travel to the stone surfaces and adhere to them by means of mucilaginous sheaths. In time, cyanobacteria would become dominant, along with lichen development, which was here less conspicuous than in strictly rural contexts (Tomaselli et al. 2000c; Lamenti et al. 2000b).

In the case of Venetian villas and, in particular, the Villa Cordellina (Vicenza) or the Villa Pisani (Stra, Venice), significant lichen colonization takes place in addition to the abundant algal patinas. In the study on the considerable lichen colonization on the statues in the garden of the Palladian Villa Cordellina it was possible to show how the use of agricultural fertilizers had encouraged the development of nitrophilic lichens and how the resulting rich mosaic pattern of lichens was so abundant that it formed the basis for a secondary development of bryophytes (Plate 3). Furthermore, the study made it possible to follow the process of recolonization, which took place only eighteen months after the conservation intervention, and to evaluate the negative impact of the soil in such close contact with the statuary (Seaward et al. 1990). The development of lichen species (*Caloplaca decipiens, Lecanora muralis, Phaeophysica orbicularis,* and *Xanthoria parietina*) facilitated the development of nooks and niches in which it was possible to observe accumulations of dust and earth particles —rich in nutrients—which provided a sort of microsoil for the establishment of biodeteriogenic bryophytes such as *Grimmia pulvinata*.

5.2.3 Monuments and Artifacts in Coastal Environments

by Antonella Altieri and Daniela Pinna

5.2.3a *Characteristics of the Environment and of the Materials*

With the term "coastal" we refer here to the environment represented by the strip of land directly influenced by the action of the sea; it is subdivided in two main environmental types: sandy coastline and maritime rocks. The common factor for both is the high concentration of marine salts in the air and in the substrate, which constitutes one of the main causes of stone deterioration in coastal areas, in addition to being a limiting factor for the development of plant and animal species. These coastal areas can obviously include both urban and rural environments, creating complex situations due to the multiple influences involved; however, here we will be analyzing the specific effects resulting from the action of salt and wind.

The concentration of soluble salts, and specifically that of sodium chloride, is a limiting factor selecting living organisms that are able to colonize and develop in a saline environment; these are called halophilic or halotolerant. These organisms exhibit particular structural adaptations: many plants, for instance, are succulents and have water storage tissues with cells rich in mucilage that retains water, or else they have leaves with an extremely limited surface area to reduce water loss to a minimum. Certain halophilic vascular plants grow on substrates with saline concentrations of 2–6% and are able to accumulate up to 10% of sodium chloride in their vacuolar sap.

In halophilic and halotolerant microorganisms, the osmotic equilibrium with the external environment is maintained by means of two main strategies for adaptation: 1. the accumulation of high concentrations of the intracellular K^+ ion, instead of the Na^+, and a modification of the enzymatic system, or 2. the accumulation in the cytoplasm of compatible organic solutes, such as glucosylglycerol.

The concentration of soluble salts in the substrates is subject to variations that are dependent on environmental factors as well as on the characteristics of the constituent materials.

The environmental factors that most influence the variations in the concentrations of salts on the materials are:

—distance from the sea;
—height above sea level;
—exposure to winds carrying marine aerosol;
—topographic characteristics, and in particular:
  a) an open environment, and hence exposure to the action of winds and marine aerosol;
  b) an environment protected by natural barriers (trees and shrubs) or artificial ones (constructions, buildings), resulting in a reduction in the deposit of salts on the surfaces.

The salt deposition can occur either within the wall structures (where the foundations are in contact with seawater) or on the surface of the constituent materials (as a result of the transport and deposit of marine aerosol). The main chemical elements found in marine aerosols are sodium, chlorine, magnesium, calcium, and sulphur. Among the crystalline forms, the most frequently occurring are sodium chloride (NaCl), potassium chloride (KCl), and calcium sulphate (CaSO₄). Sodium chloride is the most abundant soluble salt in the aerosol; it is the main cause of the breakdown processes that take place in stone—especially in limestone, where this particular salt has a synergic action when in combination with other soluble phases, such as, for instance, calcium sulphate dihydrate ($CaSO_4 \cdot 2H_2O$).

The presence of sodium sulphate (Na_2SO_4), on the other hand, seems to be the result of the reaction between the sodium chloride from the sea and sulphuric acid (H_2SO_4), which is largely an anthropogenic product.

In addition, thermal and rainfall factors must be taken into account as well as season changes, their effects being the modification of the humidity levels or the washing away of the salts present on the surfaces.

Also, the characteristics of the lithotype, in particular its mineralogical composition and porosity, contribute to differences in suscep-

tibility to chemical, physical, and biological deterioration.

5.2.3b *Problems of Biodeterioration*

In coastal areas, the degradation of monuments and stone artifacts located outdoors is mostly due to the crystallization of the soluble salts contained in seawater. The damage is attributable to the corrosiveness of sodium chloride and is accentuated or accelerated by wind and drought conditions.

When analyzing biodeterioration caused by microorganisms on stone artifacts, an often encountered problem is the difficulty in distinguishing deterioration due to abiotic causes from that caused by microbial action. For example, monuments damaged by salt efflorescence are an ideal habitat for the development of halophilic and/or halotolerant microorganisms (Chapter 2, section 2.4), which, in turn, participate in biogeochemical processes and in salt crystallization phenomena. In this instance, cause and effect overlap and may thus be confused when the damage is analyzed.

Some of the most common microorganisms identified on stone in coastal regions are: among the archaeobacteria, the genera *Halobacterium* and *Halococcus*, among heterotrophic bacteria the genera *Halomonas* and *Bacillus*, among cyanobacteria *Gloeocapsa* and *Myxosarcina*, among the eumycetes *Penicillium* and *Aspergillus*, and among the algae *Chlorella* and *Dunaliella* (Salvadori et al. 1994; Incerti et al. 1997; Saiz-Jimenez and Laiz 2000). *Bacillus cereus*, in particular, is able to grow in cultures with a 15–20% concentration of NaCl. With regard to the specific composition of halophilic bacteria communities found on artifacts, it should be emphasized that there may be variations depending on which diagnostic techniques are employed: for example, it has been noted that the bacteria isolated and identified on wall paintings using culture analysis were different from those identified using molecular techniques (Rölleke et al. 1998).

Lichens are also frequently found on rocks and monuments in coastal areas; however, they are often cited without precise systematic identification (Fig. 5.13).

Fig. 5.13 *Lichens on the Mohai, Easter Island*
– Photo O. Salvadori

The lichens of coastal regions are specialized organisms with a strong preference for marine aerosol rich in salts and for the basic pH of seawater (8.0–8.3). Coastal areas can be distinguished into three zones (coastal, semicoastal, and terrestrial), and to each of these correspond typical species of lichens; for example, *Verrucaria maura* is a species characteristic of the coastal zone and has often been confused—because of its black thallus—with the deposits caused by oil pollution on the rocks in contact with the waves (Gilbert 2000). The terrestrial zone, on the other hand, begins at the point where the lichens receive most of their water and nutrients from rainfall and from the substrate rather than from the sea spray. This is accompanied by a reduction in the pH of the water originating from the environment.

Dirina massiliensis is a lichen characteristic of coastal areas and has been observed on several monuments (Seaward and Giacobini 1988; Nimis 1995; Di Francesco et al. 1998; Carballal et al. 2001); it has a white thallus and a chalky-floury appearance and can extensively colonize both calcareous and siliceous rocks. On calcareous rocks, it forms rather thick thalli (up to 2–3 mm) and has strong deterioration power, as it produces abundant quantities of oxalic acid, which combines with the calcium contained in the stone to form calcium oxalate. *Roccella phycopsis* is another lichen linked to coastal environments and frequently found in Mediterranean areas; along with *Dirina massiliensis*, it characterizes the association *Dirinetum repandae* (Clauzade and Roux 1975). The latter has been found in archaeological sites in the Latium region, especially on vertical walls composed of alkaline tuffs and brick with northern exposure, predominantly on the upper parts of these walls, which are more exposed to the moist winds from the sea (Nimis et al. 1987).

From the point of view of biodeterioration, it should be noted that when environmental factors are particularly aggressive it seems that lichens can play a protective role; particularly in certain coastal regions it has been observed that on the most eroded granite there was no lichen cover, which was instead only found on more compact substrates (Carballal et al. 2001). For the interpretation of this phenomenon, which is open to a variety of explanations, see Chapter 1, section 5 and Chapter 4, section 3.

Similar differences in resistance to salinity are also found in mosses and higher plants, which result in their precise spatial distribution according to gradients of salt concentration (halophilic seriation). It is obvious that when the salt concentration is particularly high, as occurs on stone materials in the coastal areas (not semicoastal or terrestrial), the species found are specific to coastal rocky environments.

As far as mosses are concerned, although they are not reported in specific studies on monuments in coastal areas, *Pottia crinita*, *Tortella flavovirens*, and *Grimmia maritima* are typical halophilic species that grow on rocks—the first two are widespread in Italy, but *Grimmia maritima* has been found only in northwestern Europe (Great Britain) (Cortini Pedrotti 2001).

Among plants, the sea fennel (*Crithmum maritimum*) and numerous members of the genus *Limonium* are halophiles characteristic of the area of land that is directly reached by sea spray; *Daucus gingidium* and *Suaeda fruticosa* are also strongly halotolerant, while the species *Matthiola incana* and *Lotus cytisoides* colonize rocky habitats that are more sheltered from sea spray. Among the less halophilic plants frequently found on stone in coastal environments we must also note *Lobularia maritima* and *Hyosciamus albus*. A study of the colonization of monuments by halophilic vegetation in the coastal areas of southern Italy has shown that these species form specific communities in relation to the degree of salinity and can therefore be used as bioindicators of the salinity levels (Caneva et al. 1990). In relation to the coastal habitat, a study of the ruderal vegetation of Venice (Gamper and Bacchetta 2001) showed that certain halophilic and subhalophilic species were widespread throughout the city and linked to salinity and water gradients, in strict correlation with the bioclimate (temperate-oceanic in the sub-Mediterranean variant) and with the biogeographic location (on the southern limits of the Euro-Siberian region) of the investigated contexts.

In these environments, the salts act as limiting factors and select not only the species, but also the dynamic evolution of the plant population. Plant communities in coastal environments often exhibit strong similarities in the composition of the species and in their structure, independently of the geographic area. Indeed, floristic similarities are preserved not only because of the highly selective environment, but also—in the case of vascular plants—thanks to the transportation of seeds, fruits, and vegetative propagules across tides and currents, in addition to the dispersal facilitated by sea birds. Ruderal coastal communities consist mainly of pioneer plants with a grassy and suffrutescent habitus, which colonize the surfaces of walls but with a very low percentage cover; an evolution toward more complex structural stages is rarely observed (Caneva et al. 1990; Altieri et al. 1999a, 2000b).

5.2.4 Fountains and Nymphaea

by Anna Maria Pietrini and Sandra Ricci

5.2.4a *Characteristics of the Environment and of the Materials*

These artifacts exhibit microenvironmental characteristics that define them as systems with a certain level of differentiation compared to their surrounding context and will therefore be treated separately.

The conservation of artistic and monumental fountains raises, and always has raised, a series of highly complex problems to which no single, definitive solution can be found. The damage to the constituent materials—generally stone but sometimes also metal—is mostly due to the action of water, which, if on the one hand it represents the characteristic element and the main attraction of fountains, on the other constitutes its weakness and the main cause of its physical, chemical, and biological deterioration. The view of the artifact as a whole, as it was intended to be, is often obstructed by calcareous incrustations and biological patinas that can alter or even completely mask its three-dimensional articulation, resulting in repeated maintenance operations.

As a type of artifact, the fountain is an ideal growth substrate for many organisms and microorganisms—both animal and plant—that require a continuous and constant supply of water for their metabolic activities. In addition, fountains are usually located outdoors, and direct sunlight favors the development of photoautotrophic organisms: plant species—including photosynthesizing microflora and bryophytic and phanerogamic flora—thus play a role of primary importance in the colonization and, hence, the deterioration of these works.

Microenvironmental conditions of great diversity can coexist in fountains: there are almost always areas that are constantly wet, others that are only sporadically reached by the water supplying the fountain, and still other areas that remain completely dry. The parts that are constantly wet are the basins and pools, or also the architectural elements that serve a continuous water flow; the latter are distinguished according to the quantity and to the turbulence of the water that runs over them. These diverse situations determine different distributions of organisms on the fountain according to the specific ecological requirements of each (Fig. 5.14).

The presence of nighttime illumination and the addition of disinfecting substances or substances sequestering ions in solution can influence the system and modify the overall ecological conditions.

Nymphaea are particular types of artifacts that share many similarities with fountain environments. Gardens of *domus* or patrician villas have had nymphaea since ancient times; these sometimes rather mysterious places were constituted by artificial caves or grottoes in which the natural appearance of rocks was recreated with wall decorations composed of calcareous incrustations (tartars), scales of pumice stone, gravel from rivers, fragments of colored stone, shells, branches of coral, terracotta, and glass mosaic *tesserae*. The theme of the grotto was intimately linked with that of water, which represented the dominant element of the place; the water flow could come from narrow tubes hidden among the rocks in the greenery, or gush vigorously from either the pavement or the ceiling, or else feed little waterfalls and other surprising water play systems designed for "effect."

Both an architectural and a naturalistic element, at times embellished with frescoes, festoons, statues, and symbolic elements, the grotto was ubiquitous in villas of the Roman epoch, as shown in documentation from the first century B.C.E. (Bragantini 1999). These architectural works were rediscovered and given prominence by the Italian architects of the Renaissance who then exported them throughout Europe, with precious examples dating to the seventeenth century. This was followed by a long period of decadence during which many nymphaea were either abandoned or destroyed. Today, these precious works are often found in a state of advanced deterioration and have become true caves in the process of becoming ruins, exhibiting highly interesting and unique natural biocoenoses, due precisely to the peculiar growth habitat.

Water and light are the most important environmental parameters in the selection of the plant species able to colonize these environments. With regard to the supply of water,

Fig. 5.14 *Different types of biological colonization as related to their microhabitats (Fountain of the Dragons, Villa d'Este, Tivoli)* – Photo S. Ricci

in most nymphaea the original installations are no longer functional; when planning conservation interventions it is therefore necessary to evaluate the other sources of humidity, such as condensation phenomena, capillary water rise, infiltrations from the roof of the cave, and also the high levels of relative humidity in the air. Generally speaking, the higher the water content of the substrate—also taking into account the porosity of the materials—the more conspicuous the biological developments.

The intensity of the illumination and the duration of the period of exposure are also important factors in the selection of the biological species able to grow in a nymphaeum. Low levels of light, for example, discourage the development of several plant organisms and select sciaphilic species (that prefer shady places), similarly to what takes place in semienclosed environments.

5.2.4b *Problems of Biodeterioration*

The frequency and the extent of biodeterioration on fountains are well known, as is reflected in the many historical sources that document conservation efforts directed to the removal of such phenomena.

From an ecological standpoint, we can say that the aquatic communities found in the basins of fountains are fairly similar to those populating natural freshwater bodies, such as lakes and ponds, i.e., planktonic microorganisms, floating in the water, and benthic organisms, adhering to the substrate. The life forms found on the surfaces over which the water flows are instead composed of specific and specialized microorganisms, mostly belonging to the benthos; these are furnished with minuscule structures of attachment, such as peduncles, which allow anchorage to the surfaces that keep them from detaching under the pressure of the water flow. The factor of water speed takes on particular importance in encouraging or limiting the growth of microflora, because it alters temperature values, the availability of nutrients, and the quantity of oxygen and carbon dioxide present in the water.

The areas that are only sporadically moistened by spray or by small spurts of water allow the development of poikilohydric organisms, which are able to survive for variable periods of time in the absence of water.

And last, the portions that are permanently dry represent a subaerial environment in which the supply of water comes only from rain or from condensation phenomena. In this case, the biocoenoses that colonize the substrate are aerophytic, i.e., typical of substrates exposed to the air, and are mainly constituted by epilithic or chasmolithic species. In some cases, endolithic forms have also been encountered, as for instance in the Fountain of Neptune in Trento, where endolithic filamentous cyanobacteria were found to be developing inside the material to a maximum depth of 1.9 mm (Salvadori 2000). The development of endolithic biodeteriogens is probably more widespread than is generally realized, the infrequent isolation being due mostly to the choice of sampling methods. In order to find endolithic forms it is indeed necessary to remove scales from the stone, beneath any calcareous incrustations that might be present, and not limit the sampling to the surface (Chapter 9).

The frequency and distribution of plant organisms in the fountains do not only depend on the presence of water, but also on other factors such as temperature, the availability of nutrients, and the constituent material of the work. In the great majority of cases, the latter is natural or artificial stone, which is a particularly favorable substrate for the establishment and the successive development of many organisms. Indeed, the various lithotypes offer a rough surface that facilitates the adhesion of vegetative structures; they also provide micronutrients in the form of mineral ions, which then become part of cellular metabolism. Some lithotypes are more easily colonized than others: travertine is one example, since its large number of structural cavities allows pioneer organisms to establish themselves and organic substances and humus to be deposited; the latter, in turn, favor the onset of secondary colonization by mosses and vascular plants. Highly porous stone materials, such as tuffs, peperino, and sandstones, are naturally among the easiest to colonize (Chapter 4, section 3.1).

In time, the stone surface can acquire new chemical substances constituted by the salts in the water (calcium, magnesium, etc.), which are deposited on the surfaces through wetting and subsequent evaporation, forming adhering incrustations mainly consisting of calcite. This mineral is easily colonized, not so much because of its chemical characteristics, but because of its lack of compactness and, hence, great porosity.

Artistic fountains made out of metal alloys are, at least in theory, less subject to biological colonization than those made out of stone, because the majority of metals (e.g., copper, lead, etc.) release ions that, in large concentrations, are toxic to living organisms (Chapter 4, section 2). In reality, however, a calcareous incrustation made up of calcite crystals quickly forms on the metallic surfaces of the fountain, which de facto eliminates the inhibiting effect of the underlying metal and therefore allows microbial growth to occur in an analogous way to stone surfaces. A good example of this is the Fountain of the Naiads in Rome, made of bronze, but almost always entirely covered by a carbonate layer that favors the development of bacteria, algae, and mosses (Plate 24).

As far as the quality of the water is concerned, it contains organic nutrients, mostly consisting of nitrates and phosphates; the latter are sometimes added to limit the deposit of carbonates and subsequent incrustations, but they also influence the growth of organisms, making it very rapid and substantial. Also, these substances may already be present in the water supply to the fountain, often as a result of faulty purification or an imperfect filtration, or else originating in the surrounding environment, especially if the work is located in a non-paved area or in a park.

To counteract biodeterioration in fountains for conservation purposes, it is very important to ensure that the water supply is of good quality. A recent study carried out on some of the fountains located in the sixteenth-century garden of the Villa d'Este in Tivoli showed that the water from the Aniene River, although it undergoes a process of purification through a technologically advanced system that provides partial filtration, decalcification, and disinfection by means of ultraviolet radiation (UV-C), nevertheless still has a sufficient enough amount of particles in suspension to favor the growth of cyanobacteria, algae, and mosses. In particular, two fountains (the Fountain of the Dragons and the Rometta Fountain) were already showing signs of new colonizations on the basins and the sculptural groups only three months after completion of the conservation intervention.

Fountains fed by drinking water exhibit, on the whole, only modest colonization phenomena on the surfaces, since the processes that make the water drinkable considerably reduce the number and the vitality of the microorganisms present. The different methods of disinfection employed vary according to the characteristics of the water supply and may include the successive use of various procedures such as prechlorination, clariflocculation, rapid filtration, ozonation, filtration over active carbon, and final chlorination (Anastasi et al. 1984; Gucci 1989).

As far as underwater night illumination in fountains is concerned, it should be noted that the process of photosynthesis using radiation from lamps is considerably lower than solar radiation; however, the presence of spotlights—either submerged or positioned on the edge of the basin—constitutes an added risk for the work, partly due to the temperature increase of the water, and also intensifies the levels of colonization by photosynthesizing microflora in the areas immediately adjacent to the lamps and even the glass of the lamps themselves.

Overall, the fountain environment is characterized by a considerable floristic variety, and the most represented taxonomic groups are cyanobacteria and algae (Chlorophyceae, Diatoms, and Rhodophyceae) (Bold and Wynne 1978), although we cannot ignore the role played by mosses and lichens (Table 5.4) as well as by vascular plants, among which more or less hygrophilic species are selected according to their capacities of adaptation.

In general, the most abundant growth takes place in the areas where water flow is limited or nonexistent (because this guarantees the stability of the microclimate factors), while a more stringent selection of organisms occurs in areas with a more rapid water flow, since here only some of the taxa—with a filamentous organization—succeed in resisting the leaching action of the water by placing themselves in the direction of the current.

Investigations carried out on the Fountain of the Four Rivers and on the Trevi Fountain in Rome as well as on the fountains by Bernini in Ariccia (near Rome) (Pietrini 1991; Ricci and Pietrini 1994a; Altieri et al. 2000a) have shown a strong correlation between the structural characteristics of the algal species present and the morphology of the biological alterations on the works: a predominance of coccoidal forms, both solitary and colonial, usually results in colored patinas and gelatinous deposits, while a predominance of filamentous forms results in thick felt-like layers that can sometimes trap planktonic species inside.

The portions of the fountains above water—which are also colonized by lichens and ruderal plants, depending mostly on the gradients of hydrophilia and nitrophilia—are subject to a very similar type of deterioration as that found on works in stone outdoors, as discussed earlier in this chapter (section 2.2b). Lichen genera most often encountered on stone fountains

Table 5.4 List of the genera of cyanobacteria and algae frequently found on monumental fountains (from Foged 1983; Pietrini 1988, 1991, 1995; Bolivar and Sanchez-Castillo 1997; Altieri et al. 2000a)

Cyanobacteria	Bacillariophyceae	Chlorophyceae
Aphanocapsa	Achnanthes	Apatococcus
Aphanothece	Amphora	Chlorella
Chamaesiphon	Cocconeis	Chlorosarcinopsis
Chlorogloea	Cyclotella	Cladophora
Chroococcus	Cymbella	Cosmarium
Gloeocapsa	Diploneis	Crucigenia
Hyella	Epithemia	Euastrum
Lyngbya	Eunotia	Mougeotia
Microcystis	Fragilaria	Oocystis
Myxosarcina	Frustulia	Pediastrum
Nostoc	Gomphonema	Poloidion
Oscillatoria	Hantzschia	Pseudopleurococcus
Phormidium	Navicula	Scenedesmus
Pleurocapsa	Neidium	Scotiellopsis
Pseudoanabaena	Nitzschia	Staurastrum
Rivularia	Pinnularia	Stichococcus
Schizothrix	Surirella	Tetracystis
Scytonema	Synedra	Tetraedron
Symploca		Tetraspora
Synechococcus		Trentepohlia
Synechocystis		Ulothrix

are *Aspicilia, Caloplaca, Candelariella, Lecanora, Lecania,* and *Verrucaria* (Bartoli 1990; Altieri et al. 2000a).

As far as mosses are concerned, those present on the wet structures of the fountains belong, for the most part, to hygrophilic or aquatic species. Many are incrusting and bring about the formation of carbonate layers—sometimes of considerable thickness—as a result of the deposit and fixation of $CaCO_3$ on the vegetative structures. This process occurs mainly on structures subjected to the falling of water, in a similar way to what takes place in natural environments beneath waterfalls and trickles of water. The carbon dioxide produced by the metabolic activity of the plants contributes to the transformation of the calcium bicarbonate present in the water into calcium carbonate (Charrier 1960). The portions that are intermittently wet or constantly damp are colonized by species that are sensitive to the presence of water. Among these are: *Barbula subulata, Brachythecium rutabulum, B. plumosum, Bryum argenteum, B. bicolor, B. donianum, Eurhynchium praelongum, Gymnostomum calcareum, Fissidens taxifolius, Philonotis fontana,* and *Tortula subulata* (Bolivar and Sanchez-Castillo 1997; Altieri et al. 2000a). On the dry portions of the artifacts, a bryophytic flora develops that is represented by mesophilic and xerophilic biocoenoses (Smith 1993), with *Tortula muralis, Grimmia pulvinata,* and *Bryum capillare* as the most frequently encountered species.

When describing the state of conservation of a fountain, it is necessary to take into account that the prevalent form of alteration is constituted by calcareous incrustations, which are formed, almost without exception, on the surfaces as a result of the flow, the stagnation, and also the evaporation of the water. These incrustations can be of varying thickness, appearance, and color, depending on the action of the water. Further surface deposits such as dust, soil, guano, and biological patinas can be added to the incrustations.

The damage incurred by a type of work such as a fountain from the growth of epilithic organisms is mostly of an aesthetic order: depending on the size of their spread as well as their thickness and color, biological patinas and moss pillows can alter—sometimes profoundly—the overall appearance and visual harmony of the monument.

The physico-chemical damage produced by epilithic and chasmolithic biodeteriogens is of lesser consequence because it occurs mostly on the calcareous deposits covering the original stone or metal surface; these deposits are a sort of "sacrificial layer" in the context of biodegradation, paradoxically assuming a protective role. As far as endolithic microalgae are concerned, *Hyella fontana* is of particular importance because of its perforating action, observed, for example, on the fountain in the Alhambra in Granada (Spain) (Bolivar and Sanchez-Castillo 1997). But, as already mentioned, the development of endolithic forms directly on the substrate may be underestimated.

It should be noted that biological growth can also cause damage to the water supply systems of the fountains: the presence of a considerable biomass on the structures of the fountain can reduce the water flow inside the piping and quickly render inoperable the filtration units that may be present.

In addition, from a health and hygiene point of view, the biological patinas present in the aquatic environment of a fountain may give rise to the establishment and then proliferation of other microorganisms such as bacteria, fungi, protozoa, metazoa, sometimes with pathogenic or toxic effects. These may also induce processes of putrefaction of organic substances, releasing unpleasant odors in the water and in the air.

Nymphaea are also mostly vulnerable to colonization by photosynthetic organisms and microorganisms, even though the presence of soil and organic nutrients may encourage the development of heterotrophic bacteria and fungi as well. The prolonged neglect of many nymphaea has resulted in the establishment of considerable amounts of ruderal flora, prevalently with herbaceous habitus, but at times also woody, typical of rocky environments. Among the most commonly found are ferns, in particular *Adiantum capillus-veneris, Phyllitis scolopendrium,* some phanerogames (*Cymbalaria muralis, Parietaria diffusa, Umbilicus rupestris*) and tree species, such as *Ficus carica,* or climbing species, such as *Hedera helix, Ficus repens, Parthenocissus quinquefolia* (Caneva and Cutini 1999).

It is interesting to note that, in the past, the presence of mosses and plants in these places was seen as a positive ornamental element that made nymphea look more similar to natural caves. For a long time this concept prevailed over the notion that artifacts must be free of vegetation because of the well-known deterioration of stone structures induced by roots.

Although the debate is still open, it is generally thought that if the work was conceived to reproduce a natural environment—and especially if the colonization has reached a stage of maturity (e.g., the Nymphaeum of the Rains on the Palatine)—certain biocoenoses can be established that are in and of themselves of particular interest from a naturalistic point of view. This is the case, for example, of the *Eucladio-Adiantetum,* which in the context of natural environments is defined as a priority habitat by European directives for nature

Fig. 5.15 *Nymphaeum of the Rains on the Palatine (Rome)* – Photo A. Merante

conservation (Fig. 5.15). In these cases, removal of the flora (as is sometimes carried out in a drastic fashion) is therefore not recommended, and it is undoubtedly preferable to simply limit the control to any woody plants that may be present.

5.3 SEMIENCLOSED ENVIRONMENTS

by Ada Roccardi, Sandra Ricci, and Anna Maria Pietrini

Semienclosed environments are characterized by spaces that are partially circumscribed by walls or by natural materials such as soil or rock; they exhibit intermediate characteristics between outdoor environments and enclosed spaces, discussed in preceding sections.

These types of environments include natural cavities, for example, small caves that contain artistic expressions such as graffiti or paintings, or man-made architectural structures that differ from those discussed earlier in this chapter (section 1.4) in that they are in close contact with the outdoor environment. To this category

belong rock churches, porticos, loggias, and all outdoor spaces with a roof cover. The protective coverings over archaeological structures and monuments may be of a temporary nature or erected as provisional structures during the excavation; or else, they may be permanent elements of a plan to turn the archaeological site into an "open air museum." The microclimate characteristics found in the environments under these coverings vary according to the size and type of roofing. The various types of coverings will be discussed in Chapter 7 (section 1.2) where the systems aimed at the prevention of deterioration are treated.

Semienclosed environments exhibit particular microclimate conditions that depend, to a varying degree, on external climate and exposure and are also conditioned by the architectural characteristics of the building, the constituent materials, and the construction techniques, in addition to the size of the interchange area with the outdoor environment. The microclimate of these environments is different if the site is small and regular in shape rather than having a more highly articulated structure. In the first case, the values for temperature, light, and humidity can be rather uniform, while in the second they tend to exhibit a more heterogeneous behavior, which varies mostly according to the distances from the openings to the outside. Even though they do not exhibit the high selectivity of biodeteriogenic organisms characteristic of enclosed spaces, semienclosed spaces—because of their microenvironmental variability—influence the phenomena of biodeterioration in relation to the ecological requirements of the organisms. It is clear then, how artifacts may be colonized by different plant associations, depending on the various levels of humidity, light, and temperature.

5.3.1 Loggias and Porticoes

5.3.1a *Characteristics of the Environment and of the Materials*
Loggias and porticoes are covered environments characterized by broad frontal and lateral openings; depending on the architectural structure, the microenvironmental conditions

within them can either be uniform or exhibit strong fluctuations.

They can be situated at different heights from the ground; at ground level, we have the examples of the porticoes of Roman villas, the cloisters of churches and abbeys, and the porticoes of historic palaces. These environments can suffer from water rising from the soil by capillary action—especially when there are gardens and fountains in the immediate vicinity—and can therefore provide very favorable habitats for the development of biodeteriogens. Biological deterioration is further facilitated if ventilation is limited and shading is high.

When porticoes and loggias are situated at a higher level, the ventilation is generally higher and the influence of humidity from the soil is reduced. A clear example of the difference in conditions can be found in the small loggias of the Leaning Tower of Pisa, in which a progressive increase in ventilation is registered when progressing up the tower, along with a substantial reduction of the water levels in the structure, resulting in a different distribution of biodeteriogenic organisms. In this instance, a primary role is therefore played by the factor of exposure, which influences the degree to which the surfaces become wet through rainfall and their exposure to winds (Vedovello 2000).

In addition, the roof cover of the loggias can considerably influence the microclimate depending on its extensiveness, inclination, and height from the pavement. In environments near the sea (for example, the cathedral of Santa Fosca on the island of Torcello, Venice), the marine aerosol can also play an associated role, selecting from among possible biodeteriogens the species with halophilic behavior (section 2.3).

5.3.1b *Problems of Biodeterioration*

In areas of shade or those reached by rainfall, in which there are high levels of humidity both in the walls and the air, it is possible to observe the development of photoautotrophic flora constituted by cyanobacteria and algae, heterotrophic microflora, bryophytes, and also lichens; the growth determines the extent of the alterations according to the level of water content in the walls.

Among the most relevant studies of biodeterioration of loggias and porticoes—and also because of their historical and artistic importance—we would like to mention the cases of the monastery of Los Jeronimos in Lisbon (Ascaso et al. 2002) and of the monastery of the Benedictines in Catania (Poli Marchese et al. 1998). In both cases, the biological growth, which varied in relation to exposure and to edaphic conditions, was particularly rich and, in some areas, resulted in considerable damage.

When porticoes and loggias have good exposure to sunlight and are well ventilated, stone surfaces can be free of biological colonizations; in some cases, where there is infiltration of water or else driving rain, extensive blackish patinas will be present, constituted by associations of photosynthesizing microorganisms and microfungi. The still ongoing study on the Tower of Pisa shows how the most exposed stone surfaces—such as those of the belfry and of the higher loggias—exhibit low biological growth, with limited colonization by crustose lichens, cyanobacteria, and a few xerophilic bryophytes that are strongly adapted to hydric stresses. In the lower portions of the tower, on the other hand, large areas are colonized by mixed biocoenoses with a prevalence of cyanobacteria, green algae, and mosses, while lichens are mostly present in the *craquelures* of the columns. Among the cyanobacteria and the algae, the most commonly found species on this site are *Gloeocapsa*, *Scytonema*, *Calothrix*, *Pleurococcus*, and *Trebouxia*, because they are able to adapt to the lack of water by forming durable spores or with cellular dehydration, without this causing damage to the protoplasmic structures and, therefore, without losing vitality. Bryophytes are limited to xerophilic or mesophilic species, such as *Grimmia pulvinata*, *Tortula muralis*, and *Bryum* sp., and their development has been observed exclusively on surfaces facing north, northeast, and east, which are the shadiest areas and those receiving the majority of the rainfall. Similarly in lichens, moving from north to east, a gradual decrease in the number of hygrophilic species can be observed, while the nitrophilic species are present on the horizontal surfaces of the upper floors.

Fig. 5.16 *(a) The loggia of the Farnese Palace (Caprarola, near Viterbo); (b) development of* Dirina massiliensis *on the frescoes* – Photo A. Roccardi

In general, lichen colonization in loggias and porticoes tends to take place on the most exterior portions, where environmental conditions are similar to those found outdoors; it is rare to find them on internal surfaces that receive little ventilation or rainfall, nor a sufficient quantity of light (Nimis 1993). In these environments, a strong selection of lichen species takes place

(Piervittori and Roccardi 1998); among the rare instances of colonization, the most important one—both because of the beauty of the site and the extent of the damage inflicted—is that of the Farnese Palace at Caprarola (Viterbo). On the mural paintings of the loggia, an extensive attack of *Dirina massiliensis* f. *sorediata* was observed; because of its peculiar characteristic of developing on vertical walls that are not reached by either rainfall or direct sunlight, it is particularly aggressive and able to colonize artifacts that are placed under shelter (Seaward et al. 1989; Roccardi and Bianchetti 1989; Altieri et al. 2000b) (Fig. 5.16).

Only rarely do plants colonize loggias, since they require a direct supply of water or, at least, high levels of humidity in the substrate; in addition, their development depends on their root system being able to penetrate the walls. The most widespread forms are those of the grassy-ruderal type and, among these, *Parietaria diffusa*, *Cymbalaria muralis*, and *Sonchus tenerrimus* seem to be the most common.

5.3.2 Rupestrian Environments

5.3.2a *Characteristics of the Environment and of the Materials*

The cultural heritage most often encountered in this unique habitat is that of dwellings and churches dug into the rock. These are compounds in which the activity of man closely overlaps that of nature; examples of this are the hundred and more churches near Matera (Italy), in which are preserved notable pictorial testimonies of the artistic/cultural movement that spread through Southern Italy up to the Norman-Swabian epoch. Rupestrian civilizations of considerable size and consequence have been found in several areas of the Mediterranean, in Sicily (e.g., Pantalica), Greece, the Anatolian Peninsula (Cappadocia), as well as in the Middle East and Central and North America (for instance, the Indian *pueblo*).

Inscriptions on rocks dating from prehistoric times are even more widespread on all continents (from the rock art of the Australian aborigines to that of the desert nomads of the Sahara, from the inscriptions in the

Valcamonica in the Alps to those of Central Asia and North America).

Because they are located within a natural habitat, rupestrian settlements are generally in close contact with the surrounding and overlying soil, which is often enriched with organic nutrients linked to the use of fertilizers in agriculture, to the dung from grazing animals, or else to any industrial activities that may discharge pollutants.

These environments often suffer from the same damages characteristic and specific to hypogean environments: infiltrations of rainwater, condensation resulting from thermal inertia, and also humidity through hygroscopicity when there is an abundance of salts, which is sometimes caused by an inappropriate use of the caves, e.g., as animal shelters.

If the rupestrian settlements are extensive, the interior presents a variety of microenvironmental conditions that vary greatly in relation to the distance from the outside. Progressing inward, natural light gradually diminishes, as do aeration and ventilation, while the humidity levels become more constant and generally increase.

From an ecological point of view, here, too, the main limiting factor for the diffusion of organisms is water, whether from rainfall, infiltrations, or phenomena of capillary rise.

In the areas closest to the outside, the development of plant biocoenoses is similar to that on works in stone located outdoors, while what takes place on the inside is a selection of all the potential biodeteriogens that have adapted to the various microenvironmental conditions present.

5.3.2b *Problems of Biodeterioration*

The most evident examples of biodeterioration are usually found in well-lit areas in the vicinity of the entrance. These areas are usually covered by patinas of varying thickness and color, for the most part composed by photosynthesizing microflora, with populations constituted by epilithic and endolithic species of cyanobacteria and green algae, and also by lichen thalli.

In areas that are permanently damp or subjected to water infiltrations, or also in areas characterized by high levels of humidity or

phenomena of stagnation, colonizations of Chlorophyceae (which require elevated and constant supplies of water) tend to predominate, as do hygrophilic lichens, both gelatinous (*Collema* and *Leptogium*) and leprose (*Lepraria*).

The study carried out on the frescoes of the Crypt of the Original Sin—where all the surfaces receive a sufficient quantity of light—in the gorges of Matera (Fig. 5.17) has shown how the water content of the surfaces, linked with the infiltrations of water from above or with the different porosity of the substrate, selected photosynthesizing microorganisms, thus allowing an ecological interpretation of the different morphologies of the patinas. In particular, the study showed a transition from dark patinas—in which cyanobacteria predominate—to brown patinas and then to brilliant green ones, which are almost exclusively constituted by Chlorophyceae; the progression was related to rising levels of moisture (Caneva et al. 2002). Such tight correlation has made it possible to use this information to map out the different water contents of the walls at various heights (Plate 20).

When the rupestrian environment is of considerable size, light plays a particularly important role as a limiting factor for the growth of photosynthesizing organisms. In the absence of artificial lighting, the gradual reduction in sunlight determines a distribution of the species as related to the individual light requirements. The discriminating action of this factor is also related to the seasons as well as the exposure and inclination of the walls. Proceeding toward the interior, it is therefore possible to observe a reduction in the number of plant species and the disappearance of endolithic microflora and lichens. In some cases, the cover of biological patinas on the surfaces can nevertheless remain extensive, and is determined by highly selected species with a strong capacity for adaptation, on the basis of their particular affinity with the growth habitat. From the analysis of the specific composition of the biocoenoses in areas with lower amounts of light, it seems that they are mostly constituted by cyanobacteria and, to a lesser extent, by Bacillariophyceae, Xanthophyceae, and Chlorophyceae (Asencio and Aboal 2001; Caneva et al. 2002).

Fig. 5.17 *Rupestrian Crypt of the Original Sin in the gorge at Matera: (a) entrance; (b and c) interiors*
– Photo G. Caneva

Bryophytes—both mosses and liverworts—are frequent in rupestrian environments, because they do not require direct supplies of water and are able to survive even in low levels of light. Even in their case, the composition of the biocoenoses vary in relation to the light factor. With low levels of light, it is exclusively shade-loving and sciaphilic species that develop, such as *Brachytecium velutinum*, *Thamnium alopecurum*, *Eucladium verticillatum*, *Eurhynchium praelongum*, and *Rhynchostegiella tenella*, similarly to what takes place in cave environments (Berta and Chiappini 1978).

As far as higher plants are concerned, in the innermost areas it is still possible to encounter some of the species, especially *Adiantum capillus-veneris*, *Umbilicus rupestris*, and *Parietaria diffusa* (Caneva et al. 1992b).

In areas that are in almost complete darkness (5–20 lux), plants and bryophytes disappear completely, while it is still possible to observe the presence of a few sciaphilic cyanobacteria —known colonizers of the rocky surfaces of caves—such as *Geitlera calcarea*, *Leptolyngbya tenuissima*, *Scytonema julianum*, *Gloeocapsa sanguinea*, *Gloeothece rupestris*, *Aphanocapsa grevillei*, and *Chlorogloea microcystoides* (Golubic 1967; Dell'Uomo 1982; Hoffmann 1989; Pietrini and Ricci 1993; Aboal et al. 1994; Pietrini et al. 2002).

In rupestrian environments, it is also often possible to find microfungi belonging mostly to the group of Dematiaceae, because of their capacity to develop on substrates lacking in organic nutrients, as well as chemoautotrophic and heterotrophic bacteria, in particular actinomycetes, where there is contamination of the surfaces with soil substances. The microbial communities that develop are in every way similar to those referred to earlier in this chapter (section 1.4).

5.4 MARINE AND FRESHWATER ENVIRONMENTS

by Sandra Ricci and Maria Pia Nugari

5.4a Characteristics of the Environment and of the Materials

Aquatic environments, both marine and freshwater, can be the final resting place for stone, ceramics, metal, and wooden artifacts of archaeological as well as historical and artistic interest.

As far as stone works are concerned, these are usually wall structures that belong to archaeological sites built near the coastline and now submerged because of a rise in sea level. The conditions of the objects and their resting environment can vary according to the depth of the water and the possible periodic resurfacing linked to tide movements or phenomena of bradyseism.

Other metal or wood artifacts may find themselves submerged accidentally: this is frequently the case with ships wrecked in the open sea or in the proximity of ports. As artifacts, ships and the objects contained within them constitute a heritage to be preserved and cared for and often present new and difficult problems, both in the area of immediate intervention and in subsequent restoration and conservation.

In an aquatic environment, the distribution of the organisms and microorganisms that are potentially damaging for artifacts is strongly influenced by a variety of factors, such as the type of substrate, the turbidity of the waters, light, temperature, the availability of oxygen, and pH.

Analyzing the marine environment from an ecosystemic point of view, we can determine the presence of various communities of microorganisms as well as plant and animal organisms. The first are often the primary colonizers and develop on submerged surfaces forming biofilms that, in turn, lead to a variety of processes —e.g., corrosion—and encourage the establishment of other, larger organisms (macroalgae, animals). The microorganisms present in the biofilm or in the sediments in which the objects lie buried can induce conditions of anaerobiosis, which enable the development of other groups, such as anaerobic bacteria; well known, for example, are the corrosion phenomena provoked by sulphur-reducing bacteria on the surface of metallic artifacts.

The aquatic organisms that establish themselves on the biofilm are mostly benthonic; they are able to colonize materials submerged in the sea, adhering to them for part or all of their life cycle and occasionally establishing trophic relationships with these materials. Archaeological works in stone can be considered analogous to a rocky seabed, because they are similar both in nature and in exposure.

The relationship that benthonic organisms establish with the rocky, metal, or wooden substrate are conditioned by several environmental factors and are particularly important where plant species are concerned: the micro- and macroalgae that establish themselves on the submerged structures are not able to move, and therefore the environmental parameters present in the growth area play a fundamental role in selecting their development. Light is the parameter that most strongly influences the distribution of benthonic algae and phanerogams in an aquatic environment. The intensity and quality of light vary considerably with the depth and the turbidity of the water as well as with the topographic characteristics of the seabed. Light does not directly influence animal organisms, although it can influence their development indirectly by influencing their trophic relationship with plant life.

The *phytobenthos* on stone surfaces modifies its floristic composition in relation to depth. The epilittoral zone—with occasional water spray—and the supralittoral and eulittoral zones—within the tidal area—are prevalently colonized by epilithic and endolithic cyanobacteria; as the water gets deeper, eukaryotic algae begin to appear (Giaccone et al. 1994). The area affected by algal colonization usually reaches 150–200 m in depth, beyond which light does not penetrate. Since the development of algae is closely linked to the presence of light, algal fouling is more pervasive closer to the water surface.

The deeper regions are exclusively colonized by a few groups of algae, mostly Rhodophyceae; these are able to efficiently carry out the photosynthetic function, thanks to the presence in their cells of phycoerythrin (a pigment that absorbs green-blue light, which is the light with the maximum power of penetration).

Unlike autotrophic microorganisms, the distribution of heterotrophic organisms such as bacteria and fungi in the aquatic medium is mostly linked to the availability of organic substances and, since they are mainly aerobic, to the concentration of oxygen (Jordan 2001). Studies on fungi are more frequent; their presence in marine, oceanic, and pelagic environments is rare, while it is significantly more common in coastal waters, which are richer in biological pollutants from the land and in organic substances from the inflow of eutrophic waters. Saprophytic fungi are most widespread in freshwater, but they are also present in seawater, albeit to a lesser degree. Those belonging to the group of *Mastigomycotina* are adapted to living in an aquatic environment, both because of their nutritional requirements and because of the method of dispersal of their spores. The best known in this group belong to the *Saprolegnales* and, in particular, to the genus *Achlya*. Certain groups of hyphomycetes are also present in waters with an organic substrate, because they have adapted to dispersing their spores in this habitat (Cooke 1979). Ascomycetes and deuteromycetes are frequently present in marine waters because they are able to tolerate both wide ranges and small differences in saline concentration, although, strictly speaking, the definition of "marine" is applicable only to fungi that are able to live and reproduce in salt concentrations of 30‰. In addition to salinity, the other factor that strongly influences the distribution of fungal species in marine waters is temperature. These two factors—salinity and temperature—to which fungal species adapt, often induce morphological variations that make systematic identification very complex.

5.4b Problems of Biodeterioration

In the great majority of cases, investigations relating to the biodeterioration of submerged structures are directed to the identification of the algae responsible for the biofouling of shipwrecks, pipes, and industrial plants. Marine algae are considered one of the groups of organisms that are most closely involved in the deterioration of natural substrates such as rocks, but also of artificial, man-made products such as objects in stone, plastics, glass, varnishes, and metals. The benthonic algae that colonize the submerged stone in the waters of the Mediterranean belong mostly to the systematic groups of the Bacillariophyceae, Rhodophyceae, Phaeophyceae, and Chlorophyceae. Mollusks and polychaetes play a very important role in biodeterioration, but animal forms are not discussed here.

Degradation includes processes of solubilization and boring of rocks, corrosion of metals, damages due to mechanical action on the substrate, and penetration by vegetative structures (filaments or propagules) or by portions of the thallus (rhizoids) with an anchoring function (Fletcher 1988).

Among the Phaeophyceae, one of the most widespread algal genera is *Cystoseira*, which includes about fifty species and is amply distributed in the Mediterranean, where it forms different associations with other algae; the most frequently found species is *C. barbata* (Pizzuto and Serio 1995; Pizzuto 1999). *Dictyota dichotoma* is also very frequent; it lives anchored to rocks, in calm waters, and near the surface. The following are also somewhat frequent: *Dictyota linearis, Ectocarpus confervoides, Sargassum hornuschii, Stilophora rhizoides.*

Rhodophyceae are also often encountered, *Pterocladiella capillacea* being particularly amply diffused. The genus *Ceramium*—particularly the species *C. flaccidum*—represents one of the main components of biofouling of submerged works (Fletcher 1988). The genus *Polysiphonia* is often present with the species *P. sertularioides.*

Of the Chlorophyceae, the genera *Cladophora* and *Ulva* have been frequently found.

Most macroalgae often exhibit dense coverings of epiphytic diatoms, which anchor themselves by means of gelatinous peduncles.

Research carried out on the submerged structures of the Roman fish pools from the site of Torre Astura (Nettuno, near Rome)—which date from the first century B.C.E.—has allowed the study of how algal flora is highly influenced by the physico-chemical characteristics of the water (Ricci 2003). The particular topographic position of some of the walls of the site had favored the development of a great variety of algae. Indeed, although these structures were at a negligible depth (0.5–1.5 meters), they benefited from a good water exchange, which guaranteed a good supply of oxygen and nutrients to the biocoenoses present, while other structures, situated in less exposed and less oxygenated areas, exhibited a less abundant floristic wealth. Experiments carried out *in situ* with quarry samples (Carrara marble and limestone), submerged in seawater at a depth of 1–2 meters, showed how after only two months both animal and vegetable colonization had already taken hold; this made it possible to establish the dynamics and the time frame of the process of biological colonization in the particular environmental conditions and specific positioning of the site. Investigations of this kind make it possible to plan conservation of submerged structures and to define regular maintenance interventions.

The fact that algal colonization can constitute some sort of "historical archive" was emphasized on the occasion of the study of the incrusting algae (*Sporolithon* sp., *Neogoniolithon notarisii*, *Spongites fruticulosa*) found on the shipwreck of the *Iulia Felix*, offshore of Grado (Trieste). From the isotopic analysis of the stratifications of the carbonate algal incrustations, it seems that it is possible to reconstruct and date the various processes of deposition and, hence, understand the natural evolution of the underlying seabed (Bressan et al. 1994).

With regard to heterotrophic organisms involved in the degradation of immersed artifacts, most studies are directed to the biodeterioration of wooden finds, such as boat wreckages, pile dwellings, etc. Many fungal species have indeed been identified as responsible for the decay (Mouzoras 1986; Jones et al. 2001). Among the marine lignicolous fungi, the species encountered most frequently belong to ascomycetes and deuteromycetes and are responsible for soft rot. The wood surfaces involved in this form of degradation become soft, and the abrasive action of water and sand can remove the outermost layers of the wood exposing inner layers to further attack; this results in increasingly widespread colonizations with secondary attacks by wood-degrading bacteria. In additon to cellulolytic bacteria, soft-rot fungi belonging

to the genera *Stemphilium* and *Fusarium* were isolated on samples from the wreck of a Roman cargo ship recovered in the Tyrrhenian Sea a mile from the coast at Ladispoli, near Rome (D'Urbano et al. 1989). Most soft-rot fungi require relatively high levels of dissolved oxygen; however, it was shown that some species are able to remain active at depths of more than two thousand meters (Kohlmeyer 1969).

Water-logged wood can also be attacked by fungi responsible for brown rot and white rot, such as ascomycetes and basidiomycetes. For these fungi, especially for basidiomycetes, the limiting factor is the diffusion of oxygen; most of them cannot tolerate low concentrations of oxygen and cannot survive in anaerobiosis. Moreover, a fall in the partial pressure of oxygen results in a reduction in the activity of cellulolytic and ligninolytic enzymatic systems. These species of fungi are therefore rare in deep seawater, but frequent in tidal zones and especially in freshwater, rather than in the sea.

Bacteria are also commonly involved in the biodegradation of wood in a marine environment; many cellulolytic species find favorable conditions for their development in wood submerged in the sea (Jones et al. 2001). They are responsible for the erosions, cavities, and tunnels in the cell walls of the wood, especially in the innermost layers, producing either partial or total loss of the S2 layer of the secondary cell wall. The morphological characteristics of this kind of decay are documented in many SEM and TEM studies on shipwrecks (Abbate Edlemann et al. 1989; Blanchette 2000). The taxonomic position of these bacteria is still the subject of ongoing study because of the difficulty of isolating them in a pure culture; studies under the microscope have shown that they are gram-negative bacilli and that, in the case of the bacteria responsible for erosion, some species belong to the Myxobacteriales and Cytophagales (Blanchette et al. 1990; Eaton and Hale 1993).

Bacteria that degrade wood can live even with reduced oxygen or no oxygen at all. In particular, the bacteria responsible for erosion seem to be the most tolerant to this factor, since they are able to survive in conditions close to anaerobiosis. Precisely because of the greater requirements for oxygen (Fig. 5.18), on the other hand, attacks by bacteria responsible

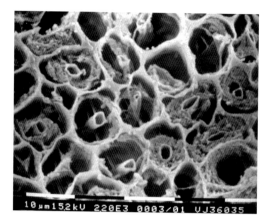

Fig. 5.18 *Effects of a bacterial attack on water-logged wood (from Blanchette 1990)*

for boring are not encountered at great depths, although they have been isolated in wood placed at depths of around 5,000 m both in the Pacific and the Atlantic oceans (Jordan 2001).

These bacteria also show a great capacity for adaptation as far as temperature is concerned; they were found on wood recovered from the bed of the Baltic Sea, which means that active biodegradation phenomena may possibly also take place at rather low temperatures (Blanchette et al. 1990).

Among the deterioration phenomena caused by heterotrophic microorganisms in a marine environment, we should also cite—although it is limited in scale—the decay taking place on limestone due to active penetration by certain species of fungi. These fungi are called marine nonlignicolous fungi and are studied mostly because of their capacity to penetrate within the tubules of perforating animals, within shells, and within calcareous incrustations of algae; but they are also able to damage limestone artifacts. Among these we find ascomycetes—with *Halosphaeria quadricornuta* and *Ramispora salina*—and deuteromycetes—with *Periconia prolifera*, responsible for the decomposition of the limestone coatings produced by the teredos during the boring of wood, or *Pharcidia balina* and certain species of *Cirrenalia* and *Humicola*, which decompose the plates of barnacles and the shells of mollusks (Cooke 1979).

Lake environments exhibit colonizations of submerged materials that differ from those found in marine environments because of the different salinity of the water. First and foremost, different systematic groups are involved. In lakes, aquatic and aerophytic species often coincide and can develop both in the aquatic and in the terrestrial environment. Benthonic algal species are less frequent compared to planktonic ones and are limited to filamentous or coccoid forms of reduced dimensions; their development takes place mostly in the form of green mucilaginous layers. Algal colonization of artifacts of historical and artistic interest in lakes is less frequent and of a lesser extent than in marine environments. Lake water often has levels of temperature that are not suitable for the development of many species of algae and levels of turbidity that can limit the development of benthonic forms, which are not able to move toward more suitable light conditions.

On the other hand, heterotrophic fungi and bacteria that degrade organic materials are relatively frequent in freshwater environments also; however, most of the species require high levels of oxygen dissolved in the water and are therefore frequently found in the vicinity of the banks and shores of lakes rather than on the beds. The stratification of sediments can indeed lead to conditions of anaerobiosis, which are incompatible with life for the majority of species.

The role of fungi and bacteria in the deterioration of objects of historical and artistic value found in freshwater has not been studied specifically. It is safe to assume, however, that the same processes of decomposition that take place on organic matter present in the waters—such as vegetable tissue or dead animals—would also occur on artifacts of an organic nature located in an aquatic environment.

5.5 EDAPHIC ENVIRONMENTS

by Claudia Sorlini, Giancarlo Ranalli, and Elisabetta Zanardini

5.5a Characteristics of the Environment and of the Materials

Examined superficially, the edaphic matrix (i.e., the ground or soil) might appear as an inert material, purely and simply a reserve of nutritional elements for the development of vegetable organisms, or as a depository for mineral and organic materials that accumulate there before being drawn up by the plant. In reality, the soil should be considered as a mature stage of development with a relatively stable equilibrium, having achieved the optimum relationship among the resources of the soil, vegetation, and climate, and based on a succession of physico-chemical and biological reactions that are synthesized in the cycles of matter (C, N, P, S, etc.) characteristic of the biosphere (Fig. 5.19).

The composition of the soil is closely related to the lithological matrix of those portions of the earth's crust that are normally exposed to the atmosphere and subject to biotic influences. The mineral component is of a lithological nature, while the organisms and their residues represent the organic part: these two components constitute the solid phase (about 50% of the volume), although they differ considerably in their proportions: in sandy soil, the mineral fraction reaches up to 99%; in cultivated land 96–98%; and between 60% and 80% in soil with an accumulation of organic substances protected from oxidation. The liquid and gaseous phases of the remaining 50% and their ratios can be subjected to rapid fluctuations in response to rainfall, losses through drainage, and absorption by vegetation, if it is present. The composition of the air present in the soil can be significantly different from the air in the atmosphere: there can be different levels of moisture, higher levels of CO_2, lower oxygen levels, and the presence of other gases produced by microbial metabolism; all of these can influence the vital activities of the microflora, which can potentially be biodegrading (in conditions of aerobiosis, microaerophilia, or anaerobiosis) for works of artistic or archaeological value typically found in edaphic environments.

The number of biological species that live in the soil is extremely high, especially in terms of microflora. The latter varies greatly, particularly in relation to the water content (both capillary and hygroscopic) and the level of organic substances present, to the pH, to the temperature, and to the aeration. However, microflora is present even in soils that are very poor in organic materials and that have extremely low levels of humidity, such as for instance desert sand, where the microflora can reach values as low as $10^3–10^4$ of Units Forming Colonies (UFC) per gram of sand. In soils that are rich in organic substances, aerated, moist, with pH levels that are close to neutral, and that are not treated

Fig. 5.19 *Cross-section reconstructing an ideal soil structure and relative distribution of microflora*

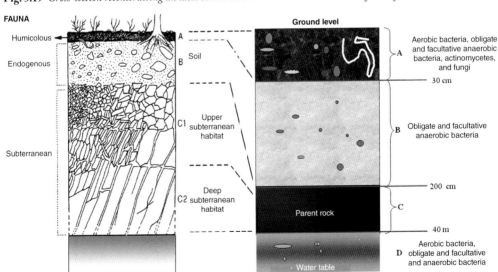

with agrochemical products commonly used in the cultivation of fields, the concentration of microorganisms can reach values of several billion per gram of fresh weight. It should also be added that most of the microbial biomass is distributed in the top 30 cm of soil, progressively diminishing farther down. Until recently, it was thought that microflora practically disappeared at a depth of two meters, while today we have evidence of the presence of microflora at much greater depths as well as in the water tables, although in lower quantities. Abundant rainfall and irrigation contribute to the percolation of microorganisms through the soil.

It should be noted that the works of historical and artistic interest that can still be found in these environments are mostly of an inorganic nature (stone, ceramic, and metal), since organic materials are quickly attacked by edaphic microflora and more or less rapidly mineralized. In some very specific cases, linked to the conditions of the resting place of the object—such as clay seams that create an anoxic environment, sandy soils with exceptionally low water content, or also permafrost—exceptional conditions are created that allow the preservation of organic materials as well.

5.5b Problems of Biodeterioration

The aerobic microbial communities operate in the aerated layers of the soil closest to the surface; overall, their metabolic versatility is significant and, therefore, their degrading and biodeteriogenic power is high. Both chemo-heterotrophic and chemoautrophic bacteria are present. Among the first, certain groups of proteolytic, cellulolytic, amylolytic, and lipolytic bacteria (*Pseudomonas, Sarcina, Bacteroides, Cytophaga, Sporocytophaga, Sorangium, Vibrio, Cellvibrio, Cellfalcicula, Bacillus, Clostridium, Alcaligenes, Staphylococcus*) contain the genetic information required to enable them to degrade numerous organic molecules for which the enzymes responsible for breakdown (hydrolase, peptidase, cellulase, amylase, lipase) are synthesized by induction. The second category includes bacteria (*Thiobacillus, Thiosphaera, Desulfovibrio, Desulfobacter, Desulfococcus, Idrogenomas*) capable of producing compounds, such as sulphuric acid

and nitric acid, that have an aggressive action on all materials, including metal and stone, and that are able to oxidize ferrous to ferric iron.

Their activity is generally greater than that of anaerobic microflora, in terms of the range of materials degraded and often also in terms of a higher speed of attack. The superficial layers of the soil also contain considerable concentrations of photosynthesizing bacteria, belonging especially to the cyanobacteria.

Progressing deeper and deeper into the soil, the quantity of oxygen diminishes rapidly until environments are reached that are completely anoxic, where the potential for carrying out oxidation/reduction reactions is extremely low. In these environments, neither aerobic bacteria nor fungi can be active; as a rare exception, it is possible to find certain yeasts that, in the absence of oxygen, are able to activate an anaerobic, fermentation-type metabolism. However, their presence deep underground is more hypothetical than proven.

Photosynthesizing organisms are also not present here, since the light indispensable for their metabolism does not penetrate. The only microorganisms that can remain active are the obligate and facultative anaerobes, provided—as already mentioned—that the level of free water is sufficiently high (Chapter 2, section 2).

Two microbial groups can be active under these conditions; the first is the group of fermenting microorganisms (bacteria, yeasts), which, if they are able to hydrolyze macromolecules (cellulose, casein, starch, etc.), increase the products of the hydrolysis. In environmental conditions favorable to them, these microorganisms can attack organic works of art composed of cellulose, keratin, fibrin, collagen, etc.; therefore it is very difficult to find paint layers with organic binders after they are buried.

The products of the metabolism of these microorganisms are acids and alcohols, both of low molecular weight; they can have a corrosive action on the materials but only if the soil is not buffered, i.e., acid and subacid soils. However, these products are important because they function as a growth substrate for the second group of microorganisms active in conditions of anaerobiosis, i.e., the bacteria that carry out anaerobic respiration. These do indeed require molecules (mostly organic ones) with low molecular weight as electron donors as well as oxidized inorganic compounds as electron receptors; of the latter, the most important ones are nitrates, sulphates, and carbon dioxide, which are converted into molecular gaseous nitrogen, hydrogen sulphide, and methane, respectively. Of the three gases produced, the one that has the most biodeteriogenic power is hydrogen sulphide, which can corrode not only the metals of the artifacts, but also those present within the stone.

What emerges clearly, in light of the above, is that the risks for interred works are greatest when these are buried near the surface, i.e., in an aerobic environment.

Among the many works of varying chemical nature subjected to attack by the aerobic microflora present in the surface layers of the soil we can cite examples where complex microbial communities colonized monuments, such as the Roman necropolis of Carmona in Spain (Heyrman and Swings 2001; Heyrman et al. 2003) and the castle of Herberstein in Austria (Heyrman et al. 2002).

The sulphuric acid produced biologically by sulphur-oxidizing bacteria—in particular those belonging to the genus *Thiobacillus*—and the nitric acid produced by nitrifying bacteria—of the genera *Nitrobacter* and *Nitrosomonas*)—act as powerful corrosives reacting with the calcium ions to form calcium sulphate (gypsum) and calcium nitrate, which are soluble and easily washed away (Agarossi et al. 1985; Bondonno et al. 1989).

At greater depths, as mentioned earlier, the most troubling action is attributable to the sulphur-reducing bacteria on metals, especially those containing iron (Little et al. 1998).

In the case of buried objects made of bronze and copper, it is important to correlate the corrosion processes with the chemistry of the soil; factors such as pH do indeed have a strong influence on the condition of the metal, since the more acid soils are more aggressive. As far as soil is concerned, the most relevant parameters are: texture, pH, concentration of soluble salts (sulphates and chlorides), content of clay and of organic substances, electrical conductivity, and buffering capacity—the values of these parameters give a measure of the "aggressivity" of the soil. To these factors, we

should add possible microbial corrosion due to the presence of particular groups of microorganisms, especially in conditions of anaerobiosis (Rosenquist 1961; Tylecote 1979). In recent years, archaeologists have noted an increased level of deterioration in metal objects, linking corrosion and the state of deterioration to the properties of the soil and to the environment in which the objects were found (Macqueen 1980; Gerwin and Baumhauer 2000). The comparison between archaeological finds brought to light in the first half of the twentieth century and those excavated more recently (1990–1995) from the same sites confirms a continuous increase in the degree of deterioration, influenced by the high content of chlorides and sulphates and also by the concentration of atmospheric pollutants that fall to earth with the rainfall or in the form of dry deposits. The increased levels of damage could be ascribed to the high levels of sulphur compounds and oxides of nitrogen in the atmosphere, to acid rainfall with the consequent acidification of the soil, and also to the high concentration of salts in agricultural soil due to intensive applications of fertilizers.

Therefore, while in the past it was thought that archaeological works in iron or bronze would not be very damaged if buried, in recent years an increased understanding of the corrosion of such materials has often led, instead, to their removal from the edaphic environment and to subsequent conservation measures. To this end, in-depth studies of the composition of the soil are necessary to be able to revise and evaluate the conditions in archaeological sites and to develop suitable measures for their protection and conservation (Gerwin and Baumhauer 2000).

On the other hand, soil can also play an important protective role for works of art that are buried in it. This occurs, for example, in desert climates where the levels of free water are so low that they do not allow microorganisms —which are nevertheless present—to carry out their metabolic activities in any evident way. For this reason, the sands of the desert have protected for centuries and at times millennia (and indeed continue to do so) the most precious vestiges of the history of human civilization, as for example the temples and, even more so, the tombs of ancient Egypt (Chapter 6, section 2.1). In these conditions of aridity, not even the mechanisms of physico-chemical deterioration can be triggered.

Based on the presupposition that deterioration was interrupted by burial, a comparison was made, in an archaeological site of a Roman settlement in ancient Palestine, between the different patterns of deterioration now present in exhibited works of art and those on objects located in the hypogean environments; this comparison has been used as a biomonitoring instrument for changes in climate (Danin 1986).

Even in temperate and humid climates, nonmetallic works of art buried in anoxic environments are generally well preserved, since deep into the soil the microflora is scarce and often finds itself in less than optimal conditions for its activities to take place, which are very weak, if not altogether latent. This is repeatedly the case in archaeological excavations where at the moment of discovery the objects are in good condition, but after a few decades the physico-chemical and biological processes lead to more or less irreversible alterations in the materials (as, for example, with the Etruscan tombs at Tarquinia, the *Domus Aurea* excavations in Rome, or the Roman garden house in Santa Giulia in Brescia). The latter, for example, when brought to light with the excavations, exhibited the splendor of its frescoed walls in excellent condition; but after twenty years of exposure to light and air, because of the less than suitable conditions of conservation that favored the establishment of high levels of humidity, the frescoes were deteriorated by an invasion of aerobic microflora with a high presence of actinomycetes (Sorlini et al. 1983).

This is why, when it is not possible to immediately carry out the necessary conservation interventions, it is often preferable to put the same soil back (if its characteristics are suitable) and cover the vestiges of the past brought to light by the excavations with it, or else with sand or some other inert material, in order to temporarily guarantee preservation conditions, while waiting for the appropriate environmental measures to be put in place (Chapter 7, section 1.2).

If this same soil is rich in organic substances and microorganisms and is sufficiently damp, the aerobic conditions created by moving it may trigger rapid processes of deterioration. It is then necessary, especially in the case of organic finds, to adopt strategies for the compression of the soil, so as to reduce to a minimum the residual levels of oxygen.

Furthermore, it should be remembered that no microorganism is able to degrade lignin anaerobically. There are known cases of wooden materials being preserved for thousands of years in anoxic environments, even in alkaline or acidic conditions, which reduce even further the number of microbial populations (the alkaline reducing fungal and the acidic reducing the bacterial populations). This is what happened in the fossil forest of Dunarobba (Terni), where trees of Taxodiaceae—which are now extinct in Europe—were buried for about two million years in a flow of clay that preserved the wooden structure in a practically intact state (which is therefore not really fossilized) up to the moment of its exposure to the subaerated environment.

Optimal preservation conditions are offered, as mentioned, not only by hot and dry environments such as those of desert sands, but also by extremely cold environments, where the activity of free water is extremely low, because the water is immobilized in the form of ice. This is why it has been possible for mummified human bodies (e.g., the mummy of Similaun, Ötzi) or for bodies of prehistoric animals buried in glaciers (the mammoths of Siberia) to be preserved, even for thousands of years, in optimal conditions.

In the preservation of buried materials, an important role is played by the clay component for two main reasons: the first is that the layers of clay are practically impermeable to water and, as such, contribute to good preservation conditions for the works of art buried within it; second, since clays are rich in charges on their surface, they easily create bonds with potential pollutants, at times even highly aggressive ones, partially bonding with them and thus preventing their deteriorating action on the works of art.

CHAPTER 6

BIODETERIORATION PROBLEMS IN RELATION TO GEOGRAPHICAL AND CLIMATIC CONTEXTS

by Giulia Caneva and Alessandra Pacini

The most important variables that affect biodeterioration phenomena have already been analyzed in the preceding chapters; however, materials and environmental conditions being equal, the particular climatic context in which the work is located as well as its environmental history and the biogeographical characteristics of the site are all significant additional factors in biodeterioration.

Indeed, the presence of potentially biodeteriogenic organisms depends not only on the environmental conditions of the surrounding area (Chapter 2), but also on phenomena that occurred in past history, giving rise to processes of speciation with the genesis of particular groups that adapted to the specific conditions of the various biogeographical regions.

While offering a general understanding of the processes of biodeterioration, this chapter has the more specific objective of highlighting the most salient problems associated with a given territorial context being examined on a large scale, while emphasizing the risk factors that may be more or less relevant depending on the geographical contexts involved. Such a perspective, if combined with the environmental typologies we have already encountered, supplies the fundamental information required to outline a large-scale "map of biological risks" of sorts, which constitutes an important instrument in preventive conservation. These issues have been emphasized by several authors and, on occasion, comparative analyses of colonization phenomena occurring in different climatic contexts have been undertaken in order to point out tendencies and correlations on a macroclimatic scale (Danin and Caneva 1990; Warscheid et al. 1990; Ortega-Calvo et al. 1993; May et al. 2000). Although interpretation of the phenomena can often be complex, given the overlap of local and topographic factors with phenomena on a much larger scale, it is nevertheless legitimate to make a few observations on some general trends. Therefore, while avoiding an unnecessary repetition of concepts that have already received ample coverage, we will primarily focus on whether or not a particular phenomenon or problem is amplified or reduced in relation to a generic environmental location.

6.1 GENERAL PRINCIPLES

The diversity of the various climates present on our planet is the result of temperature and pressure gradients present in the atmosphere; these are in turn generated by the motions of the planet Earth in its orbit round the Sun, by the various radiations emitted by the Sun, by the contours and structure of the Earth's surface, as well as by the physico-chemical and biological emissions of the planet towards the atmosphere (Odum 1988; Bullini et al. 1998).

To the main climate differences that are mainly linked to latitude and to the equally substantial differences linked to altitude, other variables—often of considerable importance—must be added, such as for instance the distribution of the oceanic masses and the emergent landmasses as well as the general orography and hydrography of the land; all these together determine the extremely varied body of climates which characterize our planet.

Because climate phenomena vary over time giving rise to the seasons, a description of climate cannot escape an analysis covering annual cycles and must take into account a time sequence of at least twenty years in order to include in the mean the least representative levels of temperature and rainfall, both the very high and the very low values.

Various climate indices have been developed —even with the awareness of their inevitable limitations—to give a concise representation of climate (for instance, Lang's rain factor, De Martonne's aridity factor, Emberger's pluviometric quotient, the oceanicity index of Amman, Rivas-Martinez's thermicity index, etc.); these are climate descriptors designed to highlight specific phenomena, and a more specialized literature should be consulted for their analysis (Odum 1988; Pignatti (Ed.) 1995; Bullini et al. 1998).

In the present context we consider that it is important to discuss bioclimates, i.e., climates in relation to the development of biological organisms (Chapter 2, section 2.6), and we include a number of bioclimatic diagrams indicative of climate type. In these graphs (for example, Fig. 6.5) we can follow parallel variations in the most significant parameters, i.e., the average temperatures (T) and the total rainfall (P) during the various months of the year, in addition to the respective average and total amount for the year; it is customary to adopt a scale of 1:2 of T in relation to P, in order to be able to visually represent the critical periods of drought (dotted area), or on the contrary the periods of hydric excess (visually represented as dark areas).

For ease of description we synthesized and grouped the main typologies, selecting them based on their representativeness of the varieties encountered on Earth, especially of the locations in which the main human settlements flourished and where we may therefore find the monuments and works that constitute the cultural heritage of humanity (Fig. 6.1). Consequently, we omitted locations that are of importance from a naturalistic point of view but are less so culturally, i.e., the Arctic tundra, the cold Boreal forests or taiga, the Asiatic and North American steppes, in which native human populations were quite low in numbers and generally nomadic, and where the material evidence linked to man is correspondingly limited.

6.1.1 Historical and Biogeographical Aspects

Although many biodeteriogenic organisms are acknowledged to be cosmopolitan, in reality there are significant floristic differences between individual geographic locations, as a function of the climate variations and the different environmental histories of the sites. The floristic differences become especially evident if the species—and, in the case of microflora, also the individual microbial strains—are characterized with precision and if the identification goes beyond the level of genus. If, on the other hand,

Fig. 6.1 *World distribution of the main sites recognized as World Heritage Sites by UNESCO*

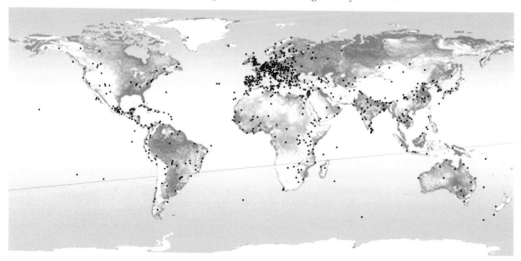

the identification does stop at the level of genus, there is less evidence of geographical differentiation because vicariant species are often present in different territories. Hence the interest in biogeographical problems, the main objective of which is to achieve an understanding of biological phenomena in their spatial manifestations (Margalef 1974) and, more specifically, to describe and analyze (in terms of causes) the distribution of living organisms, both in its present and its historical dimensions (Zunino and Zullini 1995).

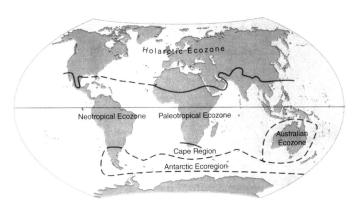

Fig. 6.2 *Main phytogeographic regions of the world (modified from Takhtajan, in Zunino and Zullini 1995)*

It is useful to remember that, with the works of Darwin and Wallace, the scientific bases were laid for an understanding of present phenomena from a historical and evolutionary perspective. The current floristic, faunal, and microbial diversity of a site cannot be explained solely in terms of present characteristics; the impact of the succession of bioclimatic events that has characterized the history of the Earth must also be taken into consideration. Ice ages, the drying up of seas and the consequent saline stresses, or else the excesses of heat or the submersion of coastlines are all key phenomena of critical importance for the understanding of present characteristics.

The geographical distribution of species on our planet presents certain recurring characteristics; various systems of biogeographical classification have been defined through a comparative analysis of the distribution areas of species. Takhtajan's model (Fig. 6.2) recognizes the existence of specific floristic areas, some quite extensive—such as the Holarctic region, which covers North America, Eurasia, and North Africa—and others that are much more localized—such as the Cape region situated in the area of the Cape of Good Hope, biogeographically very isolated, and characterized by more than 80% of endemic species. On the other hand, it is noteworthy that the tropical regions of the "New World" are different from those of Africa, South Asia, and the Pacific Islands, thus defining respectively the Neo- and Paleotropical regions.

6.1.2 Bioclimatic and Biogeographical Regions

Just as the bioclimate of a site is relevant to the particular problems of biodeterioration characteristic of certain geographical contexts, on a larger ecological scale it contributes to the rise of different biomes, i.e., biogeographical regions characterized by a particular macroclimate, in which specific plant, animal, and microbial communities are found.

Moreover, an analysis of the present climate makes it possible to characterize the most frequently recurring types of climate in relation to their influence on biological organisms. On the basis of the average annual temperature and the total average rainfall it is possible to establish the fundamental types of biomes present on Earth, which range from deserts and grasslands, to deciduous and tropical forests, to the taiga and the tundra (Fig. 6.3).

It is not easy to fit the complexity of climate types into clear-cut categories; different climate parameters and indices have been proposed in order to provide an adequate classification. According to the Russian climatologist Wladimir Köppen (1936) there are five distinct main types (tropical, arid, temperate, boreal, polar), which are primarily based on thermal values, with the exception of the arid climate, which is based on pluviometric measurements.

Subsequently, Thornthwaite and also Walter and Lieth emphasized the critical role played by rainfall levels and pluviometric

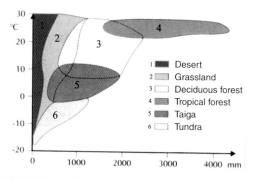

Fig. 6.3 *Distribution of biomes in relation to temperature values and rainfall (modified from Zunino and Zullini 1995)*

regimes as well and presented more articulated proposals for climate types and subtypes (Walter 1994; Table 6.1).

In the present context, it seems useful to discuss some of the broader climate typologies that recur more frequently in relation to the location of human settlements and their works and artifacts (Fig. 6.1).

It is precisely due to these inherently favorable climate conditions that one of the areas with the highest concentration of ancient human settlements—and consequently also the richest in monuments and artifacts linked to the culture of humans—is the region stretching from the Mediterranean up to Central Europe.

6.2 PROBLEMS OF BIODETERIORATION

6.2.1 Desert Climates

Desert climates are defined as those in which the annual rainfall does not exceed 200 mm, with extreme cases receiving as little as a few millimeters of water, and in which a relatively high year-round temperature leads to the more or less rapid evaporation of all meteoric waters.

Such a climate (class III) is present in various regions of the globe in the subtropical belt where downward currents create areas of high pressure, i.e.: in Africa, in those areas of the Sahara situated close to the Mediterranean coasts and in Namibia; in Asia, in the Arabic Peninsula and in areas stretching from the

Middle East towards the Iranian Sistan, the Valley of the Indus, and the Western part of China; in America, along parts of the coast of California and Chile; and in Australia, in the whole of the central area of the continent (Fig. 6.4). Desert climates are therefore characterized by great aridity throughout the year, because of the extremely low rainfall as well as the high temperatures (Fig. 6.5); as a result, biological phenomena are very limited.

The temperature variations between night and day—which can at times exceed 50°C—are also of considerable importance, as they create further stress factors for biological organisms.

In this context, aside from adaptation phenomena such as succulence or the acquisition of specific ecophysiological mechanisms, the most evident phenomenon from the floristic point of view is the reduction in the length of the life cycle —down to a few weeks or even a few days— with showy flowering taking place during the brief periods of adequate water supply.

A further limiting factor for biological growth in desert and subdesert climates is the high concentration of salts that results from intense evaporation and leaves salts to crystallize on surfaces. This phenomenon is often accompanied by a considerable increase in volume and therefore has highly destructive consequences for the substrate. From a biological point of view, salts create high levels of osmotic pressure that render the habitat inhospitable for most organisms, with the exception of halophilic ones.

In these climatic contexts, the use of earthen material (adobe) in building construction can be frequently—although not exclusively—encountered, because it does not require baking and the consequent use of wood (difficult to find in these climates) for fuel. In addition, the low rainfall reduces the necessity of frequent repairs. Entire cities in Asia—for instance, Bam in the Iranian Sistan; historic settlements in the Arabian Peninsula, among which ancient cities in Oman; several cities in India and in China—are examples of the use of this building technique, with variations in relation to the importance of the buildings. It should be remembered that the use of earthen material with a high clay content is also found in tropical and subtropical climates (for

Table 6.1 Climate classification according to Walter and Lieth (organized into classes, I–IX, and subclasses)

Class I
Equatorial zone (found between latitudes 10°N and 10°S), always humid, or with two rainy seasons, without cold periods, and with temperatures that are almost always above 20°C, with minimal seasonal variations in temperature.

Class II
Tropical zone (found between latitudes 10°N and 10°S), with a humid/wet season in the summer and with a cooler and drier period.

Class III
Desert subtropical-arid zone, with high temperatures and low rainfall, resulting from the compression of large air masses that warm up and become drier when descending; winter may be cold, but frosts are only occasional.

Class IV
Transition zone, with winter rain and dry summers; no significantly cold season, although frosts are not completely absent. Mediterranean regions belong to this zone.

Class V
Temperate-warm zone, always humid and with mild temperatures, considerable fluctuations in temperature are possible, but frost is only occasional; typical of certain parts of Australia, Tasmania, and New Zealand.

Class VI
Typical temperate zone, with a distinct, although short, cold season; present in Central Europe and on the East Coast of the United States.

Class VII
Temperate-arid continental zone, with a considerable variation in temperature between summer and winter, i.e., with markedly hot summers and cold winters, and with low total annual rainfall; to this zone belongs all of Eurasia up to the Far East, as well as part of North America, and the southern part of Argentina.

Class VIII
Temperate-boreal or cold zone, with a very long cold season (longer than six months) and a cold and humid summer, in which the average monthly temperature of the warmest month is above 10°C; this kind of climate is only found in the northern hemisphere.

Class IX
Polar arctic zone, with low rainfall throughout the year; there is a brief season without ice (warmest month with a temperature lower than 10°C) and with no night, and a long winter without daylight.

Subclass a	Relatively dry for the corresponding climate zone
Subclass h	Relatively humid for the corresponding climate zone
Subclass hh	Extremely humid for the corresponding climate zone
Subclass oc	Oceanic variant (in nontropical climates)
Subclass co	Continental variant
Subclass fr	Regular frosts in tropical climates
Subclass wr	With rainfall prevalent during winter, where this is not typical for the zone in question
Subclass sr	With rainfall prevalent during summer, where this is not typical for the zone in question
Subclass swr	Winter and summer rainfalls, where these are not typical for the zone in question
Subclass ep	Episodic precipitations in extreme desert conditions
Subclass nm	Nonmeasurable precipitation in the form of dew and fog in desert conditions

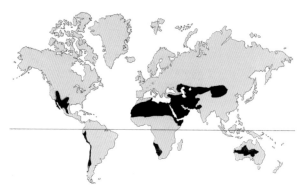

Fig. 6.4 *World distribution of desert areas (modified from Zunino and Zullini 1995)*

the beginning of systematic studies of organisms able to survive environmental conditions of extreme aridity, both in "hot deserts," as examined above, and in "cold deserts." Some studies carried out mostly in the Negev desert in Israel and Palestine (Krumbein 1983; Danin 1986) have shown how rocks are usually colonized by resistant organisms, which become visible as changes in the color of the stone (not always easy to recognize) or as other typical kinds of alterations (exfoliations, decohesions, and various morphologies of micropitting); the different colonizations, and therefore the alterations, are correlated to the climate conditions and exposure of the works. These studies have documented the survival capacity of organisms to extreme conditions, for instance to annual rainfalls of only 100–150 mm and on stone with a porosity of the order of 0.5–2.0%, displaying typical alteration phenomena related to the conditions of exposure and the degree of porosity of the various stones (Danin 1986).

example, in India, Central America, or Africa) and even in temperate ones (for example, in northern Italy and central Europe), but in these instances it is usually linked to its use in poorer dwellings or to social structures that are technologically less evolved.

Unfired clay is not in itself easily attacked by microflora, although the inclusion of organic material in the mixture (vegetable stalks, leaves, and sometimes dung) can have a considerable impact on deterioration issues. There are very few studies dealing with biological alteration in this context. This is in part due to the fact that the intrinsic building typology of these works requires ongoing maintenance operations, which limits the presence of biodeteriogens to pioneer organisms in the first place.

The environmental conditions induced by a desert climate do not imply a complete absence of life forms in desert areas. Numerous studies have shown specific adaptations of microorganisms to extreme conditions of excessive heat and severe lack of water. The 1970s saw

In these areas, the phenomena of biological colonization are much slower and less invasive than in other climate conditions; they are frequently underestimated because of the presence of alteration phenomena that are similar to those of a chemical origin (efflorescences, exfoliations, fractures, pitting, etc.). Alteration phenomena that in the past were attributed to purely chemical causes are now frequently recognized as being caused either solely or synergetically by specific biological populations, since it has been experimentally proven that in nature the processes of trans-

Fig. 6.5 *Bioclimate diagrams (with fluctuations of rainfall and temperatures over the twelve months of the year) typical of desert climates: (a) Cairo, Egypt; (b) Nukus, Uzbekistan*

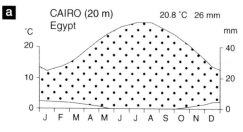
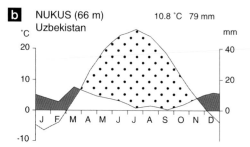

formation of the material tend to be biogeochemical (Ehrlich 1981; Krumbein 1983; Odum 1988; Bullini et al. 1998). A classic example, especially in a desert climate, is that of rock varnish or desert varnish, i.e., dark surface patinas, varying in thickness from just a few microns to 500 microns, that form on many rocks subjected to extended exposure time but not in direct contact with the soil (Fig. 6.6). At a chemical level, these patinas are composed of oxides of manganese and iron, silica, and clay minerals, but several authors—already back in the 1920s—had postulated their biological genesis in spite of their appearance being that

Fig. 6.6 *Desert varnish, with the preferential formation of dark patinas in areas with greater supplies of moisture, in the Valley of the Kings (Luxor, Egypt)* – Photo G. Caneva

of an abiotic phenomenon. Among the organisms involved in these processes are thought to be several groups of bacteria and microcolonial fungi (MCF), such as *Lichenothelia convexa*, which have the capacity to cause the deposit of the component materials of these patinas. The phenomenon does not seem to be restricted exclusively to desert climates, though, since similar patinas have been identified on works located in tropical bioclimates (Garza-Valdes and Stross 1992).

In addition, it seems that in very hot and dry areas with high levels of osmotic pressure due to saline concentrations, endolithic fungi become recurrent organisms, albeit difficult to identify (Sterflinger and Krumbein 1997). Other endolithic forms of lichens and cyanobacteria have also been frequently observed—as in an ecological seriation—with variations in stational conditions and in relation to microclimate conditions (Pohl and Schneider 2002).

In these climatic contexts, microbiological studies directed to the conservation of works of art have primarily been carried out in environments in which microclimate and topographical conditions would be more favorable to biological development, i.e., hypogea, tombs, and other damp enclosed spaces. In the case of the Egyptian tombs in the Valley of Kings and Queens (Luxor) in particular,

investigations were directed toward biological air monitoring, since the opening of the tombs and the influx of visitors could create conditions of biological colonization that might have serious consequences for the conservation of wall paintings (Chapter 5, section 1.4 and Chapter 7, section 3.1b). The studies on Nefertari's tomb showed that after the tomb was opened and conservation interventions carried out, the microbial content (bacteria and fungi) had greatly increased (Arai 1987, 1988); various bacteria of the genera *Alcaligenes*, *Acinetobacter*, *Bacillus*, *Micrococcus*, *Listeria*, *Corynebacterium*, *Arthrobacter*, as well as yeasts of the genus *Rhodosporium* and deuteromycetes (*Penicillium* in particular), were identified both in Nefertari's and Tutankhamen's tombs (Ammar et al. 1987).

Very few data are still available on the possible role played by microflora in the degradation of Egyptian monuments; in the case of the temples of Karnak at Luxor, morphological examination, in association with respirometric and histochemical techniques, has shown the presence of microflora and in particular of cyanobacteria (*Calothrix parietina*), along with other nonidentified forms (Curri and Paleni 1975). More recent publications make a brief and general mention of the possibility of biodeterioration phenomena, but without

Fig. 6.7 *Phenomena of biological colonization (not investigated on a systematic level) on monuments of ancient Egypt: (a) Temple of Edfu, near the Nile River, with circular development of microorganisms; (b) Temple of Phylae at Aswan (Egypt), which in the past was partially submerged as a result of the construction of the dam, with a clear development of dark biological patinas corresponding to areas enriched with nutrients immediately above the level of submersion* – Photo G. Caneva

supplying specific qualitative or quantitative data. Certain studies, for instance, mention the presence of biological organisms associated with the degradation of the temples of the Fayum (Helmi 1988), but the correlation between the presence of these organisms—some of which are ubiquitous—and the various forms of stone deterioration requires further investigation. A more in-depth study is also required in the case of biodeterioration phenomena in the tombs of Tell Basta (Zagazig), where seasonal fluctuations of the microbial communities were also monitored. What was found here is a certain increase in the phenom-

ena during winter months, when rainfall was greater, as well as high levels of colonization in the most deteriorated stones (10^4–10^5 CFU/g), with a fairly noticeable role being played by the actinomycetes (May et al. 2000).

Classic examples of alterations due to biological deterioration are frequently found on the temples of Upper Egypt, related to gradients of humidity and the input of organic materials, which are more substantial locally; however, such phenomena have as yet to become the object of a specific diagnostic analysis (Fig. 6.7).

In desert climates, phanerogamic plant colonization appears somewhat restricted, especially if compared to that encountered in tropical or Mediterranean climates. However, the role of certain xerophilic and phreatophytic species is not negligible; the roots can penetrate the substrate to a depth of several meters, as, for example, in the case of capers (*Capparis spinosa*, *C. decidua*) (Fig. 6.8)—which are also a distinctly calciophilic species—as well as tamarisks (*Tamarix* sp.) and figs (*Ficus carica*). During the restoration of the minarets of Samarkand (Uzbekistan), such root behavior was indeed observed with the roots of the capers (the plants sometimes looked like small shrubs) penetrating the walls to a depth of several meters (Fajrusina 1974).

Because of the still limited understanding of the problems surrounding the biological attack of materials in desert climates, it is not easy to categorize the available data, but we can at least mention the following common denominators:

—the slow establishment of phenomena of biological attack on the materials (usually on a scale of decades or even centuries);
—a reduced biodiversity in the microbial and plant composition of the community, with prevailing presence of highly adapted populations of the pioneer type, such as bacteria, fungi, lichens, and algae, and in particular of poikilohydric species with a prolonged capacity for dormancy in periods of drought as well as persistence of biological forms with short life cycles (therophytes);

—biological growth of any consequence only in the most humid enclosed environments (tombs, deposits) or in particular microclimate and topographical situations;

—development of perennial plant forms—which are in any case never very showy—only in conditions of a prolonged absence of maintenance operations;

—colonization occurring preferentially in fissures or in-depth (chasmolithic and endolithic habitus), in relation to the typology of the substrate and to microenvironmental fluctuations.

As far as the last point is concerned, it seems that in climates with strong fluctuations in the moisture levels, endolithic species do indeed tend to be favored, since they are more protected from variations in the external environment; in addition, a preferential development of filamentous and sporulating forms tends to take place (May et al. 2000).

It should be emphasized, however, that it is precisely in these environments, where the processes of deterioration are mostly due to physical agents (wind erosion, strong overheating, and consequent expansion, followed by contraction during the cooler hours) and to chemical and physico-chemical ones (strong evaporation and subsequent crystallization of salts), that careful planning of the flora and vegetation in the immediate surroundings can significantly improve the state of conservation of the site, as for instance has been proposed for the site of Moenjodaro in Pakistan (Chapter 7, section 1.2c).

6.2.2 Mediterranean Climates

Mediterranean climates are characterized by an annual rainfall that varies between 400 mm and a maximum of 1,500 mm (borderline case), by temperatures that are never too rigorous (generally above 0°C, even in winter), and especially by dry summers. What all Mediterranean climates have in common is that the lowest rainfall always occurs during the period in which the temperatures are the

Fig. 6.8 *(a)* Capparis spinosa—*a species with a distribution ranging from the arid zones of Eurasia up to the Mediterranean basin—on the minarets of Samarkand (Uzbekistan); (b)* Capparis decidua—*limited to the more arid zones of the Iranian Sistan and of Pakistan—in the archaeological site of Moenjodaro (Pakistan), where it appeared to be closely correlated to the presence of underground walls* – Photo G. Caneva

highest, thus giving rise to periods of xericity of various lengths. Average annual temperatures are above 14°C, and average temperatures during the coldest months are between 3 and 8°C (Fig. 6.9).

These conditions are not only typical of the Mediterranean basin, but can be found also in parts of California, central Chile, the Cape Province in South Africa, and along the southern coast of Australia (Fig. 6.10).

In these climates, the most limiting period is the summer when more or less extreme levels of aridity are established; during this season, biological growth may or may not be in a quiescent phase. Organisms exhibit survival adaptations that, in the case of plants, require either a large development of forms with a short life cycle or the acquisition of mechanisms

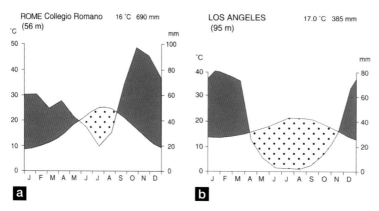

Fig. 6.9 *Bioclimate diagrams (with fluctuations of rainfall and temperatures over the twelve months of the year) typical of Mediterranean climates: (a) Rome; (b) Los Angeles*

North European climates. Such phenomena may cause indirect mechanical damage (Chapter 1) and obviously occur mostly in climates in which periods of substantial rainfall alternate with prolonged drought (Ortega-Calvo et al. 1993).

Given the high concentration of cultural heritage in these climates—and in particular in the Mediterranean basin, which saw the birth of Greco-Roman civilization—a substantial amount of literature is available on degradation, including when linked to biodeterioration. Several studies describe the most relevant problems—both for works in outdoor environments and for those located in enclosed spaces—as well as the most frequently occurring species. Many of these have been cited in the present volume, either with reference to the biological attack on materials or to case studies of colonizations in various environments.

Obviously, the various sectors of the Mediterranean basin are climatically and biogeographically diversified to a certain extent, and not only as a reflection of the gradients of rainfall and temperature.

As a general rule, we can say that in the case of works in outdoor environments, the dominant colonization is that of cyanobacteria, algae, and lichens as pioneer organisms, as has been made abundantly clear in this volume (Fig. 6.11); yet, the role of fungi—especially meristematic black fungi, as has recently been shown (Urzì et al. 2000a)—is not negligible (Chapter 3, section 2).

As is to be expected in advanced stages of

for protection against dehydration (sclerified leaves, the sinking of the stoma, etc.).

The microflora shows similar adaptations directed to the overcoming of stresses, with organisms adopting a metabolically inactive state and also acquiring specific forms of resistance in order to pull through the unfavorable season.

The effects of the cycles of absorption and desiccation of the extrapolymeric substances produced by microbial and microalgal communities are an example of degradation mechanisms resulting from the action of pioneer communities; these are different in Mediterranean climates compared to those found in more temperate and colder

Fig. 6.10 *Worldwide distribution of Mediterranean areas (modified from Zunino and Zullini 1995)*

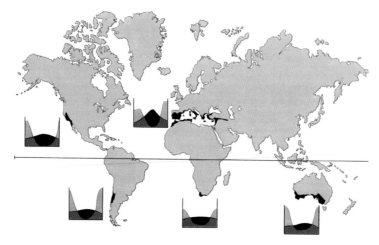

Fig. 6.11 *Pioneer stages of biological colonization in a Mediterranean environment: (a) patinas of cyanobacteria on marble fragments in the archaeological area of Ephesus (Turkey); (b) growth of lichens on megaliths in the prehistoric area near Evora (Portugal)* – Photo G. Caneva

Fig. 6.12 *Evolved stages of biological colonization, with large developments of shrub and tree communities in a Mediterranean environment in Italy: (a) Tiberius's Roman Villa in Capri; (b) bridge covered in ivy at Canale Monterano (Rome), in transitional areas with temperate climate* – Photo G. Caneva

development, communities of bryophytes and especially vascular plants with a woody structure become established as secondary colonizers (Chapter 2, section 2.3); these can exercise a mechanical action, producing an alteration of the substrate that can at times be quite significant (Fig. 6.12).

A comparison between the processes of alteration of stone materials in Jerusalem—which is at the limit of the Mediterranean climate and tends toward a subdesert climate—and stone materials in Rome—which has typical Mediterranean values in the city center, but with transitional elements toward the temperate Central European climate—has revealed a difference between the dominant communities: aside from the epilithic forms with dominant cyanobacteria that are frequently found in both contexts, the chasmolithic

forms with a more rapid growth seem to be more frequent in the Roman context, while the euendolithic and slower-growing forms seem to be more prevalent in Jerusalem (Danin and Caneva 1990). The length of time required for limestone to be completely covered with biological patinas is of the order of a century in Jerusalem, while it is relatively shorter in Rome. Here, the progression of pitting phenomena of a biological nature can lead to a loss of 1 mm of material in the span of forty years, while on the average in Jerusalem progressions of 1 mm over a period of two hundred years have been observed (Danin and Caneva 1990).

Further studies of an ecological nature on the biodeterioration of Minoan monuments on Crete and of the ancient fortifications of Khania did not reveal any clear correlations between biological growth and the fluctuations in local temperatures and rainfall. Rather, the development of microflora seemed to be linked to the topographical and exposure conditions of the sampled surfaces, and the most important microbial alteration phenomena seemed to be located in areas that were least subject to strong limiting conditions (such as extreme overheating, exposure to dominant winds, to salt spray, etc.). It should also be emphasized that fluctuations in the moisture content of the surfaces may induce significant modifications of the osmotic values, which can at times become quite selective; the presence of biofilms is thought to contribute to an increased stability of the environment, making it less subject to variations in external conditions (May et al. 2000). Even though, as mentioned earlier, it is difficult to make statements that are valid at a general level, we can say that, on a macroscale, the biodeterioration phenomena typical of Mediterranean contexts share the following characteristics:

—the gradual progression of the phenomena of biological attack on materials;
—an average biodiversity in the microbial and plant composition of the communities, with dominant presence of pioneer populations (bacterial, algal, fungal, and lichenic species) and especially of poikilohydric species with the capacity for dormancy in periods of drought;
—an increase in the effects of biological attack through physical mechanisms occurring during the hottest periods due to the effects linked to the hydration-dehydration of biofilms or thalli;
—the dominance of vascular plants with short life cycles (therophytes) in the earlier stages of establishment, and their eventual substitution with perennial forms adapted to be tolerant of summer drought;
—a substantial increase in the number of colonizing organisms in the most favorable microenvironmental conditions, especially rural areas and gardens;
—a colonization that can expand either over the surface or in depth, depending on the nature of the substrate and the microenvironmental variations.

6.2.3 Temperate Climates

As mentioned earlier, temperate climates cover a considerable range of types (four climate classes, from V to VIII), from typical, to hot, arid, and cold ones. As a whole, they are characterized by more or less marked seasonal variations and, although they may have a more or less prolonged cold period, they never reach the thermal rigors of polar regions (hence their etymology); the rainfall also differs and is variously distributed.

To different areas of temperate climate correspond different biomes, such as those of the temperate broad-leafed deciduous forests, with oak forests with limes, maple, and ash mixed in among the oak; the forests of *Eucalyptus-Nathofagus*; the perennial grasslands of the steppes, which develop where the climate is very cold in winter and hot and dry in the summer so that trees are unable to grow; and finally the coniferous forests of the taiga. Among these biomes, that of the broad-leafed deciduous forest is especially significant because of its links with human settlements; it is present in various forms in central Europe, including northern Italy and considerable tracts of the Balkan peninsula, in parts of central and northern Asia (including considerable territories in China and Russia), and in the east coast of North America (Fig. 6.13).

It is to be noted that in Europe the bioclimatic transition progresses from north to south in the following way: first, it moves from the temperate-boreal climate (class VIII) to the typical-temperate one (class VI), then from this to the Mediterranean one (class IV), subsequently tending toward the desert and subdesert climates as it proceeds southward (class III); going east, the climate moves toward temperate-dry continental (class VII), while going west it moves toward the oceanic variant of the typical-temperate climates.

In Italy, the boundary between temperate and Mediterranean climates separates the regions of northern Italy from those of central and southern Italy; exceptions to this are: a narrow coastal region in the west (Liguria)—where the climate is Mediterranean and which becomes wider as it moves farther south (for instance, Tuscany and Lazio)—and a strip in

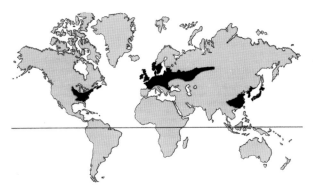

Fig. 6.13 *Worldwide distribution of temperate deciduous broad-leafed forests (modified from Zunino and Zullini 1995)*

the Apennines with temperate climate, which reaches as far down as Calabria.

In addition, the Atlantic regions of Europe—and in particular the northern shores of Portugal, Galicia, the Basque Country, Gascony, Normandy, and Ireland—present peculiar climate characteristics that are linked to the effects of oceanicity; these are reflected in significant increases of rainfall and in temperatures mitigated by the presence of huge expanses of water, with the exception of coastal areas where cold currents flow.

Even more complex and highly articulated is the climate situation of Asia, which varies greatly from the northern and western regions to those in the east and the south. Indeed, China exhibits what is essentially a continental climate, but the northwest areas are arid, while in the southeast temperate or humid-temperate climates prevail, and a limited area of the extreme south and southeast is tropical. In the northern regions, the high pressures from central Siberia exert a particular influence in winter, causing a significant lowering of the temperatures and associated dryness, while in summer warm, humid air penetrates from the Pacific Ocean. Japan is also in a temperate zone, but exhibits a marked oceanic influence, with high levels of rainfall (1,000–2,600 mm).

Climates with oceanic influence and with less rigorous temperatures are also found in the southern hemisphere, and in particular in New Zealand, where the distinctive bioclimatic characteristics, along with the absence of floristic impoverishment due to glaciations, sustain examples of primitive flora that are no longer in existence elsewhere.

With reference to typical-temperate climates such as those of Central Europe, average annual temperatures vary between 10–13°C, with rigorous winters and recurrent snowfalls; annual rainfall fluctuates between 350 mm and a maximum of 2,300 mm (in areas with oceanic influence), without any substantial reduction during the summer months. Cold-boreal climates, which are found in the northernmost areas of Eurasia and America, are characterized by much shorter summers and longer winters; more specifically, we refer to this climate type when the average air temperature falls below 10°C for more than 120 days a year and the cold season lasts longer than six months (Fig. 6.14).

From a biological point of view, the most limiting factor is the winter cold, which selects species adapted to survive this critical period in quiescent form. Extremely cold environmental conditions are particularly inhospitable for woody vegetation, which have altitudinal and latitudinal limits that are clearly influenced by the thermal parameter.

Fig. 6.14 *Bioclimate diagrams (with fluctuations in rainfall and temperatures over the twelve months of the year) characteristic of typical-temperate climates: (a) Luxembourg; (b) Moscow*

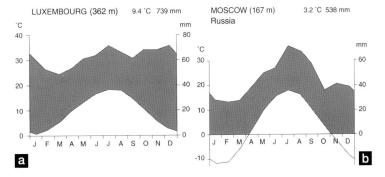

The moisture levels, which in other contexts are frequently the most determining factors, approach the threshold values for survival only rarely in temperate conditions. Furthermore, considering the relatively high levels of rainfall associated with low temperatures, temperate climates usually show relatively high hygrometric values (RH); this is a positive factor for the growth of biological populations, which are, however, often limited by the low temperatures. The phenomena of winter frost are indeed not negligible, as they can have serious consequences for the materials due to physico-mechanical damage. According to some authors, in these climates (as opposed to Mediterranean climates) the principal mechanism of biodeterioration of stone in outdoor environments resulting from the action of microbial and microalgal communities is not so much that of absorption-dehydration as that of freeze-thaw cycles of the extrapolymeric substances produced, which causes indirect mechanical damage as a result of the changes in volume (Ortega-Calvo et al. 1993). However, this does not mean that other mechanisms typical of the action of microbial groups do not exist, such as the emission of specific substances—in particular that of acid and complex-forming ones (Chapter 1)—or alterations due to mechanical effects.

In conclusion, if in the Mediterranean the main and most often recurring limiting factor is water, in temperate climates equally important limiting factors are low temperature values, although there is, of course, a reciprocal interaction between the two. Seasonal fluctuations are present in these climates as well, but they are less marked than those encountered in Mediterranean climates (May et al. 2000).

In certain cases, it has been shown how the extremes of environmental factors, especially of temperature but also moisture, favor the development of endolithic forms (both algae and lichens).

As always, the time factor determines the progression from pioneer to more evolved stages (Fig. 6.15), but with slower dynamics than those encountered in Mediterranean climates if the cold is a strongly limiting factor, and faster dynamics if the water supply becomes significant. In temperate climates

Fig. 6.15 *(a) Pioneer stages of lichenic colonization in Normandy* – Photo F. Sacco; *(b) evolved stages of development of* Hedera helix *on the remains of the church of Viller la Ville (Brussels)* – Photo J. Fox

with oceanic influence—such as northern Portugal and Galicia, for instance—the high levels of rainfall result in rapid and widespread biological development, undoubtedly much more critical than in Mediterranean regions where it is limited by summer drought. Unfortunately, present knowledge on biodeterioration in these geographical areas is still lacking, although the adoption in building techniques of certain practices for the protection of exposed surfaces demonstrates how prevalent this problem is.

An additional element that is different from other climatic contexts is the fairly common use of wood—as well as other materials of vegetable origin (for instance, bamboo)—for traditional construction, especially in Northern Europe but also in China and Japan; this is due to the ease of availability of this type of material in the forests surrounding the settlements. As a consequence, the use of wood in exteriors finds its maximum potential in these climates, with the associated problems (although here reduced by the cold climate) of durability and biodeterioration linked to it.

Limiting ourselves to the results of the investigations carried out in a Central European environment, we can thus summarize the specific problems of biodeterioration encountered:

—the gradual establishment of biodeterioration phenomena on the materials involved;

—an average biodiversity in the microbial and plant composition of the communities, with a dominance of pioneer populations from bacterial to algal, and with a fairly relevant presence of bryophytes and lichens in areas that are less subject to pollution;

—synergy of the phenomena of microbial attack primarily with phenomena of a physical type linked to freeze-thaw cycles of the materials during the winter months;

—slow but constant development of colonizing organisms, and a dynamism of the biodeteriogenic communities in relation to seasonal conditions that is not particularly marked;

—detectable biological colonization both on exposed surfaces and in the deeper layers of the material.

In regions with oceanic influence, on the other hand, where the levels of rainfall and the hygrometric values are much more favorable, the development speed of the biodeteriogens and the global dynamism of the communities seem much faster. From an ecological point of view, it is primarily hygrophilic species that are selected—from meso- to microtherms—and colonizations by bryophytes and ferns as well as lichens (normally of reduced proportions in other climatic contexts) are quite frequent.

6.2.4 Tropical Climates

Tropical climates are encountered from the tropical to the equatorial belt of the planet (including therefore classes I and II), where they give rise to the biomes of the evergreen tropical rain forests (Fig. 6.16) and to less markedly humid variants (such as the savannas, for instance).

Tropical regions occupy a vast area of the planet adding up to about 20% of the land mass and, because of their latitudinal positions, they are characterized by year-round high levels of solar radiation. As a result, the average annual temperatures are always high (varying between 24 and 27°C), variations in temperature are relatively small over the span of the year, and there is no real winter period (the average temperature for the coldest month being above 18°C).

Depending on which climate classification system is used, tropical zones can be further subdivided into those that are markedly humid,

Fig. 6.16 *Localization of humid tropical forests (modified from Zunino and Zullini 1995)*

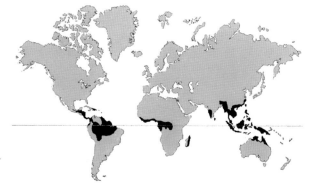

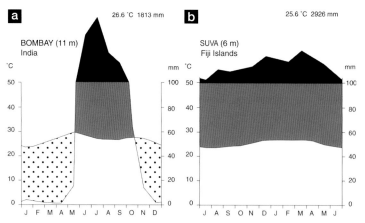

Fig 6.17 *Bioclimate diagrams (with fluctuations of rainfall and temperatures during the twelve months of the year) typical of tropical climates: (a) Bombay (India); (b) Suva (Fiji Islands)*

those with a monsoon climate, and those with alternating humid and dry periods.

Wet tropical zones are located between 10 and 15° latitude and the equator; they have a rainfall in excess of 1,500 mm and sometimes higher than 3,000 mm, with minimum monthly levels that never fall below 60 mm (Fig. 6.17b).

Tropical monsoon zones are located in an area between 10 and 20°; they have very high levels of rainfall (2,500–4,000 mm), but an alternation of wet and dry seasons is possible, and monthly rainfall can fall below 60 mm.

Tropical zones with alternating dry and wet periods are also located in the area between 10 and 20°, but they have rainfalls that vary between 760 and 1,500 mm; they are characterized by a more or less marked dry period during the cooler season, while the summer season is wet (Fig. 6.17a).

Even in the least favorable seasons, these climate types are optimal for biological growth, and often proportionately so to the levels of rainfall; these are the biomes that generally present the maximum levels of biodiversity as well as biomass and productivity (speed of development of the biomass per unit of time). In these conditions, vegetation reaches its maximum complexity, forming multilayered forests with many species of lianas and epiphytes; they host rich and complex microbial and animal communities, which, however, have not been sufficiently studied yet. It is these ecosystems that provide a source of new species for science as well as genetic resources with great potential, including for the biomedical field.

There are no periods of interruption of biological activity here, because there are no critical environmental conditions, and microbial activity proceeds rapidly throughout all seasons. Similarly, the growth of woody plants does not exhibit the characteristic winter interruption or the slowing down in the summer that take place in temperate and Mediterranean climates. It is well known that this alternation gives rise to the formation of annual growth rings, which are the result of a different density between the wood produced during spring and the one produced during the period of resumption of vegetative growth in late summer—a fact used as a base for dating in dendrochronology. Tropical woods do not exhibit alternating growth rings, and the porosity is continuous and diffused throughout the wood; the production of biomass and the speed of production are parallel and much more consistent.

As a result of the intense general development of all the different biological species, the problems of biodeterioration of materials in these contexts are obviously more serious and widespread, both for works in outdoor environments and for those in enclosed spaces, which are clearly influenced by the surrounding bioclimate. This optimum growth can be explained by the absence of limiting environmental factors, both in terms of temperature and rainfall. The high levels of rainfall result in consistently high levels of relative environmental humidity, with average readings of 60–65% in dry periods and 75–85% in the wet season, and with values that are close to condensation levels in cooler locations that receive less sunlight, such as the conditions often found not only in undergrowth, but also in enclosed environments such as storage areas and hypogea.

At present, several studies deal with the biological attack of works of artistic and historical or archaeological interest in tropical Asia (in particular India, Sri Lanka, Indonesia, Thailand, and the Philippines) (Hyvert 1972a, 1972b; Lee and Wee 1982; Riederer 1984; Sadirin 1988; Agrawal and Dhawan 1989; Saxena et al. 1991; Aranyanak 1992; Garg et al. 1994, 1995; Uchida et al. 1999; Kovacik 2000; Chung et al. 2003) and in Central and South America (in particular Mexico, Guatemala, Honduras, Cuba, and Brazil) (Hale 1973; Cepero et al. 1992; Martinez et al. 1994; Videla et al. 2000; Caneva et al. 2004), while there are very few dealing with tropical Africa.

Beginning in the 1970s, a number of contributions were published—often as a collection of conference papers—that looked at problems of biodeterioration in tropical climates as a general theme (for example, Agrawal et al. 1987; Agrawal and Dhawan 1989; Toishi et al. 1992). However, much of the data in these publications is given mainly in the form of preliminary reports or as expert technical accounts of site visits related to conservation projects of specific sites, which does not make it easy to access the gathered information.

There is still a great paucity of data relating to biological colonization in enclosed environments such as libraries and museums; recent surveys of the literature available emphasize the shortage of detailed information (Bisht 1985; Zyska 1997). Information is often limited to general, if alarming, mentions of bacterial, fungal, or algal attacks and to a variety of descriptions of the associated degradation phenomena. Among the cited case studies of biological attacks in enclosed environments, fungal growth has a significant presence in these climates as well and is noted in the case of attacks on organic materials—such as wood, paper, paintings, leather, or other samples of ethnographic interest—as well as in attacks on stone, ceramic, and glass materials.

Some authors (Chapter 7) reiterate the importance of prevention techniques and measures used in local traditions, where certain precautions aimed at reducing the risk of biological deterioration of the materials are taken, such as the reduction in the levels of sugar and humidity in the raw materials, especially those of vegetable origin. This was carefully carried out—even more than in other climatic contexts, where many similar precautions are also taken—by gathering the plants during the periods when these parameters were at their lowest, then drying them out over a long period of time with appropriate seasoning, before the start of the manufacturing process. As far as wood is concerned, much attention has always been given to choosing the most durable species—both for building technologies and for carpentry—because these are rich in tannins and have low levels of residual organic material in the parenchymatic portions of the wood. Moreover, the practice of charring—known among peasants throughout the world since time immemorial—is frequently carried out for the parts that are most exposed to the risk of biodeterioration, especially if they are immersed in water or buried in the ground.

In the context of external environments, authors have emphasized the rapid biological colonization of exposed monuments as well as the richness and variety of the colonizing organisms and microorganisms. In addition to the identification of species, which of course vary according to the different biogeographical contexts, the various studies show how the exposed materials quickly pass from exhibiting a biological covering due to the attack of algae, cyanobacteria, and fungi (which causes a blackening that spreads over the exposed surfaces) to a covering of lichens and mosses, which produce thalli with different morphologies and pigmentation.

An exhaustive floristic list of the microflora found in tropical areas is not really possible because of their extremely high biodiversity. As an example, the analysis of the various phenomena of algal colonization on painted plaster in Latin America revealed the presence of more than 1,300 morphotypes; of these, only a fraction was identifiable at the level of species, although it was possible to recognize a distinct prevalence of populations of cyanobacteria and green algae (Gaylarde and Gaylarde 2000). At a quantitative level also, it has been shown how populations of the magnitude of 10^2–10^5 UFC g^{-1} are common. According to the abovementioned literature, the following genera are found most frequently: among cyanobacteria,

Anabaena, Anacystis, Aphanothece, Calothrix, Chroococcus, Chlorogloea, Gloeocapsa, Lyngbia, Microcystis, Nostoc, Oscillatoria, Phormidium, Schizothrix, Scytonema, Tolypothrix; and among the green microalgae, *Chlorella, Chlorococcum, Hormidium, Oocystis, Pleurococcus, Trentepohlia,* in addition to diatoms such as *Navicula, Nitzschia, Pinnularia.*

Even in the case of attacks by lichens and bryophytes, the species identified on works in both archaeological and monumental contexts amount to several hundreds. Among lichens, frequent reference is made both to crustose species belonging to the genera *Aspicilia, Bacidia, Blastenia, Caloplaca, Candelariella, Chiodectron, Disploschistes, Endocarpon, Lecanora, Peltigera, Placynthium, Verrucaria,* and to foliose species belonging to the genera *Candelaria, Collema, Dirinaria, Heterodermia, Parmelia, Parmotrema, Peltula, Physcia,* as well as to tufted ones such as *Roccella* and *Usnea.* Much more abundant than in other climates are also the mosses and liverworts belonging to the genera *Barbula, Bryum, Frullania, Lejeunea, Marchantia, Papillaria, Plagiochila, Rhacomitrium,* and *Stereophyllum.*

In these climatic contexts, problems associated with attacks by bacteria, both autothrophs and heterotrophs, are also very significant—among the bacteria are *Bacillus, Pseudomonas, Nitrosomonas, Nitrosococcus, Micrococcus,* etc. (Guiamet et al. 1998; Videla et al. 2000; Videla and Saiz-Jimenez 2002)—but even more significant are attacks by fungi and actinomycetes, both in outdoor and enclosed environments. Among the fungi most frequently observed on stone monuments are the following: *Alternaria, Aspergillus* (*A. flavus, A. elegans, A. fumigatus, A. niger*), *Aureobasidium, Cephalosporium, Cladosporium* (*C. cladosporioides*), *Curvularia* (*C. lunata, C. verrugulosa*), *Fusarium* (*F. roseum*), *Penicillium* (*P. multicolor, P. crustosum, P. frequentans, P. glabrum, P. notatum*), *Phoma;* while among the actinomycetes we note *Nocardia, Micropolyspora, Micromonospora, Microellobiosporum, Streptomyces* (Hyvert 1972a, 1972b; Lim et al. 1989; Warscheid et al. 1990; Cepero et al. 1992; Kumar and Kumar 1999). Obviously, the environments in which the greatest development of heterotrophs is observed are subterranean ones such as caves and grottoes, which are in contact with the soil and with

Fig. 6.18 *Colonization by phanerogamic vegetation in a tropical environment: (a) site of Ta Prom (Cambodia); (b) trees of the species* Ficus elastica *on Khmer monuments in Cambodia –* Photo F. Prantera

organic materials. In the study of the painted caves in Ajanta in India where, for example, more than fifty different species of fungi were identified (Tilak 1991; Dhawan et al. 1992), various authors have emphasized the aggressive action of these populations, in particular of lichenic populations, which have the capacity to penetrate deeply into the substrate with the fungal hyphae (Hale 1973; Kumar and Kumar 1999) and can also produce high levels of chelating substances, which have the power to extract ions from the substrate. It should be mentioned that especially with reference to porous stone exposed to outdoor environments, some authors have spoken of the potential bioprotective qualities of biological patinas against other aggressive environmental factors (particularly wind and rain); nonetheless, the effects of the physical and chemical activity of these organisms cannot be ignored (for the discussion of this complex problem we refer back to the observations made in Chapter 4, section 1).

Lastly, as examples of the more aggressive types of biodeteriogens—if only because of the serious physico-mechanical damage they can cause—we would like to mention vascular plants, which develop with considerable speed and vigor if not kept under control by maintenance operations. With the appropriate edaphic conditions, a rapid colonization by the phanerogamic vegetation takes place, ranging from ferns to grasses and trees; the neglect of a site then leads to its colonization by the forest within a few short decades, as has been described in the case of the Khmer monuments in Cambodia (Fig. 6.18) and the Mayan monuments in Mexico, Guatemala, and Honduras (Plates 25 and 26).

Indeed, these conditions do not produce plant communities of the short-cycled pioneer type, but within a brief time period after the development of mosses and lichens, forest communities are quickly developed; these are characterized by large-size woody species with extensive columnar roots that englobe the monuments, becoming themselves part of the structure and, therefore, difficult to remove (Shah and Shah 1992–93; Mishra et al. 1995b). In addition to the "monumentalization" problem (in the sense that the trees themselves become objects to be protected) of the arboreal elements, some of which had been originally planted deliberately for symbolic and religious purposes (for example, *Ficus religiosa*, sacred to Buddha in the East, or *Ceiba pentandra*, similarly sacred in the Mayan world), even the removal of these tree elements proves difficult to carry out. Indeed, when such a close interpenetration between plants and monuments is reached (Fig. 6.19), it is necessary to proceed with great caution in the dismantling and putting back together of the disconnected elements, carefully calibrating the structural consolidation (Chapter 8, section 4.2).

In conclusion, although it is difficult to summarize in a few words the characteristics of biological attack in tropical climates, we can summarize what has been expounded above in the following points:

—rapid establishment of the phenomena of biological attack on both organic and inorganic materials (in the space of a few years, months, weeks, or even days in the case of organic materials);
—great biodiversity of the microbial and plant communities, which add up to hundreds of different species;
—easy attack by microorganisms even on materials that are the least susceptible to biodeterioration, such as glass;
—considerable development on stone and plaster of the more hygrophilic plant groups, such as mosses, liverworts, and pteridophytes;
—high biomass growth of the colonizing organisms and high dynamism of the communities, with a rapid passage from pioneer to evolved forms;
—easy colonization even by heterotrophic organisms on inorganic materials and in outdoor environments;
—frequent colonization even in the deepest layers of the materials.

Fig. 6.19 *Details of roots of* Ficus elastica *on the site of Ta Prom (Cambodia)*

PLATES

Plate 1 Effects of water as a limiting factor on biological growth. Biological colonization limited to zones that receive a sufficient supply of water through percolation: (a) growth of algae, mosses, and lichens in areas deprived of cover (Roman Forum) – Photo A. Merante; (b) growth of algae and lichens in relation to percolation (Trani) – Photo G. Caneva; (c) green patinas of Chlorophyceae, alternating with black crusts from pollution, as related to the different levels of water supply (Venice) – Photo O. Salvadori

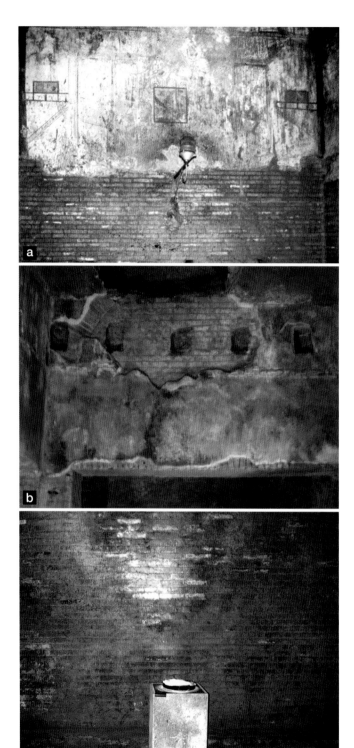

Plate 2 Effects of light as a limiting factor on biological growth.
Biological colonization limited to areas that receive sufficient light in a hypogean environment (*Domus Aurea*, Rome), with favorable values of both temperature and humidity: (a) algal patinas near the old lighting installations with incandescent bulbs – Photo ICR; (b, c) further algal development near the new lighting installations
– Photos G. Caneva

Plate 3 Effects of eutrophication on biological growth. Colonization by nitrophilic species, especially lichens with orange thalli, limited to areas with sufficient supplies of water and nitrogenous substances: (a) marble statue in a park in Lisbon – Photo O. Salvadori; (b) marble statue in Villa Cordellina (Vicenza) – Photo ICR; (c) marble stele in Ostia Antica – Photo O. Salvadori; (d) brick column in Ostia Antica – Photo O. Salvadori; (e) cement wall in the suburbs of Rome – Photo G. Caneva

Plate 4 Effect of the porosity of materials on biological growth. Differential growth of biological populations in relation to the porosity of the substrate and to the resulting ability to retain water: (a) development of mosses limited to porous carbonate rocks on buildings in Galicia – Photo J. Izco Sevillano; (b) development of mosses only on calcarenites with high porosity on a corner of the Venosa Castle – Photo G. Caneva; (c) growth of lichens limited to the mortar used in the remounting of a marble column in Ostia Antica – Photo G. Caneva; (d) development of lichens and cyanobacteria (with a greater tolerance for lack of water) on travertine (a less porous material) and of mosses (with greater requirements for water) on tufa (a more porous material) with equal conditions of exposure in the Roman Forum – Photo G. Caneva

Plate 5 Bacteria— Morphology of the colonies and microbial strains seen under the optical microscope (OM). (a) Colony of the strain *Modestobacter* sp., isolated from carbonate stone with outdoor exposure; (b) same strain seen under the microscope with epifluorescence and stained with Acridine Orange; (c) colony of *Streptomyces* sp., isolated from hypogean environments; (d) same strain observed under the OM in phase contrast, showing a clear filamentous structure (with a diameter between 0.5 and 1.0 μm); (e) refractile endospores in cells of *Bacillus* sp. observed in phase contrast; (f) cells of *Escherichia coli*, stained with Gram, observed in phase contrast; (g) cells of *Candida albicans*, stained with Gram, observed in crossed-polars; (h) cells of *Staphilococcus* sp., stained with Gram, observed in phase contrast

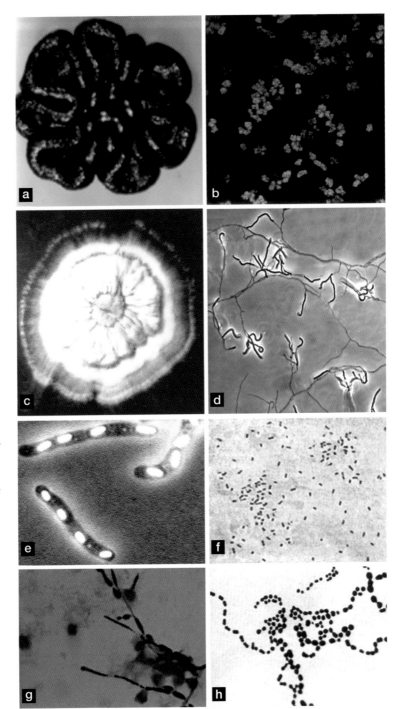

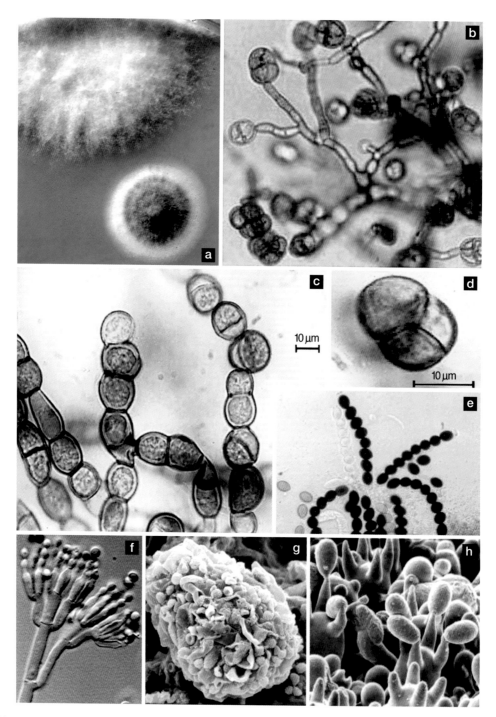

Plate 6 Fungi—Morphology of the colonies and of different fungal groups seen under the optical and the electron microscope. (a) Colonies of hyphomycetes in a culture, with emission of pigments – Photo G. Caneva; (b) conidiophore and conidia of *Ulocladium* sp.; (c) mycelium and (d) meristematic cells of *Coniosporium uncinatus* under the OM – Photo C. Urzì and F. De Leo; (e) asci with ascospores; (f) conidiophore and conidia of *Penicillium* sp.; (g) sclerotia that formed on marble; (h) basidia and basidiospores under the SEM – Photo O. Maggi and A. Persiani

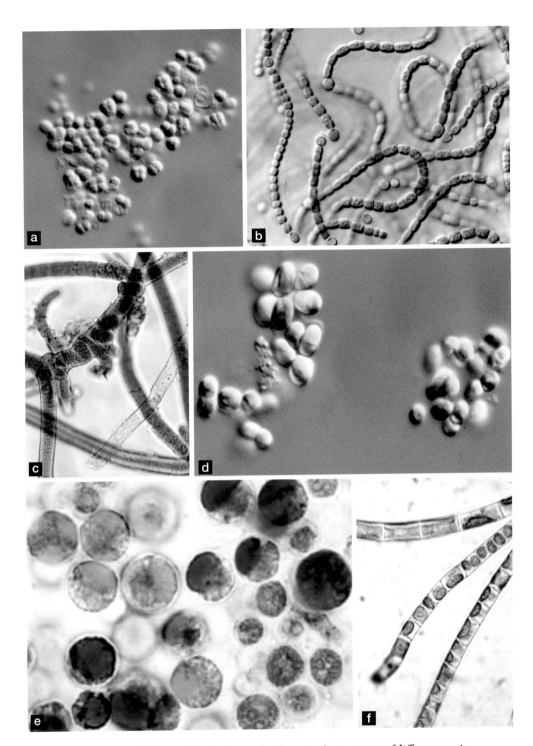

Plate 7 Cyanobacteria and algae—Morphology under the optical microscope of different species of cyanobacteria (a–b) and green algae (c–f). (a) *Chroococcidiopsis* sp.; (b) *Nostoc* sp.; (c) *Scytonema* sp.; (d) *Stichococcus* sp.; (e) *Haemotococcus* sp.; (f) *Ulothrix* sp. – Photo M. L. Tomaselli

Plate 8 Lichens—Main growth forms of lichens. Endolithic crustose, (a) *Rinodina immersa*; epilithic crustose, (b) *Ochrolechia parella*, (c) *Caroplaca saxicola*, (d) *Diplotomma venustum*; foliose, (e) *Xanthoparmelia conspersa*; fruticose, (f) *Roccella phycopsis* – Photos M. Tretiach

Plate 9 Bryophytes—Morphology of vegetative and reproductive structures. Image under the stereomicroscope of (a) gametophytes and (b) sporophytes of moss; image under the optical microscope of protonema and gametophytes of (c) moss and (f) a foliose liverwort; (d, e) cushions of moss. – Photos (d, e, f) S. Ricci

Plate 10 Higher plants—Morphology of different ruderal plants on Roman walls. (a) *Sonchus tenerrimus*;
(b) *Capparis spinosa*; (c) *Ficus carica*; (d) *Hedera helix*; (e) *Antirrhinum majus* subsp. *tortuosum*; (f) *Trachelium
caeruleum*; (g) *Parietaria judaica*; (h) *Umbilicus rupestris* and *Asplenium trichomanes* – Photos G. Caneva

Plate 11 Wood. Different types of rot resulting from fungal attack. (a) Example of white rot; (b) example of brown rot; (c) example of soft rot – Photos A. Gambetta and E. Orlandi

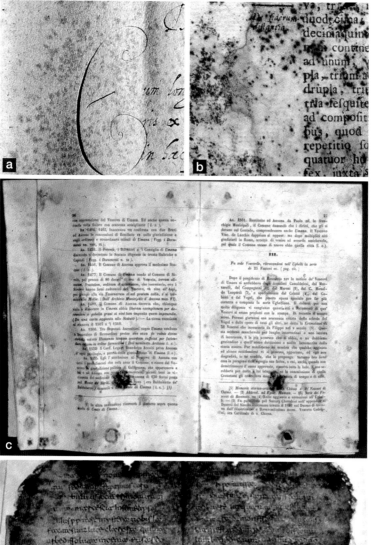

Plate 12 Paper—Foxing and chromatic alterations. (a) Typical foxing stains on the page of a book – Photo F. Gallo; (b) various typologies of stains linked to chromogenous organisms (pink) and Dematiaceae (black) – Photo ICPL; (c) chromatic and structural alterations resulting from microbial attack on a book – Photo ICPL; (d) extensive damage by cellulolytic organisms with the typical destruction of the marginal areas of the pages – Photo ICPL

Plate 13 Textiles—Stains and biological alterations. (a, b) Selective fungal attack on a silk woven textile with a cotton and linen weft; (a) gray stains due to proteolytic strains (*Penicillium chrysogenum* and *Aspergillus flavus-oryzae*) and small black stains due to a cellulolytic strain (*Chaetomium* sp.); (b) detail of the attack by *Chaetomium* sp.; (c) different types of stains from fungi on a hemp fabric; (d–f) images showing fibers stained by an attack of chromogenous fungi; (d–e) under the stereomicroscope; (f) under the optical microscope; (g) extensive fungal colonies on a linen fabric – Photos ICR

Plate 14 Parchment, leather—Fungal attack. (a) Abundant development of mycelium on leather bindings kept in a damp environment; (b, c) details of stains and fungal colonies on the spines of the books; (d) fungal attacks on ancient parchment; (e–g) details of the stains produced by the attack of different microbial species in relation to the development of the mycelium and the production of exopigments – Photos ICPL

Plate 15 Stone—Development of endolithic forms. Grayish pigmentation induced by the presence of photo-synthesizing endolithic microorganisms: (a, b) tombstones in the Jewish cemetery in the Venice Lido—in (b) a cleaning test induced the emission of photosynthetic pigments; (c, d) balustrade of Villa Manin in Passariano (Udine) – Photos O. Salvadori

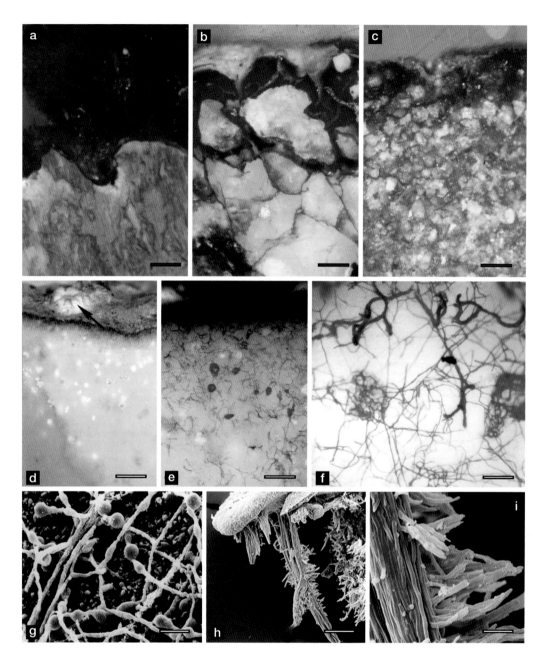

Plate 16 Stone—Alterations induced by lichens. Thallus-substrate relationship in epilithic crustose lichens *Lecanora sulphurea* (a) and *Diploschistes actinostomus* (b, c), and in endolithic ones *Petractis clausa* (d–f). Microphotographs under SEM: (g) hyphae of *Verrucaria baldensis*; (h, i) rhizines with lateral ramifications of *Physconia*; the foliose lichens anchor themselves to the substrate thanks to structures such as these – Photos O. Salvadori and M. Tretiach

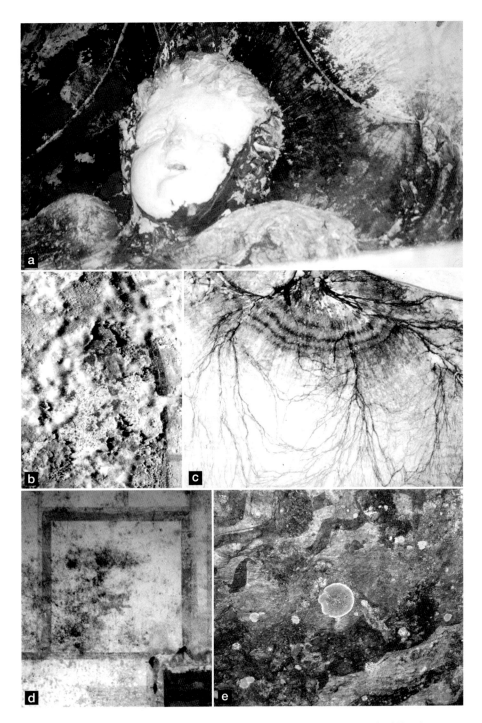

Plate 17 Frescoes and stuccos—Alterations induced by various biological agents. (a, c) Basidiomycetes attacking the wooden beams behind the plaster and piercing through to the front with rhizomorphs through a hole (c) developing on the stuccos (Church of the Spinsters, Venice) – Photo O. Salvadori; (b) fungal attack on frescoes (Crypt of the Cathedral of Anagni) – Photo ICR; (d) algal attack on frescoes of the *Domus Aurea* – Photo G. Caneva; (e) lichen attack on frescoes of the Crypt of the Original Sin (Matera) – Photo G. Caneva

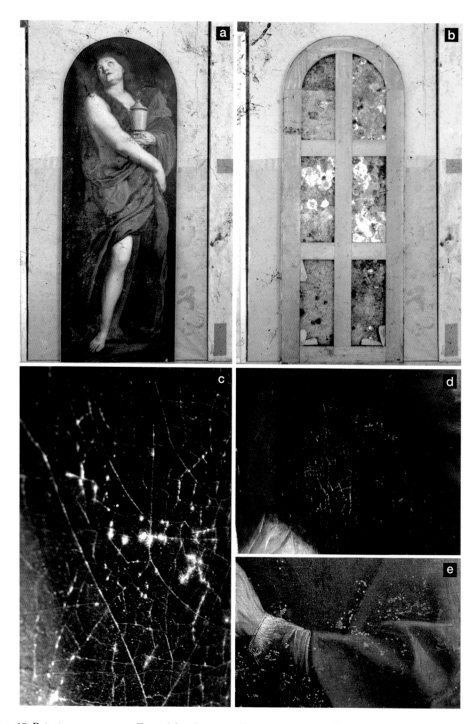

Plate 18 Paintings on canvas—Fungal development. Canvas paintings under severe attack by microfungi:
(a) only traces of the attack are visible on the paint layer; (b) massive development on the canvas support
(Cathedral of Rieti); (d) penetration of the fungal hyphae from the reverse to the front of the painting with a
detachment of the paint layer; (c) image under the stereomicroscope; (e) rare example of the colonization by
fungi on the finishing varnish of an oil painting: small colonies linked to the deposit of dust and dirt are visible
– Photo ICR

Plate 19 Photographs—Various alterations on different supports. (a) American daguerreotype, one sixth of the plate mounted in a gilt metal frame and glazed, and inserted into a partial mount—the microfungal colonies are visible beneath the glass (collection D. Matè) – Photo R. Bianchi; (b) gelatin/silver bromide glass plate, belonging to the Fondo Peliti (ING) – Photo R. Manni; (c) end of nineteenth-century albumen print, "visiting-card" format, with damage from foxing (collection D. Matè) – Photo M. C. Sclocchi; (d) gelatin/silver bromide print with destruction of the emulsion (Fondo Ente EUR ACS concession) – Photo M. C. Sclocchi; (e) early nineteenth-century albumen print with microfungal attack (collection D. Matè) – Photo M. C. Sclocchi; (f) black-and-white film from the 1930s with biological damage (Fondo Ente EUR ACS concession) – Photo M. C. Sclocchi

Plate 20 Biological alterations in caves and hypogean environments. (a) Development of actinomycetes with the appearance of whitish patinas in the Etruscan Tomb of the Leopards (Tarquinia) – Photo ICR; (b) algal and cyanobacterial patinas in the rock church of the Original Sin (Matera) – Photo G. Caneva; (c) patinas induced by actinomycetes and algae and superficial development of roots of *Ficus carica* on the frescoes of the *Domus Tiberiana* (Rome) – Photos G. Caneva

Plate 21 Biological alterations in urban environments. (a, b) Widespread presence of black patinas induced by cyanobacteria and lichens as shown in photographs dating from the early twentieth century: (a) Cestia Pyramid (Rome) – Photo ICCD, and (b) Trajan's Column (Rome) – Photo Soprintendenza Archeologica Roma; (c) patinas of green algae on the Stairway of the Giants (Venice) – Photo O. Salvadori; (d) patinas of cyanobacteria in the areas subject to percolations in the Tower of Los Jerosolimos (Lisbon) – Photo G. Caneva; (e) black crusts due to pollution, which deposited—in contrast to biological patinas—in the areas where no water has percolated (Torlonia Palace, Rome) – Photos G. Caneva

Plate 22 Plant communities present in dynamically evolved stages in archaeological sites in Rome.
(a, b) Aspects showing a dominance of *Spartium junceum* and *Ulmus minor* (class Rhamno-Prunetea) in the
Massentium Circus; (c) subnitrophilic communities of the Urtico caudatae-Smyrnietum olusastri in the
Palatine; (d) communities that tend toward the class Quercetea ilicis on the Palatine; (e) hygrophilic ruderal
aspects, with a dominance of species belonging to the class of Querco-Fagetea on the Palatine – Photos G. Caneva

Plate 23 Biological colonization in rural environments. Park of the Monsters of Bomarzo (Viterbo):
(a) Sphinx statue; (b) Ceres; (c) Ogre; (d) family coat of arms on the leaning house; (e) Proteus-Glaucus; (f) partially worked house. The widespread growth of mosses (b, c, e, f), lichens (a, d, e), and ferns (f) on the statues,
although causing an alteration of the substrate (in [e] especially), nevertheless contributes to the creation of a
bond between nature and the work of man – Photos G. Caneva

Plate 24 Fountains—Examples of biological colonization.
(a) Fountain of the Naiads (Rome) with abundant carbonate incrustations covered by algae and mosses; (b) Fountain of the Hundred Fountains (Villa d'Este, Tivoli) with an abundant development of mosses, ferns (in particular *Adiantum capillus-veneris*), and higher plants; (c) Fountain of the Ovato (Villa d'Este, Tivoli) with widespread biological growth, which contributes to give the impression of a natural habitat
– Photo P. Piccioni

Plate 25 Pioneer stages of colonization in tropical climates. (a–c) Development of algae and lichens on works in Southeast Asia: (a) church in Bombay (India); (b) statue in Bali (Indonesia); (c) Temple of Borobudur (Indonesia); (d–h) development of cyanobacteria, algae, and lichens on Mayan monuments in Central America: (d, e) black patinas induced by cyanobacteria at Chichén Itzá (Yucatán, Mexico); (f–h) development of algae and lichens at Copán (Honduras) – Photos G. Caneva

Plate 26 Evolved stages of colonization in tropical climates. (a, b) Development of trees on monuments in Southeast Asia: (a) building in Bombay (India); (b) Khmer temple in Ta Prom (Cambodia) – Photo F. Prantera; (c–f) similar phenomena on Mayan monuments in Central America: (c, d, f) archaeological site of Copán (Honduras); (e) archaeological area of Cobá (Mexico) – Photos G. Caneva except (b)

Plate 27 Prevention of biodeterioration in museum environments. (a) Control of the relative humidity and temperature of the air in proximity of painted panels by means of battery-operated data loggers (Civic Museums of Pesaro); (b) fungal attack on the reverse of a painting with a gradient of development from the bottom to the top, correlated to the capillary rise of humidity in the wall on which the painting was hung; (b–e) severe fungal colonization on the reverse of a painting kept in permanently damp conditions; (d) chromogenous fungi; (e) sterile mycelium; (f) cellulolytic fungi (*Penicillium* and *Chaetomium*) – Photo ICR

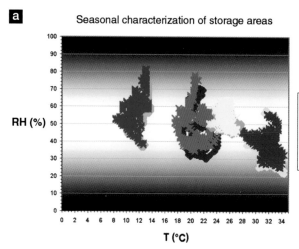

Plate 28 Prevention of biodeterioration in underground storage areas and other underground environments.
(a) Representation by points dispersal of the seasonal temperature and hygrometric variations in two Italian museum storage areas: both show considerable fluctuations over the span of a whole year (elaboration by E. Giani); (b–c) timed lighting system in a hypogean environment: (b) green monochromatic light in the absence of visitors; (c) light with high chromatic definition and automatic switching on in the presence of visitors
– Photo B. Mazzone

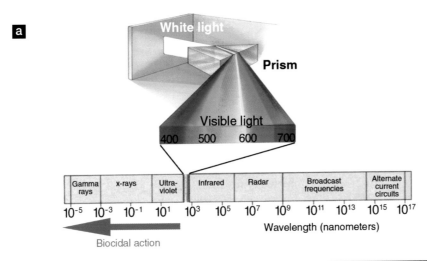

Plate 29 Physical methods of disinfection. (a) Electromagnetic spectrum, showing the increase in biocide efficacy in correspondence to the increase of the frequency of the radiation; (b–d) irradiation with UV-C of algal patinas present on a marble altar (b), and results before (d) and after (c) the treatment (Mythraeum of San Clemente, Rome) – Photo ICR

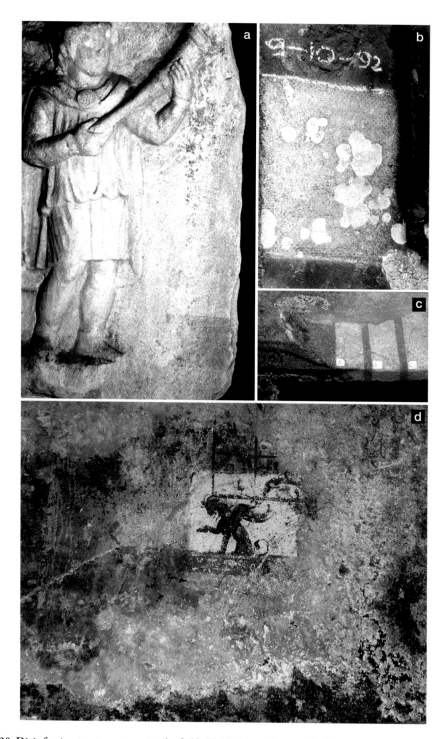

Plate 30 Disinfection treatment tests in the field. (a) Marble surface attacked by algae (portion still visible in the lower test areas protected from the radiation) after treatment with UV-C; (b) stone surface attacked by lichens after treatment with biocides – Photo L. Zambon; (c) test areas with comparison of three biocide products; (d) test on a fresco after disinfection and cleaning – Photo ICR

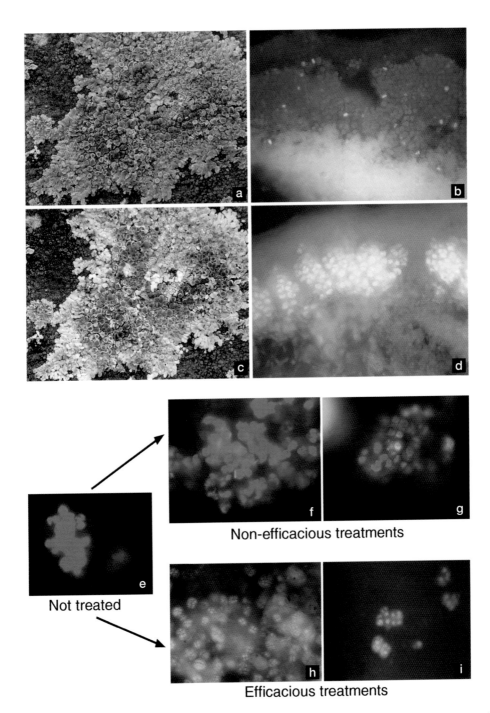

Non-efficacious treatments

Not treated

Efficacious treatments

Plate 31 Tests with biocides. (a–d) Treatment with biocides on lichens: *Lecanora muralis* after treatment with biocides, in which it is possible to distinguish macroscopic modifications in the thallus and different behavior under fluorescence, as can be seen in the section of the lichen observed after one day (a, b) and after eight days (c, d) – Photo O. Salvadori; (e) image under fluorescence of nontreated algae and cyanobacteria, and (f–i) different morphologies in relation to the differing efficacy of the biocide treatments – Photo ICR—it should be noted that, in general, bright red shows an optimal photosynthetic capacity, while devitalization results in an alteration in color

Plate 32 Bioremediation. Details of the fourteenth-century fresco by Spinello Aretino, *The Conversion of Saint Ephisius and the Battle* (a) before and (b) after biological cleaning with the use of living bacterial cells of *Pseudomonas stutzeri* – Photos G. Ranalli

CHAPTER 7

PREVENTION OF BIODETERIORATION

In the past, the field of conservation of cultural heritage was focused almost exclusively on the study of surfaces and the alterations encountered on them. A critical change in the planning of conservation intervention was brought about with the decision to take into consideration not only the surface on which the diagnostic tests and the subsequent cleaning and control treatments are carried out, but also the environment in which the work is kept. The environment itself, as the "container" of the work, can indeed be the cause of serious alterations; through the study of environmental parameters it is possible to determine the causes that led to its degradation and take the appropriate preventive and control measures. In 1987, the *Carta del Restauro* (Conservation Bill) already included the concepts of preventive conservation—directed both toward the object and the environment—and maintenance, which, again, involves both the work and the environment in which it is kept: the work of art does not exist in isolation but constitutes, along with the environment, a physical system in evolution.

7.1. GUIDELINES FOR PREVENTIVE CONSERVATION

Along with specific conservation interventions, a sensible program of conservation of cultural heritage must include an overall reevaluation and upgrade of the premises in which the works of art are kept, in order to preserve them from physical, chemical, and biological degradation. Preservation from biodeterioration phenomena entails constant maintenance of the works as well as a particular attention to the parameters of the surrounding environment. It is well known that climate and microclimate conditions (humidity, temperature, and light) as well as levels of chemical and biological pollution in the environment can greatly influence the development of biodeteriogens on works (Chapter 2). Among the environmental factors, humidity is undoubtedly the one that mostly influences the development of autotrophic and heterotrophic microflora; in the case of photosynthetic micro- and macroflora, light (both natural and artificial) constitutes an additional parameter that is essential for their growth. Thermal conditions can also play a selective role for biodeteriogenic species. If the moisture content of the materials is low, and the RH of the air is lower than 65–70%, the conditions required for the growth of biodeteriogenic microorganisms are generally not present. Exceeding these conditions is often sufficient to trigger colonization by microorganisms with a prevalence of cyanobacteria, algae, mosses, and liverworts when the levels of moisture and light are high. The presence of carriers of microbiological pollution and of situations that favor the activation of dormant spores (Chapter 2, section 2.6) are equally dangerous conditions, and prevention must therefore be directed to control all the possible factors that might induce them. Variations in microclimate conditions are often the most dangerous because they can also lead to changes of a physico-mechanical nature (expansions, contractions, etc.), which then result in the instability of the work's constituent materials, thus creating a risk situation.

Visitors cause significant variations in environmental conditions, especially if their numbers are large and the space is small; to the thermal, hygrometric, and carbon dioxide contribution that is linked to the respiration or transpiration of the human body and to the type of clothing, other elements may be added such as variable quantities of biological particles, dust and spores brought in on shoes or clothes, aerosol droplets from coughing or sneezing, and even flakes of skin. In premises in which cultural heritage is housed, the specific conservation exigencies of the materials should be the priority when choosing environmental parameters; and if the conditions for the well-being of visitors and staff are not optimal, then compromise solutions need to be sought, such as limiting opening hours, or separating spaces devoted to conservation from those used for other work activities.

In indoor environments, the factors favoring biodeterioration are generally easier to control. Here, the biological pollution of the air is also part of the parameters that require monitoring for conservation purposes. Indeed, the spreading of biodeteriogenic microorganisms occurs through the dispersal of spores and vegetative propagules that can come in from the outside, transported by air currents and by water or visitors; or on the other hand, they may be of internal origin.

Therefore, measures have to be taken to determine the sources of pollution and to control, where possible, the access routes as well as the deposits of dust and other particulates on valuable surfaces. A chemical study of the composition of the air is also desirable, since gaseous and particulate pollutants can not only cause abiotic types of damage, but can also favor, select, or inhibit colonization by deteriogens. Certain substances can provide a nutritional substrate for heterotrophic microorganisms encouraging their development, as is the case of noncombusted derivatives of oil (saturated and unsaturated hydrocarbons), which are utilized by specific fungi. In contrast, sulphur products generally inhibit microorganisms, with the exception of chemoautotrophic bacteria. Others again can prove toxic, such as for example derivatives of chlorine or lead, etc.

The methods for the prevention of biodeterioration, also called indirect methods, are therefore designed to slow down or inhibit biological growth, by acting not directly on the biological agent but rather on the factors that determine, or at least could determine,

its development. Consequently, the tools on which prevention must be based are the studies directed to characterizing the environment (microclimate and aerobiological monitoring) and the substrate (constituent materials, history of conservation of the work, alteration products, etc.), but also those directed to an in-depth knowledge of the ecological requirements of the different systematic groups of biodeteriogens. A detailed preliminary diagnostic investigation to determine the chemico-physical characteristics of the substrate, the alteration products, and the organisms and microorganisms responsible for the degradation is therefore of crucial importance.

When defining the methods of prevention best suited to a particular case, one cannot ignore the differences in environmental typologies, which—with their individual characteristics—present specific risk situations and inevitably condition the possible corrective strategies to be adopted.

7.1.1 Enclosed Environments

7.1.1a *Museums, Archives, and Libraries*
by Giovanna Pasquariello, Maria Carla Sclocchi, Donatella Matè, and Paola Valenti

The maintenance, care, and conservation of historical, artistic, and documentary heritage are the main objectives to be pursued in order to pass on cultural vestiges to future generations. To this end, the best strategy consists in the elaboration of a theoretic system of reference along with practical directives relating to the conservation and exhibition environments and to the different types of works in them; such directives must also be compatible with an optimal use of objects and premises.

In the Italian context, technical-scientific criteria and standards for the functioning and development of museums were recently defined in order to guarantee an adequate level of general access to and use of their contents as well as their security and risk reduction. The decree *"Atto di indirizzo sui criteri tecnico-scientifici e sugli standard di funzionamento e sviluppo dei musei"* (Act addressing the technical-scientific criteria and the operational and development standards of museums) (Suppl. G.U. n. 244 of 19 October, 2001) covers a variety of areas and contains operational instructions regarding the maintenance and care of museum collections in particular, and also regulations for conservation and restoration that include exhibition and transport issues. As far as libraries and archives are concerned, specific national recommendations have not been developed yet; however, provisions contained in the above-mentioned document represent an essential point of reference for library and archive materials as well.

To conserve objects in museums, archives, and libraries means to prevent and limit both the damage caused by the natural degradation of the constituent materials of the work and the damage linked to its use. At its core, a successful approach to the protection of the different types of artifacts from biodeterioration must look at the environment and the work of art as a system, i.e., two interacting elements of a whole.

The elements of this system are: the characteristics and the conditions of the environment and of the work; the biological agents involved; and the physical, chemical, and anthropic factors. An accurate analysis of the interaction among all these elements is crucial to define potential risk situations. The term environment is understood here as place of conservation, which includes macroenvironments—storage areas, consultation rooms, exhibition spaces—and microenvironments—cupboards, drawers, shelves, and display cases.

An accurate overview and detailed knowledge of both the object and its environment are the first steps in order to assess biodeterioration risks and identify which verifications and measures are necessary to reduce them. To this end, two report forms (conservation and facility-environmental reports) have been developed, which must be updated regularly (Marcone et al. 2001; Giani et al. 2002; Accardo et al. 2003). The conservation report includes information regarding the techniques of execution (constituent materials and procedures), the "history" of the work in question, information on any accessories such as frames, containers, etc., if present, as well as the state of conservation. The facility-environmental report contains information regarding the

characteristics of the various conservation environments, the microclimate and lighting conditions, the quality of the air, as well as the type of use of the facility, and the number of visitors. At present, there is no single model of such a report form that is applicable nationally; however, the above-mentioned document does provide an environmental and conservation reference base for museums. Many libraries and archives use report forms they have developed on their own; these are also often computerized.

Table 7.1 summarizes the guidelines for preventive conservation against the risks of biodeterioration. Three key objectives are identified: reducing to a minimum the influx of biodeteriogens into conservation spaces; limiting the opportunities for contact between biodeteriogens that have penetrated the environment and the works conserved within it (since it is impossible to reduce their influx to zero); controlling the conditions that favor the development of biodeteriogens that have entered the conservation environment and that may have come in contact with the works. It is possible to achieve an appropriate and complete prevention of biodeterioration by planning and applying the verifications and measures recommended in Table 7.1.

Analyzing the composition of the air—both of its biological and of its chemical atmospheric pollutants—contributes to a better assessment of actual risk situation for the objects. The gaseous and particulate chemical pollutants that favor or inhibit the growth of microorganisms (Saiz-Jimenez 1995a) are known, as are the threshold values relative to the various types of chemical pollutants that are dangerous for cultural heritage (Suppl. G.U. n. 244, 2001). Because of the risk of chemical damage, it is advisable to monitor chemical and biological pollutants contemporaneously; as regards the latter, we should bear in mind that not all microorganisms present in the air are potential biodeteriogens; it is therefore essential to know which species have the capacity to degrade which kinds of works. The combination of high concentrations of microorganisms dispersed in the air that may be potential biodeteriogens and favorable microclimate conditions—specific to the various types of object—is essential in

Table 7.1 Guidelines for the preventive conservation of biodeterioration

Objective	Monitoring	Operations
Reducing the number of biodeteriogens entering the environment where the works are kept	• Periodic analysis of the composition of the air • Periodic inspection of the structures of the building • Periodic inspection of the microenvironments	• Maintenance of the structures of the building • Limiting the opening of the windows • Regulating opening hours and the number of visitors • Control of materials on entry • Filtering of the air with appropriate systems
Limiting the contact of biodeteriogens with the works	• Periodic analysis of the composition of the air in the microenvironments containing the works • Analysis of the state of conservation of the works • Determination of the presence of dust • Analysis of the state of conservation of the containers of the works	• Protection of the works in appropriate containers • Maintenance of air circulation in the space with appropriate installations and/or layout of the building • Regulating the handling of works by staff members • Definition of suitable procedures for the movement, storage, packing, transport, and exhibition of the works • Periodic dusting with suitable implements
Prevention of the development of biodeteriogens in the environment and on the works	• Microclimate monitoring of the macro- and micro-environments of the environments where the works are kept • Lighting control • Verification of the materials and the products employed in conservation	• Maintenance of the climate control of the environment (temperature, relative humidity) • Maintenance of the circulation of air • Maintenance of lighting installations with appropriate systems and regulations • Periodic dusting with suitable implements • Definition of appropriate conservation procedures, including materials and products to be employed

order to trigger the biodeterioration process (Plate 27).

As far as the historical, artistic, and documentary heritage is concerned, no recommended threshold levels for biological pollutants are available at the present time (below in this chapter, section 3.1). The working group called *Aerobiologia e Beni Culturali* (Aerobiology and Cultural Heritage)—AIA (Italian Association of Aerobiology) is at present conducting studies for a definition of biological risk levels for the various categories of cultural heritage and also for the development of health risk indices for conservation staff (Mandrioli et al. 1988;

Gallo 1993; Singh et al. 1995; Gallo et al. 1996, 1998; Pasquariello et al. 1998; Montacutelli et al. 2000; Pasquariello 2000, 2001; Gallo et al. 2002; Giani et al. 2002; Sclocchi et al. 2002; Tarsitani and Trama 2002; Micali et al. 2003).

To optimize the quality of the air, it is thus necessary to adopt all those passive and active maintenance measures that limit the concentrations of airborne pollutants. When an air-conditioning system is present, or the installation of one is being planned, it is imperative that it be provided with a filtering system to remove both biological and chemical pollutants in order to avoid an increase in their concentrations in the indoor environment. When planning exhibitions, or in the case of extended opening hours or of other events resulting in an increased number of visitors, the influx must be rigorously limited and continuous monitoring of the levels of the above-mentioned pollutants must be in place.

An adequate air circulation in the various conservation areas constitutes one of the most effective measures aimed at limiting the risks of biodeterioration. Indeed, it has been demonstrated that adequate ventilation within a space will considerably reduce the possibility of deposit of airborne microbial spores and therefore of contact with the works (James et al. 1991; Florian 1997). In addition, good air circulation helps to control water condensation phenomena, which can often occur on cold surfaces such as furniture, glass, and metal containers (Florian 1997; Pasquariello 2001; Matè and Sclocchi 2003).

Air circulation can take place either naturally or by mechanical means. Natural ventilation, obtained by opening windows and doors, is not usually advisable as it allows the entry of atmospheric pollutants (both biological and chemical) into the macro- and the micro-environments and is potentially dangerous to cultural heritage; nevertheless, natural ventilation is preferable to a total absence of ventilation. Mechanical ventilation, also called forced ventilation, is produced by using different systems and installations and is undoubtedly the most suitable solution when it is carried out correctly.

The best means of containing the risks of biological degradation are air-conditioning and climate-control systems (Filippi 1987; Filippi et al. 1994; Ashrae 1997a, 1997b; Pasquariello 2001; Matè et al. 2003). It is essential, however, that they be working correctly and that their maintenance be well scheduled and especially that it includes the regular substitution of filters, which ensure air purification (Raffellini and Cellai 1997; De Santoli and Fracastoro

1998). When this does not take place, the air-conditioning systems can actually function in the opposite direction and spread into the air the microorganisms that have deposited onto the filters and that have developed there; this, of course, increases the level of biological pollution within the conservation areas.

Dusting includes a series of operations that must be carried out respecting specific criteria for the works of art, the furniture, and the containers, so as to drastically reduce the quantity of microorganisms carried by the dust (Arruzzolo et al. 1997). Dusting can be carried out either manually or mechanically, using different procedures and implements (Matè et al. 2004), as can be seen in Figure 7.1.

Generally, there are two types of dusting: the one that takes place periodically, called ordinary, and the extraordinary one. The latter has to be performed in the following cases: after special operations or events (for example, remodeling or upgrade operations); on works that have not been "handled" in any way for more than five years; and, when an object is first introduced into a conservation space.

Ordinary dusting must be carried out regularly every year; both ordinary and extraordinary dusting must be carried out by specialized personnel. During these regularly scheduled operations the objects are moved, which is useful in order to be able to check on their state of conservation.

The design and production of furniture and containers (cupboards, cases, drawers, display cases, and shelving) must include a

Fig. 7.1 *Different types of dusting and their applications*

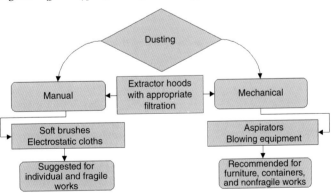

careful evaluation of technical, conservation, and exhibition requirements that vary according to the type of object. As a general rule, the most suitable materials are stainless steel and anodized aluminum, with finishes resistant to corrosion. Furniture and containers must not be placed in front of windows, nor indeed in the proximity of heating or conditioning systems, nor against walls to guarantee a sufficient air circulation.

Special attention should be given to the constituent materials of containers and protective covers (cases, boxes, folders, envelopes) used to store collections of works, particularly if these are of an organic nature and therefore even more fragile. The recommendation is to use durable, acid-free paper and cardboard products (UNI 10586, 1997; UNI 10829, 1999) as well as adhesives that are chemically inert, stable, and bioresistant. Such containers constitute a barrier against dust as well as against biological and chemical pollutants and also protect the works from light and from impact. When works are frequently consulted, and especially if they need to be viewed, plastic envelopes (polyethylene or polyester) are at times used for practical reasons. These types of coverings are usually not advisable: in the case of hygroscopic materials, processes of condensation may well occur, in addition to electrostatic phenomena—the recommended alternative is the use of paper envelopes, which allow the object to "breathe."

The above-mentioned legal decree related to museum environments (Suppl. G.U. n. 244, 2001) includes a set of reference recommendations that may also be useful for libraries and archives. International regulations specific to individual categories of cultural heritage, such as for instance for photographic material, are now also available (ISO 18902, 1999; ISO 18911, 2000; ISO 18918, 2000; ISO 18920, 2000).

A different approach must be taken when museums, libraries, and archives have furniture and containers—almost always made of wood—that are in and of themselves of historical and artistic merit because they are from the same period as the building in which they are housed. In such circumstances it is only possible to perform certain updates in the areas of security and conservation and to make specific adjustments to some of the situations in order to contain the biological risk.

The display case is a container for the permanent or temporary exhibition of a wide variety of artistic and archaeological objects and works of art. Display cases can be made of various materials, such as glass, crystal, acrylic resins, wood, and metal, and can be of different shapes, depending on the exhibition space available, on the nature of works to be exhibited, or else on aesthetic parameters (Accardo et al. 1995; Pasquariello and Maggi 1998, 2003; Giani 2002).

In the past, display cases were primarily used to protect works from theft, acts of vandalism, improper manipulation, and dust. Only recently has the display case been conceived as bringing together all the elements that must satisfy the objectives of prevention, conservation, and viewing of the works of art and of the archaeological objects they contain (Fig. 7.2).

There are two basic kinds of display case: aerated (i.e., not sealed) and sealed (Fig. 7.3).

The nonsealed display case allows microclimate exchanges with the outside environment. This type of display case—without RH regulation capacity and with external lighting—can

Fig. 7.2 *Different kinds of display cases and their uses*

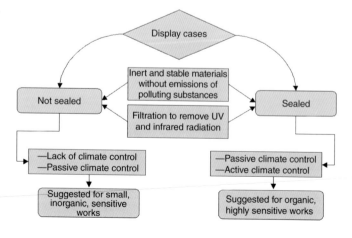

also have internal lighting and is suitable for the exhibition of works that are not highly sensitive to microclimate fluctuations (for example, works in stone).

The sealed display case, if it is well sealed, has almost no exchanges with the outside environment and can have a variety of climate control systems that can regulate both the microclimate and the quality of the air. Different systems, such as for instance the clima-box, are employed to achieve stable thermohygrometric conditions suitable for the conserva-

Fig. 7.3 *Examples of display cases: (a and b) not sealed; (c) sealed* – Photo G. Berucci

tion requirements of the works to be exhibited (Accardo et al. 1995). Essential prerequisites are a high overall sealing capacity and the use of materials (joints, seals, etc.) that are compatible with the nature of the works to be installed inside the display case. There are essentially two ways to regulate the internal microclimate: passive and active.

With passive climate control, also called stabilization, a preselected RH level is maintained within the display case; a variety of absorbent materials are used for this purpose, such as silica gel (preconditioned to the required level of RH), Art Sorb, and semipermeable membranes (Metro and Grzywacz 1992; Pasquariello and Maggi 1998, 2003; Tétreault 1999; Giani 2002). Museums, libraries, and archives also make use of nonsealed display cases that have been adapted and upgraded with the use of such materials.

The display cases with active climate control can have an interior atmosphere that has been either modified or is regulated. In the latter case, the regulation of the microclimate is based on a central air-conditioning system that depends on external elements such as radiators, dehumidifiers, etc. In the display cases with modified atmosphere, the air inside has been substituted by inert gases. Display cases with

a mixed system for stabilization also exist, i.e., with both active and passive climate control (AA.VV. 2002b).

As far as the quality of air inside the display case is concerned, the choice of construction materials is crucial: materials must be chemically stable and inert and not release polluting substances. Different active control systems are used for the elimination of polluting or contaminating components, and especially of those that have been preventively identified as critical for the various categories of works to be conserved: forced ventilation, with the air being recirculated through appropriate filters (active carbon ones), or chemical traps that have the capacity to absorb pollutants (Brimblecombe 1990; Pasquariello and Maggi 1998; Putt and Menegazzi 1999; AA.VV. 2002a; Giani 2002; Lord and Lord 2002).

The choice of lighting system and the levels of light in the display case are also important elements to be taken into consideration, both for the physical safety of the work and for maintaining the stability of the climate inside (Accardo et al. 1995). The light source, which of course must guarantee a satisfactory level of illumination of the objects, must not be a heat source and must include selective filters that reduce UV and infrared radiation. Today, fiber optics seem to offer the safest method of illumination; also, they can be positioned both outside and inside the display case. Recommended maximum values of illumination are available for each of the different categories of object and art works, subdivided according to their photosensitivity (Suppl. G.U. n. 244, 2001).

The display cases with climate-control systems are suitable for the permanent and the temporary exhibition of works of art that are particularly sensitive to microclimate fluctuations, to light, and to biological and chemical pollutants. These display cases must be carefully controlled with a system of ongoing monitoring of both the thermohygrometric parameters and the levels of polluting substances, so as to be able to intervene quickly should any anomaly arise. It is advisable to exhibit works that belong to the same category of cultural heritage together in the same display case, endeavoring to recreate the microclimate conditions of their place of origin.

So far, no recommendations are in existence with regard to the main technical requirements for display cases used for permanent and temporary exhibition. There is, however, an exhaustive bibliography on the subject, which can be of assistance when choosing the most suitable model for the different conservation and exhibition requirements (Burge 1990, 1995; Accardo et al. 1995; Pasquariello and Maggi 1998; AA.VV. 2002a, 2002b; Giani 2002).

Nevertheless, to stop the development of biodeteriogens, it is critical to keep the environmental macro- and microclimate under control, not only by monitoring thermohygrometric parameters and controlling light, air circulation, etc., but also by contemporaneously taking into account the various spatial and temporal situations and their correlation with the various typologies of the works conserved, exhibited, or consulted. In addition, appropriate choices must be made in the selection of the materials employed in conservation, storage, packing, transport, and exhibition; these must all be chemically inert and stable, bioresistant and only minimally hygroscopic.

For the preventive conservation of artistic, historical, and documentary heritage to be effective, the environmental thermohygrometric values must remain as stable as possible and must not fluctuate over the length of the day, nor indeed seasonally, whether in a room or in a container.

Table 7.2 shows the thermohygrometric conditions recommended for the conservation of materials and objects of an organic or mixed nature that are particularly susceptible to biological attack (Suppl. G.U. n. 244, 2001).

We should also point out that the parameters considered "optimal" for the conservation of the various kinds of objects often do not coincide with those adopted for the comfort of the public; and, similarly, the optimal values adopted from the perspective of the prevention of biodeterioration do not always coincide with those that are optimal from the perspective of the physical and mechanical properties of the work (ICCROM-ICR, 1983; Thomson 1986). Particular care should therefore be devoted to those environments that are open to the public and that also house objects, because it is here that the differing requirements for conservation, consultation, and exhibition coexist.

Regulations are available that give an indication of the general principles for the choice and the control of the microclimate in environments designated for conservation (UNI 10586, 1997; UNI 10829, 1999; UNI 10969, 2002).

The control of lighting (natural light, artificial light, or a combination of the two) is important in the various environments of conservation, consultation, and exhibition. When natural light is present, direct light falling on the objects should be avoided; it is advisable, for instance, to screen windows where possible with blinds or curtains, or to apply the appropriate films to the window glass in order to reduce UV and infrared radiation. For artificial illumination, the use of "cold," filtered light sources is recommended. At present, the light provided by fiber optics, with UV and infrared radiation removed, is considered the most suitable among artificial lights. Calculating of the total annual radiation to which the works are exposed is essential to be able to determine the length of rest periods from exhibition or which corrections need to be carried out. And finally, it is of great importance that the recommended levels of illumination be respected for each of the categories of object (Suppl. G.U. n. 244, 2001).

Table 7.2 Recommended thermohygrometric values for the prevention of microbial attacks on organic artifacts (Suppl. G.U. n. 224, 2001)

Organic Objects		Relative Humidity (%)	Max. Daily Fluctuation of RH	Temperature (°C)*	Max. Daily Fluctuation of T
Paintings	On canvas	40–55	6	19–24	1.5
	On panel	50–60	2	19–24	1.5
Wood	All types	50–60	2	19–24	1.5
	Archaeological	50–60	2	19–24	1.5
	Water-logged			< 4	
Paper	All types	40–55	6	18–22	1.5
	Pastels, watercolors	< 65	5	< 10	3
	Books and manuscripts	45–55	5	< 21	3
	Graphic material	45–55	5	< 21	3
Leather, skins, and parchment		45–55	5	4–10	1.5
Textiles	Cellulosic	30–50	6	19–24	1.5
	Proteinaceous	> 50–55		19–24	1.5
Ethnographic collections		20–35	5	15–23	2
Stable materials		35–65		15–30	

*The temperature must not reach 0°C.

The movement of works covers a series of operations—handling, packing, and transport —that can lead to situations of potential physical, chemical, and biological risk. These operations must be carried out following specific guidelines, both for the different kinds of movement and the different types of works.

New acquisitions or bequests and objects that are to be exhibited, whether temporarily or permanently, require special attention.

The use of protective gloves is recommended for the handling of each object, including for the consultation of original material such as books, prints, drawings, and archival documents. Access to particularly fragile works (e.g., ancient incunabula, etc.) should be limited in order to reduce potential physical, chemical, mechanical, and biological risks; it is for this reason that visitors and students are supplied with reproductions or digital images of the works requested. Both public and private institutions have their own internal regulations regarding access and consultation; in addition, computerized data banks are available for consultation, structured into sections and subsections according to type of cultural heritage.

Transport can be internal or external (movement from one space to another, loans for exhibitions, etc.) and includes packing and transport procedures. A conservation report is required for each individual object; it should record the state of conservation before and after transport and be accompanied by an accurate photographic record, so as to be able to identify and document possible damages incurred.

As a general rule, the following recommendations should be remembered as far as packing is concerned:

— the use of suitable containers with characteristics that protect from physical, chemical, and biological damage;
— the use of buffering material inside the container;
— utilization of thermoisolating materials and immobilizing systems;
— continuous monitoring of T and RH.

As far as transport is concerned:

— utilization of climate-controlled means of transport;
— storage of containers in suitable environments (far from sources of potential danger).

7.1.1b *Churches, Crypts, and Subterranean Environments*
by Maria Pia Nugari and Anna Maria Pietrini

Although not originally designed to satisfy conservation requirements, churches, crypts, and subterranean environments often house works of great historical and artistic value, which must be protected and preserved from degradation. These environments, especially subterranean ones, are characterized by microclimate and environmental conditions that are highly favorable to the development of biodeteriogenic microflora (Chapter 5, sections 1.3 and 1.4); nevertheless, it is often impossible to install air-conditioning, heating, or ventilation systems that would allow the correction of the environmental parameters and thus provide appropriate conservation conditions.

To establish guidelines with absolute validity across the wide range of structural and architectural spaces is a complex affair; it is therefore indispensable to identify the different factors that could induce biological attacks on works of art on the basis of criteria of a general nature and then determine for each individual case if it is possible to intervene.

In churches, for example, we find a wide variety of architectural characteristics that range from small chapels to other larger and more imposing structures, such as cathedrals, or else complicated, articulated buildings such as Baroque churches. Because of the wide range of architectural characteristics, churches represent environments that may differ considerably from a microclimate perspective and may exhibit significant differences even within one single building, as, for instance, in the case of a side chapel versus a central nave or a sacristy or crypt. The latter, especially, often exhibits characteristics that are typical of subterranean environments, with steadily high levels of RH. Indeed, clearly substantial differences can be identified between churches and subterranean environments in general. Even with the possible variations indicated above, churches usually exhibit microenvironmental conditions that are comparable in many respects to those described for historical buildings transformed into museums, libraries, and archives (earlier in this chapter, section 1.1a).

In churches, therefore, the prevention of biodeterioration can be carried out following general criteria directed to the containment of the conditions that would favor the development of biodeteriogens. Such an objective can at times be reached with minimal interventions that act on critical elements; these must be identified on the basis of microclimate and aerobiological studies as well as on evaluations of the facilities management of the church. Possible measures may be: the choice of type and positioning of entranceways for visitors and/or churchgoers; the aeration of the space during carefully chosen periods; the regulation of heating systems (temperature, length of operation), if present.

For example, in the Chapel of Crucifixion of the Sacred Mount of Varallo Sesia, Vercelli, in which sixteenth-century frescoes are preserved, it was possible to reduce the phenomena of biological and chemical degradation on the wall paintings by setting up the appropriate opening and closing of the chapel doors, depending on the season. A series of physical, chemical, and biological investigations made it possible to identify which among the four access doors to the chapel should be kept permanently closed—because they constituted the main routes of entry for airborne biodeteriogenic microflora—and which showed different behavioral patterns depending on the season and therefore had to be kept either open or closed, depending on the requirements for a correct air exchange and to give access to churchgoers (Nugari and Roccardi 1996, 2001) (Fig. 7.4).

In the case of the Cathedral of Anagni (Frosinone), it was thanks to a microclimate and aerobiological study that it was possible to identify the critical elements for the conservation of the priceless thirteenth-century frescoes in the crypt. The study showed that the airborne microflora traveled from the church floor to the crypt through one of the access ramps

and that the risk of contamination was highest during the summer period. On the basis of this, it was suggested that carpets be used to absorb the dust and that one of the ramps be closed during the summer, as useful measures for reducing microbiological contamination of the frescoes (Nugari and Roccardi 2001; Nugari et al. 2003).

In subterranean environments, the moisture content of the walls and the relative humidity are generally high. Although even here the first prevention strategy is to intervene in any way possible to reduce the availability of water, such environments almost always present substantial obstacles to the implementation of interventions that would lower the hygrometric levels of the site to within safety thresholds. Microbial development on the materials is therefore very frequent, and the moisture content of the structures is often such as to even induce developments of photoautotrophic flora and microflora as soon as the light conditions (natural or artificial) allow photosynthetic processes to take place (D'Urbano et al. 1998).

Temperature is one factor that limits biological colonization, and in hypogean environments it is usually somewhere between 10 and 18°C, but this does not really constitute a safety element, because of the existence of species that have adapted to these temperatures, and also because the presence of the public can easily raise the temperatures to close to 20°C, due to the limited air exchanges with the outside environment and to the often limited dimensions of these spaces. Particularly precarious conditions of ecological equilibrium are thus created, in which even minimal modifications in microclimate—induced by the presence of visitors, by maintenance work, or by other uses of the area—become factors that trigger the onset of microbial attacks, even when there had been no previous biological colonization (Giacobini et al. 1986; Laiz et al. 2000; Nugari and Roccardi 2002). The high levels of humidity also produce condensation phenomena on the surfaces—even when the differences in temperature between the surfaces and the air are small (1–2°C)—favoring the development of bacteria and algae (Hoyos and Soler 1993; Pantazidou et al. 1997; Ariño et al. 1997b).

Fig. 7.4 *Fluctuations in the microbial charge before and after the controlled management of the entrance doors of the Chapel of the Crucifixion of the Sacred Mount of Varallo Sesia (modified from Nugari and Roccardi 1996)*

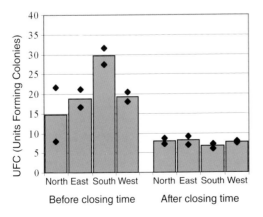

Among the prevention strategies in underground environments, controlling the influx of visitors thus becomes critical, as is also the monitoring of any other intervention that may induce a rise in temperature, relative humidity, and carbon dioxide. Stable conditions of humidity and temperature seem indeed to be useful in reducing the development of heterotrophic microorganisms, especially fungi, while fluctuations in these parameters can induce the germination of spores and favor the colonization of the surfaces. The influx of air currents from the outside, the opening to the public of spaces that are normally closed, the increase in the number of visitors, or the presence of conservation teams and activities can all become dangerous elements that upset the equilibrium and that, if not carefully controlled and managed, can increase the risk factors for biodeterioration (Arai 1974, 1983, 1988; Agarossi et al. 1985; Tilak 1991; Florian 1993; Agarossi 1994; Deshpande and Gangawane 1995; Pantazidou et al. 1997; Nugari et al. 1998).

During the conservation operations at the Tomb of the Ogre in the necropolis of Tarquinia, an increment of fifteen times the microbial concentration in the air (from 686 to 10,950 UFC/m^3) was registered. Such a situation had made the environment dangerous not only for the preservation of the paintings, but also for the health of the conservation team. With the agreement of the safety officers, it was therefore deemed necessary to limit both the number of people working in the area at any one time and the length of their stay (Nugari and Roccardi 2002).

It should also be emphasized that: 1. in addition to encouraging the transport of spores, pollen, and vegetative propagules from the outside to the inside, the entry of visitors in hypogean environments also induces a considerable increase, over time, in the numbers of airborne microorganisms, especially fungi and bacteria (through the raising of dust from the ground, the emission of biological aerosol, etc.), and 2. the autopurifying mechanisms that normally eliminate microbial pollution are usually considerably slowed down in these environments because of the lack of any efficient air exchanges with the outside. At the end of visiting time, a doubling in the microbial load of the air compared to that registered prior to opening to the public has frequently been registered (Monte and Ferrari 2000; Nugari and Roccardi 2002). In addition, dangerous, stagnant pockets of airborne contaminants are often created, and therefore attention must be concentrated on the deeper and most internal areas. In the cave of Basura in Toirano (Savona), for example, the fungal and bacterial load registered in the depths of the cave was ten to twenty times greater than near the entrance (Nugari and Roccardi 2002).

All these elements need to be considered case by case using appropriate microclimate and aerobiological studies that assess seasonal fluctuations and the exchange of air with the outside. Sometimes not even the controlled influx of visitors is sufficient to prevent biodeterioration; then the only possible preservation strategy is the total closure of the hypogean space to the public. This solution has been adopted for many caves and tombs; one of the first examples of this is the caves of Lascaux in France, which contain paintings dating from Paleolithic times. In addition to blocking visitor access, numerous disinfection operations of the surfaces had to be carried out here in order to remove biological patinas that had survived in a quiescent phase for a long period of time (Lefevre 1974). For small spaces—as for instance the Etruscan tombs in the necropolis of Tarquinia—the solution of hermetically sealing these environments has been adopted, using glass doors and timed illumination. This makes it possible to view the paintings from the outside without modifying the conditions of humidity, temperature, and air quality on the inside, all of which remain stable, thus reducing the biodeterioration risk.

As far as the proliferation of photoautotrophic biodeteriogens is concerned (cyanobacteria, algae, mosses), the measures are aimed at inhibiting photosynthetic processes and consist mostly in regulating the lighting. As a general rule, the levels of lighting should be as low as possible, and, if necessary, can be reduced from one space to the next to allow the human eye to adapt gradually. The planning of a lighting installation must therefore try to achieve a compromise between good visibility and appropriate preservation of the works. The balance can vary depending on the different planned uses of the space, on their microclimate parameters, on the objects displayed, and on their state of conservation.

As for any other photosynthetic organism, the growth process of cyanobacteria, algae, and mosses is subordinated to the quality of light, i.e., to the spectral characteristics of the light source, the quantity of energy radiated in a unit of time onto a unit of surface area (illuminance), and the length of the irradiation period. On the basis of a comparison between the absorption spectra of the main photosynthetic pigments (these range between 400 and 700 nm) and the emission spectra of some of the light sources commonly used in the field of light technology, it has been determined that, in order to limit photosynthetic activity and therefore the development of autotrophic microflora on illuminated walls, it could be useful to use lamps that emit radiation with a wavelength of 500–600 nm—corresponding to emissions in the yellow-green zone—because this is the portion of the spectrum that is the least active in the photosynthetic process. Aside from chromatic considerations, it has already been observed in Chapter 2 (section 2.2) that, although this green light is the least useful for photosynthesis, there are organisms that have nevertheless adapted to using its energy and are selected according to how best they use the wavelength of a light source (Fig. 2.6).

With the present state of knowledge, incandescent lights—which emit increasingly intense radiation in the band between 400 and 700 nm in the visible part of the spectrum, in a similar way to solar radiation—seem to be those that most favor the growth of photosynthetic microflora, also because of the temperature increase they induce on the illuminated surfaces. Very low levels of illuminance (10–15 lux) at this wavelength are sufficient for the growth of certain cyanobacteria to occur; also, given the same illuminance, the nearer the light source the greater the development of photosynthetic organisms on the surfaces (Plate 2).

The best results are achieved with the adoption of light sources characterized by an emission of radiation concentrated in the central zone of the visible wavelength range.

More specifically, some researchers who studied artificial lighting in caves open to tourists analyzed the influence of light sources with different spectral compositions on the growth of photosynthetic organisms and noted the effectiveness of high- and low-pressure sodium vapor lamps (Imprescia and Muzi 1989). On the basis of such information, a number of experiments were carried out in hypogean environments in Rome; they showed that low-pressure sodium vapor lamps—which emit a yellow monochromatic light, in a zone of the spectrum (589 nm) where the absorption of energy by photosynthetic pigments is the lowest—successfully inhibited the development of photoautotrophic microflora on wall surfaces saturated with water for relatively long periods of time (about two years), even with high levels of illuminance. However, the poor color resolution of these lamps is totally inadequate for the illumination of works of art such as frescoes, rock paintings, etc., for which the required color rendering should be close to that of natural light. These lamps could therefore be used in areas where the visual requirements are limited to providing adequate lighting for visitors to circulate among structures and where distinguishing colors is not necessary.

High-pressure sodium vapor lamps provide a better quality of light, allowing a better appreciation of chromatic qualities. From a biological point of view, at average illuminance levels these lamps inhibited the colonization of damp walls by photosynthesizing microorganisms for a period of several months.

These results lead to the conclusion that, although there is no artificial light source able to completely stop the process of photosynthesis in chlorophyll, it is nevertheless possible to inhibit and slow down the processes of growth of cyanobacteria, algae, and mosses by acting on the light source in all its various aspects of quality and quantity as well as on the duration of lighting time. For example, high chromatic definition lamps could be used but at reduced levels of illuminance, or, conversely, lamps emitting radiation that is little used by photosynthesizing organisms could be used in combination with higher illuminance levels and also calibrating lighting duration to adequate levels (Chapter 2, section 2.2).

In environments where the risk of development of photosynthetic microflora is particularly high, timed lighting installations should be planned as well as the use of lamps with low PAR (Photosynthetic Active Radiation) efficiency and, at the same time, the levels of illuminance of the surfaces should be low. To compensate for the not quite satisfactory viewing conditions offered by these light sources, various suggestions have been made, among which the combined and alternating use of sodium vapor lamps—for general and continuous illumination of a space—and of high chromatic definition lamps (e.g., halogen lamps) that offer a detailed and punctiform but temporary viewing; the second would switch on automatically when sensors detect the presence of visitors (Plate 28) (Mazzone 1999).

It is clearly impossible to extend the results of a few specific studies to every kind of subterranean environment; every biocoenosis requires an individual analysis, and every site has its own unique microclimate situation. Nevertheless, given that the growth of photosynthesizing microflora is always influenced by a combination of environmental and photometric parameters (Imprescia 1995), the most successful strategy would be for the planning of a lighting installation to be considered an interdisciplinary concern with the figures of the curator, the biologist, and the lighting engineer all working together.

From all of the above, it is clear that in the area of control of biodeterioration in subterranean environments it is difficult to find solutions with overall applicability. Relative humidity and water content of the walls are elements that are indeed difficult to control and that represent the main risk factor. Solutions have to be sought and tried out on a case-by-case basis, identifying where possible the limiting factor on which to act.

At a more general level, the following recommendations may serve as guidance:

—reduce the required lighting to the lowest levels possible, adopting lamps emitting cold light;
—avoid increases in temperature;
—maintain stable microclimate parameters;

—limit the number of visitors and the length of their stay;
—guarantee a good exchange of air, within limits so as to avoid dangerous daily fluctuations in temperature and an excessive evaporation from the surfaces, which could entail problems of chemical deterioration through the formation of salts.

It is also important to remember that underground environments often achieve an equilibrium in which the presence of microbial colonization can be minimal or confined to small areas, but that this equilibrium is very unstable, and small environmental variations—for example, increases in temperature of only 2–3°C—can trigger serious biological damage. Any intervention (opening or closing of passage routes communicating with the outside environment or with adjacent spaces, changes in the management of access to visitors, presence of conservation teams, and maintenance operations, etc.) on a preexistent consolidated equilibrium must be carefully examined and monitored so as to give the possibility for a rapid and appropriate response.

Decisions should always be assessed at an interdisciplinary level and should take into consideration all the various aspects of the problem to ensure that the measures taken to prevent biological deterioration do not themselves cause the manifestation of other degradation factors. The fundamental premise of a unified and organic project of conservation is the knowledge and awareness of the different problems relevant to the different areas of specialization. On a number of occasions, several studies have produced positive results with multidisciplinary diagnostic contributions, as for instance in the hypogean complex of San Clemente in Rome, where the results of an extensive campaign of diagnostic investigation —carried out by biologists, physicists, and chemists—were used to identify a series of measures aimed at the prevention of biodeterioration phenomena and at the reduction of the speed at which the frescoes and the walls were degrading (Chapter 5, section 1.3). To be more specific, the main interventions were: the creation of an extensive drainage cavity along the external walls with canalization of rainwater; the revision of the network for the collection of the water originating from existing rainwater pipes; the creation of a new pavement in the external areas adjacent to the church; the planning of a visiting route with the building of an internal walkway that prevented the raising of dust and particles; the regulation of the influx of visitors and of the length of their stay; and, finally, the installation of a new lighting system (Giani et al. 2004).

7.1.2 Outdoor Environments

by Antonella Altieri and Daniela Pinna

In outdoor environments, the possibilities of an intervention to bring the levels of humidity, temperature, and light to below the threshold values necessary for the growth of biodeteriogenic organisms are generally limited and difficult to put into practice, and in any case, they are most often part of an overall conservation plan that requires an assessment of the variables concerning the work in question, of the environmental conditions surrounding it, and of the type of use and presentation that is intended for it.

Architectural and sculptural works and, more seldom, natural objects of historical interest located outdoors are all exposed to degradation by atmospheric agents and are susceptible to biodeterioration. The possible conservation and prevention measures present considerably varying degrees of complexity.

The conservation problems of a historical building that has preserved the integrity of its structural elements are indeed very different from those of a structure that is reduced to ruin and that has, over time, assumed a form and a structural state that no longer correspond to the building's original functions; as a result, this type of structure presents problems pertaining not only to the degradation of its constituent materials but also to the presentation of the ruins. Indeed, in archaeological contexts the *in situ* conservation of architectural structures and decorative surfaces may become not only a necessity—due to their extensiveness and size—but also the final objective

on which to base conservation principles and methods (Melucco 1996; Laurenti et al. 1998; Michaelides 2001).

And finally, in the case of archaeological finds made of wood not specifically treated to resist outdoor conditions—such as prehistoric material (villages of pile-dwellers), ritual monuments (e.g., totems), and, more seldom, fossilized forests with remains of individual trees that are no longer living and composed of more or less degraded but not yet fossilized wood (for example, Dunarobba, Italy) (Biondi and Brugiapaglia 1991; Berti 2000)—the problems associated with their *in situ* conservation are even more complex, because of the even greater susceptibility to biological degradation of wood, its contact with the soil, and often also because of the large dimensions of the structures.

7.1.2a *Direct Interventions on Materials in Use*
The growth of biological organisms on buildings and works of historical interest, whether these are ruins or structurally whole, is mostly encountered on surfaces where both moisture and nutrients accumulate, for example on roofs or at the top of structures, on vertical surfaces exposed to driving rain or near eaves or gutter systems, or on surfaces in contact with the soil (Plate 3).

As mentioned, the moisture present in architectural and also natural structures can be due to rising capillary action or to infiltrations, percolations, and condensation phenomena. It is also closely related to the porosity of the constituent materials, to their hygroscopicity, to the nature of the soluble salts present both in the materials themselves and in the water, and to the dimensions of the structures involved.

In some cases, the prevention of biological colonization is easily accomplished with periodic maintenance that removes accumulation of debris and repairs accidental damage causing water leaks. But in the case of improvement works due to and protection from rising humidity in structures, the interventions tend to be more complex.

In these situations, along with traditional remedies such as drainage systems, cavities, aeration systems, and the cutting of walls to insert isolating materials, the injection of materials with an organic silicon base (siloxane) to provide a chemical barrier against capillary rise of water has also been experimented with as an alternative method (Bartolini et al. 2000b). The increased hydrorepellence of the constituent materials through the application of protective products can contribute to slowing down new biological colonizations on conserved works, but the choice of the type of materials to be used to this end must be evaluated very carefully. Tests should be carried out—both in the laboratory and *in situ*—to verify the effectiveness and also the chemical, physical, and biological effects produced by the protective materials. It has been established that certain synthetic polymers (Chapter 4, section 6) can be colonized by microflora (bacteria, fungi, and lichens) in the presence of favorable environmental conditions and high levels of microbial contamination, thus canceling out the effects of the conservation treatment (Koestler et al. 1988; Leznicka et al. 1991; Nugari and Pietrini 1997; Nugari and Bartolini 1997; Mansch et al. 1999; Pinna and Salvadori 1999; Bartolini et al. 2000a; Koestler 2000). In the area of protective substances used in conservation, some companies are at present investigating the possibility to develop formulations, to be mixed in with the plaster, that have both a hydrorepellent and an antivegetative action; however, the production of these types of materials is still at the experimental stage.

When planning interventions using protective surface coatings it is nevertheless important to take into consideration factors such as the topological conditions and the environmental context in which the work is located (whether or not it is in an urban area; the hydrogeological conditions of the site; exposure to sunlight; ventilation; the presence of elements such as trees or structures up against the work or in its immediate vicinity) because they influence the success of the conservation intervention.

An additional type of treatment that may be considered as a prevention measure against the growth of biodeteriogens is the application of biocides in the final phase of a conservation intervention; this treatment is only carried out if the environmental conditions continue to be favorable to a biological recolonization of the work and if there is no risk of interaction with the substrate. If such conditions are present, biocides can be applied once the intervention is completed; if necessary, further applications can be planned and repeated at appropriate intervals (Nugari and Salvadori 2002). Such a procedure must take into account the possibility of interference with other materials employed in the intervention, and in particular with the consolidants, the application of which can either precede or follow that of the biocide. Schnabel (1991) describes the degradation of limestone, resulting from the reaction between the chlorine ions of the calcium hypochlorite (used as a biocide) and one of the components of the consolidant. The possible interaction between biocides and protective materials has also been observed in laboratory investigations, carried out under controlled conditions (Malagodi et al. 2000); these showed that the sequence in which the two categories of product are applied (the biocide applied either before or after the protective) modifies the effectiveness of the protective material on the substrate.

7.1.2b *Protective Interventions in Archaeological Sites*

In archaeological contexts and, on occasion, when elements of statuary are present, temporary or permanent protective shelters are used. The first are constructed to provide temporary shelter for the structures and the working personnel during excavation or conservation campaigns and are not always sufficient to satisfy the conservation requirements of ancient structures. Permanent protective shelters, on the other hand, are planned to provide stable and ongoing protection for ancient structures that have become fragile with time as well as to create visiting conditions of an archaeological site that are more similar to a museum environment (Prosperi Porta 1996; Laurenti 2000; Laurenti, forthcoming). Sometimes the concept and design of such structures also attempt to suggest ancient forms and volumes that have been lost (Fig. 7.5b). There is a wide range of structures in existence, created with different forms and materials, which can be subdivided into two main categories: overhanging open

Fig. 7.5 *Examples of protective shelters: (a) Shelters—erected in the 1960s—covering the entirety of the archaeological structures (Imperial Villa of the Casale, Piazza Armerina, Enna)* – Photo ICR; *(b) roofs protecting plasters and floor mosaics (Domus, Corfinio, Aquila, first century B.C.E. to first century C.E.), which suggest ancient forms and volumes* – Photo ICR; *(c) roofs protecting the remains of a fossil forest (Dunarobba, Terni)* – Photo S. Berti

structures and structures that essentially cover the entirety of the work (Figs. 7.5a–c).

The materials used and the architectural structure of the protective covering, as well as the installations for the collection and the drainage of rainwater and also the screenings, all determine the effectiveness of the conservation project (Santoro and Santopuoli 2000). In planning these structures, some of the guiding principles are: simplicity of the coverings, low construction costs and use of traditional materials that are easily available, and least possible interference with the ancient works requiring protection. An example of structure of the utmost simplicity built to protect a heritage site is the one constructed for the forest of Dunarobba, where the trees are only partially fossilized and are very sensitive to thermohygrometric changes; here, simple metallic roofing has been erected to protect the remains of the trunks from the direct action of atmospheric agents. These roofs, usually single elements except where a series of very large and adjacent trunks were present, are covered with reed matting in order to avoid excessive heating and are oriented so as to protect the surface of the trunks from the prevailing winds (Berti 2000) (Fig. 7.5c).

However, the installation of any sort of roofing, whether temporary or permanent, results in the creation of a new set of microenvironmental conditions, which at times can trigger—instead of preventing or inhibiting—chemical, physical, and biological degradation. Several examples of protective structures are cited in the literature, in which the use of unsuitable materials resulted in degradation phenomena in the works to be preserved. An interesting case is the report by Child (1998) on the structure erected in Cyprus in the 1950s to protect a royal tomb found at a depth of about three meters, which contained skeletal remains, textiles, and a funeral chariot. In an attempt to preserve the finds *in situ*, the excavation was first covered with glass, and then a roof was erected over the glass, preventing air circulation and creating a greenhouse effect. The dramatic result was an accelerated degradation of the materials (textiles, bones, wood) due to water infiltrations, the presence of salts, intensified fluctuations in thermohygrometric levels and, predictably, growth of plants.

Another resounding example is that of the theater of Heraclea Minoa in Agrigento, where the protective cover built in the 1960s (De Miro 1965)—with seat coverings made of perspex in the shape of those thought to have been originally present and placed over

Fig. 7.6 *Greek theatre of Heraclea Minoa (Agrigento): (a) Perspex coverings dating from the 1960s, made to protect the seating (the plant growth is visible); (b) the seating after the removal of the covers, with very evident plant growth; (c) construction in progress of a new covering for the* cavea – Photo C. Caldi

levels of irradiation within different plant species, within the same species when a plant is adapted to full sun as opposed to when it is adapted to shade, and even within the same plant between leaves exposed to the sun and leaves in the shade (Larcher 1975; Bullini et al. 1998).

In addition, while the presence of protective structures may eliminate damage caused by rain, it has no effect on damage caused by the capillary rise of water from the soil, and provides only a partial solution to the problems induced by light and particularly by temperature, which, in the case of transparent covers, can instead reach higher levels than outside temperature due to the greenhouse effect.

the remaining existing ones—created a warm-humid microclimate ideal for plant growth, which almost completely obscured the view of the underlying seats (Fig. 7.6a–b). Only recently, in connection with the conservation and restoration operations of the *cavea*, was the old covering removed and the *cavea* protected with a roof-like structure (Meli and Alongi 1995) (Fig. 7.6).

In the planning of a permanent protective structure, one of the goals should be to guarantee microclimate conditions that, if not optimal, at least do not accelerate the process of degradation of the work. The use of transparent instead of opaque materials tends to favor the development of photosynthesizing organisms. There is indeed a strong correlation between illuminance and photosynthesis: in the shade, reduced illuminance drastically reduces photosynthetic activity. Still, it is difficult to establish what light levels are limiting for the colonization by photoautotrophic organisms because of the high adaptive capacity of plant species: the points of compensation and of light saturation vary in

The coexistence and interaction of certain microenvironmental factors can trigger and encourage the development of organisms beneath the shelter (Chapter 2, section 2). Among photoautotrophic organisms, cyanobacteria, algae, liverworts, and, to a lesser degree, mosses are in general more sensitive to the humidity factor, while lichens and plants are also selected according to the intensity of the light, the ventilation, and the availability of nutrients. Poor ventilation and a reduction in thermal fluctuations can instead trigger the development of heterotrophic microorganisms on stone materials contaminated with some form of organic matter, such as that provided by deposits of soil or of organic atmospheric pollutants and by metabolic products from previous biological colonization.

Consequently, the main objectives in the planning of a protective shelter that is efficient and that prevents biodeterioration should be:

—an adequate protection from the action of rainwater;

—construction of channeling and drainage systems for rainwater;
—protection from condensation phenomena;
—possibility of regulating the aeration and the natural lighting of the spaces protected by the sheltering structure;
—careful evaluation of how to enclose the site or artifact and ongoing subsequent monitoring of the enclosure boundaries.

With regard to this last factor, it is well known that the fewer the openings to the exterior, the smaller the risk of deposit of particles on the surfaces. On the other hand, if the protective structures enclose the work and the influx of visitors is considerable, problems of aerobiological pollution may arise, resulting in conditions favorable to microbial colonizations similar to those found in enclosed spaces (Chapter 5, section 1).

In addition, particular attention should be paid to setting up excavation sites in the proximity of artifacts, avoiding embankments near archaeological structures; these can cause moisture stagnation and contamination by organic and biological material, which, in turn, are factors that act synergistically in triggering biodeterioration phenomena linked to the development of fungi and—among the bacteria—of actinomycetes (Laurenti 2006).

In addition to protective structures, the practice of temporary reburial is used for the *in situ* conservation of the more fragile archaeological artifacts that are more susceptible to degradation, such as mosaics, plasters, and mud structures, as well as shipwrecks. This system of physical protection is based on employing materials placed in direct contact with the archaeological structures and is adopted during excavation or when it is otherwise difficult to build protective systems able to guarantee microenvironmental conditions suitable for conservation. Temporary reburial should be used only as a seasonal or annual solution.

Since the 1970s and the beginning of the 1980s, separating materials have been used in temporary reburial, as for example, sheets of polyethylene, builders' netting, or geosynthetic (planar material, made from synthetic fibers, usually employed for geotechnical engineering purposes) (Menicali 1993; Cazzuffi 1999); these are combined with inert materials such as sand, pozzolan, or expanded clay. The ease with which the materials used in temporary reburial can be removed allows the work to be used and viewed, while also offering the possibility of regularly monitoring its state of conservation. The nature of the artifact to be protected; its surroundings; the duration of the proposed period of reburial; and the presence of other systems of protection and safeguard of the archaeological remains are all factors that determine the modalities of reburial (Laurenti and Altieri 2000).

For example, in a reburial system consisting of protection of the ancient surface with a nonwoven geosynthetic textile material that is in contact with it, which is then covered with a 10–20 cm layer of expanded clay (Fig. 7.7), the microclimate in contact with the surface of the pavements or walls will be characterized by:

—reduced daily thermal fluctuations compared to the outside environment (Fig. 7.8b);
—a considerable reduction of the irradiation of the pavement surfaces, comparable to 0.1% of the solar radiation outside, measured at midday (the Taurine Baths, Civitavecchia, Rome);

Fig. 7.7 *Temporary reburial of mosaic paving with a nonwoven geotextile ("Reemay 2033") and expanded clay (the Taurine Baths, Civitavecchia, Rome)* – Photo ICR

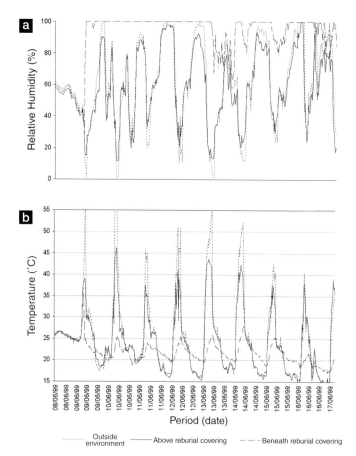

Outside environment ....   Above reburial covering ———   Beneath reburial covering – – –

Fig. 7.8 *Microclimate fluctuations measured beneath the nonwoven geotextile and expanded clay used for reburial and in the outdoors: (a) relative humidity; (b) temperature*

conditions of high humidity, poor ventilation, and contact with the soil can trigger the development of such microorganisms. Yet, to date no such phenomena have been reported on stone materials; on the other hand, in the case of organic materials such as the wood of shipwrecks or of architectural structures—more susceptible to degradation by heterotrophic microorganisms—the choice of temporary reburial must be made with great care, especially in terms of the selection of geosynthetic materials.

7.1.2c *Interventions on the Environment*

Among the interventions on the environment that can prove effective in the prevention of degradation, we would like to cite the effects resulting from the planting of vegetation in the proximity of a heritage site or work of cultural interest.

In some cases, the presence of plants can provide a protection system against environmental factors that could otherwise cause degradation. The planting of trees and shrubs can be used to change the microclimate in proximity of the work: the shade provided by the leaves of the shrubs or trees can reduce the temperature fluctuations on the surfaces of the materials as well as the water movement within the structures (Caneva 1999b); when they act as windbreaks or barriers, trees and shrubs will instead reduce the damaging effects of wind erosion (Fig. 7.9) and, in coastal areas, of the deposit of marine aerosol on the surfaces. In addition, the presence of arboreal or shrubby species, or also of a grassy carpet in proximity of the work, can protect the surfaces from the deteriorating action of certain atmospheric pollutants and particles that are in suspension in

—reduced daily fluctuations in relative humidity, with higher levels than those found outside (Fig. 7.8a; Altieri et al. 1999).

In similar conditions, the growth of photoautotrophic organisms is inhibited by the absence of light, but it is possible to observe the development of root systems of plants that are present in the vicinity of the structure or that grow in the layer of inert material. In this case, the roots tend to grow horizontally in the interface between the pavement and the geosynthetic textile, probably thanks to the high and constant levels of humidity. Colonization by heterotrophic microorganisms, such as fungi and bacteria, could also be favored; it is known that

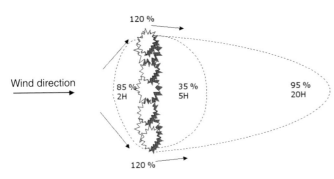

Fig. 7.9 *Example (top view) of the wind-breaking action of a hedge, the effectiveness of which can reach a distance of up to five times its height (drawing adapted from Chiusoli 1999)*

the air and that, when deposited on the leaves, remain entrapped there.

The interception of solid particulates by plants can occur by sedimentation, by impact due to air flow, or by deposition through rainfall (Bussotti et al. 1995). In an urban environment, lead concentration measurements were ten times lower in the leaves of trees bordering urban parks than in the leaves of trees bordering roads (Capannesi et al. 1981). This can be ascribed not only to the lower concentrations of pollutants as you move away from their source (Alessio et al. 2002), but also to the filtering action of the grassy carpet in parks, a filtering that does not occur near asphalt or paved roads. The filtering action of vegetation is proportional to the diameter of the particles and is particularly effective for those particles with dimensions of > 5 μm; it is also dependent on the shape and characteristics of the plant surfaces.

Finally, the presence of plant cover on the ground—especially in environmental contexts that are more at risk, such as those of certain archaeological sites—can also offer a mechanical protection, by reducing landslides and soil erosion caused by rainwater, as well as physical protection through the processes of the evaporation and transpiration of plants that regulate the groundwater (Pirola et al. 1980). In the project for the conservation of the archaeological site of Moenjodaro in Pakistan (De Marco et al. 1990), for example, it was proposed that the development of spontaneous vegetation be encouraged, utilizing

the plants as a natural pumping system to lower the water table level in proximity of the site, which was the main cause for the rapid degradation of the ancient brick walls after they had been excavated.

In the planning of the vegetative landscape of an archaeological site, however, questions do arise as to the compatibility of the selected plants with the conservation of the works and as to their historical coherence (Caneva 1999b). Already in the early twentieth century, when analyzing the possible roles of vegetation in the conservation and presentation of archeological sites, the architect and archaeologist Giacomo Boni—who also supervised the Antiquities and Fine Arts Department in Rome—had made the distinction between "classical flora" and "ornamental flora, to gladden the ruins and veil the restorations" on one side and the "parasitic flora that upsets the order of the structures of antiquity" (Boni 1917). The study of classical literary sources, of iconographic representations that recur in wall paintings, and of archaeobotanical finds (Ciarallo and De Carolis 1998; Caneva 1999a) can provide useful clues to make historically appropriate choices in the selection of the species and in their spatial and formal positioning in environments such as *viridaria*, gardens, and *horti*.

In archeological sites, the introduction of plant species should also be historically compatible, so as to recreate a landscape in which the vegetation is coherent with the site (Caneva 1999b). This kind of planning is in contrast to some solutions found in certain Mediterranean sites where "modern" hybrids were used, such as the genus *Rosa* for example, or exotic species of the genera *Acacia, Robinia, Opuntia, Agave, Yucca, Eucalyptus, Pittosporum, Ailanthus*, which have on occasion become dominant.

Nevertheless, with regard to ornamental flora in archeological contexts or in the vicinity of historical buildings, the possible negative effects, both direct and indirect, must be taken into consideration.

Direct effects include those of a mechanical nature, caused by the development of root systems inside the ancient structures, and those of a more aesthetic nature; these are produced by the growth of vegetation that hides the work, which is then no longer visible as a whole.

Among the indirect effects is the localized growth of biodeteriogens on stone surfaces that are shaded under trees and shrubs and are therefore damp. Such an indirect effect must, however, be evaluated in the context of the climate conditions of the site (Chapter 6). In geographical contexts with a temperate climate, for example, when planting evergreen trees in the proximity of a building, minimum distances have been estimated for the location of trees in relation to their height and the latitude of the site, so as to favor sunlight during the winter season, which is the period with the greatest risks for damage to the structure due to humidity. If, on the other hand, deciduous trees have been chosen, there is no need to maintain minimum distances (Massari and Massari 1993).

7.2 MICROCLIMATE MONITORING

by Elisabetta Giani

When studying the environmental characteristics of a site— be it an exhibition space or a museum warehouse, an archeological site, a cave, or a hypogeum—it is essential that planning and implementation of a microclimate investigation be included (Camuffo 1998; Bernardi 2003).

The knowledge of the microclimate conditions of the spaces under consideration and the climate conditions of the surroundings forms the obligatory basis for an understanding of the physical system artifact-environment and, consequently, for the identification of the causes of degradation.

Indeed, processes of deterioration on works of art can be caused directly by inappropriate microclimate conditions—suffice to think of the alterations caused by the presence of excessive moisture in walls or by condensation on surfaces—but they can also be determined by causes that are only indirectly linked to the microclimate: for instance, the formation of

salt efflorescences resulting from excessive ventilation, or, vice versa, the growth of bacteria, algae, fungi, etc., favored by stagnant air. In the latter situations, microclimate conditions that in and of themselves do not appear to be critical may instead lead to the establishment of both chemical and biological alterations (Chapter 2).

The setting up and fine-tuning of the most appropriate conservation conditions cannot do without an understanding of the actual situation, which will then serve as a base for the planning of both active and passive interventions on the environment in question.

7.2.1 Thermohygrometric Parameters

The physical parameters that are measured during a campaign of investigation of a microclimate are: the temperature of the air, the surface temperature, the wet-bulb temperature, relative humidity, air velocity, and illuminance (irradiance). At the same time, other derived values are also calculated, such as specific humidity, absolute humidity, and the dew point; at times, the microclimate investigation also includes a measurement of the moisture content of the walls (Accardo and Vigliano 1989).

The knowledge of how these values evolve in space and over time makes it possible to understand the microclimate behavior of the environment under consideration and, in particular, to evaluate the exchanges of temperature, moisture, etc., between the environment and the objects contained within it.

We will now give a brief description of the significance and importance of the values indicated above, along with their units of measurement. For temperatures, the unit of measurement in Europe is the degree Celsius (°C), in Anglo-Saxon countries it is the degree Fahrenheit (°F), and the MKS system (based on the meter, kilogram, and second) uses the Kelvin (K).

The indoor air temperature is measured with one or more thermohygrometric probes variably positioned depending on the specific requirements; for instance, if the goal is to study the thermal behavior of an exhibition space, with particular attention being paid to a specific painting, one or more probes will be placed near the object as well as in its surroundings. In an external environment, on the other hand, the temperature is read by placing the probe in a bell that is artificially or naturally ventilated at a height of two meters from the ground, which must be covered with grass.

The surface temperature is measured at the point of contact between the probe and the object; at times, this is a delicate operation because of the difficulty in achieving good contact between the probe and the surface of the object. The temperature measured with the wet bulb of the psychrometer depends on the degree of saturation of the air; along with the reading of the dry bulb, which instead measures the temperature of the air, it makes it possible to establish the relative humidity of the air; relative humidity is the value that indicates the degree of saturation of the air, which indicates therefore how far we are from conditions of saturation. RH is an expression of the percentage ratio (%) between the quantity of water vapor in the air and the maximum quantity of water vapor that the air could contain if it were saturated, at the same temperature and pressure. It is a value that depends on the temperature and on the quantity of vapor (moisture) present in the air; it is of considerable importance in the study

of the interaction between works of art and their environment, since the exchanges of moisture, and hence the behavior of materials, depend heavily on this parameter (Chapter 2, section 2.6) (Figs. 2.16 and 7.10). For any given temperature and relative humidity of the air, every material has its own specific moisture content at equilibrium, which is often referred to as EMC (equilibrium moisture content). As environmental conditions change, so does the moisture content of the material, leading to evaporation or absorption phenomena, which are at the root of degradation processes.

Internal air velocity is measured in the vicinity of the objects and represents the air flow that laps around them. The measurement is taken with specialized anemometers called "hot wire anemometers"; these are based on the principle of the cooling of the sensitive wire by the flow of air around it. The measurement is expressed in meters per second (m/s). It is possible to measure movements of air as small as a few centimeters per second (cm/s).

Illuminance indicates the luminous flux falling on the object; it is measured with a

Fig. 7.10 *Psychrometric diagram (modified from Accardo and Vigliano 1989)*

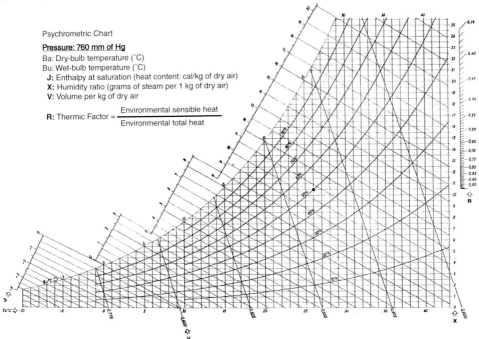

Psychrometric Chart

Pressure: 760 mm of Hg

Ba: Dry-bulb temperature (˚C)
Bu: Wet-bulb temperature (˚C)
  J: Enthalpy at saturation (heat content: cal/kg of dry air)
  X: Humidity ratio (grams of steam per 1 kg of dry air)
  V: Volume per kg of dry air

R: Thermic Factor = $\dfrac{\text{Environmental sensible heat}}{\text{Environmental total heat}}$

luxometer, and the unit of measurement is the lux. More sophisticated instruments are able to discriminate and quantify the UV component of the luminous flux. When measuring light it is a good idea to bear in mind that the effects of irradiation (illuminance) are cumulative and that it is therefore important to calculate not only the spot measurements but also the annual exposure, both in terms of the lux and the yearly dosage of UV radiation.

The derived values, calculated by means of mathematical algorithms, are those obtained from the measured levels of temperature and relative humidity.

Specific humidity represents the ratio between the number of molecules of water vapor present in a given volume of air and the total number of molecules, i.e., it is equal to the sum of those present in the vapor and those in the dry air. This is a parameter that does not depend on the temperature nor on the volume; in environments in which water vapor (moisture) is neither introduced nor removed, its fluctuations are therefore only linked to vapor exchanges between the environment and the object. The unit of measurement for specific humidity is g vapor/kg of dry air.

Absolute humidity, on the other hand, represents the ratio between the number of molecules of water vapor present in a given volume of air and the volume of air itself. Since the volume occupied by a certain mass of air also depends on its temperature, it is easy to understand why this parameter is different from the previous one. For this very reason, this parameter is also less significant; its unit of measurement is g/m^3. It is, however, often used also because of its greater intuitiveness.

The dew point represents the temperature at which water vapor in an environment reaches saturation, thus forming either fog in the air or condensation on a surface. In order to obtain the temperature at which fog is formed in the air, it is necessary to know the temperature of the air and its relative humidity, whereas if the measurements of the surface temperature of an object and the relative humidity of the air are available, it is possible to calculate the temperature below which surface condensation will take place.

7.2.2 Measurement Campaigns

The planning of a campaign of acquisition of microclimate data should be defined case by case, depending on the situation at hand and on the finances available for the project. One of the first aspects is to determine the scope of the campaign, i.e., the choice of the amount of measuring points to be used and of their positioning. The number of probes must be related to the type and size of the environment in question as well as to the specific problems requiring resolution; when a particular environment is being checked, it is in any case always recommended to install more than one probe for the reading of one and the same parameter; in addition, when possible, it is advisable to choose their location according to a grid, which means identifying verticals and horizontals in order to be able to study the spatial gradients of the measured values. When studying the particular object-environment interaction, it is also essential that the sensors—whether these are electronic probes or hair thermohygrographs—be located in the immediate vicinity of the artifacts.

In addition to data relating to indoor environments, it is also necessary to collect data from the outdoor environment, so as to be able to evaluate the repercussions it may have on indoor microclimate conditions. Whenever possible, it is therefore advisable to install a meteoclimatic weather station in the proximity of the environment being studied, which will register the values of T, RH, wind velocity, solar irradiation, and rainfall. Alternatively, it should be possible to refer to data collected by the appropriate institutions, such as astronomical observatories, the Regional Agency for the Protection of the Environment (ARPA—*Associazione Regionale Protezione Ambiente*), military meteorological institutes, etc. In every situation, there must always be the opportunity to compare the meteoclimatic data with the microclimate ones.

A second aspect to be considered is the type of instrumentation to be used, i.e., whether linked to the main electric grid or running on batteries (the latter having appeared on the market in the last few years). Campaigns of continuous measurement with instruments linked to the main grid usually provide readings for air temperature, contact temperature, relative humidity, air velocity, and illuminance; battery-run instruments, or data loggers, will continuously register the air temperature and the relative humidity, but more rarely the contact temperature (Plate 27a). Obviously, the data recording time is limited in battery-run instruments—from a few months to a few days, depending on how frequently readings are taken—because it is dependent on the life of the batteries. However, this type of instrumentation proves very useful for a quick monitoring of conditions, for all those situations in which it is difficult or impossible to connect to the electric grid, or also when the installation of monitoring equipment capable of transmitting data by radio is too expensive. Regardless of the type of equipment used, this type of continuous data acquisition envisages the recording of physical values over a period that can range from a few days (a minimum of two weeks per season is generally recommended) to a solar year, with a time interval between measurements spanning from a few minutes to a half hour, depending on the phenomena under investigation.

Spot analysis is another kind of data acquisition: it provides on-the-spot measurements—usually by means of some kind of portable instrument—and is particularly suitable for preliminary on-site readings. Naturally, if the goal is to study the evolution of the microclimate over time, a series of measurements will have to be taken repeatedly during the day and will have to be recorded (on magnetic tape if possible) in order to allow subsequent data analysis.

Whatever the choice of method for the acquisition of data, it is important that a monitoring campaign be carried out for each of the four seasons, in order to be able to characterize the environment over its annual evolution (Plate 28a).

Another important aspect to be considered in the planning of a microclimate monitoring campaign for spaces that are open to the public is the inclusion of measurements taken during closing days. This makes it possible to evaluate the impact of visitors on the indoor microclimate (Thomson 1986).

In the case of churches (or, in general, of spaces that are not heated such as hypogea, caves, crypts, etc.) readings should be taken both in the hours of worship and when the spaces are empty. Indeed, it is often this alternation of presence and absence of people that leads to conditions of risk for the works. As already mentioned several times, the presence of people increases not only the temperature but also the amount of water vapor in the air (at a rate of about 50 g/h per person) as a result of breathing. A certain amount of time after worship (or after a visit) has ended and people have left the premises, the temperature decreases again; at the same time, the increase in water vapor in the air brought about by the presence of people determines an increase in relative humidity, which may, in turn, give rise to the formation of condensation on the surfaces, especially at the end of the winter season and during spring, when the walls—even if they are very thick—are cold.

From all that has been briefly described above, it follows that, although some principles are applicable in all cases, each monitoring campaign must be custom designed for each specific situation. In essence, what the study must be capable of illustrating at the end of the campaign is what exchange mechanisms take place between the environment and the objects housed within it. This means that it must be possible to explain how the fluctuations occur in connection to the following:

—day/night and seasonal cycles;
—presence of churchgoers or visitors;
—regulation of the openings (doors, windows, etc.);
—operating functions of the installations (heating, air-conditioning, lighting, etc.).

The subsequent goal is, obviously, to be able to make suggestions for improvements on the basis of the described phenomena.

7.2.3 Data Analysis

The recorded data are then elaborated using mathematical software and the results are displayed with tables, in graphic form, or with maps illustrating the spatial distribution of both the recorded and calculated values.

The analysis of the elaborated data shows the progression of microclimate values in space and over time, highlighting areas of instability, spatial gradients, and fluctuations over time that can be at the root of not only physical and mechanical but also of biological and chemical alterations. There are indeed threshold values beneath and above which specific biological attacks may occur. The graph in Figure 7.11 shows the absence, on closing days, of the double peak that is always present on opening days and that corresponds to visiting hours (in this case 9:30 A.M.–12 P.M. and 3–6 P.M.). Opening and closing times for excavations determine the daily fluctuations both for temperature and relative humidity values, and these fluctuations can be a trigger for the occurrence of biological alteration phenomena.

An additional step in the study of microclimate can be taken by using calculation models that, on the basis of the recorded values, simulate the different situations that might occur and then predict thermal, hygrometric, radiating effects, etc., depending on the type of simulation.

As a final phase of a campaign of microclimate monitoring, it is advisable to constitute

an electronic archive of the recorded data and of the data analysis, in order to have easy access to them in the future.

The stored results can be used either for further in-depth study, for chemical and biological investigations, or else for the study of air quality.

7.3 AEROBIOLOGICAL MONITORING

by Paolo Mandrioli, Giovanna Pasquariello, and Ada Roccardi

The air, whether in enclosed spaces or in the open, contains in suspension innumerable particles and droplets of various origin, form, and size. These constitute what is called atmospheric aerosol and can be of inorganic, organic, and biological origin; their effect on the surfaces on which they are deposited can vary greatly and be strongly correlated to the nature of the surface concerned.

In general, the main goal of aerobiological monitoring is to establish the nature of the atmosphere, from a biological point of view, in terms of quality (types of biological particles) and quantity (how their atmospheric concentration varies) (Mandrioli and Caneva 1998). Airborne biological particles consist mostly of viruses, bacteria, and fungal spores, as well as the spores of bryophytes and pteridophytes, algal cells, lichenic propagules, grains of pollen, protozoic cysts, fragments and eggs of insects, and small insects. Often, these biological components are found as aggregates or else included within other solid particles (for example, accumulated in dusts) or in liquids. The term biological aerosol refers to the ensemble of particles of biological origin. Its concentration in the atmosphere varies according to its nature: it is estimated, for example, that the presence of cyanobacteria and algae is limited, while the concentration of fungal and bacterial spores is higher and varies with meteorological conditions, seasons, climate, and the presence of local sources. The number of vegetative forms is scarce because these have low resistance to UV radiation, dehydration, and thermal shock.

Fig. 7.11 *Fluctuations in T and RH in a hypogean environment (the gray band corresponds to the day of closure to the public)*

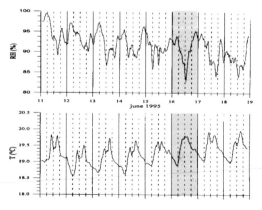

From a numerical point of view, biological aerosol constitutes only a fraction of the total population of airborne particles in the atmosphere, while its mass concentration is comparable to other kinds of particles because of its larger average diameter. In open urban environments and at certain seasonal points of the year it is possible, for example,

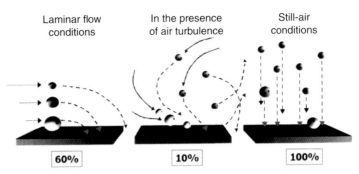

Fig. 7.12 *Different deposition mechanisms of airborne particles (elaboration by P. Mandrioli)*

to observe an average concentration of $10–10^3$ pollen/m^3, $10^3–10^4$ fungal spores/m^3, $10–10^3$ bacteria/m^3 (Maggi et al. 1998). The main sources for biological aerosol are: the soil, bodies of water, agricultural activity, industry, and natural climate phenomena.

Airborne biological particles range widely in terms of size, from viruses (0.003–0.05 μm), to insects, seeds, and fragments of lichens. The phenomenon of deposit of these particles takes place through a variety of mechanisms (Fig. 7.12) (gravitation, turbulent diffusion, inertial impact, thermo- and electrophoresis) as well as through precipitation, whereby the tiniest particles are transported by the diffusion of molecules of water vapor inside droplets of fog or in ice crystals and then dragged down to the ground. In gravitational deposition, how long the particles remain in the air is determined by the speed of sedimentation, a value that is not always easy to calculate because it depends on a number of parameters, such as the form, the surface, and the density, the latter often deriving from the atmospheric humidity value (Mandrioli and Caneva 1998).

Regardless of the type of mechanism that deposits the aerosol, the atmosphere constitutes the main dispersal and transmission method between source and substrate. The aerobiological studies applied in the area of cultural heritage contribute to the knowledge and identification of the biological components of air—both in indoor and outdoor environments—and make it possible to improve our understanding of the relationships among such components and of the degradation processes occurring on the artifacts. Because the live aerobiological particles are potential agents of biodeterioration, it is imperative to have an understanding of the risk of degradation to which the work of art is subjected in relation to the different conservation environments, the various constituent materials, the level of chemical pollution, and the relative microclimate conditions (Blomquist et al. 1984a, 1984b; Mandrioli and Caneva 1998).

To evaluate the biological risk factor in the air within enclosed spaces, it is essential to carry out a quantitative analysis—to indicate the degree of contamination in the environment—and a qualitative analysis—to determine which potentially biodeteriogenic species are present and to identify the sources emitting the particles—in order to take the appropriate preventive measures.

Aerobiological monitoring must therefore be included within a program of preventive conservation in order to be able to contain any possible risks of biodeterioration. In parallel to the aerobiological campaign, it is important to carry out a microbiological check of the surfaces of the artifacts to identify the impact of the deposition processes and any possible correlations between the potentially dangerous microorganisms isolated from the air and those present on the works of art. Several types of sampling equipment are available commercially—in use especially in the medical field—for the testing of the quality of the air and of the surfaces, and several nondestructive methods have entered common usage, also in the sector of cultural heritage (Pitzurra et al. 1997; Sbaraglia et al. 1999; Maggi et al. 2000).

In the evaluation of air quality, the choice of the method to be used for the investigations is closely linked to the proposed goals within the overall intervention program. For the choice of sampler it is important to know in advance which particles or cells are to be isolated and what is the hypothetical atmospheric load of the bioaerosol to be sampled.

On the basis of the experience acquired by researchers in the field of cultural heritage, the selected sampling techniques are those that best meet the requirements in the field of works of art, bearing in mind that bacteria and fungi are the most frequently encountered biodeteriogens in this area.

Passive methods of collection of the particles include gravitational deposit onto sedimentation plates. This is a simple form of sampling that provides the number of particles deposited per surface unit (UFC/cm^2). It is a commonly used technique that allows contemporaneous sampling in several different locations of the space under investigation and the subsequent ample identification of the species present (potentially or actually biodeteriogenic).

Among cascade impactors, the most widely used is the Andersen instrument (Fig. 7.13). This is a volumetric sampler, able to provide both qualitative and quantitative data through the aspiration of given volumes of air (UFC/m^3). The SAS (Surface Air Sampler) impactor may offer an alternative to the Andersen instrument for the evaluation of concentrations of bacteria and fungi, although the very short sampling times do not always produce significant results.

Indoor biological pollution can have external sources (soil, vehicular traffic, industries, vegetation, etc.) or internal ones (biological attacks to furnishings, containers, materials employed during conservation, presence of people and related activities, etc.). Outdoor pollution must in any case be taken into account when assessing the risk, since the majority of biological pollutants come from the outside.

In the past, not much attention was devoted to the phenomenon of atmospheric dispersal of the organisms responsible for the degradation of the materials and, in particular, of the stone surfaces of monuments in archaeological sites, and studies were limited to making a system-

Fig. 7.13 *Aerobiological survey with an Andersen (Doria Pamphili Gallery, Rome)* – Photo M. Romolaccio

atic survey of the alterations already present (Caneva et al. 1998; Piervittori and Roccardi 2002; Piervittori et al. 2002). Present knowledge of the means of dispersal of biodeteriogenic organisms in outdoor environments is therefore scarce, while in recent years there has been a great increase in the number of studies applied to the conservation of cultural heritage in enclosed spaces. On the whole, sampling of the outdoor environment has been limited to the areas adjacent to the enclosed spaces, with the aim of comparing the results of the qualitative and quantitative analyses of the aerobiological particles indoors and outdoors, of identifying their provenance, the different circulation of the pollutants, and their correlation. This kind of applied research has mostly been carried out with the goal of discovering the concentrations of allergens, as these play an important role in the epidemiological field (e.g., surveys of the transport of pollen grains), or to trace the movement of air masses (e.g., through the measurement of the concentrations of specific natural biological particles such as pollen, fungal spores, and algae).

7.3.1 Measurement Campaigns

To plan a monitoring campaign, it is necessary to make a preliminary study of the characteristics of the environment to be analyzed. The information form used for the aerobiological survey of enclosed environments can offer guidelines for the collection of data relevant to this end (Mandrioli and Caneva 1998). This form includes the following headings: goals and duration of the monitoring campaign; usage or intended purpose of the space; whether or not the space will be open to the public; its size and architectural structure; whether the monitoring is partial or global; exposure of the building and of the external walls; openings to the outdoor environment (doors, windows); transit areas (hallways, elevators); thermoregulation, air-conditioning, humidification systems; light sources (natural and artificial); type of surrounding environment; plan of the site; and any additional notes.

For a better comprehension of the phenomena of diffusion of the particles, it may prove very useful to include a comparison between the measured concentrations of aerobiological pollution and those found in spaces that are adjacent to the ones under investigation. It has often been noted that, within one and the same space, there can be areas with different microclimates and different levels of aerobiological contamination, as for instance in access and transit areas such as around doors, in hallways, on stairs, and in areas on different levels, especially if these have different usages, such as archives, storage areas, and attics (Ranalli et al. 1995).

The choice of the sampling method must be based mainly on the type of the space and on the nature of the aerobiological particles to be monitored; these parameters must be compatible with the chosen methods of calculation and analysis (Chatigny et al. 1989).

When establishing a measurement strategy, the main factors to be considered are: placement of the sampling equipment; number of samples to be taken and their duration span; physical and chemical fluctuations in the environment during sampling; analytical techniques to be used for identification; and quantification of the material isolated through sampling. The last two parameters are particularly important for the live material sampled because of its high susceptibility to environmental and sampling stresses. When choosing the location of the areas to be sampled, it is necessary to assess the microbial load of the environment both in proximity of the work of art under investigation and at a distance from it. However, it can prove very difficult and arbitrary to set up and standardize in advance the criteria and methods to be used when the characteristics of the environment to be analyzed are still unknown. A correct approach in this case would require that a series of explorative measurements be taken in conditions that are defined "normal" and "extreme": the first refer to seasonal or daily conditions that fall within the average characteristic situations of the site under investigation, the second refer to what are considered "threshold" conditions because they occur only rarely. The same criteria should be used for explorative measurements placing the sampler in several different locations within the site to be analyzed (horizontally and vertically when necessary). The results of this campaign of preliminary measurements should provide sufficient indications to draw up a protocol of the procedures to be adopted.

Microbiological identifications carried out on microfungi in nonheated environments show the presence of a higher number of fungal spores at lower sampling heights (50 cm rather than 150 cm from floor level), with an increase of 50–60%. This can be largely attributed to the process of sedimentation of the fungal spores, which occurs more easily in the absence of the air movement caused by air-conditioning and heating systems. On the contrary, the presence of sources of heat or ventilation systems can provoke an increased mixing of the particles in the environment with a canceling out of the vertical concentration gradient. The measurements carried out in heated environments—at heights of 50 cm and 220 cm from floor level—have shown essentially equal and uniformly distributed loads of bacteria and fungi. Further experiments carried out in nonheated rooms—50 cm and 150 cm from the floor—showed that the ratio between the bacterial and the fungal load, using selective culture media, is little more than one (Nugari and Roccardi 1996, 2001, and forthcoming).

The programming of a campaign of aerobiological monitoring must also take into account the temporal conditions in which the measurements are taken, which means that the time of year in which to begin the measurements as well as their frequency and duration must be established. The reason for this is that growth and development of microorganisms are inherently seasonal and closely linked to the alternation of microclimate conditions existing both in enclosed and open environments.

In addition to seasonal variations, which are generally observable over the span of a year, the higher-frequency fluctuations of the day-night cycle must be taken into consideration. As is well known, this frequency is determined mostly by day-night fluctuations in the temperature of the air and, hence, of the substrates as well as by variations in other parameters, such as humidity and luminous flux. Although with different modalities, these fluctuations are also observed in enclosed environments, as a result of the programming of heating and air-conditioning plants. It is indeed rare, even in situations with rigorous climate control, to find that the same levels of temperature and humidity are maintained stable over the twenty-four hour span.

In addition to seasonal and daily variations, another factor that characterizes fluctuations of environmental parameters in enclosed environments such as museums, galleries, churches, etc. is human activity, with regard to both visitor and churchgoer influx and the presence of personnel fulfilling various tasks. Human presence in an enclosed environment induces not only increases in temperature and humidity, but also the emission of biological aerosols such as droplets and minuscule fragments of organic tissues, which generally constitute the vehicle for the transmission of microorganisms. It is therefore important to take into account the time of day in which sampling takes place in relation to visiting hours, in order to be able to evaluate to what extent the microbial component is due to human activity in the space under investigation. Sampling carried out in the morning hours, before any activity takes place, can provide information on the basic microbial concentration; while measurements taken after the presence of a considerable number of visitors or after a period of intense activity—such as movement of dusty objects or the cleaning of the areas—will give a good indication as to the maximum microbial concentrations that might be found in that particular environment.

7.4 THE PREVENTION OF BIOLOGICAL RISK: HEALTH ASPECTS RELATED TO MICROFLORA

by Gianfranco Tarsitani

The conservation of cultural heritage covers a wide range of works, from historical buildings to textiles or furnishings. This great variety means the working environments of conservation and restoration personnel are unique and that people working in these contexts are exposed to a variety of health risks.

Because of this, aerobiological studies applied to this field have important hygienic as well as health and safety implications. As a result, the study of the degradation of cultural heritage, and especially of biodeterioration—which in recent decades has blossomed—has, at the same time, promoted a better understanding of the consequences of exposure to factors of biological risk for people operating in this sector.

As mentioned, this area of study is particularly complex because of the extremely diverse types of cultural heritage materials involved, i.e., archaeological finds, ancient books, paintings, textiles, furnishings. In such contexts, people may find themselves working on an archaeological excavation outdoors or in an enclosed environment, on scaffolding or ladders at the various sites, in conservation studios, in libraries or museums, and so on in a great variety of different situations with varying degrees of comfort. The working conditions for conservators and restorers are often inevitably precarious, since their activities are often undertaken in environments that are characterized by specific and unique structural and hygienic conditions.

Given the great heterogeneity of the possible dangers and the difficulty in defining their specific action in many cases, the effects on health can manifest themselves as a simple irritation perceived at a sensory level, or with an array of symptoms with immediate, short-, middle-, or long-term onset. In the context of such effects, it is possible to encounter pathologies with a specific recognizable etiological agent as well as nonspecific pathologies for which it is not possible to identify a cause-effect relationship with any particular contaminant or situation. Very often, the damages to health are intensified by the sum and the synergistic effect of the various danger factors (physical, chemical, and biological).

However, there is no particular awareness, nor even an instinctive perception, of the connection between potential biological risks and specific work activities among people in the field of conservation—contrary to what one might think based on widespread beliefs such as "the curse of Tutankhamen" (the unexplained deaths of archaeologists and workmen who had violated the tomb of the Pharaoh). Nevertheless, taking the example above, it seems evident that little known or unknown biological agents may survive in specific climate conditions (a tomb, a hypogeum, a sealed artifact) and, subsequent to these conditions being altered, may rapidly find a way to reactivate themselves and give rise to new colonies.

7.4.1 A Health-conscious Approach to Microbiology

Even though its methods are comparable to those used in studies directed to the safeguard of cultural heritage, the health-conscious approach to microbiological study actually operates with a logic that is diametrically opposed to them, in the sense that it places at its core of interest the health of the people employed in the various activities in question.

Let us say first that putting such precautions in place is actually a fulfillment of a series of legislative measures that are, in turn, fulfilling decrees, the most relevant of which is the legislative decree of 19 September 1994 n. 626 (G.U. 104 of 6 May 1996, Suppl. Ord.). This decree enforces various European directives regarding the improvement of workplace health

and safety measures. In identifying the general measures for the protection of the health and safety of workers, the guiding principles on which the decree is based are first and foremost an assessment of the risks involved, followed by the elimination (or reduction) of the risks at the source. Based on the knowledge gained through the technical processes involved, it is possible to program the prevention by: substituting what is dangerous with what is not, or less, dangerous; defining working methods that are risk free; giving priority to measures of collective, rather than individual, protection; limiting the number of workers exposed to danger; reducing to a minimum the use of physical, chemical, and biological agents in the workplace; keeping the worker away from the sources of risk, for inherent health reasons; setting up health, protection, and emergency measures; and providing the workforce with adequate information on issues regarding health and safety in the workplace.

Section VIII of the legislative decree 626/94 cited above refers to the issue of protection from biological agents and is therefore of specific interest to our topic. This section is applicable to all work situations in which there might be a risk of exposure to biological agents, i.e., any microorganism, even if genetically modified, or a cellular culture or a human endoparasite that could provoke infections, allergies, or be toxic. Biological agents are subdivided into the following four categories according to the risk of infection:

1. An agent that has little chance of infecting a human being;
2. An agent that can provoke serious illness in humans and pose a threat for the workers; it is unlikely that it could spread into the community; effective prophylactic and therapeutic measures are usually available;
3. An agent that can provoke serious illness in humans and pose a serious threat for the workers; the biological agent can spread in the community, but usually there are effective prophylactic and therapeutic measures available;
4. A biological agent that can provoke serious illness in humans and pose a serious

threat for the workers and that also has a high risk of propagating in the community; there are usually no effective prophylactic or therapeutic measures at hand.

The above-mentioned legislative decree cites the list of biological agents referred to in subsections 2, 3, and 4; it also states that when the biological agent cannot be unequivocally classified into either group 3 or 4, it must be listed under the higher-risk group (4).

The health risk can be particularly relevant in conservation and restoration operations, since ancient and deteriorated works are almost without exception in such conditions because of the action of living organisms. Depending on the nature of the constituent materials of the works, on the microclimate conditions, and on the pollution of the environment in which they are preserved, spores and airborne vegetative forms can find various matrices on which to develop and constitute elements of potential degradation. An example of ideal habitat for the development of fungi is the one created on the reverse of paintings on canvas that hang on damp walls; here, an organic substrate (the canvas support) is in conditions of high humidity and low air exchange.

A highly topical health problem is air pollution, which rises to the level of public health issue when considering both the length of exposure periods and the type and extent of population at risk. Air contains biological particles from a variety of different sources that can accumulate in dust and then become suspended again in the atmosphere. Among other components are fungal and bacterial spores as well as viruses. The formation of aerosol is one of the most frequent causes of contamination; aerosol (composed of minuscule droplets) may contain dispersed microorganisms, possibly pathogenic, that spread everywhere transmitting diseases. Research in libraries and museums has confirmed the presence of bacteria (*Bacillus cereus*, *B. circulans*, *B. subtilis*, *B. anthracis*, *Micrococcus luteus*, *Streptococcus pyogenes*, *Staphylococcus aureus*, *Corynebacterium pyogenes*, *Micrococcus* sp.) and microfungi (*Aspergillus*, *Penicillium*, *Trichoderma*, *Alternaria*, *Stachybotrys*, *Trichotecium*, *Rhizopus*, *Mucor*, *Cladosporium*, *Monilia*). Biological contamination within an environment can cause not only infectious diseases, but also symptoms such as irritation of the mucous membranes, headaches, and fatigue.

The phenomenon of microorganisms transported by bioaerosol is a danger and an often underestimated risk, which should instead be taken into serious consideration especially in enclosed environments and also during any activity that might place the operator in contact with microorganisms potentially harmful to his health. For example, the inhalation of high concentrations of fungal spores can cause hypersensitivity and produce allergic reactions and asthma in humans.

To evaluate the biological risk in enclosed spaces, it is important to know the concentration of the total load of bacteria and also to identify the bacterial and fungal species present in the indoor air. The total bacterial load gives an indication of the general degree of contamination of the environment in question, while the identification of the species present provides information on the existence of specific biological risks, thus offering the possibility of identifying the specific sources of their emission and of resolving the issue. No specific threshold limits have been adopted officially for indoor environments; some authorities have proposed threshold values, as for instance OSHA in the USA, which set as thresholds 750 UFC/m^3 for total bacterial load, 150 UFC/m^3 for the load of microfungi, and requires a complete absence of allergens.

The decree mentioned earlier (626/94) prescribes the adoption and use of devices for individual protection when working conditions do not allow the adoption of collective protection measures, or else when these are not sufficient. Such devices include both technical equipment (ranging from extractor hoods to protection on scaffolds) and working procedures (carrying out dangerous work when the premises are less crowded, etc.). Helmets, shoes, and—more specifically for biological risks—masks, glasses, gloves, overalls, caps, and similar items are an important addition to personal equipment, which people working in the cultural heritage sector often neglect to use.

7.4.2 Diseases of Cultural Heritage Personnel

A person working in conservation or restoration can contract a wide variety of diseases caused by biological agents. The most common symptoms found among people working in the area of cultural heritage are: allergies, skin diseases, illnesses of the respiratory tract, eye diseases, as well as possible risks for procreation.

Various types of dermatitis, whether allergic or not, are among the most common diseases attributed to the action of biological agents encountered on objects or finds. Diseases such as scabies are also a type dermatitis caused by biological agents; fibrous materials such as textiles, furs, or paper are possible vehicles for the transmission of some of these organisms. Diseases of the respiratory tract, such as bronchitis and bacterial pneumonia, can be caused by the numerous agents present in the work environment, and especially on the objects themselves. Frequent eye disorders due to the presence of biological agents are conjunctivitis but also more serious eye diseases. In addition, there are numerous examples of professional diseases that afflict conservators in charge of interventions on works that simply originate from infected areas or environments, or that carry agents that are not in and of themselves particularly pathogenic in the widest sense of the term. These agents may have remained isolated for a long time, and the organism of modern man—living in other regions and used to different conditions of hygiene and sanitation—may now have trouble reacting against them, because it has never had the opportunity to develop specific immune defenses.

Diseases linked to enclosed environments deserve a special mention. In this context, it is necessary to make a distinction between "sicknesses linked to buildings" and "sick building syndrome." The first refers to pathological conditions determined by readily identifiable causes; they apply to a limited number of people within a building and can be correlated to the presence of a specific biological contaminant. Bioaerosol plays an important role in the etiology of the different forms of "sicknesses linked to buildings" with three distinct mechanisms: infectious, allergic, and toxic. "Sick building syndrome" consists of a body of multisensory perceptions, or of a nonspecific sense of malaise, or else of a precise set of symptoms: eye and nose symptoms, muscular and joint pains, respiratory disorders, skin conditions, violent headaches, and states of anxiety. The symptoms usually appear on the first working day of the week, returning from holiday, or after any other kind of work interruption; they increase as the week progresses and also with the years at work. This syndrome strikes the majority of people working inside a building, although it does seem to be more frequent in women. It is characteristic of modern buildings with total air conditioning and, in the context of a multifactored etiopathogenesis, its main determining factor seems to be the poor quality of the air-conditioning system. There are also descriptions of "library sickness" syndromes that seem to affect both librarians and patrons.

7.4.3 Air-transmitted Diseases

In addition to providing information on the particular characteristics of the materials and the environment, the qualitative and quantitative evaluation of the biological component of the air performed with targeted analysis campaigns also contributes to a definition of actual risk situations in specific work environments and provides information on the sources of contamination essential for the selection of suitable corrective measures and appropriate interventions.

Microorganisms present in the air are incorporated into solid particles (dust of mineral, vegetable, or animal origin or deriving from the desiccation of saliva or other secretions) or liquid ones (aerosol from coughing), which protect them from environmental stresses, enabling them to survive for long periods of time. Pathogenic microorganisms present in the air reach the human body mostly through inhalation and, to a lesser degree, through contact and ingestion, causing damage at various levels.

For an evaluation of the level of pollution of indoor air, the ratio of the concentration of biocontaminants in the indoor/outdoor environments is of the utmost importance (earlier in this chapter, section 4.2). In the case of pollution caused by bacteria and viruses, this ratio is usually greater than 1, because the majority of these microorganisms are introduced from the human body into the atmosphere with speech, coughing, sneezing, and the shedding of tiny flakes of skin to which the microorganisms adhere. In the case of fungal spores, instead, the ratio is less than 1, because they are predominantly introduced from outside; however, when conditions of hygiene are not optimal, or in the presence of damp walls or stagnant water, the concentration of fungal spores can increase.

Allergic symptoms may be manifested with different degrees of severity, starting with a simple inflammation of the eyes and of the upper respiratory tract (rhinitis, sinusitis) and developing into much more serious symptoms such as allergic alveolitis and bronchial asthma. In an indoor environment, the allergic symptomatology is triggered by the presence, in the air, of acarians (which are not discussed here), epidermal elements from domestic animals, molds, and fungal spores, while the risk linked to the presence of pollens is low, since their indoor concentration is lower than outdoors.

The fungal species present in nature are numerous and ubiquitous; their development is favored by conditions of high relative humidity and air temperatures between 18 and 32°C; for this reason, they are present in large quantities in damp buildings or in air-conditioning systems. The fungi most frequently responsible for allergic reactions belong to the following genera: *Alternaria, Cladosporium, Aspergillus, Candida, Penicillium, Mucor, Fusarium,* and *Rhizopus. Alternaria* is present mostly in the outdoor environment, while *Aspergillus, Penicillium, Mucor,* and *Rhizopus* are more prevalent in damp indoor environments. Fungal spores are responsible for allergic reactions of the skin and of the respiratory system that can have continuous or seasonal manifestations.

With regard to toxic effects that may be produced by biological agents, indoor air may contain fungi that produce mycotoxins (*Fusarium, Trichocene, Trichoderma*). At low concentration levels, some mycotoxins are responsible for gastrointestinal troubles and for damage to the haemopoietic and genital systems, as well as for less specific symptoms such as asthenia and nausea, similar to those that characterize "sick building syndrome."

As far as the risk of infection is concerned, we should remember once again that the air in enclosed environments can constitute a vehicle for contagion for diseases of the respiratory tract and, more generally, for air-transmitted diseases.

Numerous kinds of microorganisms are present in the air outside, but in low concentrations and subject to mechanisms of autopurification (desiccation, action of UV rays and of oxygen, cleaning of the atmosphere through the action of rainwater). In enclosed environments, the concentration levels of germs are noticeably higher than outside, both because of the lack of natural mechanisms of purification and because of overcrowding. In addition, these environments may also present microclimate conditions (ideal humidity and temperature) that encourage the development of mycetes, bacteria, protozoa, and acarians. In particular, humidifiers, vaporizers, poorly functioning air-conditioning systems, and indeed any systems or situations that allow water stagnation, all represent an excellent *pabulum* for the development of microorganisms.

Microorganisms that penetrate through the air-conditioning system can either be transported on dust particles and on detritus of building materials or they can proliferate inside the ducts—the latter provide an excellent culture ground because they offer a protected space, with suitable temperature and humidity, as well as nutrients in the dirt. In addition, since microbes and spores are small and light, they are carried along in the flow of air introduced into the buildings. We should also take into consideration that in the case of air-conditioning systems that recirculate the air, the polluting agents may be diffused throughout all the premises that are fed by the same system.

Because airborne pathogenic microorganisms enter the human body mostly through inhalation, we can say that, in a way, the human respiratory ducts are a sort of "air sampler," in which the particles are captured through sedimentation according to their size (although their particular shape and chemical composition are also of importance). The greater the diameter of the particles, the lesser the damage; the largest ones (3–30 μm) remain in the upper respiratory tract and are therefore eliminated during expectoration. Particles with a diameter of 0.5–3 μm are those responsible for more severe damage to the respiratory tract, because they are able to reach the pulmonary alveoli, while the even smaller ones are not deposited at all and are eliminated when breathing out.

The risk of infection linked to the inhalation of microbial aerosol is related to factors inherent to both the microorganism and the host. The first include pathogenicity, virulence, and bacterial concentration. Pathogenicity is the intrinsic capacity of a microbial species to provoke an infectious illness in a specific animal species; virulence is the degree of pathogenicity that a specific strain of a pathogenic microbial species develops toward the host animal species (for example, there are strains of the diphtheria bacterium that are more virulent toward man than others). Obviously, virulence levels being equal, it is the concentration of bacteria that is most important; the higher the load, the higher the probability that it will cause a state of infection and disease. This state can be produced either by the presence of a massive microbial dose, or else by repeated penetrations of a small number of germs at close intervals. Overcrowding in enclosed environments favors the spread of infection because of the higher concentration of sources of infection within a single space and, therefore, a higher exposure to contagion. On the other hand, the penetration of small and repeated doses of microorganisms at distant intervals will instead favor the development of a state of acquired immunity.

As far as the host is concerned, we distinguish between intrinsic factors—such as age, gender, constitution, and ethnicity—and contingent factors—such as lifestyle, working conditions, nutritional state, and health. People at either extreme of the age band (earliest infancy and advanced age) have a higher degree of sensitivity to infection; those with either congenital or acquired states of lowered immunity, or those affected by chronic debilitating illnesses are also more sensitive. Indulgence in habits such as smoking or the abuse of alcohol and conditions of malnutrition can also make the organism more prone to contracting infectious illnesses.

In accordance with legislation (earlier in this chapter, section 4.1), an adequate health supervision of the personnel must be guaranteed based on the results of the risk assessment. The doctor and/or the authority responsible of this supervision must be aware of the circumstances and conditions of exposure of each member of the workforce.

Although the main sources of danger for people working in the area of cultural heritage seem to be generally well known, a precise quantification of the risk of the pathologies associated with them has not been defined. There is a real need for specific research on these subjects in order to identify strategies and measures that will reduce the risks and improve safety and working conditions.

One of the major obstacles to the improvement of working conditions is the general lack of attention for even the most elementary safety measures in daily activity, resulting from a lack of understanding of the potential dangers and a lack of good practices in adopting appropriate safety-conscious behaviors. People working in the area of cultural heritage, who often did not receive specialized training, must be given the correct tools to be able to intervene appropriately in risk assessment and management and to succeed in changing methods of operation rooted in long-term working habits.

To summarize, an educational campaign should be undertaken with regard to all forms of risk, including those of a biological nature, with the aim of establishing an awareness of the importance of a correct risk evaluation and encouraging the acquisition of operational autonomy. An effective and successful educational campaign must produce such levels of understanding and awareness necessary that choosing models of behavior aimed at maximum levels of safety will become second nature, on the basis of the following three cardinal points:

—Communication of the dangers inherent in the risk, aiming to provide the capacity for a correct perception of the risk

factors involved and of the gravity of the consequences of a possible incident;
—Analysis of the risk through formalized procedures, which—once the potential dangers are identified—will make it possible to analyze the modalities and probabilities of an incident occurring;
—Management of the risk that enables its reduction within levels of sustainability and provides parameters and procedures for the management of inevitable residual risk.

The most difficult and complex stage is undoubtedly the communication of the risk. Contrary to the two subsequent points, in the first there is no recourse to formalized procedures; it relies only on communication techniques and is strongly tied to subjective perceptions in the exchanges between the transmitter and the receiver of the message. It is at this level that it is possible to win a great battle for civilization, pursuing those objectives fundamental to a democratic approach to health.

CHAPTER 8

CONTROL OF BIODETERIORATION AND BIOREMEDIATION TECHNIQUES

The control of biodeterioration of our cultural heritage includes all the measures used to eliminate degradation by microorganisms and organisms and, whenever possible, also those used to delay their reappearance. These types of measures are referred to as "direct methods" of elimination of biological growth, in that they foresee a direct intervention on the biodeteriogens with the application of mechanical, physical, and chemical methods with biocidal action.

The efficacy of these operations depends on a number of factors such as the methods adopted, the products used, the organisms targeted, the state of conservation of the work, etc.; but it is important to emphasize that continued growth of undesirable organisms is inevitable if the environmental conditions favorable to them remain in place. Hence, the importance of all the activities outlined in Chapter 7, aimed at creating—whenever possible—environmental and substrate conditions that are unsuitable for the life of biodeteriogens (indirect methods).

In this chapter we will illustrate the general methodological criteria of the control of biodeterioration, distinguishing them on the basis of the constituent materials of the works of art, namely organic or inorganic. In terms of the latter, we will be looking in particular at the topics of disinfection and herbicide treatments in works in stone, since an extensive literature as well as many case histories are available in this area. On the other hand, treatments on other kinds of inorganic materials (e.g., metals, glass) are carried out very rarely, partly because as a substrate they are less widespread than stone, but mostly because of their lower bioreceptivity and the difficulty of interpreting the biological alterations.

There is also a section dedicated to bioremediation, in which the development of biological agents is actively encouraged because it fulfills a positive function instead of a negative one: thanks to some specific metabolic characteristics, some microbial species are utilized in order to treat the chemical and physical damage on certain materials.

8.1 METHODOLOGICAL ASPECTS OF THE TREATMENTS

by Giulia Caneva, Maria Pia Nugari, and Ornella Salvadori

Above and beyond the chosen methodology and the nature of the biodeteriogenic agent to be eliminated, the methods employed in the control of biodeterioration still foresee interventions on the surface of the work; these, although selected with the utmost respect for the constituent materials, will nevertheless always involve either a mechanical, a physical, or a chemical action on the works.

The decision to perform an intervention or not and the choice of the best procedures must be considered with the greatest care, evaluating the different aspects involved in the problem (Caneva et al. 1994a, 1996; Koestler et al. 1997; Warscheid 2000, 2003).

Even though the elements requiring assessment are complex, the main points to be considered when setting up a treatment methodology can be summarized as follows:

—identification of the etiology of the damage;
—suitability of the intervention;
—risks linked with the intervention itself.

The first consideration must be whether or not the problem is of biological origin. This point has priority over others and might seem obvious, but the fact is that such a question does not always have an easy answer. A biological attack is not always readily recognizable as such to the naked eye, nor does it always have such a distinct and unique morphology as to exclude the possibility of being confused with degradation phenomena of purely chemical origin. Biocide treatments—whether employing chemical or physical methods—should therefore only be chosen after an accurate diagnosis aimed at the identification of the colonizing macro- and microflora and of its relationship with the alteration phenomena present.

The appropriateness of an intervention must always be evaluated, even if the existence of a biodeterioration problem has been determined. The need for intervention must take into account several factors that, as a whole, contribute toward the formulation of the most accurate answer:

- Identification of the biodeteriogens and quantification of the damage. The knowledge of the ecological and physical as well as the morphological and structural characteristics of the species present makes it possible to understand what conditions favor their development, their life cycles, their relationship with the substrate, and what particular mechanisms of biodeterioration are in action. Clearly, a treatment is only considered appropriate if the information gathered points to a high potential of biodegradation;
- Results of the treatment over time. If the effects of the treatment can only be short term, the decision must depend on the actual gravity of the alteration and how long it will take for recolonization to take place. In the case of severe damage, treatment will of course prove a necessity, but it is essential that periodic maintenance be included as part of the overall plan. In the case of inconsequential damages, if there is no possibility of providing a plan for subsequent maintenance, treatment becomes superfluous.

- Ecosystemic consequences of the intervention. In this case there may be a variety of problems:
a) A change in the equilibrium of the microbial communities: a biocide treatment directed toward the eradication of only some of the species may induce the proliferation of others, which find themselves without competition. In some hypogea, for example, new, more aggressive and more resistant species developed after a biocide treatment with a limited range of action (Agarossi et al. 1988).
b) Indiscriminate destruction of the biodiversity: interventions involving nonspecific methods can result—especially in the case of outdoor works in stone—in the destruction of all life forms in the treated area, with the consequent reduction of floristic elements that might otherwise have been of interest (Nimis et al. 1992; Ariño and Saiz-Jimenez 1996a; Celesti Grapow and Blasi 2003). This is often the case with total herbicide treatments; in many contexts—such as wall enclosures, or architectural structures where it is not absolutely necessary to preserve the surface finishing (for example, the *nuraghi*)—a complete removal of all the plant communities should not be performed, and the interventions should instead be confined to the eradication of only the most dangerous species. The appropriateness of interventions must be carefully evaluated in other situations also, such as for instance in caves, nymphaea, or monuments in rural environments, which were intrinsically conceived as expressions of the bond between nature and the work of mankind (Plate 23).

If the damage and the lesions caused by growth processes do indeed justify an intervention, an additional aspect to be evaluated is what level of risk the removal of the biodeteriogens may entail for the work of art or artifact. The result of such an evaluation may be to decide against any intervention, or else to intervene but by adopting appropriate methodologies and precautionary measures.

The decision to avoid an intervention may, for instance, be taken in the case of certain slow-growing endolithic species (e.g., lichens): their elimination might involve the removal of the superficial layers of the stone, which, in turn, would expose particularly fragile surfaces to new kinds of damage, increasing the dynamics of deterioration. But not intervening is certainly not a choice in the case of microorganisms with direct degrading action on the substrate; here, blocking the enzymatic attack on the constituent materials is a necessity.

Preliminary operations of various kinds must be carried out when treatment is unavoidable, and the physico-chemical conditions and the state of conservation of the work are critical. Among these preliminary operations are: the preliminary consolidation of the substrate to avoid further worsening of the state of conservation as a result of subsequent treatments; the partial drying out of materials to improve the effectiveness of the treatments; and also the removal of elements that are not directly pertinent to the work and that have facilitated its biodeterioration, such as the linings of canvases applied with starch-based glues. Similarly, in the removal of trees that are solidly rooted in walls further stresses for the substrate are unavoidable, but the operation must be performed to prevent a total collapse of the structure in the future; in such an instance, it is advisable to add structural consolidation to the intervention.

Once the need for an intervention has been established, the best method for carrying out the treatment must be chosen. Depending on the kinds of organisms present, on the types of materials of the work to be treated, and on its state of conservation, the decision has to be made whether mechanical, physical, or chemical methods of treatment are most appropriate, and—in the case of the latter—which products and techniques are the most suitable. The advantages and disadvantages of each method, along with the limits of their application and the costs involved, must be evaluated for each individual case. Also, it is important not to forget that the treatment itself may have an interfering effect on the constituent materials of the work. Interventions with inappropriate materials and methods can sometimes cause worse damage than the one induced by the

biodeteriogens, such as for instance irreversible changes to pigments.

The characteristics of the various applicable methods must therefore be known or be investigated experimentally to establish the following:

—the efficacy in relation to macro- and microorganisms that deteriorate the various materials;
—the possible interactions with the substrate;
—the best method of application in relation to the different kinds of biological colonization, the nature of the materials, and the state of conservation of the work;
—the long-term effects and the possible interactions with other products used during conservation;
—innovative methods and alternatives to the traditional methods of disinfection and disinfestation.

The efficacy of a treatment is often determined by the specific phase of the life cycle of the species to be eliminated. In the case of species with a marked seasonal rhythm—such as, for example, vascular plants—it is necessary to plan for interventions to be performed at the best moment for the treatment, i.e., when the species are most vulnerable. Microorganisms and lichens must also be treated during their metabolically active phases, or else the result will not be satisfactory.

Alongside the treatments, it is essential to plan and carry out operations that have an impact on the environmental conditions that allowed the biological development to occur in the first place, with the objective to modify, at the very least, the levels of the most influential parameter (Chapter 7): only if this is accomplished will the results of the treatment be long term.

In the case of works located outdoors, or if it is impossible to intervene on environmental parameters, a plan for periodic maintenance is essential and should be programmed taking into account the growth rhythms of the species involved and the dynamics of recolonization; this would make it possible to intervene in the initial phases of the developments or even preventively.

8.2 **MECHANICAL METHODS**

by Giulia Caneva, Maria Pia Nugari, and Ornella Salvadori

Mechanical methods consist of the physical removal of the biodeteriogens with manual instruments such as scalpels, paint brushes, brushes, spatulas, small scrapers, microaspirators, etc. These methods are widely used because of their simplicity and the immediacy of their results.

It should be pointed out, however, that it is very difficult to achieve a complete removal of the vegetative or reproductive structures of the species present, especially if the colonization has gone beyond the initial phases, without seriously damaging the substrate—the results are therefore generally short-lived. Indeed, whether it is a fungal mycelium, a lichenic thallus, or, even more so, the root of a plant, colonization is never limited to a superficial phenomenon; thus, if damage to the substrate is to be avoided, mechanical removal cannot ever be anything but partial. The substrate of a biological colonization is usually altered in its physical and chemical characteristics and therefore exhibits a greater fragility toward the action of mechanical instruments. It is fairly common that damage is caused by imprudent operators during the manual extirpation of plants anchored in walls or in the cleaning of surfaces covered with biological patinas performed with stiff brushes (Fig. 8.1); or, during the removal of algal patinas, for instance, stains are frequently formed due to the emission of photosynthesizing pigments onto the substrate because of the rupturing of cell walls.

However, there are occasions when mechanical removal can be the answer: for instance, in the removal of gymnosperms (for instance, cypresses, pines, etc.) from archaeological sites or near monuments, the felling at the base is sufficient to kill these species, so that it is not necessary to totally extirpate the plant (unless there are structural requirements), nor to employ biocides in order to avoid new shoots.

In addition to ease of applicability, mechanical methods have one more advantage in that they do not add anything that may cause further degradation; and, when executed expertly, they can prove quite useful, especially when combined with chemical methods (Fig. 8.2).

A reduction in the biomass may prove useful as a preliminary treatment to the application of a biocide: in the case of highly sporified fungal colonizations, the removal of the spores could be accomplished with aspirators or microaspirators (Fig. 8.3) or with damp or electrostatically charged materials, while removal of thick mucilaginous layers of cyanobacteria and algae could take place with blotters or brushes. Such operations must be carried out with the utmost care: if a simple dusting is performed, it may instead be contributing to the dispersal of the spores or vegetative propagules into the air, thus encouraging the diffusion of biological damage (Schulz 1994). A microaspiration executed with precision using a stereomicroscope can prove very useful for the removal of fungal colonizations in their early growth stages, if, however, the object is monitored periodically after the operation to allow an early identification of any new developments.

Mechanical methods are also often used for the removal of devitalized biodeteriogens after treatment with biocides.

Fig. 8.1 *Damage to the substrate through erosion of the surface, due to the use of stiff brushes*

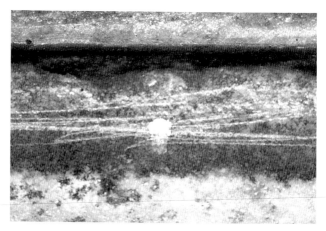

Fig. 8.2 *Elimination of a fungal attack from a fresco using a variety of methods (Giotto, Lower Basilica in Assisi)—the darker square in the center shows the condition prior to the intervention* – Photo ICR

8.3 **PHYSICAL METHODS**

by Stefano Berti, Flavia Pinzari, and Piero Tiano

The physical methods illustrated here perform a biocidal action on the biodeteriogenic organisms either directly—by interacting with the genetic cell material, as in the case of ionizing radiations, or through other processes of physical alteration of the cells themselves (use of intense heat or freezing processes)—or indirectly, through the reduction of the constituent elements of the atmosphere that are indispensable to the biodeteriogens for their metabolic processes (e.g., modified atmospheres).

Mainly, physical methods make use of electromagnetic radiations (UV, gamma, or beta waves), microwaves, low-tension electric currents, or low and high temperatures (Figs. 8.4 and 8.5). These methods have been prevalently used on organic moveable artifacts and against insect infestation; in some cases, they were even effective against birds (neither insects nor birds are discussed here) as well as in the elimination of biological patinas from stone materials.

The use of physical methods is not yet widespread for a variety of reasons relating to applicability, cost, the risk for the operators if not adequately protected, and to the issue of possible interference with the constituent materials of the works of art. To determine the level of danger of the treatment, it is often necessary to establish numerous factors relating to the operation; for instance, in the case of radiations, their danger level depends on the intensity (the dosage) and the duration of their application as well as on the distance of the object from the emitting source (Brokerhof 1989; Pointing et al. 1998). At present, however, the various effects on materials of cultural and historical interest are not known for all the types of radiation, nor is there complete information on the duration of exposure or dosage necessary for these radiations to have the desired biocidal effect. In addition, some treatments require the use of very costly and complex equipment.

Fig. 8.3 *Aspiration of the microbial load from the reverse of a painting attacked by fungi*

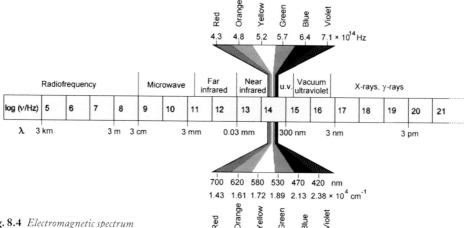

Fig. 8.4 *Electromagnetic spectrum*

In any case, all the interventions using physical methods have absolutely no preventive power; they can only eliminate living organisms and, sometimes, also the resistant and reproductive forms present in the treated objects. Indeed, the objects may find themselves immediately recolonized after the treatment, if they are not preserved in suitable and protected environments.

Ultraviolet radiation (UV) has germicide effects between 200 and 280 nm, with maximum effectiveness in the region of 230–275 nm. The application of UV radiation is very simple and not costly, even though it is limited due to its very low penetration. In addition, the efficacy of the treatment is dependent on its duration, on the type of substrate, and on the physiological state of the microorganisms. UV radiation can also induce photooxidation phenomena in organic materials and interact with some pigments, effects that considerably limit the field of its application.

Of all the electromagnetic radiations available, gamma rays are the most studied as a possible method of disinfection for materials of cultural interest. They are widely used for the control of microorganisms in other sectors, such as the food and cosmetics industries. Gamma rays are produced by the disintegration of the nucleus of radioactive isotopes (isotopes of Cobalt-60 are generally used) carried out with irradiators that can regulate the dosage of the emitted radiation. The cost of this equipment and the necessity of having locations suitable for the use of radioactive sources make this method a difficult one to be adopted by institutions dedicated to conservation of cultural heritage. Advantages offered by this treatment are the absence of any residual radioactivity in the materials that have

Fig. 8.5

Classification of electromagnetic waves based on frequency and wavelength

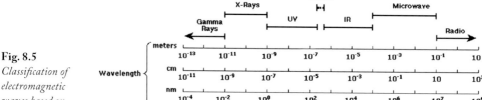

been exposed to the gamma rays as well as the sterilizing and disinfesting efficiency, which, although it diminishes in relation to the thickness and density of the treated object, nevertheless has an excellent penetration. Gamma rays are indeed more penetrating than UV rays and are therefore useful for in-depth treatment of materials.

Beta rays also have disinfecting powers. These are high-energy electrons produced by a strongly heated cathode. The results are comparable with those obtained with gamma rays, although their penetration is not so good, and they are more difficult to obtain.

The use of high temperatures (55°C) or very low ones (–20°C) is generally limited to the elimination of insects and is not usually applied to microbial attacks. Still, we should mention the application of hot vapor (even recently) to stone materials (Formica Castaldi 1988; Orial and Marie-Victoire 2002), even though this use can have certain drawbacks. Intense heat can also be induced by microwaves, which have a rather high wavelength (Fig. 8.4) and therefore a relatively low energy, but are able to increase the movement of dipolar molecules (e.g., water). Most living organisms die at a certain temperature (50–55°C), which is easily attainable using microwaves. Yet, microwaves have limited penetration and can never be used in the presence of metal elements.

8.3.1 Organic Materials

There are few physical methods of disinfection of organic materials capable of controlling the action of microorganisms responsible for biodeterioration, and they are not always easy to apply in the field of cultural heritage for several reasons. Some methods that are widely used for the control of insect infestations (Gilbert 1991; Wellheiser 1992) do not always have a biocidal effect on fungi and bacteria: among these are freezing, modified atmospheres, microwaves, and infrared radiation (Wellheiser 1992).

Atmospheres modified with inert gases (with nitrogen, argon, or simply with chemicals that "sequester" all available oxygen) are highly effective against insects and can sometimes be used to slow down or stop fungal activity (Macauley and Griffin 1969; Walsh and Stewart 1971; Gilbert 1989; Hocking 1990); but for most species, their toxic effect is limited, although further study is required in this area (Tavzes et al. 2001). Recently, it was shown that anoxic atmospheres created with argon are effective in the control of some brown-rot fungi (*Antrodia vaillantii*, *Coniophora puteana*, *Serpula lacrymans*), but not of *Trametes versicolor*, exemplifying how survival in low concentrations of oxygen is species-specific (Tavzes et al. 2003). Exposure to argon and nitrogen does not seem to have any effect on pigments (Suzuki and Koestler 2003).

The use of other gases—such as carbon dioxide—to obtain very low levels of oxygen is generally discouraged due to the greater reactivity of these gases in relation to the materials. For example, in the presence of moisture in the materials, CO_2 produces carbonic acid, which in turn has the capacity to alter the substrate. However, all treatments using modified atmospheres require the monitoring of the relative humidity of the enclosed environments into which the gases are introduced as well as of the moisture activity in the treated materials. Condensation phenomena on the materials, on one hand, and excessive dehydration, on the other, must both be avoided. The most widely used containers are built using sealed multilayered plastics as an effective barrier against both gases and water vapor.

For small or medium-sized objects, it is possible to use heat-sealed bags made of a transparent material impermeable to oxygen and containing oxygen absorbers (Fig. 8.6). Several products are available commercially that contain substances capable of developing exothermic reactions, which absorb the oxygen in the prepared bag. It is always necessary to monitor both the temperature and the relative humidity during the whole length of the treatment.

In general, UV treatments that are effective on microorganisms are also harmful to the organic materials on which these organisms grow. Pigments and also the main structural macromolecules (cellulose, collagen, keratin) are quickly altered by these radiations, which therefore find little application for the disinfection of cultural heritage material of an organic

Fig. 8.6 *Heat-sealed bag containing an oxygen absorber* – Photo R. Gasperini

nature. In the case of paper, for instance, a sterilizing treatment using UV is comparable to years of aging, which includes the depolymerization of cellulose chains and the paper turning to a yellow color before its time. However, the danger of these treatments obviously varies according to the chemical nature of the materials. Recently, the proposal was made to reduce the amount of airborne fungal and bacterial spores inside museums, archives, and libraries by continuously recycling the air through a UV room. The higher the intensity of the UV lamps, the more the microbial contamination in the air is reduced (Rossmoore et al. 2002).

The use of gamma rays is limited due to the damaging effects these have on organic materials, which are all the more serious the higher the dosage of radiation used. As in the case of UV radiation, the damages are comparable to an early aging of the materials and a loss in mechanical strength. According to Wellheiser (1992), a dose of gamma rays of 4.5 kGy for a duration of 48 minutes is sufficient to kill most species of fungi, although other authors indicate different minimum dosages (for the same exposure): 8 kGy (Hanus 1985) and 10 kGy (Bonetti et al. 1979). The temperature of the material influences the treatment, in the sense that for objects that undergo preliminary heating, it is possible to greatly reduce the dosage of radiation necessary to produce a lethal effect: for objects heated to 60°C, a dose of 0.5 kGy is sufficient to obtain complete sterilization. Low levels of relative humidity seem to have a nega-

tive effect on the efficacy of the treatment. Also, a number of studies carried out on ancient materials showed that sensitivity to gamma radiation varies from material to material; for example, wood (in contrast to paper and textiles) does not show a great sensitivity. Indeed, while paper suffers negative effects that are especially visible over time (Butterfield 1987), wood seems to have a greater resistance, as shown by Pointing et al. (1998), who did not find any negative effects on archaeological wood even at overall doses of 100 kGy. To inactivate the biodeteriogens in the wood of a Tudor ship, doses between 20 and 100 kGy—depending on the degree of degradation and the thickness of the wood—were used recently (Jones et al. 2003).

The fungicidal properties of beta radiation and their influence on the physico-chemical characteristics of paper were also tested recently; the experiments showed negative effects on the materials similar to those of gamma radiation (Flieder et al. 1995; Rakotonirainy et al. 1999a). With regard to possible damage to the materials, Wellheiser (1992) cites studies conducted by Brokerhof (1989) on the effects of beta radiations on cellulose, cotton, leather, and various synthetic materials; these show that any structural changes are in any case proportional to the applied dosages.

Microwaves have not yet found wide application in the area of disinfection of organic materials; however, some authors (Flieder et al. 1995; Rakotonirainy et al. 1999a) have proven their disinfectant properties and the lack of negative effects on paper materials.

Heat-based treatments (both wet and dry) kill microorganisms only if they are applied over long periods of time and with high temperatures; this is prohibitive for most organic materials such as wood, paper, and leather.

Freezing is used almost exclusively for the elimination of insects in all their stages (adult, larvae, and eggs) from wood and paper materials. The exposure of living organisms to temperatures much lower than zero causes their water content to be transformed into ice, with the consequent rupture of the cells. This phenomenon depends on both the level of low temperature achieved and the duration of exposure; it is recommended that the object to be treated remain at −20°C for at least seventy-two hours (Florian 1997).

8.3.2 Inorganic Materials

Photosynthesizing organisms require a source of light in order to live; their numbers can therefore be reduced by covering the object that acts as the substrate to these organisms (where this is possible) with wrappings or canvases that are impermeable to light. The duration of the wrapping can be between two weeks and two months, depending on the species present and the density of the development (Fig. 8.7). However, this method does not guarantee the death of all algae, since many of them have the capacity to live as heterotrophs in the dark. The possible biocidal effect of two months' darkness was tested on lichens with positive results (Prieto et al. 1996), even though it is known that some lichens can retain vitality even after being kept in the dark for several years in a herbarium. It was observed, for example, that the construction—during the Second World War—of a protective anti-aircraft structure over Trajan's Column in Rome, which kept the monument in the dark for six years, had a biocidal effect on the biological populations that had been present previously (Caneva et al. 1994b).

As mentioned in Chapter 7, especially with regard to rock paintings or petroglyphs, a simple covering can substantially reduce the supply of rainwater that would otherwise be absorbed by the rocks. This makes it possible to significantly reduce epilithic biological growth (Young and Wainwright 1995).

The use of ultraviolet radiation—both because of its lack of penetration and because of possible negative effects—has been limited to the elimination of algal patinas from the surfaces of stone and of nonpainted plaster (Plate 29) (Van der Molen et al. 1980; Petushkova et al. 1988; Bartolini et al. 1999). Positive results were obtained from tests carried out with lamps 25 cm from the stone surface, for three cycles of ten hours each, with low-pressure Phillips lamps (TUV G36T UV-C long life) (Bartolini et al. 1999).

Fig. 8.7 *Roman statue of the biblical Judith (Boboli Gardens, Florence) before and after intervening to reduce the biological patina by wrapping the statue with a canvas that is impermeable to light, for a period of approximately one month* – Photo P. Tiano

Infrared lamps can be used to apply local heat and dry out stone artifacts attacked by biodeteriogens, but the temperature of the object should never exceed 70°C.

Tests have also been carried out using laser for the removal of lichens and biological "crusts": an energy of 0.5 J/cm², and slightly more in the case of lichens (0.8 J/cm²), is sufficient to obtain their removal (Maravelaki et al. 1996). Cleaning tests have been carried out using laser on archaeological metal artifacts (Pini et al. 2000).

Systems based on the use of electric current are employed to drive away birds from monuments and are therefore not discussed in this context.

8.4 CHEMICAL METHODS

by Giulia Caneva, Maria Pia Nugari, and Ornella Salvadori

So far, the control of biological growth on organic and inorganic materials has been largely carried out using biocides or pesticides. The term pesticide refers to any chemical substance able to eliminate unwanted biological species. The root of the term (pesti-) indicates the target organism—the object to be controlled—and the second part of the term (-cide) indicates the lethal action. So far, a vast range of both organic and inorganic products has been used for the disinfection (i.e., the killing of pathogenic microorganisms and the reduction of saprophytic ones) or the disinfestations (i.e., the control of plants and animals) of objects of cultural interest; these products are called disinfectants and disinfestants, respectively. The terms bactericide, algicide, fungicide, lichenicide, and herbicide indicate the type of target organism, i.e., bacteria, algae, fungi, lichens, and plants, respectively.

In general, biocides are employed to eliminate the macro- or microorganisms responsible for the degradation, although recently some products were proposed that should have a preventive action on biological growth.

Commercial products or formulas contain the active principle (AP), which carries out the biocide activity, and other ingredients (coformulates), which generally are included to improve the performance. Commercial brand names are often constituted by names followed by letters or numbers, and only the correct denominations of the products will ensure a correct identification. For example, Preventol®, Hyamine®, and Metatin® are only the registered trademarks of some of the manufacturing companies and do not correspond to one single AP, while Preventol R80, Preventol O, and Preventol PN allow for an immediate and unequivocal identification of the AP (alkyl-dimethyl-benzyl-ammonium chloride, ortho-phenylphenol, and sodium pentachlorophenate, respectively). Unfortunately, much of the literature regarding conservation interventions or product testing often cites incomplete data in this area and is therefore of very little use.

Biocides can be divided into various categories: 1. according to their chemical nature (organic and inorganic compounds); 2. on the basis of the presence of characteristic functional groups (borates, chlorates, copper compounds, nitroorganic compounds, heterocyclic compounds, phosphoorganic compounds, etc.); 3. according to the type of formulation (liquid, solid, or gaseous compounds); and 4. according to the type of action (oxidizing agents, tensio-active ones, alkalizing agents, etc.).

Microbicides can be divided into two main categories: 1. those that act through contact—for example, oxidizing agents such as hypochlorite, or disinfectants acting through the interaction with the membranes, such as the salts of quaternary ammonium—and 2. those that inhibit certain specific metabolic activities (Denyer and Stewart 1998) (Fig. 8.8). The time required for the biocide to become effective is clearly linked to the nature of its action.

The products employed in the field of conservation should fulfill the following criteria:

—high efficacy against biodeteriogens;
—absence of interference with the original materials;
—low toxicity for human health;
—low risk of environmental pollution.

The efficacy is the measure of the biocidal action against the targeted organisms. Its evaluation takes into account: the dose of the

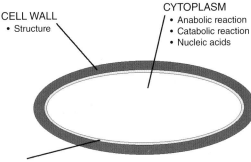

CELL WALL
• Structure

CYTOPLASM
• Anabolic reaction
• Catabolic reaction
• Nucleic acids

PLASMA MEMBRANE
• Structural integrity
• Membrane-bound enzymes
• Transport mechanisms
• Processes that need energy

Fig. 8.8 *Schematic representation of the action sites of biocides (modified from Denyer and Stewart 1998)*

product (expressed as the quantity of biocide/unit of surface or volume of air), the spectrum of activity (the range of the action against the target organisms), and the persistence of the action (duration of the treatment in relation to the permanence of toxic residues).

Therefore, the lower the dose required, the broader the spectrum of action, and the longer the continuing duration of its action, the more effective the biocide is judged to be. The first parameter is usually the one most taken into consideration, because often a specific and selective action is required and the removal of all residues of the biocide is recommended in order to avoid interactions with the substrate, which inherently makes any long-term action impossible.

In addition to being influenced by the intrinsic characteristics of the product, the efficacy of a biocide on a specific target organism can be determined by other parameters such as: the concentrations employed; the type of solvent utilized; the pH of the solutions; the duration of the application; the stability of the product; the microclimate parameters; the spread of the colonization; the nature and the state of conservation of the substrate; and its water content. These factors can at least partially account for the different levels of performance demonstrated by one and the same active principle in different experimental treatments (Caneva et al. 1996).

It is of the utmost importance to first verify that there is no interference, i.e., to determine that the use of a specific biocide is safe for the constituent materials of a work of art. The recent application of various scientific methods of analysis has shown that many biocides may have collateral negative effects—both short and long term—which are not always visually perceptible (Koestler et al. 1993; Nugari and Salvadori 2003a, 2003b; Suzuki and Koestler 2003). Effects on a macroscale are sometimes visible, such as stains, yellowing, bleaching, opacifications, modifications in the colors of pigments in paint layers, or even more extreme reactions (for example, corrosions, oxidations, solubilizations) in some materials, or also the presence of unpleasant residual odors after treatment, all of which therefore advise against the use of such products. The interfering action of biocides with the substrate is a function not only of the chemical characteristics of the product, but also of: the concentrations utilized, the contact time, the methods of application and, last but not least, the different types of substrate, which behave differently according to their various natures. The possibility should also be taken into consideration that a biocide may react with other products used before or after the intervention, during the various conservation phases. If there is any doubt concerning the possible incompatibility between biocides and any conservation products (generally listed in the technical sheets), or with regard to possible secondary interference with the materials, or if there are any issues regarding toxicological risks, all residues of the biocide product employed must be removed once its action is completed.

Whenever the available information is not sufficient with regard to the efficacy of the different products on the relevant species as well as their possible interference with the substrate, it is good practice to carry out preliminary tests (Chapter 9, section 2.1).

Knowledge of the toxicological characteristics of the biocides is essential in the choice of products to be used and must be considered with care. There are various levels of toxicity: acute (the effects are evident after a single application or after a brief exposure); subacute (repeated, short exposures); chronic (continuous and prolonged exposure of low doses over a long period of time); and sub-chronic (continuous exposure at a low dosage and for a relatively short period of time).

The toxicological properties of a specific active principle are usually determined with testing on laboratory animals (in particular, rats and rabbits), carried out by the manufacturers and submitted to the relevant health authorities to receive authorization for sale. Acute toxicity is expressed as the dose (DL_{50}, lethal dose) or concentration (CL_{50}, lethal concentration) of a product able to cause the death of 50% of the animals employed in the tests. DL_{50} refers to oral or cutaneous intake and is expressed in mg of substance per kg of body weight, while CL_{50} refers to inhaled intake and is expressed in mg/l or mg/m^3 of air.

It is of great importance to identify the concentration or the dose at which no toxic effects are observed; this level is expressed using a variety of indices such as LOEL (Lowest Observed Effect Level)—i.e., the lowest concentration or dose of substance at which an effect was observed—and NOEL (No Observed Effect Level)—i.e., the highest concentration or dose at which no effects are displayed. The toxicity threshold is therefore situated between the values of these two indices. These values can be extrapolated from animals to humans, generally dividing the obtained result by a security factor, thus obtaining the ADI (Admissible Daily Intake); this is the maximum daily dose (expressed in mg/kg) that humans can absorb without risk (Catizone and Zanin 2001).

As evidence of a more recent concern, toxicological experimentation has also been directed to the definition of mutagenicity (capacity to induce mutations at a genetic level), carcinogenicity (capacity of inducing cancer), teratogenicity (capacity of inducing anomalies in embryonic development), and embryotoxicity (with toxic effects on the embryo).

In the past, active principles were classified on the basis of toxicological characteristics into four toxicological classes; as a result of the European Union Directive 94/494, a new law (D.L. n. 194 of 17/03/1995) came into force in Italy in June 1995, which reclassifies all active principles into "highly toxic/toxic," "harmful," and "irritant" and includes the respective danger symbols and abbreviations in the labels; all other substances are defined as "nonclassified" (Fig. 8.9).

On May, 14 2000, a directive of the European Community (European Biocidal Products Directive 1998/8/EC) for the regulation of the use of biocides came into operation; this directive regulates the market and sanctions the application of biocides in Europe. It also differentiates active substances, biocide products, and basic active substances and addresses the areas of application of twenty-three types of products, such as various disinfectants, preservatives, insecticides, and antiadherent substances. However, although much of the necessary data for the assessment of biocides is already available in the literature, for many of the compounds it is still necessary to carry out further tests and experiments. The direct consequence of this is that the most commonly used compounds are those that are already authorized as well as their mixtures.

The methods of use of the product also influence the toxicological risk; operators must therefore take adequate precautions on the basis of the level of toxicity of the product. To avoid exposure of the skin and of the respiratory tracts, it is advisable to wear protective clothing, and

Fig. 8.9 *Correspondences between old and new classifications and danger labeling (modified from Catizone and Zanin 2001)*

in particular gloves, glasses, and masks. Other protective coverings (for instance, overalls made of a material that is impermeable to acids and solvents) and boots may also prove necessary when highly toxic products are being sprayed. In addition, it is important to avoid eating, drinking, or smoking while using these pesticides, so as to reduce the risk of oral exposure.

The assessment of the risk of pesticide use to the environment is an additional factor that must be taken into account, especially in the case of operations involving herbicides, during which a potentially substantial dose might be applied, or at least dispersed, in the environment (below in this chapter, section 4.2.b). Current regulations include an ecotoxicological assessment that evaluates both the intrinsic danger level of the product and its dispersion as well as the consequences for the environment. This risk can be linked to a variety of factors, such as drifting, i.e., the traveling of the product from the treatment site to adjacent areas, and undesirable secondary effects on other plants, animals, and microorganisms in the soil. In addition to the nature of the products employed and the frequency of the treatments, the risk is also a function of other parameters such as the chemical composition and the lay of the soil, its water content, the pH, the climate characteristics, the microflora present, etc.

The methods of application of biocides vary according to: the materials composing the work of art, its state of conservation and the size of the area to be treated, the specific target organism, its spread and diffusion, and the selected product. Biocidal products can be applied through spraying, brushing, application of compresses, and fumigation. Injection and dispersal of granular formulations (below in this chapter, section 4.2.b) are sometimes also options, but these are limited to herbicides. Biocides can be diluted or dispersed in water or in other organic solvents; concentrations vary: generally between 0.1 and 3%, and up to 10% for injections. Depending on the type of

biological growth present, it may sometimes be appropriate to carry out a preliminary partial removal of the biomass.

Spraying and brushing are the most frequent methods of application. The application with a brush has the advantage of allowing a deeper penetration of the biocides (Fig. 8.10); spraying is preferable for the treatment of large surfaces or, generally, when the state of conservation of the material is poor or uncertain.

The application of compresses is carried out to increase the time of contact of the biocidal solutions and to exploit the lytic action of water; carboxymethyl cellulose and paper pulp are the materials employed for this type of application. The compress is then covered with polyethylene or a similar material, to stop the solvent from evaporating too rapidly (Fig. 8.11).

Fumigation—i.e., the dispersal of biocides in gaseous or vapor form in the air and in the materials—is widely used for the treatment of organic materials and only rarely for stone, especially since it must be carried out in enclosed environments and is therefore limited in its use to objects that can be transported into the laboratory or, at least, isolated from the external atmosphere (for example, caves and tombs). The treatment is usually carried out in an autoclave (Fig. 8.12), in which it is possible to regulate the pressure in order to facilitate the penetration of gas or vapor, in fumigation chambers, or

Fig. 8.10 *Brushed biocide treatment on the reverse of a canvas* – Photo ICR

Fig. 8.11 *Treatment with biocide compresses on stone* – Photo ICR

in other completely sealed environments (these are sometimes created *in situ* with sheets of polythene).

The chemical substances that are most frequently used in the area of conservation vary according to the regulations in force in the different countries. Due to a recent and better understanding of toxicological characteristics, and especially of chronic and environmental toxicity, several countries have prohibited or strictly limited the use of several products that were used in the past (for example, pentachlorophenol, tributyl tin oxide, ethylene oxide, methyl bromide, formaldehyde, prussic acid, arsenic, mercury derivatives, etc.) (Nugari and Salvadori 2003b).

Fig. 8.12 *Autoclave for fumigation treatments with biocides* – Photo ICPL

8.4.1 ORGANIC MATERIALS

by Maria Pia Nugari, Corrado Fanelli, and Sabrina Palanti

As with all artifacts of historical and artistic interest, the two main strategies for controlling biodeterioration of organic materials are: prevention and cure. In the case of artifacts of an organic nature—usually preserved in indoor environments—prevention can be realized by controlling the environmental conditions favorable to the development of biodeteriogens (temperature, environmental humidity, light) and also by using natural or synthetic compounds with strong antimicrobial action (both antibacterial and antifungal) in the preservation spaces; this makes it possible to slow down or even stop the development of many of the microorganisms that can deteriorate the various materials of organic origin.

As an alternative to synthetic products, the use of natural plant extracts has been suggested because of their antifungal action; the vapors of certain essential oils from plant extracts (clove, absinthe, eucalyptus, lavender, etc.) (Fig. 8.13) do indeed exhibit a fungistatic action and can be employed to limit the development and spread of cellulolytic fungi in libraries, archives, and bookstores (De Billerbeck et al. 2001). Other essential oils were tested directly on fungal cultures, but the concentrations required for sufficient efficacy were very high; the recommended use is therefore for the control of biocontamination (Rakotonirainy et al. 1998). The drawback in the use of these substances is that the terpenoid structures of most essential oils tend to oxidize and can cause retention phenomena in the treated object; systems of microencapsulation of the substances to avoid their rapid oxidation are being tested to attempt to counter these effects (Henry et al. 2001).

For the disinfection of libraries and archives, it was recently pro-

Fig. 8.13 *(a) Plants of* Artemisia absinthium *(absinthe) and (b) detail of its heads, which are used for their antibacterial and fungistatic action* – Photo P. M. Guarrera

posed to use electrically generated aerosols (Electrofog®), which allow the fumigation of the premises with fungicides. This method makes it possible to kill the airborne fungal spores as well, without altering temperature or relative humidity in the environment. The biocide product that has proven most effective with this method is thiazolyl-4-2-benzimidazole, which successfully decontaminated both the air and the surfaces; however, for toxicological reasons, the system can only be used by competent personnel (Rakotonirainy and Flieder 1994; Rakotonirainy et al. 1999b).

The chemical methods proposed for the direct treatment of organic artifacts are mostly based on the use of compounds that are primarily designed to control the growth of deteriogenic microorganisms by slowing it down and ultimately stopping it. The selection of the compounds for the protection or treatment of the materials must be made on a case-by-case basis according to the type of biological attack, the type of deteriorated material, and also the level of alteration. It bears repeating that, in the case of organic artifacts, it is best to limit the treatment with chemical products as much as possible. The risk of interference during the disinfection of organic materials is particularly

problematic in this area. Quite often, organic artifacts are made up of several components, each of which can have a different chemical reactivity to the biocidal products used. In this context, and precisely to be able to intervene successfully while avoiding "treatment damage," it is extremely important to set up a system of early detection of biodeteriogenic microorganisms, which can sometimes be present without being visible to the naked eye. For the early diagnosis of fungal colonization on paper materials, an electronic sensor was recently developed; this system, called "electronic nose," was able to detect the volatile chemical substances produced by the metabolism of fungi that are biodeteriogenic of book materials (Canhoto et al. 2004).

Since the biodeteriogens of organic materials are for the most part fungi and bacteria, the search for disinfectant products for these organisms also examines highly specific substances that target the biosynthetic processes of molecules found in these microorganisms but without interfering with the materials to be protected. Similar strategies have also long been used in the fight against phytopathogenic bacteria and fungi as well as against dermatophytes (the fungi that attack human skin), given that fungi and bacteria are among the main pathogens of plants and are also etiological agents for human and animal skin diseases. In fungi for instance, the most vulnerable target compounds are the components of the cell walls (constituted in large proportion of chitin and kitosane), which exhibit a structure that is quite different from the cell wall structure of plants (Chapter 4, section 1.a). In addition to substances that inhibit the synthesis of chitin, other substances exist that are active specifically within the synthesis processes of other molecules peculiar to fungi, such as ergosterol, the most abundant sterol present in fungal membranes. Biocide compounds with this action (econazol, strobilutrin A, triazine, fluorouridine) have been tested on different strains of fungal biodeteriogens of paper. An assessment of the efficacy of these treatments on various kinds of paper showed that the triazines give the best results (Ricelli et al. 1995, 1999; Rossi et al. 1998, 1999; Fanelli et al. 2001).

However, while the most interesting antifungal compounds have these two targets, there are others in use that have other targets,

and often more than one. Indeed, the use of compounds that have only one specific target has the disadvantage that, in time, their use can induce the selection of populations that have become resistant to such compounds, which therefore lose their efficacy. Consequently, the use of mixtures of compounds that work on several targets (integrated approach) is increasingly common since this makes it more difficult for mutant resistant strains to form. It is therefore more than evident that the choice of chemical compound to be used for the control

Table 8.1 Some of the main toxic gases and their negative effects on materials (modified from Sclocchi 2002; Nugari and Salvadori 2003b)

Fumigating Gases	Microbicidal Action	Insecticidal Action	Toxicological Problems	Effects on Materials
Cyanohydric acid (HCN)	Weak fungicide	YES	Highly toxic	Alters the color and shine of metals. Creates residual odors in ethnographic material.
Ethylene oxide (C_2H_4O)	YES	YES	Highly toxic and carcinogenic. Flammable, to be used only in sealed systems (autoclaves and sterilizing units)	Reacts with hydrogen sulphide groups in proteins and other polymers. Can induce polymerization. Can also oxidize copper and brass. Causes loss of resistance in paper, cotton, and silk.
Formaldehyde (CH_2O)	YES	NO	Toxic and potentially carcinogenic	Has a tendency to polymerize and to precipitate as a thin whitish film. Cannot be used on proteinaceous substances. Causes the oxidation of lead in conditions of high humidity.
Methyl bromide (CH_3Br)	Doubtful	YES	Highly toxic; suspected of being carcinogenic. Being phased out	Materials containing sulphur cannot be treated (such as wool, skins, etc.). Gives rise to the formation of mercaptans, which give a bad odor to leather and parchment. Interacts with varnishes, natural resins, and lead-based pigments.
Thymol ($C_{10}H_{14}O$)	YES	NO	An irritant and an allergenic	Causes inks, paints, and varnishes to dissolve. Softens acrylates, glues, and skins. Causes yellowing in paper and acrylic resins.

of biodeterioration in different matrices is an exceedingly complex problem, which must be tackled case by case because of the large number of variables involved.

The most widely used products for the disinfection of organic materials are substances with a wide spectrum of action and, in particular, those that are active in gaseous or vapor form. These substances—although limited to treatment of materials that can be transported to the laboratory or at least isolated from the outdoor environment—guarantee a good penetration of the product into all parts of the artifact (for example, in the treatment of book collections) and also the possibility of working on particularly fragile materials (easel paintings). However, most of the products active in this form—such as methyl bromide, ethylene oxide, and formaldehyde, some of which are already banned in certain countries—are highly toxic; their use, if it is absolutely necessary in the absence of any other practicable alternative, must be limited and/or only carried out by specialized personnel. In addition, even though these substances are still widely used, they are not entirely innocuous for the works of art, and numerous studies have shown the negative effects induced by them (Daniels and Boyd 1986; Florian and Grimstad 1987; Baker et al. 1990; Schulz 1994; Raychaudhuri and Brimblecombe 2000; Sclocchi 2002; Nugari and Salvadori 2003a, 2003b).

Because insect damage plays a significant role in the biodegradation of works of art of an organic nature, the most widely used biocides tend to have a dual action, as microbicides and as insecticides. However, the microbicide activity of some of these products, such as thymol and methyl bromide, has not always been verified (Florian and Grimstad 1987; Nyuksha 1994; Schulz 1994; Sclocchi 2002). Table 8.1 lists the main gaseous substances in use, their range of action, and any possible negative effects on the materials treated.

With regard to ethylene oxide—which is among the most widely used gases thanks to its broad range of action—it is important to remember that it is an explosive gas and must therefore be used mixed with inert gases in appropriate airtight spaces. At present, it is mixed with carbon dioxide, as the previously used Freon is now included in the list of substances that damage the ozone layer and, hence, the environment. The lower overall efficacy of this mixture imposes specific methods and length of treatments (Table 8.2).

As far the use of biocides in solution is concerned, there are few studies that deal in any systematic way with their applicability on materials of an organic nature and that evaluate their efficacy and interference.

The most frequently recurring classes of biocides in the literature are essentially organic compounds and, in particular, products with quaternary salts of ammonium as a base or else phenyl derivatives (Nugari et al. 1987; Koestler et al. 1993; Nugari 1995; Nugari and Salvadori 2003a, 2003b; Suzuki and Koestler 2003). The inherent fragility of organic artifacts—and the fact that they are often composed of many different materials—requires a greater attention on issues of interference and often prevents the use of products requiring application of water-based solutions with a brush.

In general, for the disinfection of paintings—especially of those with a textile support—action is taken on the support rather than on the painted surfaces. Nebulization is the most widely used method, and in many cases the preferred formulations are those soluble in organic solvents because they minimize the contact between water and pigments or other potentially soluble elements. Precisely because of its solubility in organic solvents, o-phenylphenol is widely used for easel paintings and ancient

Table 8.2 Conditions of treatment of paper with ethylene oxide (modified from Sclocchi 2002)

Ethylene oxide + CO_2 (g/m³)	
Temperature	20–30°C
Relative humidity	50–60%
Humidity of the paper	6–7% by weight
Time in the autoclave	> 44 hours
Air washing after treatment (complete exchange of air in the chamber)	No less than 10 treatments

textiles, even though its efficacy is limited, but still greater for fungi than for bacteria (Giuliani and Nugari 1993; Koestler et al. 1993; Nugari and Salvadori 2003a, 2003b). Thymol has been used on the reverse of pastel paintings (Schulz 1994). Ethanol in a 10% water solution has also been proposed for the disinfection of fungal attacks (Szczepanowska and Cavaliere 2003). Benzisothiazolone in ethanol has been adopted instead to control fungal development not on the paintings themselves but on materials in their proximity, such as glass, frames, and stretchers (Berovic 2003). Because of their good microbicide efficacy and persistence over time, other derivatives of isothiazolone have been used in the treatment of wooden materials and are still employed today in the disinfection of water-logged archaeological wood during conservation and consolidation treatments (Unger et al. 2001).

Among inorganic biocides, we should mention boron derivatives, which however are not sufficiently effective against microfungi (Unger et al. 2001); sodium borohydrate has recently been proposed, but rather with the function of eliminating the dark stains caused by dematiaceous fungi than as a disinfectant (Szczepanowska and Cavaliere 2003).

8.4.2 **STONE MATERIALS**

by Maria Pia Nugari and Ornella Salvadori

Treatment with chemical substances is by far the most frequently adopted method for the elimination of biodeteriogens from stone artifacts, and numerous biocides have been in use for this purpose (Caneva et al. 1996) (Table 8.3).

Many studies have been conducted on commercial products to identify the most suitable biocides in this particular area and to determine both how effective and how harmless their use is on stone surfaces (Koestler and Salvadori 1996; Wakefield and Jones 1996; Tiano 1998; Warscheid and Braams 2000). Unfortunately, the results of these investigations—conducted either in the laboratory or *in situ*—often furnish information only on specific cases or problems without discussing the more general aspects

relating to the overall conservation project of the monument under consideration.

Tests to determine the efficacy of treatment on the biodeteriogens should be carried out both in the laboratory and *in situ* (Chapter 9, section 2.1a) (Fig. 8.14). The investigation methods employed can strongly influence the results; moreover, strong discrepancies have been noted between results obtained in the laboratory and those obtained outdoors. Laboratory analyses have therefore only limited validity in predicting the effect of biocidal products *in situ* (Koestler and Salvadori 1996). The microorganisms used in the efficacy tests in the lab are often more sensitive than those present in the biofilms on stone surfaces, which are usually much more resistant to the biocide action. Thus, the active range of chemical products may be significantly influenced by the specific actual conditions of application. In addition, biocides are often applied in much stronger concentrations than the minimum inhibitory concentration (MIC) established in the lab; this implies a substantial risk of interference with the materials but often still with disappointing results,

Fig. 8.14 *On-site comparative testing of biocides* – Photo ICR

Table 8.3 Most frequently used biocides in the conservation of stone (modified from Caneva and Pinna 2001)

Chemical Classification	Chemical Composition	Commercial Name	BF	CA	L	P
Inorganic compounds	Sodium and potassium hypochlorite		•	•	•	
	Lithium hypochlorite				•	
	Sodium sulphite				•	
	Hydrogen peroxide		•	•		
	Sodium octaborate	Polybor		•		
Phosphoorganic compounds	Glyphosate	Roundup, Spasor, Rodeo			•	
Alcohols	Ethanol		•			
Phenol derivatives	Thymol		•			
	o-phenyl-phenol	Lysol	•		•	
	p-chloro-m-cresol		•			
	Chlorinated and phenolic compounds	Panacide, Halophane, Thaltox C		•	•	
	Sodium pentachlorophenate			•	•	
Nitroorganic compounds (ureic and carbamates)	Diuron	Karmex	•	•	•	
	Diuron	Diuron	•	•	•	
	Chlobromuron	Maloran			•	
	Fluometuron	Lito 3			•	
	Sodium dithiocarbamate					•
Quaternary ammonium salts	Alkyl-benzyl-dimethyl-ammonium chloride	Preventol R50	•	•		
		Preventol R80	•	•	•	
		Neo Desogen	•	•	•	
		Hyamine 3500		•	•	
		BAC	•	•	•	
		Catamin AB		•	•	
	Benzyl-dodecyl-bis(2-hydroxyethyl)-ammonium chloride	Bradophen	•			
	(Diisobutylphenoxyethoxyethyl)dimethyl-benzyl-ammonium chloride	Hyamine 1622				
	Dodecyl-benzyl-trimethyl-ammonium chloride	Gloquat C		•		
	Lauryl-dimethyl-benzyl-ammonium chloride	Cequartyl		•	•	
Organic metal salts	Tri-n-butyl tin oxide	TBTO	•	•	•	
		Thaltox		•	•	
	Tri-n-butyl tin naphthenate	Metatin N58-10	•		•	
Pyridine	2,3,5,6 tetrachloro-4-methyl sulfonyl pyridine	Algophase		•	•	
	picloram	Tordon 22K, Uniran				•
Heterocyclic compounds (diazines and triazines)	bromacil	Hyvar X			•	•
	hexazinone	Velpar, Velpar L		•	•	•
	terbutryn	Igran			•	
	terbutylazine	Primatol M50				•
	secbumeton	Primatol 3588				•
Mixtures	quaternary ammonium salt + tri-n-butyl tin naphthenate	Metatin N58-10/101	•	•	•	
	quaternary ammonium salt + tri-n-butyl tin oxide	Thaltox Q		•	•	
		Thaltox 20			•	
		Murasol 20			•	
	Dimethyl-thio sodium carbamate + 2-mercaptobenzothiazole	Vancide 51	•	•	•	

Key: B = bacteria; F = fungi; C = cyanobacteria; A = algae; L = lichens; P = higher plants

since higher concentrations do not necessarily result in greater efficacy.

How long a biocide takes to act depends on the selected product and on the nature of the organisms and microorganisms to be removed. The oxidizing agents, for example, act rapidly but may prove highly aggressive toward the substrate and are often not very effective. In the case of biocides that inhibit a metabolic process, the efficacy is closely correlated to the duration of the treatment. In this instance, the time span required for the action varies depending on the nature of the microorganisms to be eliminated, between three and seven days for bacteria, fungi, or algae (Gomez-Alarcon et al. 1999; Pietrini et al. 1999), and up to ten days in the case of lichens (Nimis and Salvadori 1997). A second application of the product often proves necessary when thick biofilms or sporulating microorganisms are present.

Other factors that may influence the efficacy of biocides on stone materials are:

—the kind of substrate (for example, the porosity of the stone and the presence of some minerals, such as illite and smectite, increase the quantity of product absorbed, thus increasing the efficacy of the treatment) (Grant and Bravery 1985; Young et al. 1995);
—the presence of organic material (dust, pollen, etc.) or of pollutants on the surface;
—the contact time;
—meteorological conditions (temperature, wind, rain) and light intensity.

Some of these conditions can be simulated in the laboratory, but it is impossible to recreate a complex ecosystem such as that of a colonized stone in the open. Preliminary *in situ* tests are therefore extremely useful to study, at the same time, the biocide-microorganisms and the biocide-stone substrate interactions, both in the short and the long term.

There are many factors that may cause a negative interaction between the biocide and the stone. As in the case of efficacy, the most important ones are: the actual chemical composition of the product, the conditions of its usage (concentration, method and duration of the application, etc.), the mineralogical and petrographic characteristics of the substrate, and the environmental conditions (Nugari 1999).

The susceptibility of the different lithotypes varies according to the biocides used; consequently, it is possible, by means of specific studies, to select biocides with the least amount of contraindications. From experiments carried out by Altieri et al. (1999c), for instance, it emerged that, of four commercially available formulations, only one biocide (based on the quaternary ammonium salt + the naphthenate of tri-n-butyl-tin) yellowed and darkened travertine even when applied in low concentrations, while the other tested products did not interfere with this lithotype. Marbles and calcareous stone are particularly sensitive to acid substances, and it is therefore advisable to use substances with a pH near to neutral, or even to add neutralizing substances (Mouga and Almeida 1997). The corrosion of calcite and the formation of rusty stains resulting from the use of hydrogen peroxide are well known (Nugari et al. 1993a, 1993b; Kumar and Kumar 1999). Despite this, there are many products with known negative effects that still appear on many lists of recommended products and are still used.

The biocide-stone interaction can be influenced by environmental factors (light, rain), but also by substances used before or after the biocide during the course of the conservation project. The products based on alkyl-dimethyl-benzyl-ammonium chloride are incompatible with anionic surfactants, nitrates, hydrogen peroxide, and many other substances that may reduce the efficacy of the treatment. In general, reducing the contact time of the biocide with the stone to the absolute minimum necessary will also reduce any possible negative interactions to a minimum; rinsing with water after the product has carried out its action will also avoid possible long-term interactions with the substrate, although it will of course also reduce any long-term residual action over time.

We should also mention that the operators working in this sector need to acquire a better knowledge of the chemical and physical characteristics (for instance, solubility, stability, incompatibility with other chemical substances, etc.) as well as the nature of the action of the active principles and of the commercial products they are using. It is not rare to find that the nature of the action of a product is not known and that, consequently, the duration of application is incorrect, or again that the advised concentrations for usage are doubled or tripled in an attempt to increase the efficacy, often with a negative impact on the final result. Higher concentrations are sometimes less effective than more diluted solutions (Pinck et al. 2002) and can frequently give rise to negative collateral effects.

In the conservation of stone, many different operations are carried out (e.g., cleaning, consolidation, protection, etc.) utilizing a variety of products and methods. The mixing of biocides with cleaning substances, consolidants, and/or water repellents, or the sequence of their application should all be evaluated with extreme care because of the possible negative interactions among the various substances, which can cause the loss of the very properties for which they were originally chosen (cleaning power, water repellence, etc.). The investigations conducted so far in this area to define the most appropriate application procedures have yielded useful but not exhaustive information (Leznicka 1992; Tiano et al. 1997; Malagodi et al. 2000; Ariño et al. 2002).

In recent years, it has become common practice to apply water repellents and biocides at the end of a conservation intervention, in order to protect the stone and inhibit recolonization for as long as possible. The application of the biocide before the water repellent seems to alter the characteristics of the latter, diminishing its water-repellent properties. In contrast, the application of the biocide after the protective treatment has been carried out does not seem to influence the characteristics of the protective treatment. However, these interaction effects between the biocide and the protective substance cannot be generalized and, in any case, vary according to the products used (Balzarotti Kämmlein et al. 1999; Malagodi et al. 2000). A study conducted on mortars *in situ*, treated either with silicon-based products or with silicon-based products followed by the application of a biocide, showed that microbial growth is inhibited for a period of 5–8 years on all the surfaces, with the exception of the areas affected by capillary water rise; in

these areas, the development of photoautotrophs was present only on the mortars treated solely with silicone resins (Ariño et al. 2002).

The need to develop resin-based products that also include biocides (i.e., polymers with antimicrobial action) is strongly felt in several industrial sectors and represents a new and very interesting area of research (Leeming et al. 2002). The aim would be to have an "on-demand" biocide that is liberated from the supporting resin when microorganisms are present (Denyer and Stewart 1998). Various methods can be used to obtain products with these characteristics, depending on the nature of the resins and the biocides used (for example, whether or not they are water soluble), and the result must be proven analytically—it is not simply a question of mixing two products. Quaresima et al. (1997) have polymerized certain monomers, to which Cu^{2+} has been added, controlling the release of the copper over time, thus demonstrating that these materials could constitute a model for the development of new products for the control of biodeterioration (Corain et al. 1998). Products with these characteristics (resin + biocide) have been commercially available in the conservation sector for a few years now, but preliminary analyses in the laboratory have not shown an increase in their biocide efficacy when compared to the same biocides and resins employed separately (Blazquez et al. 2000).

Recently, special mortars have been developed for use in conservation; these have antimicrobial properties to inhibit colonization by biodeteriogens, which otherwise takes place very rapidly because of the general composition and porosity of mortars. These special mortars have been obtained by adding 5% zeolitized tuff (pumice) previously subjected to an exchange with copper (Ferone et al. 2000), or else by adding sepiolite, which acts as a carrier for the biocide (Martinez-Ramirez et al. 1998a, 1998b). The biocide action is due to the release of Cu^{2+} ions or to the presence of pentachlorophenol adsorbed onto the sepiolite, respectively. Although these investigations are still at a preliminary stage and require a careful assessment (for example, the use of pentachlorophenol is banned in many countries for toxicological reasons), they are certainly promising as a potential and interesting new field of research.

8.4.2a **Disinfection**

The term disinfection (first coined in the medical field) means the elimination of harmful microorganisms and the reduction of saprophytic ones. In this section, we are also going to include lichenicide treatments, because they are generally carried out using similar methods and often also with the same products as those employed in the elimination of biodeteriogenic microorganisms (bacteria, algae, fungi).

In the past, the disinfection of stone was not the object of much study; indeed, the elimination of biodeteriogenic microorganisms would often simply be part of the cleaning operations, frequently with negative consequences for the work of art. The tools and the cleaning substances employed often produced little more than a reduction of the biomass, but not the killing of the species present; in addition, a propagation of the vegetative and reproductive structures often occurred as a result, thus increasing the risk of contamination of other surfaces. In the case of colonization by heterotrophic microorganisms in enclosed environments, on the other hand, indirect measures were often thought to be sufficient, such as the repair of walls or the application of consolidating and water-repellent substances, which, by reducing the water supply, were thought to limit microbial development and diffusion. Such treatments should however be appropriately assessed, since it is well known that some consolidating and protective substances will actually favor the establishment of specific biological attacks (Chapter 4, section 6).

The killing of microorganisms and lichens must therefore be considered an operation in itself that has its own specific requirements in terms of time frame, materials, and methods, and must be planned and executed before embarking on cleaning operations. Careful consideration must be given even to the practice of reducing the volume of the biomass through brushing prior to disinfection, intended to improve the efficacy of the treatment. Indeed, in the case of artifacts altered by the presence of algal patinas, this operation can cause the rupture of the microbial cells and the release of photosynthesizing pigments with consequent staining that is sometimes difficult to remove; in the case of sorediate or isidiate lichens, the brushing action itself induces countless vegetative propagules to spread into the atmosphere. This was observed in the instance of the lichen attack on the frescoes at Caprarola (Chapter 5, section 3.1b), where the operations of surface dusting carried out to remove the thalli had, in fact, clearly contributed to the propagation of the attack.

The disinfectants generally used in the conservation of stone were not originally developed for this particular field of application but for the medical sector, and consequently the efficacy tests generally relate to microbial species that are different from those that are biodeteriogenic for works of art; lab and *in situ* testing is therefore necessary. It is important that the tests be carried out on known species to allow comparisons between results from different laboratories; indeed, lab analyses have shown that different species will respond differently to one and the same biocide (Caneva et al. 1996; Monte and Nichi 1997; Nimis and Salvadori 1997; Pietrini et al. 1999). It should also be remembered that the capacity to resist or tolerate biocides also depends on a variety of factors related to the species; the fungus *Alternaria*, for example, exhibits a particular resistance to certain biocides (Bassi et al. 1984), depending on the stage of its development and the growth conditions. Also, it is known that microorganisms in the vegetative phase, without spores, are more sensitive to biocide treatments.

It is essential that laboratory results be verified *in situ* because the efficacy of the biocide can be strongly influenced by the type of microbial association, the resistance forms (e.g., spores), the characteristics of the microorganisms that determine the alteration (e.g., the presence of mucilaginous sheaths), and also by the presence of biofilms. Among algal patinas, for example, the black, thick, and mucilaginous ones that are composed primarily by cyanobacteria are generally less susceptible to the action of biocides than patinas constituted of biocoenoses with a predominance of green algae (Ricci and Pietrini 2004; Bartolini and Ricci 2004; Gasperini and Salvadori, unpublished). In the case of mixed populations, it is necessary to adopt products with a wide spectrum of action, active against all the colonizing species present on the substrate; if this is not done, the resistant species can indeed increase in number because of the elimination of the competition. And it is for this very reason that these types of products are used in the great majority of the cases.

The efficacy of the *in situ* treatments must be verified with appropriate tests in the laboratory; being able to see the macroscopic changes in the biological patinas or the lichenic thalli induced by the treatment is not always proof that the devitalization of the colonizing species has actually taken place. For example, the efficacy of an algicide treatment in inhibiting algal growth is often assessed *in situ* with the naked eye, noting any changes in color in the patinas and/or the disappearance of all pigmentation once cleaning has been completed. In addition to often providing misleading results (apparent success of the treatment because of the partial discoloration of the patinas), such an empirical method is not sufficient to obtain scientifically valid comparisons of the results from different investigations. In order to evaluate *in situ* efficacy, a recent development has been the application of colorimetric measurements and measurements of chlorophyll fluorescence by means of a portable fluorimeter and a fluorescence LIDAR spectrometer (Wakefield and Jones 1996; Tomaselli et al. 2003). Laboratory observation of algal patinas and of lichenic sections under the fluorescence microscope allows rapid and objective evaluations of *in situ* treatments (Nugari et al. 1993a; Monte and Nichi 1997; Nimis and Salvadori 1997; Bartolini et al. 1999) (Plate 31).

Some authors have shown that biocides can have different degrees of efficacy depending on the nature of the substrate on which they are applied, e.g., its porosity and mineralogical composition (Grant and Bravery 1985; Young et al. 1995). Thus, the substrate can play a determining role in the efficacy of a treatment, despite the fact that this aspect is by and large neglected.

Treatments can be differentiated according to the product used, the type and/or degree of colonization, and the extent of its cover; the type of artifact, its size, and its state of conservation can further influence the choice.

The customary application methods for biocides in solution are brushing or spraying (nebulizing) (Fig. 8.15). Spraying must be executed with care if fungi are in their sporulation phase, because it will spread the extremely easily dispersed fungal spores into the environment, which involves risks not only for the works, but also for the operators if they are not adequately protected. The application of compresses (Fig. 8.11) is chosen mostly for colonizations of endolithic microorganisms (cyanobacteria, algae, fungi, lichens), which can reach considerable depths within the material.

Only for certain substances that are active in vapor form—such as thymol and formaldehyde, neither of which is much used today because of toxicological concerns—has vaporization or fumigation been used as a method. Elmer et al. (1993) experimented, and obtained good results, with fumigation using ethylene oxide in enclosed chambers on objects in stone (marble, limestone, quartzitic sandstone) that could be transported easily. However, since the toxicological characteristics of this gas are known (Caneva et al. 1996), it is recommended to make limited use of it.

The number of applications and the contact times depend on the product chosen as well as on the targeted organism and its metabolic phase. A short application time (from a few hours to a day) can be effective only with biocides that act through contact, interacting with the structure of the cell walls or membranes, but not for products that act by interfering with metabolic reactions; for these, a longer time frame is required and sometimes also repeated applications. In general, a second application is

Fig. 8.15 *Spray treatment* – Photo ICR

always necessary when biofilms or spores are present. In the case of poikilohydric organisms, such as lichens and cyanobacteria, it is advisable to first nebulize with water the surfaces to be treated—at least a few hours before the application of the biocide—because this enables the organisms to return to an active metabolic state and therefore increases the efficacy of the treatment.

Once its action has been carried out, the biocide must in any case be carefully removed from the substrate using water or the appropriate dilution solvent. It is also often necessary to complete the cleaning operation with gentle systems of mechanical cleaning to remove the residues of the devitalized biomass. Sometimes the biocide is instead deliberately left on the surface or reapplied after final cleaning, in an attempt to stave off further colonizations (earlier in this chapter, section 4.2).

The products used in the disinfection of stone are composed of a wide range of chemical compounds, which include inorganic substances and commercial formulations containing one or more active principles of an organic nature (Table 8.3). The first reports detailing the use of microbicides in conservation date from the end of the 1960s; Caneva et al. (1996) contains an exhaustive historical bibliography on the subject, while several publications regarding products that are still in use today will be mentioned in the following paragraphs.

Some inorganic compounds were proposed in the past, and in some cases are again put forward today, for the reduction of biological growth (Wessel 2003). Historically, the first to be used were the borates and the fluorosilicates (fluorosilicates of zinc, aluminum, magnesium, and sodium), which were used against algae and lichens—these are almost entirely abandoned today. These substances have not always been very effective and, more importantly, they can induce alterations in the substrate: the borates can give rise to soluble salts, while the fluorosilicates produce hard and compact crusts on the surface of stone, which eventually detach. Despite this, the borates (disodium octaborate in combination with zinc, to form zinc borate on the substrate) are still used today in the United Kingdom, with the recommendation to rinse well and repeatedly after usage to avoid

the formation of sodium salts (Wakefield and Jones 1996).

Among the inorganic compounds, some metal derivatives have been suggested for use in conservation, e.g., zinc and magnesium chloride, copper sulphate or carbonate; these, however, have often proven harmful to the works since they can induce colorations in the substrate. Ammonia has also been suggested, but it is effective only as a degreaser. The most frequently used substances are hydrogen peroxide and the hypochlorites (sodium hypochlorite and, to a lesser extent, the hypochlorites of calcium and lithium). These have been used for the control of algal, bacterial, and fungal colonizations (especially algal patinas and the black stains caused by Dematiaceae) and also for lichenic thalli. Their widespread use in conservation is probably due to their ease of accessibility, low cost, very low levels of toxicity, as well as their bleaching and cleaning properties. However, their efficacy is limited and is greatly dependent on the method of application: if the concentration is not suitable and the time of contact with the biodeteriogenic agent is not sufficiently long, there is the risk that the results may not be positive. As a general rule, we can say that hydrogen peroxide has little or no effect on sporal cells, while the hypochlorites are markedly more effective and are therefore also used in the industrial sector, in the disinfection of water, etc.

Numerous studies have shown the existence of interference problems between these compounds and the substrate. Especially hydrogen peroxide, since it is a highly unstable molecule, is often stabilized in commercial products with the addition of acids, which cause a substantial and very dangerous drop in pH leading to corrosion effects on certain minerals (calcite, for example); because of this, this particular property should always be checked before use. Moreover, as it is an oxidizing substance, it is able to oxidize metal ions, especially iron, and also the organic substances present in some components of stone, giving rise to red, rusty, or black stains. Hydrogen peroxide can also provoke the rapid denaturation of chlorophyll with the consequent emission of carotenoids, which can irreversibly stain the stone with a reddish coloring (Bernardini 1993; Nugari et al. 1993a, 1993b; Kumar and Kumar 1999). In the case of the hypochlorites, the residual chlorine has an oxidizing effect on iron, and soluble salts may also be formed by the chloride ions and sodium ions; interactions with consolidants applied after the treatment are also a possibility (Tudor et al. 1990).

One of the first organic products to be utilized was formaldehyde or formalin, especially for the treatment of algal patinas and lichens, but also of fungal and bacterial colonizations in hypogean environments, such as caves with paintings or inscriptions and burial chambers. This substance is also active in vapor form and can therefore be used to disinfect an enclosed environment as a whole. Formaldehyde was also employed against some mosses and lichens. Its high toxicity, in addition to its now recognized carcinogenic properties, strictly limits its use today.

Phenol derivatives were widely used in the past and are partly still used today. The chemical, physical, and toxicological properties of these substances vary greatly according to the product under consideration; their antimicrobial action is consequently also very variable. In general, they possess good fungicidal and bactericidal properties, while being less effective against algae and lichens. Pentachlorophenol and its sodium salt—which were effective and widely used in the past—are now not available commercially in most countries because of their toxicity. Orthophenyl-phenol and its sodium salt are particularly successful against fungi and less so against algae and bacteria. However, a chromatic alteration was reported on a fresco treated with this product, due to the interaction with a fixative applied during an earlier intervention (Bettini et al. 1988). In the case of other phenol derivatives, such a p-chlorometacresol, which is widely employed for the disinfection of fungal, bacterial, and algal attacks, studies show conflicting results; the treatments also seem to induce changes in the color of the stone, probably as a result of reactions with the iron in the substrate. Thymol—both in liquid and gaseous form—has been widely used as a bactericide and fungicide, especially in the past; however, its range of action is unclear, and the assessments of its efficacy are conflicting. The incomplete disinfection linked to its usage may cause the development of resistant bacterial species that are particularly harmful (e.g., the genera *Pseudomonas* and *Bacillus*) (Petushkova and Lyalikova 1994).

The most widely used class of products for the control of microbial colonizations is that of the salts of quaternary ammonium; these surfactants constitute a vast group of substances that have found wide use in the field of conservation because they combine a detergent action and wide-range efficacy with a middle-to-low level of toxicity. However, both efficacy and toxicity can vary according to the product under consideration.

A great number of salts have been tested for possible use in the field of conservation (alkyl-aryl-trimethyl-ammonium chloride; lauryl-dimethyl-benzyl-ammonium chloride; lauryl-dimethyl-benzyl-ammonium bromide; dodecyl-dioxetyl-benzyl ammonium chloride; p-tolyl-dodecyl-trimethyl-ammonium oxetyl-sulphate, etc.), but among them all, alkyl-dimethyl-benzyl-ammonium chloride is certainly the most widely employed product in workshop practice in Italy and abroad (Fig. 8.16).

It has been applied in various formulations and with good results as a bactericide, algicide, fungicide, and lichenicide (Caneva et al. 1996; Nimis and Salvadori 1997; Pietrini et al. 1999; Tomaselli et al. 2003). Choosing the appropriate concentration for each product is essential for the success of the treatment; unsatisfactory results seem indeed to be attributable more to an incorrect use of the biocide (concentration, duration of application, etc.) than to a lack of efficacy of the product itself. Generally speaking, the biocidal action does not persist over time (Pietrini et al. 1999), yet the literature contains references to contradictory results that show, for example, an absence of recolonization one year after a treatment of lichens on granite (Prieto et al. 1996).

Fig. 8.16 *Chemical formula of alkyl-dimethyl-benzyl-ammonium chloride*

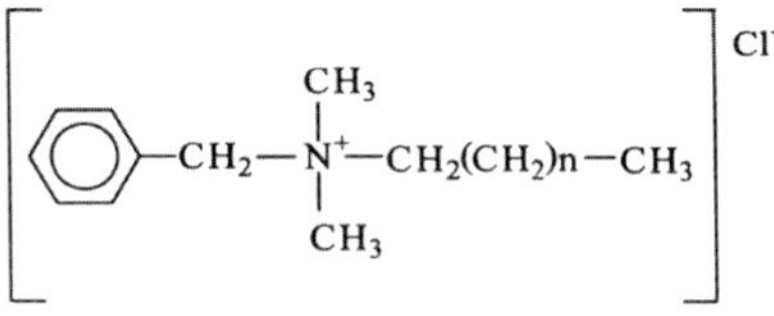

Concentrated solutions (10%) can cause irreversible changes in the color of the stone, as demonstrated by tests carried out both in the laboratory and *in situ* (Nugari et al. 1993b; Nugari and Salvadori 2003a). However, these effects are never encountered when the product is used at the recommended lower concentrations. The reported appearance of pink or orangish stains after the treatment of algal patinas—caused by the release of photosynthetic pigments from the cells—can be attributed to certain solvents present in the formulations (for instance, iso-propanol): the alcohol fraction does indeed facilitate penetration into the cells and thus increase its efficacy, but it can also cause the extraction of the pigments (Bernardini 1993). These stains are generally easy to remove only if action is taken quickly, and it is therefore always advisable to check the applications of the product.

Treatments have also been successfully carried out on both indoor and outdoor wall paintings against heterotrophic and autotrophic bacteria as well as algae and fungi; no problems of interference with the colors of the frescoes have been reported (Pietrini and Ricci 1991; Nugari et al. 2003; Ricci and Pietrini 2004).

The salts of quaternary ammonium have also been widely used in formulations containing other active principles, especially in mixtures with organotin. These products are usually quite effective, especially against lichen colonizations, both in the short and long term. Recently, though, new and stricter toxicological laws are in the process of reducing use of organotin compounds.

Organometallic compounds—made of an organic molecule linked to a heavy metal—are another group of substances with bactericidal, fungicidal, algicidal, and lichenicidal properties. Derivatives of mercury, zinc, and tin were among the most widely used; the limitations on the use of the mercury and tin derivatives are linked to their high volatility, extreme instability to light, as well as their ecotoxicity (Paulus 1990; Petersen and Hammer 1992). The derivatives of mercury (phenylmercury acetate, ethylmercury thiosalicylate of sodium), which were frequently used in the past, are at present almost never used because

of their neurotoxicity, especially at a chronic level. Certain derivatives of tin, on the other hand, are still being used today—with good results—as bactericides, fungicides, algicides, and lichenicides, both on frescoes and on stone materials. The oxide of tri-n-butyl tin, which has short- and long-term effectiveness against bacteria, algae, and fungi, has also been used in the treatment of lichens, but its use is now prohibited in many countries. The naphthenate of tri-n-butyl tin, which is less toxic, is at present still used with success as a bactericide, algicide, fungicide, and lichenicide, especially in formulations mixed with the salts of quaternary ammonium for increased efficacy (Tiano et al. 1994; Rebrikova and Ageeva 1995; Nimis and Salvadori 1997; Monte and Nichi 1997; Pietrini et al. 1999; Ascaso et al. 2002; Tomaselli et al. 2003). This is one of the few biocides that seems to protect stone surfaces from recolonization for any length of time, sometimes even for several years (Bernardini 1993; Pietrini et al. 1999). This may be due not only to its insolubility in water, but also to the fact that this product seems to form an amorphous layer on the surface of the stone that appears to make the stone hydrophobic (Altieri et al. 1997).

A derivative of pyridine—2,3,5,6 tetrachloro-4-meth-sulphonyl pyridine—has given positive results particularly with algal colonization, but also with colonizations by bacteria, fungi, and lichens (Tiano et al. 1994; Nimis and Salvadori 1997; Urzì et al. 2000b; Pinck et al. 2002; Ascaso et al. 2002; Hernandez-Marine et al. 2003b). It can be used directly as a biocide or, in the case of algal patinas, applied after disinfection with alkyl-dimethyl-benzyl-ammonium chloride, with the aim of protecting the surface against recolonization over time. A recent experiment carried out in a hypogean environment showed that it offered protection against new algal colonizations for a period of about five years (Pietrini et al. 1999).

Derivatives of urea have been widely used as lichenicides, but almost exclusively in Italy; among these are chlorobromuron, diuron, and fluometuron. Diuron is the compound that proved to be most effective in the treatment of lichens and is still used as such (Prieto et al.

1996; Monte and Nichi 1997; Tomaselli et al. 2003); overall, trials on algae have given unsatisfactory results instead.

For the treatment of algal patinas and lichens, some products have been used that are based on the same substances generally employed as herbicides, such as the diazines (bromacile), the triazines (terbutryn, hexazinone) (Monte and Nichi 1997), and also dichlorophene, paraquat/diquat, MCPA, 2,4 D, simazine, and chlorotiamide; the results were rarely satisfactory, both because of low efficacy and problems of interference.

Fig. 8.17 *Trees of heaven (*Ailanthus altissima*): although appearing to be individual trees, they are in reality interconnected because they are suckers that have grown out after the felling of the mother plant –* Photo G. Caneva

8.4.2b **Herbicide Treatments**

by Giulia Caneva, Maria Pia Nugari, and Ornella Salvadori

Treatments with herbicides (or weed killers) are those carried out against vascular plants, from Pteridophytes to Spermatophytes, and against both herbaceous and woody plants.

The use of herbicides is particularly suitable when:

—unwanted species are growing directly on the works in stone and are damaging them;

—prior to undertaking an archaeological excavation;

—particularly invasive species hinder the access to archaeological sites; for instance, the tree of heaven (*Ailanthus altissima*), false acacias (*Robinia pseudo-acacia*), or brambles (*Rubus ulmifo-lius*), all of which exhibit an extremely well-developed capacity for producing suckers, or also ivy (*Hedera helix*), which possesses a massive development of adventitious roots (Caneva 1991; Almeida et al. 1994).

Indeed, with the exception of a few species (gymnosperms, for example, among which are pines, cypresses, fir trees, etc.), simple mechanical removal is not in itself sufficient to resolve the problem (unless it is possible to remove the entire root system) because the residual presence of vegetative portions can quickly trigger the regrowth of the unwanted plant (Fig. 8.17).

However, when the aim is to selectively control the woody species, while other forms of colonization are allowed to continue to grow for naturalistic and environmental reasons, such a treatment may be carried out in the full awareness of its limitations (Fig. 8.18). Direct removal of living plants, especially if they have woody and deep roots, can prove damaging for the substrate because of the resistance of the roots themselves.

The use of herbicides is, on the other hand, not recommended for routine control of spontaneous vegetation in monumental and archaeological contexts where periodical mechanical mowing can be effective and seems more correct both in its health and safety and its naturalistic aspects (Caneva and Galotta 1994). In this way, the needless utilization of products with possible toxic residues is avoided, the wholesale removal of spontaneous flora that may include elements worthy of preservation is limited, and possible effects of soil erosion are not facilitated.

Fig. 8.18 *Mechanical and selective removal of the biological colonization present on the city walls of Lucca*
– Photo G. Caneva

Leaves and roots are the two main routes for penetration for herbicides in post-emergence treatments (against fully developed plants) and pre-emergence treatments (active as antigermination agents, i.e., before the development of the plants). In addition to the intrinsic properties of its active principle, the efficacy of an herbicide is linked to numerous external factors: meteorological conditions (temperature, sunlight, wind, RH, etc.) before, during, and after treatment; biological factors (plant species, vegetative stage, conditions of growth, etc.); and factors linked to application (concentrations used, methods of spraying, etc.).

Penetration through the leaves includes two distinct phases: retention of the droplets of the sprayed product on the surface of the leaves, and penetration inside the leaf tissues, through the integument or the stoma. Retention by the leaves is a complex phenomenon that includes various phases—adhesion and flow of the droplets on the surface of the leaf, wettability of the leaves, evaporation of the droplets—and is linked with the characteristics of the formulation employed, the methods of treatment, as well as the specific characteristics of the leaf surface (morphology, wettability, orientation, hairiness, roughness) (Catizone and Zanin 2001). Penetration through the roots varies according to the species of plant and the type of root.

Herbicides can be distinguished into contact herbicides—which exercise their phytotoxic action in the proximity of the areas of entry—and systemic herbicides—which, after penetration, are transported to other parts of the plant where they exercise their action (Fig. 8.19). These are the ones most commonly used.

Most herbicides are of an organic nature, but they also include some inorganic compounds, salts, or acids (e.g., ammonium sulphate, borax, ferrous sulphate, sodium chlorate, copper sulphate, sulphuric acid, phosphoric acid); these were used in this capacity since the earliest times (for example, kitchen salt), but today have been largely abandoned, either because of problems of toxicity, impact on the environment, or low herbicidal efficacy.

Based on their chemical structure, organic herbicides can be classified into: aliphatic carboxylic acids, aromatic carboxylic acids, nitro-organic compounds, heterocyclic compounds, phosphoorganic compounds, and derivatives of cyclohexane. A different criterion for classification is the one based on the nature of the action (for example, photosynthesis inhibitors, inhibitors of the biosynthesis of carotenoids, inhibitors of the synthesis of cell walls, inhibitors of specific enzymes, or also phyto-hormonic action, etc.).

Weed-killing interventions must be differentiated according to the species present and the general conditions of the area, after having completed a survey of the flora and vegetation present and after having ascertained the negative effects that each one may have on the substrate. The planning of these interventions should take into account the methods and the duration of the applications, which vary depending on the physiology and the biological characteristics of the species to be treated as well as the climate conditions of the area.

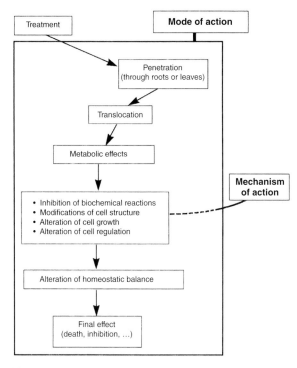

Fig. 8.19 *Schematic representation of the action of a systemic herbicide (modified from Catizone and Zanin 2001)*

For a correct planning of an herbicide operation, it is also important to take into account—in addition to the general climate conditions—the meteorological conditions that may create seasonal or topographical variations that may or may not be favorable to the treatments in question. In temperate and Mediterranean climates—which exhibit strong seasonal variations, with an alternation of favorable and unfavorable periods for biological growth—the operations should be programmed for the periods that are favorable to vegetative growth. Dormancy periods should be avoided because they limit the biocidal efficacy of the products, unless the plan is to use antigermination products, which are active in the pre-emergence stage. In general, the best periods are spring and early autumn—especially in the case of herbicides absorbed by the leaves—

although this would eliminate the possibility of controlling annual species. In tropical climates, there are no seasonal restrictions—with the exception of monsoon climates, which have substantial differences—but specific local climate conditions need to be evaluated in order to avoid unsuitable meteorological conditions. In any case, treatments should not be carried out during periods of rainfall—to avoid the washing off and dispersion of the biocides by the rain—when strong winds are blowing, or when the surfaces reach excessively high temperatures.

As mentioned, where the vegetation has undermined wall structures by penetrating them deeply, it is necessary to include consolidation operations into the overall plan, because these plants, while causing substantial damage, also function, in a way, as an element of "structural cohesion." It is therefore necessary to proceed to the partial dismantling of these stone structures to remove the lignified roots (Fig. 8.20); indeed, not only can their presence hinder the normal repositioning of the walls, but it can also lead to subsequent structural damage resulting from the decrease in the volume of the roots after the cutting. If these operations of consolidation have not been included in the planning, it is better not to carry

Fig. 8.20 *In-depth penetration of the roots of a tree of heaven, which required the dismantling of a balcony to remove the roots and avoid subsequent collapse* – Photo G. Caneva

Fig. 8.21 *(a) Incorrect treatment with pressurized pumps, with which it is difficult to control the dosage as well as the dispersion of the product (archaeological remains along the Via Sacra, Rome)* – historical photograph ICR; *(b) herbicide treatment on the* Domus Tiberiana *(Rome) with the use of a lifting device* – Photo G. Caneva

Herbicides can be applied by spraying, injection, compresses, or by scattering of granules. Spraying onto vegetative structures (in the case of products absorbed by the leaves) or on the ground (in the case of products absorbed by the roots) requires the use of low-concentration solutions (generally 0.1–1%) of the biocide in water. This type of treatment is the most common; it is quick and easy to carry out, and it can be used both for cryptogamic vegetation (mosses and ferns) and for grassy plants and trees. Although spraying with pressurized pumps is widely practiced for budget reasons when the areas to be treated are not easily accessible (Fig. 8.21a), such a practice should be avoided, because it can have negative collateral effects on the stone treated surfaces and because the dispersion of the biocide is not easily controlled. It is preferable to try to reach the areas to be treated with the help of lifting devices or, if possible, with scaffolding (Fig. 8.21b).

Treatment by injection involves the introduction of concentrated solutions of the biocide (1:10) directly into the vascular system of the plant. Such a method is only used on woody plants of a certain size; generally, it is performed after the plant has been cut down at the level of the root collar—thus saving the subsequent removal of the desiccated epigean portions—and after having drilled the stump (Fig. 8.22a) along the perimeter, or else after having made holes at an angle of 45° into the trunks (Fig. 8.22b). This method has the advantage of avoiding the dispersion of herbicides in the treatment area, which is good both from a health and hygiene point of view and also because it avoids possible problems of interference with the stone substrate.

Application by means of compresses involves the use of herbicides in the form of crystals or powders dispersed in an inert material and applied to the root collar of a freshly felled tree

out the herbicide treatment, but to defer it until it is possible to intervene at the same time on the structure of the wall itself.

Archaeological sites do not necessarily require total eradication of the vegetative species present; it is possible instead to plan selective treatments, calibrated to kill certain species while sparing others, by exploiting the variable sensitivity of each species to the different biocides, by being knowledgeable about the selectivity of the different compounds, and by choosing appropriate dosages as well as methods and techniques of distribution (Casadei and Dalla Pozza 1989; Catizone 1990).

stump. As in the previous example, this treatment is applied only to woody plants and has the advantage of limiting the dispersal of the product. It should, however, be noted that, in general, this operation has a more limited effect, because the biocide does not travel very far.

Treatment carried out by means of granule dispersal involves the use of granular formulations that are dispersed on the ground and, through the action of rain, slowly release the active agent. This treatment is therefore usually carried out in the winter time to avoid the pre-emergence of the vegetation and inhibiting the germination of the seeds. It is only used for horizontal surfaces (pavement areas, excavations, areas adjacent to walls). It is preferable to close the area to the public for a period that may vary between fifteen and thirty days, depending on the rainfall of the area, since the release of the active agents and their action take place slowly over a period of time.

Especially in the more complex cases of herbicide treatment in archaeological areas, a long-term maintenance plan is required that is subdivided into seasonal phases, including a first "correction" phase, during which an initial one-time operation is performed, followed by a "conversion" phase, which involves the progressive elimination of the most aggressive species, and concluded by an ordinary "maintenance" phase (Miravalle 1990). Such operations—also called by other authors "first aid," "maintenance," and "retouching"—usually cover a period of two years and, according to some authors, as much as three or four years (Bettini and Cinquanta 1990). In addition, these types of interventions must be adjusted to local requirements, which may suggest complete eradication, a selective or a temporary treatment, or else may be limited to the removal of shrubs, which is a kind of selective action (Bettini 1988).

Among the most frequently cited herbicides in the literature, not only with reference to the treatment of important archaeological and monumental complexes in Italy and Europe, but also in South America and Southeast Asia, we would like to mention some derivatives of urea (monuron, diuron, fluometuron, sulphonylurea), diazines (bromacyl), and especially triazines (atrazine, simazine, prome-

Fig. 8.22 *(a) Injection treatment on a tree stump; (b) injection treatment on ivy stalks that have been previously drilled to facilitate penetration of the product* – Photo G. Caneva

trine, terbutylazine, secbumeton, terbutryn). Among the urea-based compounds, diuron has had positive results although it is not much used; also, bromacyl is employed less and less, because of its high levels of persistence and ease of translocation.

Among the triazines, we would like to mention hexazinone, which carries out its action both through contact and root absorption and exhibits good biocide efficacy against herbaceous vegetation and mosses as well as shrubs. This compound has limited toxicity and is therefore still frequently employed. In addition, it does not seem to exhibit interference problems with the stone materials tested (Mambelli et al. 1989). Within this category of products, the use of terbutryn in several commercial formulations is also of interest, as has been demonstrated in various experimental treatments in areas of central and southern Italy (Caneva, unpublished).

Picloram—a derivative of pyridine absorbed both through the leaves and the roots—shows good results in the treatment of grassy, shrubby, and woody vegetation; however, negative side effects on the substrate are a possibility and must therefore be evaluated preventively.

Products based on imazapyr, a derivative of nicotinic acid, have been employed against woody plants in archaeological sites in Rome and Pompei and also in tropical areas. The active principle proved to be highly effective even against very resistant flora (such as the tree of heaven, *Ailanthus altissima*, and trees of the genus *Ficus*, in Mediterranean areas and in the tropical belt) and also had good results in localized spray treatments of tree stumps (Caneva 1991). Its use is at present limited in Italy, awaiting further study on the dispersal characteristics of the product, but it is commonly used in the USA and in other countries in Europe.

For health and safety reasons, low-toxicity products with low persistence in the soil are preferred in the area of herbicide application as well, despite their lower efficacy in prevention treatments; the selection of the compounds is led by the breadth of the spectrum of action.

Among the most widely used products today, we would like to mention those based on glyphosate, an organic compound with leaf absorption, chosen largely for its good toxicological characteristics, both in terms of low acute toxicity and low persistence in the soil (Mouga and Almeida 1997). Under normal conditions of usage, the efficacy of the product limits its field of application to mosses and herbaceous vegetation. Interference phenomena are cited in the literature, but, even though significant, they can be attributed to difficult experimental condition, and are limited to some types of substrates (Mambelli et al. 1989; Nugari et al. 1993b).

For injection applications against tree forms (*Ailanthus* in particular) ammonia solutions have been proposed for use as a systemic biocide, i.e., able to induce a gradual necrosis consequent to its traslocation into various vegetative parts (Almeida et al. 1994). However, the results of this experiment are only preliminary, since neither the dosage nor conditions of usage are specified.

8.5 BIOREMEDIATION

by Giancarlo Ranalli and Claudia Sorlini

8.5.1 Microorganisms Employed in Cleaning

The increased levels of inorganic and organic pollutants registered in the last few decades are the cause of the accelerated degradation of materials, especially of those located outdoors, exhibiting widespread corrosion phenomena. Particularly in urban surroundings, the most frequent pathologies found on stone are the gray and black crusts—mostly composed of sulphates and carbon deposits—and other alterations that cause the degradation or weathering of stone and that are defined as "sulphatation" and "nitratation."

Other pathologies may be caused by organic substances that are not part of the original artifact but have found their way onto the surface of the stone, as a result of the deposit of polluting substances from the atmosphere, of the accumulation of compounds deriving from the lyses of primary colonizers, and finally as products of earlier conservation treatments. The organic substances introduced by conservation operations can therefore include natural compounds such as animal and vegetable glues, egg casein, walnut and linseed oil, but also synthetic polymers such as the resins used as protective materials, consolidants, or water-repellents, all of which are undoubtedly the most abundant components compared to the biological pollution components.

Normally, the "cure" for these pathologies of stone involves physico-chemical techniques; these, however, can sometimes be particularly invasive and aggressive for the artifact and can also cause the introduction of toxic substances into the atmosphere, putting the operators at risk during the treatments.

An interesting alternative to the use of such substances is that of biological methods that employ microorganisms and enzymes as biological cleaning agents—as agents of "bioconservation" of works of art, so to speak—which allow "soft" interventions on the objects as well as a greater environmental safety. Both microorganisms and enzymes replicate, in optimal conditions, the same biological processes they perform in nature, contributing to the closure of the biogeochemical cycles of the material and to the maintenance of the dynamic equilibrium of the ecosystems.

The notion of using whole live cells in remediation and conservation of works of art is supported by the fact that only a minority of known microorganisms has a negative role (deteriorating pathogens), while many of them are instead responsible for positive "processes," such as that of removing unwanted organic substances In addition, microorganisms sometimes exhibit definite advantages over enzymes and especially over chemical methods when the organic substance to be removed is particularly complex and incrusted. In these cases, enzymes—which are programmed to attack specific chemical bonds—find themselves in difficulty when confronted with complex molecules that require not one single enzyme but a pool of different enzymes, which are not necessarily readily available commercially. In these cases, bacteria are able—through mechanisms of genetic induction—to synthesize the enzymes required for the degradation of the material with which they are in contact, reacting in an "intelligent" manner to the environmental conditions in which they find themselves. Moreover, the excessively drastic chemical methods able to remove the most resistant incrustations also provoke irreparable damage to the artifact.

The use of these biological agents involves an initial laboratory screening phase, i.e., a selection of the microorganisms suitable for the removal of the formed accretions or the patinas of organic material. The selection of bacteria, yeasts, and microscopic fungi can be made either by recourse to international collections of microorganisms (ATCC, DSMZ, CBS, etc.) or by isolating new microorganisms from different environmental matrices, including from the works of art themselves. The main advantage of the latter option is that microorganisms originating on the base material are generally also the best suited to grow on these materials because they have already adapted to them, as opposed to microorganisms originating from different matrices, whether environmental or not. The identification and the characterization of microorganisms isolated from the environment must be carried out with great care and precision, so as to be absolutely certain that they are not pathogenic. What can be helpful in this process are biomolecular techniques aimed at the identification of the particular genes in the DNA of the microorganisms that form the code for the enzymes useful in the treatment.

The setting up and perfecting of a biological cleaning process aimed at works of cultural interest can also require the use of specific carriers. These are inorganic materials (certain minerals, for instance, reduced to sand-like granular form) onto which the bacteria adhere forming a biofilm, or else they can be gellified organic materials inside which the cells are immobilized. The advantage of the latter lies in the fact that they are able to provide—during the biological cleaning treatment—the water and nutrients that may be required by the selected bacterial cultures in use in order to function.

A particularly important aspect of the entire bioremediation and bioconservation phase is that of the monitoring of the process itself, right from the preliminary laboratory testing with proven physico-chemical and microbiological techniques; to these can be added advanced and rapid techniques—even though these are not always specific—such as, for example, the measurement of the ATP content using bioluminescence. Thanks to its being an energetic molecule common to all organisms, ATP can be a valid bioindicator of the levels of microbial activity, with analytical response times of under ten minutes. In addition, the combination of rapid acquisition techniques of low levels of light in bioluminescence (imaging) with a computerized analysis of the data can offer a further advantage for the definition of a complex monitoring strategy, able to provide answers related to both the qualitative and quantitative aspects of the biodeteriogens and to their spatial distribution (Ranalli et al. 1997, 1998, 2003).

Also, since the microbial techniques available today are much more sophisticated and advanced—such as, for instance, those offered by the more developed techniques of molecular biology, e.g., extraction of nucleic acids, amplification, PCR, analysis of electrophoretic profiles, DGGE, cloning—it is considerably easier to adopt, in each case and whenever necessary, the specific microorganisms that are most suitable for the resolution of the different problems, even in the field of cultural heritage (Rölleke et al. 1999; Abbruscato et al. 2003).

The processes of biocleaning and the advanced biotechnologies based on the usage of microorganisms and enzymes applied to works of cultural heritage are not risky operations as long as they are carried out with due care and attention. This means that microorganisms pathogenic to humans and to the environment must be excluded; that nonsporogenic strains must be adopted, i.e., strains that are unable to survive in quiescent form even after a considerable length of time (Chapter 3, section 1); and, finally, that a thorough and gentle final cleaning operation (soft cleaning) must be carried out at the end of the biocleaning treatments. These operations are of great importance in that they have to guarantee an effective and appropriate removal of organic residues, enzymes, and of dead and living cells of the microorganisms employed previously. Because of all this, the operations must be carried out by specialized and qualified personnel, competent in the use of reagents and knowledgeable of the operating procedures of the planned interventions.

8.5.1a *Bioremoval/Biocleaning of Unwanted Organic Substances*

The very first example of cleaning by removing unwanted materials from a stone matrix was obtained with Hempel's biological compress ("Bio-Pack"), based on a paste made of urea and attapulgite with the capacity to remove patinas rich in sulphates, especially from marble surfaces (Hempel 1978).

The recent success of a case of biotechnologies applied to the field of fine arts represents a promising example in the field of bioremediation of works of art: in this case, selected bacteria were applied to the fourteenth-century frescoes by Spinello Aretino in the Monumental Cemetery in Pisa; the frescoes had suffered considerable alteration from a tenacious layer of animal glue applied in the past in an attempt to remove them (*strappo*) from the wall (Ranalli et al. 2000, 2003). First,

a bacterial culture of *Pseudomonas stutzeri* (A29) was selected in the laboratory because of its high biodegrading action against many organic compounds, among which animal glue; then it was applied—in suspension—directly on the altered frescoes, in a concentration of about 10^8 of live cells per ml (Fig. 8.23) with satisfactory results (Plate 32).

To remove compounds and organic polymers, such as hydrocarbons and synthetic resins, it is possible to use microorganisms that are able to use xenobiotic molecules as growth substrates—i.e., molecules that are not natural but synthesized chemically, which are complex and difficult to attack—until they achieve complete mineralization. In this context, mixed cultures and bacteria with the capacity of using aliphatic and aromatic hydrocarbons as the sole source of carbon and energy were isolated directly from stone samples (Saltrio stone, Carrara and Candoglia marbles) that had been exposed for a few years to the atmospheric pollutants in Milan (Zanardini et al. 2000).

8.5.1b *Bioremoval/Biocleaning of Sulphates and Nitrates*

The removal of sulphates and nitrates from the surface of works of art subjected to "sulphatation" and nitratation" can be carried out using sulphur-reducing and nitrate-reducing (denitrifying) bacteria; these operate in the absence of oxygen and convert the sulphates into hydrogen sulphate and the nitrates into molecular nitrogen (denitrification), respectively. This action was proven both in laboratory tests and on actual examples of marble affected by black crusts ($CaSO_4 \cdot 2H_2O$), which are the result of gypsum deposits originating from the exposure of the carbonates ($CaCO_3$) to atmospheric pollutants, especially sulphur dioxide (Lal Gauri et al. 1989, 1992; Heselmeyer et al. 1991). In the experimental tests, twelve hours of application of *Desulfovibrio desulfuricans* resulted in a partial cleaning of the crust, contemporaneously with an initial calcification of the most superficial layer of gesso.

To create more suitable conditions for this microflora's activity, it may be necessary to apply compresses of these microorganisms in direct contact with the stone surfaces affected

Fig. 8.23 *(a) Initial phases of the biocleaning of the fourteenth-century fresco by Spinello Aretino,* The Conversion of Saint Ephisius and the Battle; *(b) bioconservation phase of the same fresco: removal of the cotton fabric impregnated with animal glue that was used in the "strappo" technique employed during the removal of the original wall surfaces of the Monumental Cemetery in Pisa (1948)* – Photo G. Ranalli

of sepiolite colonized by the bacterial cells of *Desulfovibrio desulfuricans* and *Desulfovibrio vulgaris,* a removal of the sulphates equal to 81% of the initial quantity present was observed (Ranalli et al. 1996, 1997, 2000). In other experiments, cultures of *Pseudomonas stutzeri* (GB94), selected because of their high capacity for denitrification, were used: the results showed that, after thirty hours at 28°C in controlled anaerobic laboratory conditions, an average of 90% of the nitrates had been removed (Ranalli et al. 1996, 2000).

8.5.1c *Bioremoval/Biocleaning of Calcium Oxalate Patinas*

The presence of oxalate patinas and films on stone monuments is well documented by several authors (AA. VV. 1989; Salvadori and Realini 1996; Monte 2003). Numerous hypotheses have been formulated to explain the origin of such patinas: they could be a consequence of the use of protective/decorative treatments on the surface of the works, or residues of the combustion of fossil fuels, or else they could be due to the possible biological activity of lichens, bacteria, and fungi. Despite the fact that some authors maintain that these patinas may have a protective role, their removal has been attempted for aesthetic reasons, through the application on stone surfaces of microorganisms capable of using calcium oxalate; the results, however, have not been satisfactory (Tiano et al. 1996).

8.5.2 **Enzymes Used in Cleaning**

The use of enzymes in the field of conservation of cultural heritage, which began around 1970, has been widely applied with two specific objectives in mind:

by these pathologies and then to cover the compress with a thin film of a plastic material impermeable to oxygen. Such a covering, in addition to creating conditions of anaerobiosis, also maintains the humidity levels over a period of time; this helps prolong the metabolic activities of the applied cells, while also reducing the danger of a rapid dessication of the organisms with a resulting loss in functional capacity. More recently, interesting results were obtained on a fragment of a marble field statue containing 1,900 ppm of sulphates: after a period of thirty-six hours from the application

—aesthetics, in the sense of achieving the cleaning of the image by partially or completely removing old and disfiguring patinas;

—functionality, in that such an application allows the removal and elimination of organic substances of natural origin, e.g., adhesives.

The advantages deriving from the use of enzymes, which are highly functional proteins, compared to other traditionally employed reagents, such as solvents or alkaline and acid substances, are the speed of their action as well as their intrinsic specificity and selectivity of the substrate-enzyme relationship (Chapter 1, section 4.6). During the application period (minutes, in general), each enzyme does indeed often exhibit different affinities when acting on different compounds and substances; the factors that may strongly influence the enzyme's action in suspensions in water—even at low concentrations—are: the pH values (favorable, optimal, limiting); the temperature; and the absence of inhibitors (metal ions). The earliest application of a substance carrying out its action through mechanisms of an enzymatic kind was the empirical use of saliva as a cleaning agent for paintings; the main active component of saliva being mucin, it exercised a combined action as emulsifier and surfactant. This practice was followed by applications of amylase and protease in water-based solutions on paper supports. Still later, rough preparations of pancreatin were used, but with somewhat uncertain results. Beginning in the 1980s, enzymes in gel form were formulated, which presented specific advantages in application and final removal (Makes 1988). In 1994, the first applications of enzymatic cleaning were performed on polychrome works (*The Deposition*, National Gallery of Parma) using protease and lipase in an alkaline environment; these were followed by the successful removal from paintings of a synthetic acrylic resin, Paraloid B72 (Bellucci and Cremonesi 1994; Bonomi 1994; Cremonesi 1999; Buttazoni et al. 2000; Wolbers 2000).

Proteolytic enzymes were also employed for the cleaning of lichenic incrustations on the stone surfaces of monuments in Lecce (southern Italy); the results were positive, despite some difficulties encountered in keeping the enzyme active (Fig. 8.24) (Capponi and Meucci 1987).

8.5.3 Biocalcification for the Consolidation of Stone

Calcareous stone plays an important role in the reconstruction and the conservation of historical and artistic heritage and is the object of interventions not only of conservation, but also of protection. Still, this stone is also subjected to attack when it is exposed to unfavorable atmospheric conditions in outdoor environments and when the levels of pollution are high. Generally, after their extraction as blocks and before they are cut, the stones are put out to "dry," a stage that allows any residual water to migrate toward the surface, taking along with it a portion of the mineral salts; among these, the most important is calcium carbonate ($CaCO_3$, calcite), which creates a protective layer when it is deposited. In general, these protective layers will alter with time, after long periods of exposure to atmospheric pollutants (nitrogen oxides, hydrogen sulphide, acid rainfall, etc.); this phenomenon has sometimes been further intensified by inappropriate conservation interventions.

In nature, the role of bacteria in the precipitation of $CaCO_3$ has long been known; indeed, soil, freshwater, and salt water are all environments in which the precipitation of $CaCO_3$ crystals is actively sustained by bacteria belonging to different taxonomic groups, with the formation of marine calcareous skeletons, carbonate sediments, and limestone (Boquet et al. 1973; Morita 1980; Rivadeneyra et al. 1985, 1991, 1994). It seems that in this process the bacteria carry out a dual role, both active and passive: they can themselves act as nuclei for the neoformation of calcite crystals, or they can locally modify the chemical composition of the waters by means of calcium concentrations and the production of bicarbonate ions. The mechanisms at their disposal to induce the precipitation of calcium

Fig. 8.24 *Cleaning tests on lichenic colonizations using proteolitic enzymes: (a) before; (b) after treatment (Church of Santa Croce, Lecce) (from Capponi and Meucci 1987)*

or calcium-magnesium carbonates seem to be various:

—aerobic or anaerobic oxidation of organic compounds in environments lacking nitrogen with neutral or alkaline pH values and rich in Ca or Mg;
—aerobic or anaerobic oxidation of nitrogenous organic compounds in nonbuffered environments enriched with Ca or Mg;
—reduction of $CaSO_4$ by sulphur-reducing bacteria;
—hydrolysis of urea;
—photosynthesis.

Thanks to research carried out by various authors, it became clear, around the 1980s, that it could be possible to reproduce crystals of calcite artificially in the laboratory, subsequently elaborating a process that could be applied directly on stone in historic buildings. The process involves recreating the original natural phenomenon that leads to the formation of crystals of calcite in a variety of different environments and reproposing it today for the consolidation of stone through advanced processes of biomineralization that can be obtained through:

a) Heterotrophic bacteria;
b) Natural and synthetic macromolecules;
c) The dead cells of calcinogenic strains.

In the first case, interesting results were achieved by applying live biological cultures of *Bacillus cereus* directly onto monuments (Orial et al. 1992). The treatment proved successful in reducing the superficial porosity of the stone, but there were no noticeable effects of consolidation of the stone (Castanier et al. 2000). Other experiments carried out with the bacteria *Myxococcus xantus* seemed to induce an increase in penetration, involving not just the surface layers of the materials; however, the authors emphasize that the efficacy of the biotreatment seemed to depend on a variety of factors, which require sophisticated and accurate measures of control to guarantee high-quality results (Rodriguez-Navarro et al. 2000). At present, the large-scale use of heterotrophic bacteria on stone materials in the area of cultural heritage requires further investigation, although it does so far look very promising (Fig. 8.25).

In the second case, the consolidation of stone through advanced processes of biomineralization is induced and promoted through the employment of natural organic macromolecules (organic matrix macromolecules, OMM) and synthetic polypeptides. Indeed, by using fragments of the shells of the mollusk *Mytilus californianus*—composed of glycoproteins rich in aspartic and glutamic acids and a mixture of polysaccharides—and in the presence of calcium ions, it is possible to induce a typical configuration of the beta structure, with a consequent high affinity for those same Ca^{2+} ions that actively participate in the neoformation of calcite crystals (Wheeler and Sikes 1989; Addadi and Weiner 1990). The advantages of employing such macromolecules were shown in studies and investigations carried out on stone test samples in the laboratory, which also highlighted other advantages such as a reduced capacity for water absorption and, at the same time, an increase in the resistance of the treated surfaces (Tiano et al. 1992; Tiano 1995, 2003) (Fig. 8.25). Although the systems for extraction and purification of the organic matrices from mollusks have greatly improved in recent years with the use of ion exchange resins, the production levels are still low, and the production costs too high.

As a result, an alternative has been to adopt macromolecules that are synthetic but with an analogous structure to natural ones—such as polyaspartate (Poly-A) and polyaspartate-leucine (Poly-A-L)—which exhibit similar activities in the process of nucleation of calcite crystals (Levi et al. 1998).

Recently, a third proposed way of consolidating stone via biomineralization that has shown favorable possibilities is the use of dead cells of calcinogenic bacteria. Indeed, the residual fractions from cellular lysis have shown an effectiveness in terms of productivity and speed of neoformation of $CaCO_3$ crystals that is higher than that of the original live bacterial cells. The cell fractions of bacterial origin or their structures might therefore be directly involved in the promotion of the precipitation of $CaCO_3$ by reducing the contribution of

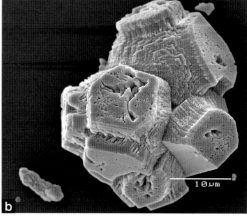

Fig. 8.25 *Biocalcification: (a) SEM image of bacterial colonies active during the biocalcification process; (b) SEM image of calcite crystals induced by bacterial fraction of the cells (BCF)* – Photo P. Tiano

organic substances on the surface of the treated stone surfaces (Tiano 2003). And finally, the analysis and the study of the genetic makeup of the bacteria involved in the biomineralization process may contribute to the development of future strategies that correspond more closely to the requirements and priorities of the conservation and restoration of works of art.

CHAPTER 9

TECHNIQUES AND METHODS OF INVESTIGATION

9.1 TECHNIQUES FOR THE STUDY OF BIODETERIORATION

by Ornella Salvadori and Clara Urzì

Identifying the organisms responsible for biodeterioration is an important step to determine their habitat and how dangerous they are, which is necessary to subsequently establish the best methods for their elimination and/or for the reduction of their deteriorating action. Clearly, the procedures for their identification and their taxonomic classification depend on whether macroorganisms or microorganisms are being studied and also vary within each taxonomic group.

However, the identification of the organism alone, if it is not associated to other investigations, often tells us little or nothing about the organism's deteriogenic action, unless it is already known. The pertinent literature clearly shows that there are many more studies of the first type, i.e., devoted to identification alone, while studies of the second type are very few and certainly not carried out routinely. Studies directed to the definition and quantification of deterioration focus on the organism-substrate relationship and on the physiological activities of the organisms. The choice of sampling method is therefore crucial and influences the various stages as well as the results obtained. It is essential to develop a plan for the investigations to be carried out, taking into consideration that the identification of a certain organism on an artifact does not necessarily mean that it causes deterioration. In addition, it is important to assess the induced decay to determine whether or not a biocide treatment is required.

9.1.1 Sampling

Sampling must be planned on the basis of diagnostic and analytical objectives (e.g., the identification of the biodeteriogens present, the relationship with the substrate, deterioration induced in the material, etc.). The techniques for sampling and for the preservation of samples vary according to the type of substrate and to whether microorganisms (not always visible with the naked eye) or macroorganisms (lichens, bryophytes, plants) are present (NORMAL 3/80, 1980). In the case of the latter, specimens are sampled if their systematic classification is doubtful, or else if they are missing from the collections in the herbarium. The samples must be as complete and representative as possible and in any case must possess all the necessary elements for diagnosis.

Sampling techniques for the study of microorganisms can be distinguished into two main groups:

— "nondestructive" or "minimally destructive" sampling techniques (for example, needle, swab, adhesive tape, etc.), with which only the biomass is removed (Fig. 9.1) or also a small part of the substrate, generally when it is already partially detached from the artifact;

— "destructive" sampling methods using instruments such as scalpels, with which a portion of the material subjected to biodeterioration is also removed.

Table 9.1, for example, illustrates the main methods to be used for the sampling of stone subjected to biodeterioration.

Fig. 9.1 *Sampling from a painting using a swab*
– Photo ICR

Depending on the kind of analysis required, sampling must be carried out using clean or sterile instruments; for all microbial investigations, sterility is a prerequisite during the sampling phase, while in other instances cleanliness of the instruments used for sampling is simply advised. The quantity of material sampled influences the number and the type of analyses that can be carried out in the laboratory and, hence, the level of identification that can be achieved. Before any actual sampling is undertaken, therefore, a preliminary choice must be made as to the sampling method that is most appropriate for the type of analysis and for the level of investigation required (Hirsch et al. 1995b; Urzì et al. 2003b).

9.1.2 Identification of Microorganisms

As mentioned earlier, the phase of the identification of microorganisms causing biodeterioration is crucial both for the definition of their habitats and of the dangers they represent for the substrate as well as to establish the subsequent course of action.

The identification procedures are different for every group under investigation, and what level of identification can be reached depends on the techniques employed (identification at the level of genus or of species). In addition to the standard techniques of observation under the microscope and the use of specific analytical keys, the following techniques can be used for the identification of the different microbial groups.

9.1.2a. *Techniques Based on Cultures*

The cultivation of the various microbial groups (microalgae, fungi, yeasts, cyanobacteria, bacteria) presupposes a knowledge of their nutritional and environmental requirements and, consequently, a correct choice of: an appropriate culture medium capable of satisfying their requirements for sources of C, N, S, O, etc.; suitable sources of energy (chemical or light); environmental conditions (pH, temperature, presence or absence of oxygen); and other factors that are essential for the growth of various organisms. It should be added that if the intention is to carry out analyses using cultures, both the sampling and the storage environments must be sterile and must be kept so until the samples arrive in the laboratory; they also must be kept at a temperature of +4°C.

Analyses of cultures can be qualitative, aimed solely at the taxonomic identification of the microorganisms, or both qualitative and quantitative, in order to also establish the number of microorganisms present/per unit of weight of the sample or unit of surface sampled.

Standard NORMAL 25/87 (1987) and 9/88 (1988) list some of the cultural media commonly used for the isolation and growth in pure culture of the biodeteriogens of natural and artificial stone. Other growth media routinely used by other researchers are also reported in the literature. The choice of culture medium may also be made in relation to the material constituting the work of art under investigation; this is especially useful for the isolation and study of heterotrophic microor-

Table 9.1 Main sampling methods applicable on stone

Nondestructive Methods	When to Use Them	Types of Microbiological Analyses
Needle	In correspondence of stains of a biological nature, growth of microorganisms in pitting, cavities, among crystals, or fibers, etc.	Microscope analysis or qualitative analysis of the culture with direct scattering on the petri dish
Swab	Surface	Observation under the microscope, qualitative or quantitative cultural analysis for the surface sampled
Adhesive tape	Surface	Observation under the microscope, qualitative cultural analysis, identification of individual cells directly *in situ*, using molecular probes (FISH)
Microscalpel	In the presence of very thick patinas	Observation under the microscope, qualitative cultural analysis
Destructive Methods	**When to Use Them**	**Types of Microbiological Analyses**
Microscalpel	Small quantities of material	Observation of the material under the microscope, qualitative-quantitative investigations, extraction of DNA for biomolecular analyses
Scalpel	A considerable quantity of material (50 mg–1 gm or more) that is already detached and nonreusable, hidden areas	Observation of the material under the microscope with or without preparation, cultural qualitative-quantitative analysis and subsequent identification of the isolated organisms, extraction of the DNA from the sample for biomolecular analyses, for the study of the microbial communities colonizing the sampled area

ganisms. The differentiation between superficial contaminants and biodeteriogens can be achieved by adopting media whose sole source of carbon is the same molecule present in the substrate: for example, in the case of wood or paper, it could be useful to choose selective growth media based on cellulose as the only source of carbon, since these would be suitable only for the growth of cellulolytic organisms (Rautela and Cowling 1966).

Yet, it is well known that many microorganisms that live in a particular environment (Chapter 3, section 1) remain vital but do not reproduce when grown in an artificial medium. Indeed, it is thought that only about 1–5% of the total number of microorganisms in existence can grow in laboratory cultures. Because of this, the cultivation of microorganisms in pure cultures is limited for the ecology and also in terms of the relationships

among the microorganisms present in the specimen. On the other hand, it does offer the possibility of an in-depth study of those microorganisms that are able to grow in artificial cultures; their isolation in pure culture allows for tests that establish their biochemical and ecophysiological properties of strains, making it possible to determine the macromolecular profile (proteins, fatty acids, and nucleic acids) that characterizes the genus or the individual species.

9.1.2b. *Molecular Techniques*
Molecular techniques are all those techniques pertaining to molecular biology that can be applied for the diagnosis of microorganisms and/or to establish the presence or absence of functional genes and proteins. There are numerous molecular techniques that can be applied

in the field of diagnostics of microorganisms that colonize works of art; these are generally based on techniques applied in various fields of microbiology (Daffonchio et al. 2000; Rölleke et al. 2000; Gonzales 2003; Urzì et al. 2003b). Molecular techniques are all based on the extraction of DNA or RNA and amplification via PCR (polymerase chain reaction) of specific portions of the DNA (for example, rDNA, operons for functional genes, etc.). The extraction of nucleic acids can be carried out either from individual isolated cells or directly from the samples to be analyzed. In the latter case, molecular study can be carried out on the entire microbial community—highlighting both the microorganisms that can be grown in cultures and those that cannot—or else on the isolated strains to identify and compare these with other sequences present in many online databases (Gonzales 2003).

Figure 9.2 summarizes the protocol in which cultural and molecular investigations are employed for the study of microbial communities that colonize works of art.

9.1.2c. *Identification of Microorganisms by Means of FISH*

FISH (Fluorescence *In Situ* Hybridization) is a technique that allows the study of complex communities as well as the identification of individual cells directly *in situ*, independently from their isolation and development in cultures. The technique consists in the use of molecular probes—with a length of 15–20 pb —linked to a fluorochrome that binds specifically to certain portions of the RNA (or DNA) of the cells that are complementary to the nucleotide sequence present in the probe. Probes are in existence, or can be designed, that are domain-specific (for Fungi, Bacteria, Archaea, etc.) or else genus- or species-specific. It is also possible to identify genetic sequences corresponding to a specific enzymatic activity. The protocol for FISH (Fig. 9.3) requires the cells to be fixed onto a support (glass, membrane, or sticky tape) followed by hybridization with the probes (Urzì and Albertano 2001; La Cono and Urzì 2003; Urzì et al. 2004).

Fig. 9.2 *Schematic representation relating to the isolation and identification of microorganisms from artifacts*

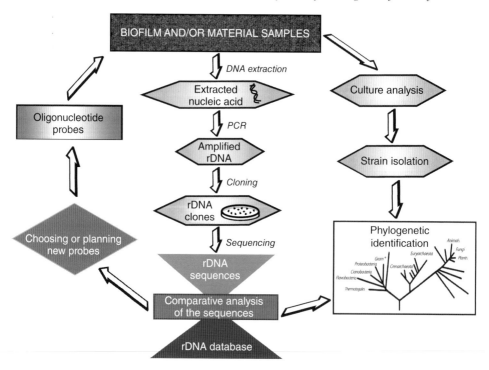

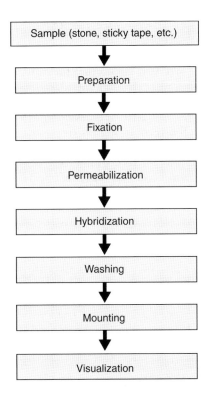

Fig. 9.3 *Schematic representation and main phases of the FISH protocol (modified from Urzì et al. 2003a)*

The diagram shows the following sequence:

Sample (stone, sticky tape, etc.) → Preparation → Fixation → Permeabilization → Hybridization → Washing → Mounting → Visualization

9.1.2d. **Detection of Microbial Activity**

The presence of microorganisms can also be detected with chemical and biochemical tests that measure their activity. Examples of this are the measuring of the ATP and of the energy charge, which is used in the study of the biodeterioration of stone, wood, textiles, and other substrates (McCarthy 1987; Stanley 1988; Tiano et al. 1989; Nieto et al. 1997a, 1997b; Ranalli 2000); the measuring of the dehydrogenase activity, which reflects the respiratory activity of microorganisms (Warscheid et al. 1990); or the measuring of the metabolic activity with a variety of colorimetric tests (McCarthy 1987; Tayler and May 1995).

Respirometric methods—which measure oxygen consumption or the emission of carbon dioxide—can be employed for a general detection of the presence of active deteriogens. The more accurate the detection system of the gases—i.e., the smaller the statistically significant measurable difference between the control and the sample—the greater the applicability of the method for early diagnosis. The measurement of the concentrations of respiratory gases can be carried out by enclosing the object under examination (a wooden artifact, a support, etc.) in a bag that is impermeable to the gases being measured and that is connected to the measuring equipment (Koestler et al. 2000).

The presence of fungi can be measured by analyzing the amount of chitin or ergosterol in the materials (Seitz et al. 1979; McCarthy 1987; Whipps 1987). Chitin and ergosterol are molecules that are part of the fungal cells and that can be used as markers for the presence of mycelium; their concentration in the substrate can be correlated with the amount of fungi present. Microcalorimetry is another useful method for the study of fungal activity in relation to water activity: it measures the heat produced by fungal metabolic action (Markova and Wadsö 1998).

The measurement of the quantity of chlorophyll pigments (chlorophylls a, b, c) with UV/Vis spectrophotometric analysis is useful to detect the presence of photosynthesizing microorganisms in samples from alterations of uncertain biological origin; this provides information relating to the vitality of these organisms and, within certain limits, can lead back to the systematic groups to which they belong (UNI 10813, 1999).

Evaluation techniques of the isotopic components might also prove useful in the field of cultural heritage since it is well known that microorganisms—especially in less than optimal growth conditions—tend to select specific isotopes (usually the lightest in weight) of certain chemical elements (S,C,N,O). With isotopic fractioning techniques it is therefore possible to acquire data relative to the biogenic or nonbiogenic genesis of certain elements and, hence, of certain compounds (e.g., sulphates, nitrates, etc.) (Ehrlich 1981).

9.1.3 **Identification of Macroflora and Ecological Analysis**

When lichens, mosses, and higher plants are present, it is advisable to supplement the floristic study (i.e., the identification of all the species found) with an ecological study of the communities they determine (NORMAL 24/86, 1986). This is carried out with phyto-sociological surveys using the methodology proposed by Braun-Blanquet (1928) in the early twentieth century and still routinely employed today for procedures of the ecological analysis of flora (Bullini et al. 1998; Pignatti 1995).

This phytosociological method is based on the following fundamental principles:

—plant communities are characterized by a particular floristic composition;
—among all the species that characterize a phytocoenosis, some express better than others the complexity of the relationships among the species, the community, and the environment; these are called characteristic species;
—characteristic species can be used to set up a hierarchical classification of plant groups, formed by progressively more comprehensive units, each one with a characteristic group of species.

Phytosociological investigation consists of an analytical phase followed by a synthesis phase. In its initial phase, the study of the plant component begins with the phytosociological surveys; these include observation and gathering of three types of data relating to the sampling site (altitude, exposure, inclination), the type of substrate, and the floristic composition. The survey must be carried out in a homogenous area that is representative of the plant typology under investigation, because the single survey must be representative of a standardized description of a sample of the vegetation in question. For each species surveyed, the relative percentage of coverage must be noted and must be expressed in terms of abundance-dominance. According to the conventional Braun-Blanquet scale, this ranges from the upper value of 5—for individuals of the species that cover more than 75% of the surface surveyed —to the lowest value (defined by +) for sporadic species with a cover of less than 1%. The entire range of the Braun-Blanquet scale is: + = < 1%; 1 = 1–5%; 2 = 5–25%; 3 = 25–50%; 4 = 50–75%; 5 = 75–100%.

In the synthesis phase, the comparison and the classification on a floristic basis of the phyto-sociological surveys carried out in similar plant contexts will lead to a definition of an average and theoretical model of the plant grouping studied, which is defined as plant association. This represents the basic phytosociological unit; for an association to be recognized as such, it must be characterized by a well-typified floristic aggregate and by characteristic species that must be variably exclusive and capable of distinguishing, floristically and ecologically, the association in question from all other associations present in the territory being studied.

On the basis of their floristic similarities, several associations may be grouped together into a larger complex to form a more synthetic classification called syntaxonomy of the vegetation. Syntaxonomy thus presents a hierarchical approach, placing the association as the category forming the basis of the classification, and then conventionally establishing higher syntaxonomic categories (literally, *syntaxa* is composed of *taxa*, which corresponds to taxonomic categories, and *sin*, which refers to the level of communities or ensembles); these can be distinguished into alliances, orders, and classes. The alliance is composed of several associations, the order by several alliances, and the class by several orders; each category is identifiable through specific suffixes (*-etum*, *-ion*, *-etalia*, *-etea*, respectively) linked to the name of the genus of the species that characterizes the grouping. The main criterion for differentiating the various groups is the floristic composition and especially the presence of characteristic species, which can therefore be distinguished into species characteristic of the association, alliance, order, and class (Fig. 9.4).

9.1.4 Analysis for the Assessment of the Relationship with the Substrate and of the Induced Deterioration

9.1.4a. *Observation of Biological Specimens under the Optical Microscope*

Observation under the optical microscope makes it possible to detect the presence of biodeteriogens, the structural and morphological characteristics useful for their taxonomic classification, the depth and means of penetration, and the effects induced by chemical and/or physical treatments directed at their elimination.

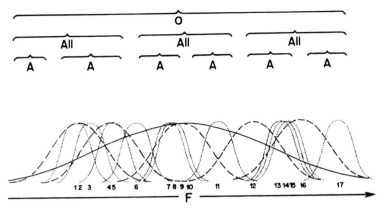

Fig. 9.4 *Ecological gradients and syntaxonomy: as a function of ecological breadth correlated to environmental factors (F), we recognize associations (A), alliances (All), and orders (O) (from Golubic 1967)*

The observations can be carried out on fresh, fixed, or appropriately stained specimens (Warscheid et al. 1990; UNI 10923, 2001). To determine what depth microorganisms have reached within the materials, it is useful to prepare and study polished and thin sections of the material, either loose or enclosed in resins. The inclusion within a resin is useful for the study of fragile materials, water-logged woods, textiles, organic archaeological finds, etc., and also to obtain sections that are suitably oriented. Thin sections of organic materials can be made with microtomes and ultramicrotomes of various kinds, while sections of inorganic materials can be obtained by polishing the sample with abrasive papers (UNI 10922, 2001).

Histochemical staining techniques are very useful in detecting the presence of microorganisms that would otherwise be difficult to see. Among the most commonly used are the staining with: methylene blue, lactophenol/blue cotton, periodic acid/Schiff's reagent (PAS), acridine orange, 4,6 diamidino-2-phenylindole dihydrochloride (DAPI) (UNI 10923, 2001).

The study of the sections—stained or not—is particularly useful when investigating the interactions between the microorganisms and the substrate (development inside the materials, preferential pathways of penetration, etc.) and to determine the damage induced deep into the structure of the materials (for example, in the study of bacterial and fungal attacks on wood). The stereoscopic microscope as well as the optical microscope in transmitted, reflected, polarized, and fluorescent light can all prove useful.

9.1.4b. Observation under the Electronic Microscope
Surface structures as well as the relationship between organisms and substrate can be studied by SEM (Scanning Electron Microscopy) (Fig. 9.5) or by ESEM (Environmental Scanning Electron Microscopy); the TEM (Transmission Electron Microscopy), on the other hand, allows the study of intracellular structures.

Samples (either biological, or in which organic materials are present) must generally first be fixed in paraformaldehyde or glutaraldehyde in a buffering solution. The protocols for SEM and TEM are different (NORMAL 19/85, 1985; Urzì and Albertano 2001).

The use of ESEM does not involve any preliminary preparation or preventive fixing of the samples; these are mounted directly onto an aluminum support for microscopy, introduced into the ESEM, and observed directly.

9.1.4c. Identification of the Induced Deterioration and of Neoformation Products
To characterize the deterioration of the substrate, many different techniques can be used depending on the materials on which the organisms have developed; we will only briefly mention some of them here.

Certain nondestructive physical methods, for example, can be employed for fungal

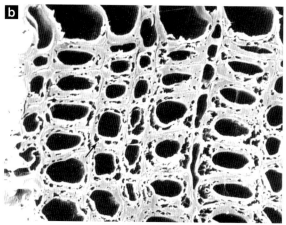

Fig. 9.5 *Interaction between microorganisms and substrate as observed under SEM: (a) textile fibers attacked by fungi* – Photo M. R. Giuliani; *(b) alterations of the cell walls in wood induced by phenomena of soft rot (from Eaton and Hale 1993)*

tometry; Raman spectroscopy; x-ray diffraction and microdiffraction; the determination of the degree of polymerization (DP); and scanning electron microscopy and energy dispersive spectroscopy (SEM/EDS).

9.1.5 Measurement of Airborne Microflora

The biological component of the air is a potential element of deterioration for works of art; with aerobiological monitoring it is possible to acquire information of critical importance for the prevention of biodeterioration (Chapter 7, section 2). To identify the microorganisms present in the air, various sampling methods can be used: gravitational deposit, impaction, aspiration, filtration, electrostatic precipitation, thermal precipitation, impingement (Mandrioli and Caneva 1998). The choice of method is closely correlated with the predetermined objectives (for example, explorative sampling, indoor or outdoor environments, types of microorganisms to be isolated, etc.).

Exposing plates containing culture media to the air is a very simple method (passive sampling); however, it only allows the gathering of purely qualitative information, since it cannot be put in relation to a known volume of air. Active sampling methods, on the other hand, involve the aspiration of known amounts of air that are then directed onto culture media or onto filters subsequently transferred to culture media; in these cases, the microbial concentration is related to known volumes of air (NORMAL 39/93, 1994; Eckhardt 1996; Mandrioli and Caneva 1998; Nugari 2003).

infections of wood only when the damage is extensive and are therefore not useful for early or preventive diagnosis. These are sonic analysis (conduction of sonic waves into the wood), radiographic methods (mainly x-rays), and electric resistance (Ross et al. 1994). Recently (as was noted in Chapter 8, section 4.1), electronic sensors have been developed that can give an early diagnosis of the presence of fungal colonizations on paper (Canhoto et al. 2004).

The use of destructive techniques always requires the removal of a certain amount, however small, of material. Among the various diagnostic techniques employed for the identification of the products of deterioration, we would like to mention FT-IR spectrophotometry;

9.2 METHODS FOR EVALUATING CONSERVATION PRODUCTS

by Maria Pia Nugari and Ornella Salvadori

The experimental assessment of the suitability of commercial products (biocides, adhesives, carriers, water-repellents, consolidants, etc.)

for conservation use is a fundamental part of the research applied to conservation, because these products are so rarely made specifically for this sector. The products used on works of art and artifacts must, indeed, not only fulfill their intended function, but they must also be capable of not modifying the chemical, physical, or biological characteristics of the constituent materials of the object. In conservation, standardized methods of investigation that would allow repeatable and comparable tests are not always available. In addition, the evaluation of the main requirements of the different conservation materials must always be carried out both in the laboratory and on the materials, in order to assess the products as used in practice. Whenever the intended use foresees leaving the products in contact with the work for a certain amount of time, it is also necessary to test not only their immediate behavior but also the reactions over time, including artificial aging tests into the plan.

Tests that are strictly biological vary according to the nature of the product to be tested. For biocides, the analyses are mainly aimed at evaluating their efficacy against biodeteriogens, while the assessment of any potential harmfulness for the artifacts is carried out with chemical and physico-chemical tests. When studying adhesive products, water-repellents, carriers, etc., biological analyses assess their degree of susceptibility to biological attack, in both the short and the long term, and also what effect any alterations induced by biological colonizations would have on their particular characteristics (adhesive power of the glues, water-repellence of the protective materials, etc.).

9.2.1 Evaluation of Biocides

9.2.1a *Evaluation of Efficacy*
The evaluation of the efficacy of a biocide involves laboratory tests as well as tests on the materials to be treated, differentiated according to the target organisms (NORMAL 38/93, 1993).

In the case of microorganisms that can be cultivated in the laboratory, the biocide—tested in different concentrations—is added to a solid or liquid culture medium, which is inoculated with the microbial suspensions under investigation to determine the minimum inhibiting concentration (MIC), i.e., the maximum dilution at which no growth occurs (Fig. 9.6). An alternative to this is the disk assay method, which involves positioning a small disk imbibed with the biocide solution in the center of a culture plate; what is measured in this case is the zone of inhibition of growth surrounding the disk. It should be emphasized, though, that such a method is influenced by the capacity of the biocide to diffuse through the agar. The results thus obtained will not always be the same when transferred into practice, since all of the following factors can considerably affect the response to the application of a biocide: the nature of the substrate; its state of conservation; the presence of dirt; the microbial associations present (usually much more complex than the microorganisms tested singly in the laboratory); the vital phase of the biodeteriogens; and the environmental conditions.

For this very reason, laboratory tests also include the use of the same type of substrate that is to undergo treatment (e.g., stone, paper, canvas, etc.) (Tiano et al. 1994, 1997; Ranalli 2000) (Fig. 9.7). Unlike the previous tests, the use of the material composing the substrate offers the advantage of being able to evaluate at the same time the biocide-substrate and biocide-microorganisms interactions. The chemical and physical characteristics of the substrate as well as its state of conservation and hydration

Fig. 9.6 *Laboratory test for the evaluation of the efficacy of biocides, comparing biological development in relation to different concentrations of biocidal substances*
– Photo ICR

Fig. 9.7 *Laboratory test to evaluate the efficacy of biocides on lining canvases (note the presence of the zone of inhibition)* – Photo ICR

In the tests on the work itself, the product to be tested is applied on areas that are limited in size, homogenous, and representative of the alteration; there should also be a control, which must be treated only with the solvent used. The microorganisms and organisms present must first be identified and their vitality verified (Fig. 9.8).

In the case of lichens, mosses, and plants, the efficacy evaluation of the biocide is carried out almost exclusively on the works *in situ*, because the growth of these organisms—typical of outdoor environments—requires time and is not always possible in laboratory conditions (Fig. 9.9).

can considerably influence the absorption and retention of the biocide, with significant consequences on its efficacy. Ideally, therefore, parallel efficacy tests should be conducted both in culture on the isolated species and on test areas on the work itself. Also, to verify the efficacy over time, the same samples can undergo accelerated aging tests that take into account the influence of temperature, humidity, and light (Tiano et al. 1994; Warscheid and Braams 2000).

After the treatment has been applied and the time span in which the product carries out the action is over, the efficacy is checked by means of:

—observation of any macroscopic alterations of the organisms or patinas in the treated areas;
—evaluation of any residual vitality by inoculating specific culture media with samples taken from the treated areas;
—assessment of the amounts of residual cellular adenosine triphosphate (ATP);
—for cyanobacteria, algae, and lichens, determination of their vitality by measurement of chlorophyll fluorescence *in situ* with fluorimeters; alternatively, observations in the laboratory under fluorescence microscope (Plate 31).

Fig. 9.8 In-situ *test to assess the efficacy of biocides on frescoes colonized by algae (Lower Basilica of San Clemente, Rome)* – Photo G. Zecca

9.2.1b *Evaluation of Biocide-Substrate Interactions*

Most of the tests aiming to evaluate the biocide-substrate interactions cannot be carried out directly on the work of art because they are often destructive or because of technical difficulties; consequently, such measurements are carried out almost exclusively in the laboratory. However, there is no standard procedure for this evaluation yet; usually, samples of the same nature

Fig. 9.9 In-situ *tests on testing strips to assess the efficacy of herbicides (Baths of Caracalla, Rome)* – Photo G. Caneva

The only measurement that can be carried out easily in the field as well, because it does not require sampling, is that of color using portable spectrophotometers or reflectance colorimeters (Nugari et al. 1993a; Tiano et al. 1994; Caneva et al. 1996; Nugari 1999).

The objective of these measurements and analyses is to attempt to establish any contraindications for the use of certain products on the substrate. The more these analyses are carried out and with increasingly sophisticated equipment, the more obvious it becomes that no kind of chemical treatment is completely innocuous for the characteristics of the substrate (Prieto et al. 1996; Sameno Puerto et al. 1996; Altieri et al. 1997).

To verify the effects of treatments over time, the same test samples can then be used to carry out accelerated aging tests.

or composition as the constituent materials are treated by brushing, capillary absorption, or immersion in biocide solutions, which are generally more concentrated than those normally used (Fig. 9.10). To perform the testing, it is necessary to know all the chemical and physical characteristics of the biocide in question; these can be drawn from the technical sheet, although in many cases it is useful to confirm them with tests, especially on the solutions that are to be tested. Before and after the application of the biocide, measurements and analyses are carried out to characterize the substrate; these may vary depending on the nature of the substrate, and among them we would like to mention:

—color measurements;
—observation of the morphological characteristics of the surface;
—evaluation of resistance to folding and/or bending;
—determination of the degree of polymerization (DP);
—measurement of porosity;
—determination of surface roughness;
—measurement of surface pH;
—assessment of water absorption over time;
—measurement of conductivity;
—quantitative/qualitative analysis of soluble salts.

9.2.2 Evaluation of the Susceptibility to Biodeterioration of Conservation Products

Materials employed in conservation (adhesives, carriers, consolidants, and water-repellent substances) can undergo a number of laboratory tests to establish:

—their susceptibility to biological colonization;

Fig. 9.10 *Tests to evaluate the degree of interference of various biocides on test samples of plaster* – Photo P. Tiano

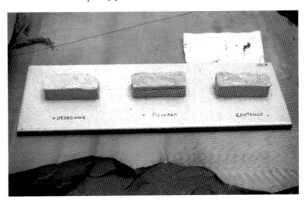

—the changes in their performance properties induced by microbial growth.

There are no products in existence for which it
could be excluded a priori that they may serve
as growth substrate for microorganisms, a
situation that could occur should environmental conditions become particularly favorable,
especially in terms of relative humidity; products should, however, remain stable during use,
even when conditions are favorable to microbial
development. There are no standard procedures
specific to the conservation sector for these
kinds of verifications, either. However, various
types of tests, on the products themselves or on
the products applied on the substrates, can be
developed, by referring to standards used for
the evaluation of materials in other fields and
adapting them to conservation.

In the tests, materials are generally subjected to the action of microorganisms with
high biodeteriogenic capacities; among these,
fungi are the most frequently used, in addition to bacteria. Indeed, because of their
decomposing ability, fungi have the capacity to metabolize the most diverse substrates,
including certain synthetic resins and the plastifiers they contain. In the standards for evaluating the resistance to biological degradation,
depending on the substance being examined,
various species are proposed that are recognized as being among the most aggressive for
the different materials; for example, in the
U.S. standards, the fungi used in testing polymeric synthetic materials are *Aspergillus niger*,
Penicillium pinophylum, *Chaetomium globosum*,
Gliocladium virens, *Aureobasidium pullulans*
(ASTM G21, 1990).

9.2.2a *Tests on the Products Themselves*

To verify the susceptibility to biodeterioration
of conservation materials, these are placed into
containers made of an inert material, usually
glass (for instance, when testing adhesives), or
they are spread on a glass slide or some other
kind of inert support (in the case of film-
forming substances or substances that require
polymerization) (Nugari and Priori 1985;
Koestler et al. 1986, 1988; Koestler and Santoro
1988; Nugari and Bartolini 1997; Koestler

2000; Tiano et al. 2000; Talarico et al. 2001).
The materials can then be:

—inoculated with a microbial suspension
(a bacterial one or, more often, a fungal
one);
—exposed outdoors to favor the natural
deposit of spores and microbial cells.

Subsequently, the samples can be incubated in
a thermostatically controlled chamber at 28°C
or, better yet, in a climate-controlled chamber
with conditions favorable to triggering microbial development, i.e., with high temperature
and RH, usually $T \geq 28°C$ and $RH \geq 90\%$.

Checks are then carried out at predetermined time intervals, registering the presence
or absence of microbial development, by means
of direct observation and under the stereomicroscope. The percentage of the cover is then
evaluated, i.e., what proportion of the surface
has been covered with microbial growth. The
standard methods—recommended for the
testing of susceptibility to biodegradation—
suggest the use of numerical conversions of
grades of susceptibility, for example from 0 to
5, for the percentage of coverage or for the time
required for the cover or for the degradation of
a material to be complete (ASTM G21, 1990).
It is also possible to determine the nature and
the number of species that have developed and,
in the case of fungi, observe if sporulation has
taken place.

This kind of test is of great interest because,
as it does not add any other nutrient substance
to the materials, it is proof of microbial growth
occurring solely at the expense of the material under examination. However, in the case
of outdoor exposure, the deposit of particles
of various nature can also contribute to the
establishment of growth and must therefore be
taken into account.

9.2.2b *Tests on the Products as Applied onto a Substrate*

All conservation products that might come
into contact with the works, even if only temporarily, should be tested, since they may leave
potentially dangerous residues behind that may
serve as growth substrates for microorganisms

(e.g., carriers for cleaning substances, precon-solidation substances, adhesives used for tem-porary facings, etc.).

The products are applied on specimens of the material prepared for this purpose (Salvadori and Nugari 1988; Leznicka et al. 1991; Sorlini et al. 1991). The application methods (e.g., immersion, capillary absorption, brush appli-cation, etc.) vary according to the characteris-tics and the properties of the product and the purpose of use. It is therefore necessary, when the product requires it, to wait for the solvent to evaporate and for polymerization to occur, which generally happens at room temperature or in ovens.

Thus prepared, test samples can further undergo the following:

—inoculation with a known microbial
  suspension;
—outdoor exposure in order to encourage
  the natural deposit of microbial spores
  and cells;
—soil burial test, i.e., immersion in
  garden soil with known microbial
  concentration.

Test samples are then incubated in conditions favorable to microbial growth, or in a ther-mostatically controlled environment (set at 28°C), or else in a climate-controlled cham-ber (usually with T ≥ 28°C and RH ≥ 90%). Checks are then carried out at predetermined time intervals, registering the presence or absence of microbial development, by means of direct observation and under the stereomi-croscope (Figs. 9.11 and 9.12). The percentage of the surface covered by microbial coloniza-tion is then evaluated, as are the morphologi-cal characteristics of the growth, the number of species that have developed and, in the case of fungi, whether or not sporulation has taken place.

In comparison to the previous tests (earlier in the chapter, section 2.2a), these tests have the advantage of more closely approaching the conditions found in practice, since the substrate on which the product is used is part of the test, and therefore allows an evaluation of the over-all effect; the amount of product—often very low in quantity—as well as the nature of the substrate can strongly influence the suscepti-bility to biological deterioration.

For those products that have to be applied in damp environments or in conditions of high levels of biological contamination—as, for example, consolidants and protective sub-stances used in hypogean environments—it might be useful to simulate the conditions of usage in practice. Treated test samples can be exposed in the same sites where the treatment will occur and periodically checked over a specific period of time (30–40 days). The con-ditions of exposure, humidity, and tempera-ture are not controlled and can fluctuate, but they constitute the actual conditions in which the products will be used and must be able to guarantee a high level of resistance to biologi-cal degradation. These tests represent there-fore a further opportunity for the verification of the products prior to direct application onto the work.

Fig. 9.11 *Marble samples treated with water repellents, showing different fungal growths according to the resins tested (far right sample is a control)* – Photo O. Salvadori

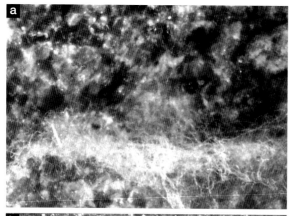

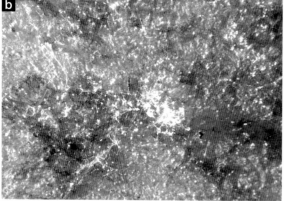

Fig. 9.12 *Fungal growth on a resin applied to (a) sandstone and (b) marble* – Photo O. Salvadori

9.2.2c *Modifications of the Performance Characteristics*

The assay of biological growth on a certain product does not necessarily imply that such a growth will interfere with its performance characteristics, i.e., its adhesive, water-repellent, consolidant powers, etc. In order to check these, it is necessary to select the most appropriate tests, which vary depending on the category of product under investigation; the samples must be tested before and after the application of the product as well as after biological growth. The investigations are essentially chemical and physical tests aimed at identifying any modifications induced by the biological growth on the properties of the product necessary for the intended use (the same commercial product can be used for different purposes, for instance, acrylic resins are used as protective coatings, adhesives, or consolidants).

Studies of this type are very rare; as an example, below here we cite a methodology for the evaluation of water-repellent substances used in the treatment of stone (Commissione NORMAL 1993; Appolonia et al. 1995). On previously treated specimens of stone—with known mineralogical and petrographic characteristics—after applying the product and after carrying out the tests evaluating the susceptibility to biodeterioration (and completely removing any microbial growth that may have occurred), the following parameters were measured:

—mass by weight;
—water absorbed through capillarity over time;
—water absorbed through capillarity in a specified period of time;
—water absorbed in conditions of low pressure over time;
—water lost through evaporation of the absorbed water;
—water-repellence (measurement of contact angle);
—permeability to water vapor;
—color of surface.

At times, the residues of biological colonization are such that they interfere with the validity of the chosen testing method; when this happens, it is necessary to adapt the investigations to each individual case, according to what can be realistically actually accomplished. For example, residues of glycopolysaccharides of biological origin—which are frequently difficult to remove from surfaces—interact with tests for water-repellence by absorbing water; or, residues of pigmented cells prevent a correct measurement of the color alterations or of the transparency of the resins being tested.

BIBLIOGRAPHY

AA.VV. [multiple authors], 1989. *Atti del Convegno—Le pellicole ad ossalato: origine e significato nella conservazione delle opere d'arte.* Centro CNR "Gino Bozza," Milano.

AA.VV., 2002a. *The Manual of Museum Exhibitions.* Ed. Altamira, London, New York.

AA.VV., 2002b. *Annali del Laboratorio Museotecnico III.* Ed. Goppion, Trezzano sul Naviglio, Milano: 255–295.

ABBATE EDLEMANN M.L., GAMBETTA A., GIACHI G., ORLANDI E., 1989. *Studio di alcune specie legnose appartenenti ad un relitto navale del VII sec. a.C., effettuato con il microscopio elettronico a scansione.* In: TAMPONE G. (Ed.), *Il Restauro del Legno*, Nardini Editore, Firenze, Vol. 1: 121–127.

ABBRUSCATO P., SORLINI C., ZANARDINI E., 2003. *Molecular characterization of lead-resistant isolates from Certosa of Pavia red stains.* In: SAIZ-JIMENEZ C. (Ed.), *Molecular Biology and Cultural Heritage, Proceedings of International Congress on Molecular Biology and Cultural Heritage,* Seville. Balkema Publishers, Lisse (NL): 109–113.

ABDEL-KAREEM O., 2000. *Microbiological testing of polymers and resins used in conservation of linen textiles.* In: *15th World Conference on Nondestructive Testing,* Associazione Italiana Prove non Distruttive, Roma.

ABDULLA H., DEWEDAR A., MAY E., 1999. *Studies on Actinomycetes as colonisers of stone in an Egyptian tomb.* International Biodeterioration & Biodegradation, *44 (4):* 165.

ABOAL M., ASENCIO A.D, PREFASI M., 1994. *Studies on cave cyanophytes from southeastern Spain:* Scytonema julianum *(Meneghini ex Franck) Richter.* Algological Studies, *75:* 31–36.

ACCARDO G., VIGLIANO G., 1989. *Strumenti e materiali del restauro.* Kappa, Roma.

ACCARDO G., GIANI E., SECCARONI C., 1995. *Evoluzione della modellistica di vetrine per la conservazione di manufatti artistici.* Materiali e Strutture: problematiche di conservazione, *3:* 115–126.

ACCARDO G., CACACE C., GIANI E., GIOVAGNOLI A., NUGARI M.P., 2003. *Museum collections: data sheets for improved management.* In: *Conservation Science 2002. Papers from the Conference held in Edinburgh 2002,* Scotland: 30–43.

ADAMO P., MARCHETIELLO A., VIOLANTE P., 1993. *The weathering of mafic rocks by lichens.* Lichenologist, *25 (3):* 285–297.

ADDADI L., WEINER S., 1990. *Stereochemical and structural relations between macromolecules and crystals in biominer-alization.* In: MANN S., WEBB J., WILLIAMS R.J.P. (Eds.), *Biomineralization: Chemical and Biochemical Perspectives.* Friburg: 133–156.

AGAROSSI G., 1994. *Biodeterioramento in ambienti ipogei: esperienze e considerazioni.* In: GUIDOBALDI F. (Ed.), *Studi e ricerche sulla conservazione delle Opere d'Arte dedicati alla memoria di Marcello Paribeni.* CNR, Roma: 1–18.

AGAROSSI G., FERRARI R., MONTE M., 1985. *Microbial biodeterioration in the hypogea: the subterranean neopythagorean Basilica of Porta Maggiore in Rome.* In: FELIX G. (Ed.), *Proceedings of 5th International Congress on Deterioration and Conservation of Stone.* Presses Polytechniques Romandes, Lausanne: 597–605.

AGAROSSI G., FERRARI R., MONTE M., 1986. *The Basilica of St. Clemente in Rome: Studies on biodeterioration.* In: *Proceeding of the Symposium on Scientific Methodologies applied to works of art.* Firenze, 1984: 52–56.

AGAROSSI G., FERRARI, R., MONTE M., GUGLIANDOLO C., MAUGERI T., 1988. *Changes of microbial system in an Etruscan tomb after biocidal treatments.* In: *Proceedings of the 6th International Congress on Deterioration and Conservation of Stone, Supplement.* Nicholas Copernicus University Press, Torun, Poland: 82–91.

AGRAWAL O.P., DHAWAN S. (Eds.), 1989. *Proceedings of the International Conference on Biodeterioration of Cultural Property,* Lucknow, India.

AGRAWAL O.P., PATHAK N., 1993. *Biodeterioration of ethnological objects: a review.* INTACH, Indian Conservation Institute.

AGRAWAL O.P., SINGH T., KHARBADE V., JAIN K.K., JOSHI J.P., 1987. *Discoloration of Thaj Mahal marble: A case of study.* In: ICOM *Committee for Conservation, Preprints of the 8th Triennial Meeting,* Sydney, Australia: 447–451.

AHMADJIAN V., 1993. *The Lichen Symbiosis.* John Wiley & Sons Inc., New York.

AKAI H., 1997. *Anti-bacterial function of natural silk materials.* Int. Wild Silkmoth & Silk, *3:* 79–81.

ALBERTANO P., 1991. *The role of photosynthetic microor-ganisms on ancient monuments. A survey on methodological approaches.* In: HICKS S., MILLER U., NILSSON S. (Eds.), *Airborne particles and gases, and their impact on the cultural heritage and its environment,* PACT, *33:* 151–159.

ALBERTANO P., 1998. *Deterioration of Roman hypogea by epilithic cyanobacteria and microalgae.* In: GUARINO A. ET AL. (Eds.), *Proceedings of the 1st International Congress on Science and Technology for the Safeguard of Cultural Heritage in the Mediterranean Basin,* CNR Editions, Palermo, *2:* 1303–1308.

ALBERTANO P., 2002. *How can cyanobacterial biofilms alter stone in hypogean environments?* In: *Art, Biology, and Conservation 2002. Biodeterioration of Works of Art*, The Metropolitan Museum of Art, New York: 100–105.

ALBERTANO P., 2003. *Methodological approaches to the study of stone alteration caused by cyanobacterial biofilms in hypogean environments*. In: KOESTLER R.J., KOESTLER V.R., CHAROLA A.E., NIETO-FERNANDEZ F.E. (Eds.), *Art, Biology, and Conservation: Biodeterioration of Works of Art*, The Metropolitan Museum of Art, New York: 302–315.

ALBERTANO P., BELLEZZA S., 2001. *Cytochemistry of cyanobacterial exopolymers in biofilms from Roman hypogea*. Nova Hedwigia, *123*: 501–518.

ALBERTANO P., BRUNO L., 2003. *The importance of light in the conservation of hypogean monuments*. In: SAIZ-JIMENEZ C. (Ed.), *Molecular Biology and Cultural Heritage, Proceedings of International Congress on Molecular Biology and Cultural Heritage*, Seville. Balkema Publishers, Lisse (NL): 171–177.

ALBERTANO P., GRILLI CAIOLA M., 1989. *A hypogean algal association*. Braun-Blanquetia, *3*: 287–292.

ALBERTANO P., URZÌ C., 1999. *Structural interactions among epilithic cyanobacteria and heterotrophic microorganisms in Roman hypogea*. Microbial Ecology, *38*: 244–252.

ALBERTANO P., KOVACIK L., GRILLI CAIOLA M., 1994. *Preliminary investigations on epilithic cyanophytes from a Roman necropolis*. Arch. Hydrobiol., Algological Studies, *75*: 71–74.

ALBERTANO P., KOVACIK L., MARVAN P., GRILLI CAIOLA M., 1995. *A terrestrial epilithic diatom from Roman catacombs*. In: MARINO D., MONTRESOR M. (Eds.), *Proceedings of the 13th International Diatom Symposium*. Biopress Limited, Bristol: 11–21.

ALBERTANO P., BRUNO L., BELLEZZA S., PARADOSSI G., 2000. *Polysaccharides as a key step in stone bioerosion*. In: FASSINA V. (Ed.), *Proceedings of the 9th International Congress on Deterioration and Conservation of Stone*, Venice. Elsevier, Amsterdam, *1*: 425–432.

ALBERTANO P., MOSCONE D., PALLESCHI G., HERMOSÍN B., SAIZ-JIMENEZ C., SÁNCHEZ-MORAL S., HERNÁNDEZ-MARINÉ M., URZÌ C., GROTH I., SCHROECKH V., SAARELA M., MATTILA-SANDHOLM T., GALLON J.R., GRAZIOTTIN F., BISCONTI F., GIULIANI R., 2003. *Cyanobacteria attack rocks (CATS): control and preventive strategies to avoid damage caused by cyanobacteria and associated microorganisms in Roman hypogean monuments*. In: SAIZ-JIMENEZ C. (Ed.), *Molecular Biology and Cultural Heritage, Proceedings of International Congress on Molecular Biology and Cultural Heritage*, Seville. Balkema Publishers, Lisse (NL): 151–162.

ALEFFI M., 1992. *Biomonitoraggio dell'inquinamento atmosferico tramite briofite: valutazione dell'I.A.P. (Index of Atmospheric Purity) in ambiente urbano. I. Fase metodologica*. Giornale Botanico Italiano, *126 (2)*: 351.

ALESSIO M., ANSELMI S., CONFORTO L., IMPROTA S., MANES F., MANFRA L., 2002. *Radiocarbon as a biomarker of urban pollution in leaves of evergreen species sampled in Rome and in rural areas (Lazio, central Italy)*. Atmospheric Environment, *36*: 5405–5416.

ALLEGRINI M.C., VITALI N., ANGELINI A., VENANZONI R., 1994. *L'uso di alcune Briofite come indicatori di accumulo nel monitoraggio dei metalli pesanti*. Giornale Botanico Italiano, *128*: 275.

ALLSOPP D., SEAL K.J., 1986. *Introduction to biodeterioration*. Edward Arnold, London: 55–61.

ALMEIDA M.T., MOUGA T., BARRACOSA P., 1994. *The weathering ability of higher plants. The case of* Ailanthus altissima *(Miller) Swingle*. International Biodeterioration & Biodegradation, *33 (4)*: 333–343.

ALTIERI A., RICCI S., 1994. *Il ruolo delle briofite nel biodeterioramento di materiali lapidei*. In: FASSINA V., OTT H., ZEZZA F. (Eds.), *Proceedings of the 3rd International Symposium on the Conservation of Monuments in the Mediterranean Basin*, Venice: 329–333.

ALTIERI A., RICCI S., 1997. *Calcium uptake in mosses and its role in stone biodeterioration*. International Biodeterioration & Biodegradation, *40*: 201–204.

ALTIERI A., PIETRINI A.M., RICCI S., 1993. *Un'associazione di alghe e muschi in un sito archeologico ipogeo*. Giornale Botanico Italiano, *127*: 611.

ALTIERI A., COLADONATO M., LONATI G., MALAGODI M., NUGARI M.P., SALVADORI O., 1997. *Effects of biocidal treatments on some Italian lithotype samples*. In: MOROPOULOU A., ZEZZA F., KOLLIAS E., PAPACHRISTODOULOU I. (Eds.), *Proceedings of the 4th International Symposium on the Conservation of Monuments in the Mediterranean*, Rhodes, Technical Chamber of Greece, *3*: 31–40.

ALTIERI A., DE PALMA G., FERRONI A., 1999a. *La vulnerabilità del sito di Tharros. Problemi conservativi e proposte di intervento*. In: ACQUARO E., FRANCISI M.T., KIROVA T.K., MELUCCO VACCARO A. (Eds.), *Studi e ricerche sui beni culturali, Tharros nomen*. La Spezia: 59–70.

ALTIERI A., LAURENTI M.C., ROCCARDI A., 1999b. *The conservation of archaeological sites: materials and techniques for short-term protection of archaeological remains*. In: MARABELLI M., PARISI C. (Eds.), *Proceedings of the 6th International Conference on Non-destructive Testing and Microanalysis for the Diagnostics and Conservation of the Cultural and Environmental Heritage*, EUROMA, Roma: 673–687.

ALTIERI A., LONATI G., MALAGODI M., NUGARI M.P., 1999c. *Colorimetric evaluation of biocidal interference with stone in UV-A aging test*. In: MARABELLI M., PARISI C. (Eds.), *Proceedings of the 6th Conference on Non-destructive Testing and Microanalysis for the Diagnostics and Conservation of the Cultural and Environmental Heritage*. EUROMA, Roma: 107–116.

ALTIERI A., MAZZONI A., PIETRINI A.M., RICCI S., ROCCARDI A., 2000a. *Indagini diagnostiche sul biodeterioramento delle fontane*. In: NATOLI M. (Ed.), *Piazza di Corte, il recupero dell'immagine berniniana*. Fratelli Palombi Ed., Roma: 34–57.

ALTIERI A., PIETRINI A.M., RICCI S., ROCCARDI A., PIERVITTORI R., 2000b. *The temples of the archaeological area of Paestum (Italy): A case study*. In: FASSINA V. (Ed.), *Proceedings of the 9th International Congress on Deterioration and Conservation of Stone*, Venice. Elsevier, Amsterdam: 433–443.

AMMAR M.S., BAKARAT K., HOGHANEM E., EL DEEB A.A., 1987. *Microflora investigations in wall paintings of the tomb of Nefertari*. Ann. Serv. Ant. Egypte: 58–63.

ANASTASI P., BORGIOLI A., MARTINI P., 1984. *Prima fase di ricerche sulla crescita algale in un impianto di potabilizzazione alimentato con acqua di lago*. Ingegneria Sanitaria, 4: 17–23.

APPELBAUM B., 1991. *Guide to Environmental Protection of Collections*. Sound View Press Madison, Connecticut.

APPOLONIA L., MIGLIORINI S., VAUDAN D., 1989. *Lo scialbo degli affreschi di epoca ottoniana della cattedrale di Aosta*. In: *Le pellicole ad ossalato: origine e significato nella conservazione delle opere d'arte*. Centro CNR "Gino Bozza," Milano: 245–254.

APPOLONIA L., FASSINA V., MATTEOLI U., MECCHI A.M., NUGARI M.P., PINNA D., PERUZZI R., SALVADORI O., SANTAMARIA U., SCALA A., TIANO P., 1995. *Methodology for the evaluation of protective products for stone materials. Part II: experimental tests on treated samples*. In: *International Colloquium on Methods of Evaluating Products for the Conservation of Porous Building Materials in Monuments*, ICCROM, Rome: 75–86.

APTROOT A., JAMES P.W., 2002. *Monitoring Lichens on Monuments*. In: NIMIS P.L., SCHEIDEGGER C., WOLSELEY P.A. (Eds.), *Monitoring with Lichens – Monitoring Lichens*. Kluwer Academic Publishers: 239–253.

ARAI H., 1974. *Microbial study on a virgin tumulus*. Kokogaku Zasshi, Journal of the Achaeological Society of Nippon, 59: 328–336.

ARAI H., 1983. *Microbiological studies on the conservation of mural paintings in tumuli*. In: *Proceedings of 7th International Symposium on the Conservation and Restoration of Cultural Property, Conservation and Restoration of Mural Paintings (I)*. Tokyo, Japan: 117–124.

ARAI H., 1984. *Antimicrobial factors found in virgin tumuli*. Biodeterioration, 4: 363–368.

ARAI H., 1987. *Biological investigations*. In: *Wall paintings of the tomb of Nefertari. Scientific studies for their conservation*, The Getty Conservation Institute, Le Caire: 54–57.

ARAI H., 1988. *On microorganisms in the Tomb of Nefertari*. Science for Conservation, 27: 13–20.

ARAI H., 2000. *Foxing caused by Fungi: twenty-five years of study*. International Biodeterioration & Biodegradation, 46: 181–188.

ARANYANAK C., 1992. *Biodeterioration of cultural materials in Thailand*. In: TOISHI K., ARAI H., KENJO T., YAMANO K. (Eds.), *Proceedings of the 2nd International Conference on Biodeterioration of Cultural Property*, Pacifico Yokohama. International Communications Specialists, Tokyo: 23–33.

ARIÑO X., SAIZ-JIMENEZ C., 1996a. *Biological diversity and cultural heritage*. Aerobiologia, 12: 279–282.

ARIÑO X., SAIZ-JIMENEZ C., 1996b. *Lichen deterioration of consolidants used in the conservation of stone monuments*. Lichenologist, 28 (4): 391–394.

ARIÑO X., ORTEGA-CALVO J.J., GOMEZ-BOLEA A., SAIZ-JIMENEZ C., 1995. *Lichen colonization of the Roman pavement at Baelo Claudia (Cadiz, Spain): Biodeterioration vs. bioprotection*. The Science of the Total Environment, 67: 353–363.

ARIÑO X., GOMEZ-BOLEA A., SAIZ-JIMENEZ C., 1997a. *Lichens on ancient mortars*. International Biodeterioration & Biodegradation, 40 (2–4): 217–224.

ARIÑO X., HERNANDEZ-MARINE M., SAIZ-JIMENEZ C., 1997b. *Colonization of Roman tombs by calcifying cyanobacteria*. Phycologia, 36: 366–373.

ARIÑO X., CANALS A., GOMEZ-BOLEA A., SAIZ-JIMENEZ C., 2002. *Assessment of the performance of a water-repellent/biocide treatment after 8 years*. In: GALÁN E., ZEZZA F. (Eds.), *Protection and Conservation of the Cultural Heritage of the Mediterranean Cities*. Proceedings of the 5th International Symposium on the Conservation of Monuments in the Mediterranean Basin, Seville: 121–125.

ARROYO I., ARROYO G., 1996. *Annual microbiological analysis of the Altamira cave (Santillana del Mar), Spain*. In: RIEDERER J. (Ed.), *Proceedings of the 8th International Congress on Deterioration and Conservation of Stone*. Berlin, Germany: 601–608.

ARTIOLI D., GIOVAGNOLI A.M., NUGARI M.P., IVONE A., LONATI G., 2000. *The Doria Pamphilj exhibition gallery: the study of environmental conditions*. In: FASSINA V. (Ed.), *Proceedings of the 9th International Congress on Deterioration and Conservation of Stone*, Venice. Elsevier, Amsterdam 1: 375–381.

ARUZZOLO G., MARINUCCI G., RICCARDI M.L., ROTILI R., RUSCHIONI E., VALENTI P., VECA E., 1997. *Istruzioni tecniche relative alle operazioni di spolveratura di materiale librario e archivistico*. Cabnewsletter-Conservazione negli Archivi e nelle Biblioteche, 6: 8–11.

ASCASO C., 1984. *Structural aspects of lichens invading their substrata*. In: VICENTE C. (Ed.), *Surface Physiology of Lichens*. Universidad Complutense, Madrid: 87–112.

ASCASO C., GALVAN J., RODRIGUEZ-PASCAL C., 1982. *The weathering of calcareous rocks by lichens*. Pedobiologia, 24: 219–229.

ASCASO C., WIERZCHOS J., CASTELLO R., 1998. *Study of the biogenic weathering of calcareous litharenite stones caused by lichen and endolithic microorganisms*. International Biodeterioration & Biodegradation, 42: 29–38.

ASCASO C., WIERZCHOS J., SOUZA-EGIPSY V., DE LOS RIOS A., DELGADO RODRIGUES J., 2002. In situ *evaluation of the biodeteriorating action of microorganisms and the effects of biocides on carbonate rock of the Jeronimos Monastery (Lisbon)*. International Biodeterioration & Biodegradation, 49: 1–12.

ASENCIO A.D., ABOAL M., 2001. *Biodeterioration of wall paintings in the caves of Murcia (SE Spain) by epilithic and chasmoendolithic microalghe*. Algological Studies, 103: 131–142.

ASHRAE P., 1997. *Air Contaminants*. In: *ASHRAE Handbook. Fundamentals*, Chapter 12. American Society of Heating, Refrigerating, and Air-Conditioning Engineers, Atlanta.

ASTM G21, 1990. *Standard Practice for Determining Resistance of Synthetic Polymeric Materials to Fungi*.

ATLAS R.M., CHOWDHURY A.N., LAL GAURI K., 1988. *Microbial calcification of gypsum-rock and sulfated marble*. Studies in Conservation, 33: 149–153.

AUGUSTI S., 1944a. *Alterazioni osservate sugli affreschi dello Zingaro nel chiostro del Platano in Napoli*. Archivio Storico Napoletano, 30: 1–8.

AUGUSTI S., 1944b. *Azione dei microrganismi e dei parassiti sui dipinti murali*. Boll. Soc. Naturalisti in Napoli, 55: 68–73.

AUGUSTI S., 1948. *Alterazioni dei dipinti murali: loro natura e cause.* Tipografia Miccoli, Napoli: 1–29.

AUSSET P., LEFÈVRE R.A., DEL MONTE M., 2000. *Early mechanisms of development of sulphated black crusts on carbonate stone.* In: FASSINA V. (Ed.), *Proceedings of the 9th International Congress on Deterioration and Conservation of Stone,* Venice. Elsevier, Amsterdam: 329–337.

BAER N.S., BANKS P.N., 1995. *Indoor air pollution effects on cultural and historic materials.* The International Journal of Museum Management and Curatorship, 4: 9–20.

BAKER M.T., BURGESS H.D., BINNIE N.E., DERRICK M.R., DRUZIK J.R., 1990. *Laboratory investigation of the fumigant Vikane®.* In: *Proceedings of the 9th Triennal Meeting ICOM,* Dresden. ICOM Committee for Conservation, Los Angeles: 804–811.

BALDI F., 2000. *Il processo di produzione delle paste chimiche e loro trattamento.* 8° Corso di Tecnologia per tecnici cartari. Scuola Interregionale di Tecnologia per Tecnici Cartari, Verona.

BALZAROTTI KÄMMLEIN R., SANSONI M., CASTRONOVO A., 1999. *An innovative water-compatible formulation of Algophase® for treatment of mortars.* In: *An International Conference on Microbiology and Conservation (ICMC '99), Of Microbes and Art. The Role of Microbial Communities in the Degradation and Protection of Cultural Heritage,* Florence, Italy: 217–220.

BARCELLONA VERO L., TABASSO LAURENZI M., 1979. *La Fontana del Tritone di L. Bernini, a Roma: un esempio di alterazione legato a fattori chimici, biologici e ambientali.* In: *Proceedings of the 3rd International Congress on Deterioration and Conservation of Stone,* Venice: 511–516.

BARCELLONA VERO L., MONTE SILA M., SILVERI A., 1973. *Influenza dell'azione dei solfobatteri nei processi di alterazione dei materiali lapidei.* In: URBANI G. (Ed.), *Problemi di Conservazione.* Edizioni Compositori, Bologna: 439–451.

BARTOLI A., 1990. *I licheni della Peschiera dei Tritoni nell'Orto Botanico di Roma, Villa Corsini.* Giornale Botanico Italiano, 125(3): 87.

BARTOLINI M., MONTE M., 2000. *Chemiolithotrophic bacteria on stone monuments: cultural methods set up.* In: FASSINA V. (Ed.), *Proceedings of the 9th International Congress on Deterioration and Conservation of Stone,* Venice. Elsevier, Amsterdam: 453–460.

BARTOLINI M., RICCI S., 2004. *Rilascio di pigmenti fotosintetici da biocenosi epilitiche trattate con biocidi.* Kermes, 56: 63–68.

BARTOLINI M., PIETRINI A.M., RICCI S., 1999. *Use of UV-C irradiation on artistic stonework for control of algae and Cyanobacteria.* In: *An International Conference on Microbiology and Conservation (ICMC'99), Of Microbes and Art. The Role of Microbial Communities in the Degradation and Protection of Cultural Heritage.* Florence, Italy: 221–227.

BARTOLINI M., NUGARI M.P., PANDOLFI A.M., 2000a. *Resistance to biodeterioration of some chemical products used for rising damp barrier.* In: GALÁN E., ŽEZZA F. (Eds.), *Protection and Conservation of the Cultural Heritage of the Mediterranean Cities. Proceedings of the 5th International Symposium on the Conservation of Monuments in the Mediterranean Basin,* Seville: 397–400.

BARTOLINI M., NUGARI M.P., PANDOLFI A., SANTAMARIA U., 2000b. *Lo sbarramento chimico all'umidità ascendente mediante prodotti silossanici: risultati sperimentali.* Bollettino ICR Nuova Serie, 1: 55–62.

BARUFFO L., TRETIACH M., ZEDDA L., LEUCKERT CH., 2001. *Sostanze licheniche: come riconoscerle e perché.* Notiziario della Società Lichenologica Italiana, 14: 5–33.

BASSI M., BARBIERI N., BONECCHI R., 1984. *St. Christopher church in Milan. 2. Biological investigations.* Arte Lombarda, 68–69: 8–12.

BASSI M., FERRARI A., REALINI M., SORLINI C., 1986. *Red stains on the Certosa of Pavia: a case of biodeterioration.* International Biodeterioration, 22: 201–205.

BAYNES-COPE A.D., 1971. *The choice of biocides for library and archival materials.* In: WALTERS A.H., HUECK-VAN DER PLAS H. (Eds.), *Biodeterioration of materials.* John Wiley and Sons, New York, 2: 381–387.

BAZZAZ F.A., 1996. *Plants in changing environments: linking physiological, population, and community ecology.* Cambridge University Press, Cambridge.

BEARLE J.W.S., LOMAS B., COOK W.D., 1998. *Atlas of fibre fracture and damage to textiles.* The Textile Institute. CRC Press, Woodhead Publishing Limited Cambridge England.

BEGON M., HARPER J.L., TOWNSEND C.R., 1989. *Ecologia: individui, popolazioni, comunità.* Zanichelli, Bologna.

BELLEZZA S., ALBERTANO P., 2003. *A Chroococcalean species from Roman hypogean sites: characterisation of* Gloeothece membranacea *(Cyanobacteria, Synechoccaceae).* Arch. Hydrobiol., Algological Studies, 109: 103–112.

BELLINZONI A.M., CANEVA G., RICCI S., 2003. *Ecological trends in travertine colonisation by pioneer algae and plant communities.* International Biodeterioration & Biodegradation, 51: 203–210.

BELLUCCI R., CREMONESI P., 1994. *L'uso degli enzimi nella conservazione e nel restauro dei dipinti.* Kermes, 21: 45–64.

BELOYANNIS N., 1985. *Physico-chemical aspects of the Acropolis problem: marble deterioration. Forms, causes and means of prevention.* In: *L'Acropoli di Atene: Conservazione e restauro.* Ed. Scientifiche Italiane: 81–84.

BERARDI C., GIULIANI M.R., 2002. *Tessuti – caratterizzazione dei materiali costitutivi e valutazione dello stato di conservazione.* In: *Il Sarcofago dell'Imperatore. Studi, ricerche e indagini sulla tomba di Federico II nella Cattedrale di Palermo. 1994–1999.* Regione Siciliana, Officine Grafiche Riunite, Palermo: 266–271.

BERNARDI A., 2002. *Metodologia per un'analisi microclimatica per la conservazione delle opere d'arte negli ambienti interni.* In: CASTELLANO A., MARTINI M., SIBILIA E. (Eds.), *Elementi di Archeometria.* Egea, Milano: 343–368.

BERNARDI A., 2003. *Conservazione opere d'arte. Il microclima negli ambienti museali.* Il Prato, Padova.

BERNARDINI C., 1993. *Biocidi e prevenzione microbiologica: alcune osservazioni di cantiere.* Kermes, 16: 12–19.

BEROVIC M., 2003. *Biodeterioration studies on pastels and oil-based paintings.* In: KOESTLER R.J., KOESTLER V.H., CHAROLA A.E., NIETO-FERNANDEZ F.E. (Eds.), *Art, Biology, and Conservation: Biodeterioration of Works of*

Art. The Metropolitan Museum of Art, New York: 50–59.

BERSELLI S., GASPARINI L., 2000. *L'archivio fotografico: manuale per la conservazione e la gestione della fotografia antica e moderna*. Zanichelli, Bologna.

BERTA A., CHIAPPINI M., 1978. *Primo contributo alla conoscenza speleobiologica vegetale della Sardegna*. In: *Atti 22° Congresso Soc. Ital. Biogeogr*. Morisia, *4*: 1– 27.

BERTI S., 2000. *The conservation of the Dunarobba tree trunks*. In: *Proceedings of the International Congress on Science and Technology for the Safeguard of Cultural Heritage in the Mediterranean Basin*, Paris 1999. Elsevier: 803–804.

BETTINI C., 1988. *Il controllo della vegetazione nelle aree archeologiche: problematiche conservative, aspetti metodologici ed esperienze applicative*. In: BISCONTIN G., VASSALLO E., VOLPIN S. (Eds.), *Le scienze, le istituzioni, gli operatori alla soglia degli anni '90, Atti Convegno Scienza e Beni Culturali*, Bressanone. Libreria Progetto Editore, Padova: 207–220.

BETTINI C., CINQUANTA A., 1990. *Vegetazione e monumenti. Esigenze e metodologie nel controllo delle infestanti ruderali*. Union Printing, Viterbo.

BETTINI C., BONADONNA L., CARRUBA G., GIACOBINI C., SCIOTI A.M., 1982. *Un'indagine relativa alla carica microbica dei dipinti murali della Cappella degli Scrovegni*. Bollettino d'Arte. Serie Speciale "Giotto a Padova," *2*: 221–233.

BETTINI C., AGAROSSI G., FERRARI R., MONTE M., 1988. *Fenomeni di biodeterioramento in ambienti ipogei dipinti. esperienze di controllo di alcune specie microbiche*. In: *Proceedings of the 2nd International Conference on Non-destructive Testing, Microanalytical Methods and Environment Evaluation for Study and Conservation of Works of Art*, Perugia, session III/1. Comas Grafica, Roma: 1–14.

BETTINI V., 1996. *Elementi di ecologia urbana*. Einaudi, Torino.

BICCHIERI M., PAPPALARDO G., ROMANO F.P., SEMENTILLI F.M., DE ACUTIS R., 2001. *Characterization of foxing stains by chemical and spectro-metric methods*. Restaurator, *22*:1–19.

BIONDI E., BRUGIAPAGLIA E., 1991. Taxodixylon gypsaceum *in the fossil forest of Dunarobba (Umbria, Central Italy)*. Flora mediterranea, *1*: 111–120.

BISHT A.S., 1985. *Conservation of ethnographic collections in museums with special reference to wooden artifacts*. Conservation of Cultural Property in India, *18–20*: 48–53.

BJELLAND T., SÆBØ L., THORSETH I.H., 2002. *The occurence of biomineralization products in four lichen species growing on sandstone in western Norway*. Lichenologist, *34 (5)*: 429–440.

BJORDAL C.G., NILSSON T., DANIEL G., 1999. *Microbial decay of waterlogged wood found in Sweden applicable to archaeology and conservation*. International Biodeterioration & Biodegradation, *43*: 63–73.

BLANCHETTE R.A., 1995. *Biodeterioration of archaeological wood*. CAB Biodeterioration Abstracts, *9*: 113–127.

BLANCHETTE R.A., 2000. *A review of microbial deterio-ration found in archaeological wood from different environments*. International Biodeterioration & Biodegradation, *46*: 189–204.

BLANCHETTE R.A., NILSSON T., DANIAL G., ABAD A., 1990. *Biological degradation of wood*. In: ROWELL R.M., BARBOUR R.J. (Eds.), *Archaeological Wood Properties, Chemistry, and Preservation*. Advances in Chemistry Series, *225*: 141–170.

BLANCHETTE R.A., HAIGHT J.E., CEASE K., HELD B.W., SIMPSON E., LIEBHART R., SAMS G.K., 1999. *Deterioration found in Tumulus MM, an 8th century BC tomb at Gordion, Turkey*. International Biodeterioration & Biodegradation, *44 (4)*: 162.

BLAZQUEZ A.B., LORENZO J., FLORES M., GÓMEZ-ALARCÓN G., 2000. *Evaluation of the effect of some biocides against organisms from historic monuments*. Aerobiologia, *16*: 423–428.

BLOMQUIST G., PALMGREN U., STRÖM G., 1984. *Improved techniques for sampling airborne fungal particles in highly contaminated environments*. Scandinavian Journal of Work, Environment, and Health, *10*: 253–258.

BLOMQUIST G., STRÖM G., STRÖMQUIST L.H., 1984b. *Sampling of high concentrations of aiborne fungi*. Scandinavian Journal of Work, Environment, and Health, *10*: 109–113.

BOLD H.C., WYNNE M.J., 1978. *Introduction to the Algae. Structure and Reproduction*. Prentice-Hall, Englewood Cliffs, New Jersey.

BOLIVAR F.C., SANCHEZ-CASTILLO P.M., 1997. *Biomineralization Processes in the Fountains of the Alhambra, Granada, Spain*. International Biodeterioration & Biodegradation, *40 (2–4)*: 205–215.

BONAVENTURA M.P., DESALLE R., EVELEIGH D.E., BALDWIN A.M., KOESTLER R.J., 2003. *Studies of fungal infestations of Tiffany's drawings: limits and advantages of classical and molecular techniques*. In: KOESTLER R.J., KOESTLER V.H., CHAROLA A.E., NIETO-FERNANDEZ F.E. (Eds.), *Art, Biology, and Conservation: Biodeterioration of Works of Art*. The Metropolitan Museum of Art, New York: 95–107.

BONDONNO A., VON HOLY A., BAECKER A.A.W., 1989. *Effects of* Desulfovibrio *and* Thiobacillus *biofilms on the corrosion of electroless nickel plated mild steel*. International Biodeterioration, *25 (4)*: 285–298.

BONETTI M.F., GALLO F., MAGAUDDA G., MARCONI C., MONTANARI M., 1979. *Essais sur l'utilisation des rayons gammas pour la stérilization des matériaux libraires*. Studies in Conservation, *24*: 59–68.

BONI G., 1917. *Flora delle ruine*. La Nuova Antologia: 27–35.

BONOMI R., 1994. *Utilizzo degli enzimi per il restauro di una scultura in terracotta policroma*. OPD Restauro, *6*: 101–107.

BOONE D., CASTENHOLZ R.W., GARRITY G.M., 2001. *Bergey's Manual of Systematic Bacteriology*, Vol. 1. Springer Verlag, New York.

BOQUET E., BORONAT A., RAMOS-CORMENZANA A., 1973. *Production of calcite (calcium carbonate) crystals by soil bacteria is a general phenomenon*. Nature, *246*: 527–529.

BOUSTEAD W., 1963. *The Conservation of works of art in tropical and sub-tropical zones*. In: *Recent Advances in Conservation*. Butterworths, London: 73–78.

BRAGANTINI I., 1999. *L'acqua, il giardino e la grotta nel mondo romano tra l'età tardo-repubblicana e la prima età imperiale*. In: LAPI BALLERINI I., MEDRI L.M. (Eds.), *Atti V Conv. Intern. su Parchi e Giardini Storici "Artifici d' acque e giardini. La cultura delle grotte e dei ninfei in Italia e in Europa,"* Firenze-Lucca, 1998: 20–24.

BRAUN-BLANQUET J., 1928. *Pflanzensoziologie.* Springer Verlag, Wien.

BRESSAN G., BABBINI L., GHIRARDELLI L., 1994. *Le alghe: archivi di storia.* In: *Operazione Iulia Felix. Lo scavo subacqueo della nave romana al largo di Grado.* Ed. della Laguna: 67–70.

BRIGHTMAN F.H., SEAWARD M.R.D., 1978. *Lichens on man-made substrates.* In: SEAWARD M.R.D. (Ed.) *Lichen Ecology.* Academic Press, London: 253–293.

BRIMBLECOMBE P., 1990. *The composition of museum atmospheres.* Atmospheric Environment, *24B (1):* 1–8.

BROKERHOF A.W., 1989. *Control of fungi and insects in objects and collections of cultural value. "A state of the art."* Central Research Laboratory for Objects of Art and Science, Amsterdam: 1–77.

BROWN P.W., MASTERS L.W., 1980. *Factors affecting the corrosion of metals in the atmophere.* In: AILOR W.H. (Ed.), *Atmospheric Corrosion.* John Wiley and Sons, Hollywood: 31–49.

BROWN S., COLE I., VINOD D., KING S., PEARSON C., 2002. *Guidelines for environmental control in Cultural Institutions.* Consortium for Heritage Collections and their Environment. Commonwealth Department of Communications, Information Technology and the Arts, Australia.

BRUNI S., CARIATI F., BIANCHI C.L., ZANARDINI E., SORLINI C., 1995. *Spectroscopic investigation on red stains affecting the Carrara marble façade of the Certosa of Pavia.* Archeometry, *37 (2):* 249–255.

BRUNO L., PIERMARINI S., ALBERTANO P., 2001. *Characterisation of spectral emission by cyanobacterial biofilms in the Roman Catacombs of Priscilla in Rome (Italy).* Nova Hedwigia, *123:* 229–236.

BULLINI L., PIGNATTI S., VIRZO DE SANTO A., 1998. *Ecosistemi artificiali.* In: *Ecologia generale.* UTET, Torino: 436–439.

BURGE H., 1990. *Bioaerosol: prevalence and health effects in the indoor environment.* Journal of Allergy and Clinical Immunology, *86 (5):* 687–701.

BURGE H., 1995. *Bioaerosols.* Lewis Publishers CRC, London.

BUSSOTTI F., GROSSONI P., BATISTONI P., FERRETTI M., CENNI E., 1995. *Preliminary studies on the ability of plant barriers to capture lead and cadmium of vehicular origin.* Aerobiologia, *11:* 11–18.

BUSWELL J.A., 1991. *Fungal degradation of lignin.* In: ARORA D.K., RAI B., MUKERJI, KNUDSEN G.R. (Eds.) *Handbook of Applied Mycology.* Marcel Dekker, New York, *1:* 425–480.

BUTTAZZONI N., CASOLI A., CREMONESI P., ROSSI P., 2000. *Preparazione e utilizzo di gel enzimatici, reagenti per la pulitura di opere policrome.* Progetto Restauro, *16:* 11–19.

BUTTERFIELD F.J., 1987. *The potential long-term effects of gamma irradiation on paper.* Studies in Conservation, *32:* 181–192.

CACACE C., 2002. *Progetto microclimatico di apertura del Sarcofago di Federico II.* In: *Il Sarcofago dell'Imperatore. Studi, ricerche e indagini sulla tomba di Federico II nella Cattedrale di Palermo. 1994–1999.* Regione Siciliana. Officine Grafiche Riunite, Palermo: 228–231.

CALDERON O.H., STAFFELDT E.E., COLEMAN C.B., 1968. *Metal-organic acid corrosion and some mechanisms associated with these corrosion processes.* In: *Biodeterioration of materials, Proceedings of the 1st International Symposium,* Southampton. Elsevier Pub., Amsterdam: 356–363.

CAMUFFO D., 1990. *Ambienti e musei. Microclimatologia di ambienti chiusi e conservazione di opere pittoriche.* In: *Atti dell'Accademia Nazionale dei Lincei.* Giornata dell'Ambiente, *82:* 157–1666.

CAMUFFO D., 1998. *Microclimate for Cultural Heritage.* Elsevier, Amsterdam.

CAÑAVERAS J.C., HOYOS M., SANCHEZ-MORAL S., SANZ-RUBIO E., BEDOYA J., SOLER V., GROTH, I, SCHUMANN P., LAIZ L., GONZALEZ I., SAIZ,-JIMENEZ C., 1999. *Microbial communities associated with hydromagnesite and needle-fiber aragonite deposits in karstic cave (Altamira, Northern Spain).* Geomicrobiology Journal, *16:* 9–25.

CANEVA G., 1985. *Ruolo della vegetazione nella degradazione di murature ed intonaci.* In: *Atti del Convegno Scienza e Beni Culturali, L'intonaco. Storia, Cultura e Tecnologia,* Bressanone: 199–209.

CANEVA G., 1990. *Il problema della crescita delle radici degli alberi nella conservazione di monumenti ipogei. Gli Orti Farnesiani sul Palatino.* In: *Roma Antica 2.* Ecole Francaise de Rome, Soprintendenza Archeologica di Roma, Roma: 687–719.

CANEVA G., 1991. *Il problema della crescita di* Ailanthus altissima *(Miller) Swingle nelle zone archeologiche e monumentali.* In: BISCONTIN G., MIETTO D.E. (Eds.), *Le pietre nell'architettura: struttura e superfici, Atti Convegno Scienza e Beni Culturali,* Bressanone. Libreria Progetto Editore, Padova: 225–234.

CANEVA G., 1993. *Ecological approach to the genesis of calcium oxalates patinas on stone monuments.* Aerobiologia, *9:* 149–156.

CANEVA G., 1994. *Il problema delle radici degli alberi nella conservazione degli ambienti ipogei.* In: GUIDOBALDI F. (Ed.), *Studi e ricerche sulla conservazione dei monumenti in memoria di Marcello Paribeni,* CNR Opere d'arte, Roma: 41–65.

CANEVA G., 1997. *A botanical approach to the planning of archaeological parks in Italy.* In: *Conservation and management in archaeological areas.* James & James, London, Vol. 3: 127–134.

CANEVA G., 1999a. *Ipotesi sul significato simbolico del giardino dipinto della Villa di Livia (Primaporta, Roma).* Bollettino della Commissione Archeologica Comunale di Roma C: 63–80.

CANEVA G., 1999b. *A botanical approach to the planning of archaeological parks in Italy.* Conservation and Management of Archaeological Sites, *3 (3):* 127–134.

CANEVA G., 2004. *Complessità degli aspetti gestionali delle comunità vegetali in aree archeologiche: il caso del Palatino.* In: *Atti del 40° Congresso della Società Italiana di Fitosociologia,* Roma: 9–11.

CANEVA G., ALTIERI A., 1988. *Biochemical mechanisms of stone weathering induced by plant growth.* In: *Proceedings of the 6th International Congress on Deterioration and Conservation of Stone.* Nicholas Copernicus University Press, Torun, Poland: 32–44.

CANEVA G., CUTINI M., 1998. *Palatino: trasformazioni ambientali ed aspetti floristico-vegetazionali legati ai problemi archeologici*. In: Giavarini C. (Ed.), *Il Palatino, area sacra sud-ovest e Domus Tiberiana*. CISTEC, Ed. L'Erma di Bretschneider, Roma: 195–258.

Caneva G., Cutini M., 1999. *La flora spontanea ed ornamentale di grotte e ninfei*. In: Lapi Ballerini I., Medri L.M. (Eds.), *Atti V Conv. Intern. su Parchi e Giardini Storici "Artifici d' acque e giardini. La cultura delle grotte e dei ninfei in Italia e in Europa,"* Firenze-Lucca 1998: 261–267.

Caneva G., Galotta G., 1994. *Floristic and structural changes of plant communities of the Domus Aurea (Rome) related to a different weed control*. In: Fassina V., Ott H., Zezza F. (Eds.), *La conservazione dei monumenti nel bacino del Mediterraneo, Atti del 3° Simposio Internazionale*, Venice: 317–322.

Caneva G., Pinna D., 2001. *Il biodeterioramento dei monumenti*. In: Catizone P., Zanin G. (Eds.), *Malerbologia*. Patron Editore, Bologna: 879–899.

Caneva G., Roccardi A., 1989. *Harmful flora in the conservation of Roman monuments*. In: Agrawal O.P., Dhawan S. (Eds.), *Proceedings of the International Conference on Biodeterioration of Cultural Property*, Lucknow 1989. Rajkamal Eletric Press, New Delhi: 315–324.

Caneva G., Salvadori O., 1989. *Sistematica e sinsistematica delle comunità vegetali nella pianificazione di interventi di restauro*. In: *Atti Convegno Scienza e Beni Culturali "Il cantiere della conoscenza, il cantiere del restauro,"* Bressanone. Libreria Progetto Editore, Padova: 325–335.

Caneva G., Roccardi A., Marenzi A., Napoleone I., 1989. *Correlation analysis in the biodeterioration of stone artworks*. International Biodeterioration, 25: 161–167.

Caneva G., Dinelli A., De Marco G., Vinci M., 1990. *Halophilous vegetation in the deterioration of stone monuments in coastal environments*. In: *Proceedings of the International Congress on Deterioration of Monuments in the Mediterranean Basin*, Bari 1989. Grafo Ed., Brescia: 231–234.

Caneva G., Gori E., Danin A., 1992a. *Incident rainfall in Rome and its relation with biodeterioration of buildings*. Atmospheric Environment, 26B: 255–259.

Caneva G., De Marco G., Dinelli A., Vinci M., 1992b. *The wall vegetation of the Roman archaeological areas*. Science and Technology in Cultural Heritage, 1: 217–226.

Caneva G., Nugari M.P., Ricci S., Salvadori, O., 1992c. *Pitting of marble Roman monuments and the related microflora*. In: Delgado Rodriguez J., Henriques F., Telmo Jeremias F. (Eds.), *Proceedings of the 7th International Congress on Deterioration and Conservation of Stone*. LNEC, Lisbon, Portugal: 521–530.

Caneva G., De Marco G., Pontrandolfi M.A., 1993. *Plant communities on the walls of Venosa Castle (Basilicata, Italy) as biodeteriogens and bioindicators*. In: Thiel M.J. (Ed.), *Proceedings of International UNESCO/RILEM Congress on Conservation of Stone and Other Materials*, Paris. E & FN Spon, London: 263–270.

Caneva G., Nugari M.P., Salvadori O., 1994a. *La biologia nel restauro*. Nardini Editore, Firenze.

Caneva G., Danin A., Ricci S., Conti C., 1994b. *The pitting of Trajan's column in Rome: An ecological model of its origin*. In: *Conservazione Patrimonio culturale II*. Contributi Centro Linceo Interdisciplinare Beniamino Segre, Accademia Nazionale dei Lincei, Roma, 88: 77–102.

Caneva G., De Marco G., Dinelli A., Vinci M., 1995a. *Le classi Parietarietea diffusae (Rivas Martinez, 1964) Oberd. 1977 e Adiantetea Br.-Bl. 1947 nelle aree archeologiche romane*. Fitosociologia, 29: 165–179.

Caneva G., Gori E., Montefinale T., 1995b. *Biodeterioration of monuments in relation to climatic changes in Rome between 19–20th centuries*. The Science of the Total Environment, 167: 205–214.

Caneva G., Nugari M.P., Pinna D., Salvadori O., 1996. *Il controllo del degrado biologico. I biocidi nel restauro dei materiali lapidei*. Nardini Editore, Firenze.

Caneva G., Nugari M.P., Maggi O., 1998a. *Il biodeterioramento dei materiali costitutivi i Beni Culturali*. In: Mandrioli P., Caneva G. (Eds), *Aerobiologia e Beni Culturali. Metodologie e tecniche di misura*. Nardini Editore, Firenze: 19–32.

Caneva G., Piervittori R., Roccardi A., 1998b. *Ambienti esterni: problematiche specifiche*. In: Mandrioli P., Caneva G. (Eds.), *Aerobiologia e Beni Culturali*, Nardini Editore, Fiesole: 247–250.

Caneva G., Nugari M.P., Pietrini A.M., Pacini A., Merante A., 2002. *Il biodeterioramento della Cripta del Peccato Originale nella gravina di Matera e sua analisi ecologica per il biomonitoraggio di parametri ambientali*. In: *Atti del 97° Congresso della Società Botanica Italiana*, Lecce, Edizioni Del Grifo: 103.

Caneva G., Pacini A., Celesti Grapow L., Ceschin S., 2003. *The Colosseum's use and state of abandonment as analysed through its flora*. International Biodeterioration & Biodegradation, 51: 211–219.

Caneva G., Di Stefano D., Ciriaco G., Ricci S., 2004. *Stone cavity and porosity as a limiting factor for biological colonisation: the travertine of Lungotevere (Rome)*. In: Kwiatkowski D., Löfvendahl R. (Eds.), *Proceedings of the 10th International Congress on Deterioration and Conservation of Stone*. Stockholm, 1: 227–32.

Canhoto O., Pinzari F., Fanelli C., Magan N., 2004. *Application of electronic-nose technology for the detection of fungal contamination in library paper*. International Biodegradation & Biodeterioration, 54: 303–309

Capannesi G., Gratani L., Amadori S., Bruno F., 1981. *Livelli di accumulo di 36 elementi in foglie di Quercus ilex nella città di Roma (Italia)*. In: *Atti del primo Congresso della Società Italiana di Ecologia*: 397–406.

Capponi G., Meucci C., 1987. *Il restauro del paramento lapideo della facciata della chiesa di S.Croce a Lecce*. Bollettino d'Arte, suppl. n. 41, Materiali lapidei, 1: 263–282.

Carballal R., Paz-Bermúdez G., Sánchez-Biezma M.J., Prieto B., 2001. *Lichen colonization of coastal churches in Galicia: biodeterioration implications*. International Biodeterioration & Biodegradation, 47: 157–163.

Carlile M.J., Watkinson S.C., 1994. *The Fungi*. Academic Press, Harcourt Brace & Company, Publisher London.

CARLILE M.J., WATKINSON S.C., GOODAY W., 2001. *The Fungi*. Academic Press, London.

CARTIER BRESSON A., 1987. *Le tecniche fotografiche storiche*. In: MASETTI-BITELLI L., VLAHOV R. (Eds.), *La Fotografia. 1. Tecniche di conservazione e problemi di restauro*. Ed. Analisi, Bologna: 49–67.

CASADEI B., DALLA POZZA G.L., 1989. *Ambiente urbano e lotte antiparassitarie*. Ed. Maggioli, Rimini.

CASATI P., 1985. *Scienze della terra*. Clued, Milano.

CASSAR M., 1995. *Environmental management*. Routledge, London.

CASTANIER S., LE METAYER-LEVREL G., RIASL G., LOUBIER J.F., PERTHUISOT J.P., 2000. *Bacterial carbon-atogenesis and applications to preservation and restoration of historic property*. In: CIFERRI O., TIANO P., MASTROMEI G. (Eds.), *Of Microbes and Art. The Role of Microbial Communities in the Degradation and Protection of Cultural Heritage*. Kluwer Academic/Plenum Publishers, Amsterdam: 246–262.

CASTENHOLZ R.W, 2001. *Phylum BX. Cyanobacteria. Oxygenic photosynthetic bacteria*. In: BOONE D.R., CASTENHOLZ R.W. (Eds.), *Bergey's Manual of Systematic Bacteriology*. Springer Verlag, New York, 2nd edition: 473–597.

CATIZONE P., 1990. *Il contenimento delle piante infestanti nelle aree di interesse archeologico*. In: MASTROROBERTO M. (Ed.), *Archeologia e Botanica, Atti convegno di studi sul contributo della botanica alla conoscenza e conser-vazione delle aree archeologiche*, Pompei. L'Erma di Bretschneider, Roma: 59–64.

CATIZONE P., ZANIN G. (Eds.), 2001. *Malerbologia*. Patron Editore, Bologna.

CAZZUFFI D., 1999. *Repertorio italiano. Geosintetici 98/99*. BE-MA Editrice, Milano.

CECCHINI G., GIORDANO G., MILANI D., 1995. *Materiali tradizionali per il restauro dei dipinti. Preparazione e applicazione secondo il manuale di Giovanni Secco Suardo*. Associazione Giovanni Secco Suardo, Lurano, Bergamo.

CELESTI GRAPOW L., BLASI C., 2003. *I siti archeologici nella conservazione della biodiversità in ambito urbano: la flora vascolare spontanea delle Terme di Caracalla a Roma*. Webbia, *58 (1)*: 77–102.

CELESTI GRAPOW L., PIGNATTI S., PIGNATTI E., 1993–94. *Analisi della flora dei siti archeologici di Roma*. Allionia, *32*: 113–118.

CENCI R.M., DAPIAGGI M., 1998. *Muschi e suoli quali indicatori delle deposizioni atmosferiche di origine antropica*. Biologi Italiani, *11*: 26–30.

CENCI R.M., PALMIERI F., 1997. *L'impiego dei muschi terrestri e del suolo per valutare le deposizioni atmosferiche di origine antropica*. Inquinamento, *1*: 36–45.

CEPERO A., MARTINEZ P., CASTRO J., SANCHEZ A., MACHADO J., 1992. *The biodeterioration of cultural property in the republic of Cuba. A review of some experiences*. In: TOISHI K., ARAI H., KENJO T., YAMANO K. (Eds.), *Proceedings of the 2nd International Conference on Biodeterioration of Cultural Property*, Pacifico Yokohama. International Communications Specialists, Tokyo: 479–487.

CESARI M.G., ROSSI W., 1972. *Le radici minacciano le tombe dipinte di Tarquinia*. Archeologia, *3*.

CESCHIN S., CUTINI M., CANEVA G., 2003. *La vegetazione ruderale dell'area archeologica del Palatino (Roma)*. Fitosociologia, *40 (1)*: 73–96.

CHAPMAN J., REISS M.J., 1994. *Ecologia: principi ed applicazioni*. Zanichelli Editore, Bologna.

CHARRIER G., 1960. *Muschi calcarizzati*. Nuovo Giornale Botanico Italiano, *67*: 263–264.

CHATIGNY M.A., MACHER J.M., BURGE H.A., SOLOMON W.R., 1989. *Sampling airborne microorganisms and aeroallergens*. In: *Air-sampling Instruments*. S.V. Hering Technical Editor, 7th edition.

CHEN J., BLUME H.P., BEYER L., 2000. *Weathering of rocks induced by lichen colonization – A review*. Catena, *39*: 121–146.

CHIARI G., COSSIO R., 2001. *Lichens on a sandstone: Do they cause damage?* In: *Proceedings of the Congress "Botany 2001 Plants and People*," Albuquerque, New Mexico.

CHILD R., 1998. *Tutela del patrimonio artistico museale*. In: *Atti del Convegno Climatologia applicata alla conservazione dei beni archeologici e storico-artistici*. Servizio Beni Culturali, Provincia Autonoma di Trento: 37–77.

CHIUSOLI A., 1999. *La scienza del paesaggio*. CLUEB, Bologna.

CHUNG Y.J., SEO M.S., LEE K.S., HAN S.H., 2003. *The biodeterioration and conservation of stone historical monuments*. Conservation Studies, *24*: 5–28.

CIARALLO A., DE CAROLIS E., 1998. *Lungo le mura di Pompei. L'antica città nel suo ambiente naturale*. Soprint. archeologica di Pompei, Electa Mondadori: 56.

CLAUZADE G., ROUX C., 1975. *Étude écologique et phytosociologique de la végétation lichénique des roches calcaires non altérées dans les régions méditerranéennes et subméditerranéennes du sud-est de la France*. Bull. Mus. Hist. Nat. Marseille, *35*: 153–208.

COMMISSIONE NORMAL, SOTTOGRUPPO SPERIMENTAZIONE PROTETTIVI, 1993. *Metodologia per la valutazione dei prodotti impiegati come protettivi per materiale lapideo. Parte I, test e trattamento dei campioni*. L'Edilizia, *VII*: 57–71.

COOKE W.B., 1979. *The Ecology of Fungi*. CRC Press, Boca Raton, Florida.

CORAIN B., FILIPPO R., QUARESIMA R., SCOCCIA G., VOLPE R., 1998. *Impiego di resine metallorganiche innovative per la protezione biologica dei beni culturali*. In: SANNA U. (Ed.), *Atti IV Congresso Nazionale AIMAT*, PTM Ed.: 769–779.

CORTINI PEDROTTI C., 1978. *La florula briologica della Grotta di Monte Cucco (Appennino Umbro-Marchigiano)*. Lavori Società Italiana Biogeografia, *7*: 759–769.

CORTINI PEDROTTI C., 2001. *Flora dei Muschi d'Italia*. Antonio Delfino Editore, Roma.

COSENTINO S., PALMAS F., 1991. *Assessment of airborne fungal spores in different industrial working environments and their importance as health hazards to workers*. Environmental Monitoring and Assessment, *16*: 127–136.

COTTER D.A., 1981. *Spore Activation*. In: TURIAN G., HOHL H.R. (Eds.), *The Fungal Spore*: 385–411.

COX C.S., WATHES C.M., 1995. *Bioaerosol Handbook*. Lewis Publishers CRC , London.

CRAGNOLINO G., TUOVINEN O.H., 1984. *The role of sulphate-reducing and sulphur-oxidising bacteria in the* localised corrosion of iron-base alloys. A review. International Biodeterioration, 20 (1): 9–26.

CRAWFORD J., 1981. L'età del collodio. Capanna Editore, Firenze.

CREMONESI P., 1999. L'uso degli enzimi nella pulitura di opere policrome. Il Prato, Padova.

CREMONESI P., 2001. L'uso di tensioattivi e chelanti nella pulitura di opere policrome. Il Prato, Padova.

CURRI S.B., PALENI A., 1975. Some aspects of the growth of chemolithotrophic microorganisms in the Karnak temples. In: ROSSI-MANARESI R. (Ed.), "The Conservation of Stone," Proceedings of the International Symposium. Bologna: 267–279.

CUTLER D.F., RICHARDSON I.B.K., 1981. Tree root and buildings. Kew Garden, Construction Press, London.

D'URBANO S., MEUCCI C., NUGARI M.P., PRIORI G.F., 1989. Valutazione del degrado biologico e chimico di legni archeologici in ambiente marino. In: TAMPONE G. (Ed.), Il Restauro del Legno, Nardini Editore, Firenze, 1: 79–84.

D'URBANO S., PANDOLFI A., PIETRINI A. M., 1998. Patologie da umidità nelle strutture murarie antiche. In: PANDOLFI A. e SPAMINATO M.L.S. (Eds.), Diagnosi e progetto per la conservazione dei materiali dell'architettura, De Luca Ed., Roma: 327–340.

DAFFONCHIO D., BORIN S., ZANARDINI E., ABBRUSCATO P., REALINI M., URZÌ C., SORLINI C., 2000. Molecular tools applied to the study of deteriorated artworks. In: CIFERRI O., TIANO P., MASTROMEI G. (Eds.), Of Microbes and Art: The Role of Microbial Communites in the Degradation and Protection of Cultural Heritage, Kluwer Academic Publisher, Dordrecht: 21–38.

DANIEL G.F., NILSSON T., 1998. Developments in the study of soft rot and bacterial decay. In: BRUCE A., PALFREYMAN J.W. (Eds.), Forest Product Biotechnology. Taylor and Francis, London.

DANIELS V., BOYD D., 1986. The yellowing of thymol in the display of prints. Studies in Conservation, 31: 156–158.

DANILOV R.A., EKELUND N.G.A., 2001. Comparison of usefulness of three types of artificial substrata (glass, wood and plastic) when studying settlement patterns of peryphyton in lakes of different trophic status. Journal of Microbiological Methods, 45: 167–170.

DANIN A., 1986. Patterns of biogenic weathering as indicators of paleoclimates in Israel. In: Proceedings of the Royal Society of Edinburgh, 89B: 243–253.

DANIN A., CANEVA G., 1990. Deterioration of limestone walls in Jerusalem and marble monuments in Rome caused by cyanobacteria. International Biodeterioration 26: 397–417.

DE BELIE N., RICHARDSON M., BRAAM C.R., SVENNERSTEDT B., LENEHAN J.J., SONCK B., 2000. Durability of Building Materials and Components in the Agricultural Environment. In: The Agricultural Environment and Timber Structures, Part I, Journal of Agricultural Engineering Research, 75: 225–241.

DE BILLERBECK G., ROQUES C., FONVIELLE J.L., 2001. Preventive treatment of the atmosphere of cultural heritage premises by volatile essential oils. Evaluation of antifungal activity of citronella essential oil. In: Fungi. A Threat for the People and Cultural Heritage through Microorganisms. München: REMA: 31–33.

DE BOCK L.A., VAN GRIEKEN R.E., CAMUFFO D., GRIME G.W., 1999. Microanalysis of museum aerosols to elucidate the soiling of paintings: Case of the Correr Museum, Venice, Italy. Environmental Science & Technology, 30 (11): 3341–3350.

DE GUICHEN G., 1980. Climate in Museums. ICCROM, Roma.

DE LEO F., URZÌ C., 2003. Fungal colonization in treated and untreated stone surfaces. In: SAIZ-JIMENEZ C. (Ed.), Molecular Biology and Cultural Heritage, Proceedings of the International Congress on Molecular Biology and Cultural Heritage, Seville. Balkema Publishers, Lisse (NL): 213–218.

DE LEO F., URZÌ C., DE HOOG G.S., 1999. Two Coniosporium species from rock surfaces. Studies in Mycology, 43: 70–79.

DE LEO F., URZÌ C., DE HOOG G.S., 2003. A new meristematic fungus, Pseudoteniolina globosa. Antonie van Leeuwenhoek, 83: 351–360.

DE MARCO G., CANEVA G., DINELLI A., 1990. Geobotanical foundations for a protection project in the Moenjodaro archaeological area. Prospezioni archeologiche, Quaderni, 1: 115–120.

DE MIRO E., 1965. Heraclea Minoa. Serie degli itinerari dei musei, gallerie e monumenti d'Italia, 110. Istituto Poligrafico dello Stato, Roma.

DE ROSSI E., CIFERRI O., 2003. The microbial flora of naturally aged silk fibroin. In: SAIZ-JIMENEZ C. (Ed.), Molecular Biology and Cultural Heritage, Proceedings of the International Congress on Molecular Biology and Cultural Heritage, Seville. Balkema Publishers, Lisse (NL): 267–270.

DE SANTOLI L., FRACASTORO G.V., 1998. La qualità dell'aria negli ambienti interni. Soluzioni e strategie. Milano: AICARR.

DEACON J.W., 1997. Modern Mycology. Blackwell Science Ltd., Oxford.

DEACON J.W., 2000. Micologia moderna. Calderini Edagricole, Bologna.

DEKKER C., 2001. Mould contamination on textile objects: consequences and avoidance. In: Fungi: A Threat for People and Cultural Heritage through Microorganisms. München Messegelände: 24–25.

DEL MONTE M., FERRARI A., 1989. Patine da biointer-azione alla luce delle superfici marmoree. In: Atti del Convegno – Le pellicole ad ossalato: origine e significato nella conservazione delle opere d'arte, Milano: 171–182.

DEL MONTE M., SABBIONI C., 1987. A study of the patina called "scialbatura" on imperial Roman marbles. Studies in Conservation, 32: 114–121.

DELL'UOMO A., 1982. Peuplements d'algues dans quelques grottes de la gorge de Frasassi. Guide-Itineraire Excursion Intern. de Phytosociologie en Italie centrale. Università degli Studi di Camerino: 207–210.

DENYER S.P, STEWART G.S.A.B., 1998. Mechanisms of action of disinfectants. International Biodeterioration & Biodegradation, 41: 261–268.

DESHPANDE J., GANGAWANE L.V., 1995. On the possibility of air-borne fungi in the biodeterioration of Buddhist paintings in Ajanta caves. In: ARAYNAK C., SINGHASIRI C. (Eds), Proceedings of the 3rd International Conference on Biodeterioration of Cultural Property, Bangkok, Thailand: 455–457.

DHAWAN S., 1987. *Role of microorganisms in corrosion of metals – A literature survey.* In: *Proceedings of the Asian Regional Seminar on Conservation of Metals in Humid Climates*, ICCROM, Roma: 100–112.

DHAWAN S., AGRAWAL O.P., 1986. *Fungal flora of miniature paper painting and land lithographs.* International Biodeterioration, *22*: 95–99.

DHAWAN S., GARG K.L., PATHAK N., 1992. *Microbial analysis of Ajanta wall paintings and their possible control in situ.* In: TOISHI K., ARAI H., KENJO T., YAMANO K. (Eds.), *Proceedings of the 2nd International Conference on Biodeterioration of Cultural Property*, Pacifico Yokohama. International Communications Specialists, Tokyo: 245–262.

DI BENEDETTO L., GRILLO M., STAGNO F., 2000. *Biodeteriorogeni dell'ex Collegio dei Gesuiti in Catania: Florula, vegetazione ed aspetti ecologici.* Archivio Geobotanico, *6*: 59–66.

DI FRANCESCO C., BEVILACQUA F., PINNA D., GRILLINI G.C., TUCCI A., 1998. *Dati storico-materiali ed orientamenti progettuali per il restauro dell'Atrio di Pomposa.* In: *Atti del Convegno "Progettare i restauri,"* Bressanone, Ed. Arcadia Ricerche: 627–635.

DIX N.J., WEBSTER J., 1995. *Fungal Ecology.* Chapman & Hall, London.

DOMASLOWSKY W., STRZELCZYK A., 1986. *Evaluation of applicability of epoxy resins to conservation of stone historic monuments.* In: *Case Studies in the Conservation of Stone and Wall Paintings*, IIC Congress, Bologna: 126–132.

DOMSCH K. H., GAMS W., ANDERSON T., 1993. *Compendium of Soil Fungi.* Academic Press, London.

DORNIEDEN T., GORBUSHINA A.A., 2000. *New methods to study the detrimental effects of poikilotroph microcolonial micromycetes (PMM) on building materials.* In: FASSINA V. (Ed.), *Proceedings of the 9th International Congress on Deterioration and Conservation of Stone*, Venice. Elsevier, Amsterdam: 461–468.

DORNIEDEN T., GORBUSHINA A.A., KRUMBEIN W.E., 2000. *Biodecay of cultural heritage as a space/time-related ecological situation – An evaluation of a series of studies.* Biodeterioration & Biodegradation, *46*: 261–270.

DREWELLO R., 1997. *Mikrobiell induzierte korrosion von Silikatglas – unter besonderer Berücksichtigung von Alkali-Erdalkali-Silikatgläser.* Ph.D. thesis, F. Alexander Universität Erlangen-Nürnberg, Germany.

DUNCAN S.J., GANIARIS H., 1987. *Some sulphide corrosion products on copper alloys and lead alloys from London waterfront sites.* In: *Proceedings of the Jubilee Conservation Conference – Recent Advances in the Conservation and Analysis of Artifacts*, London, University of London, Summer School Press: 109–118.

DURRANT L.R., 1996. *Biodegradation of lignocellulosic materials by soil fungi isolated under anaerobic conditions.* International Biodeterioration & Biodegradation, *37*: 189–195.

EATON R.A., HALE M.D.C., 1993. *Fungal decay.* In: *Wood Decay Pest and Protection*, Chapman & Hall, London: 76–97.

ECKHARDT F.E.W., 1985. *Mechanisms of the microbial degradation of minerals in sandstone monuments, medieval frescoes, and plaster.* In: FELIX G. (Ed.), *Proceedings of 5th International Congress on Deterioration and Conservation of Stone.* Presses Polytechniques Romandes, Lausanne: 643–652.

ECKHARDT F.E.W., 1996. *Microbial diversity and airborne contamination.* In: HEITZ ET AL. (Eds.), *Microbially influenced corrosion of materials.* Springer Verlag, Berlin Heidelberg: 75–95.

EDWARDS H.G.M., FARWELL D.W., SEAWARD M.R.D., GIACOBINI C., 1991. *Preliminary Raman microscopic analyses of a lichen encrustation involved in the biodeterioration of Renaissance frescoes in Central Italy.* International Biodeterioration *27(1)*: 1–9.

EDWARDS H.G.M., FARWELL D.W., JENKINS R., SEAWARD M.R.D., 1992. *Vibrational Raman spectroscopic studies of calcium oxalate monohydrate and dihydrate in lichen encrustations on Renaissance frescoes.* Journal of Raman Spectroscopy, *23*: 185–189.

EDWARDS H.G.M., EDWARDS K.A.E., FARWELL D.W., LEWIS I.R., SEAWARD M.R.D., 1994. *An approach to stone and fresco lichen biodeterioration through Fourier transform Raman microscopic investigation of thallus-substratum encrustations.* Journal of Raman Spectroscopy, *25*: 99–103.

EGGINS H.O.W., MILLS J., HOLT A., SCOTT G., 1971. *Biodeterioration and biodegradation of synthetic polymers.* In: SYKES G., SKINNER F.A. (Eds.), *Microbial Aspects of Pollution, The Society for Applied Bacteriology Symposium*, Series n. 1, Academic Press, London-New York: 267–279.

EHRLICH H.L., 1981. *Geomicrobiology.* New York, Marcel Deckker.

ELLENBERG H., 1979. *Zeigerwerte der Gefässpflanzen Mitteleuropas.* Scripta geobotanica, Aufl. Goettingen, *9 (2)*: 122.

ELLIOTT C.G., 1994. *Reproduction in Fungi.* Chapman & Hall, London.

ELMER K., ROSE E., FITZNER B., KRUMBEIN W.E., WARSCHEID T., 1993. *Sterilisation of cultural objects by ethylene oxide – Application for conservation practice by laboratory-based or mobile treatments.* In: THIEL M.J. (Ed.), *Proceedings of International UNESCO/RILEM Congress on Conservation of Stone and Other Materials*, Paris. E & FN Spon, London: 581–588.

EL-SAYED A.H.M.M., MAHMOUD W.M., DAVIS E.M., COUGHLIN R.W., 1996. *Biodegradation of polyurethane coatings by hydrocarbon-degrading bacteria.* International Biodeterioration & Biodegradation, *37*: 69–79.

ENDT D.W.V., JESSUP W.C., 1986. *The deterioration of protein materials in museums.* In: *Biodeterioration VI, Atti VI Biodeterioration Symposium*, CAB International, Washington: 332–337.

ENVIRONMENTAL HEALTH DIRECTORATE (EHD), HEALTH PROTECTION BRANCH, 1987. *Exposure guidelines for residential indoor air quality.* Minister of National Health and Welfare, Ottawa.

ERIKSSON K.E., BLANCHETTE R. A., ANDER P., 1990. *Microbial and enzymatic degradation of wood and wood components.* Springer Verlag, New York.

FAJRUSINA S.A., 1974. *Capparis spinosa L. destroyer of architectural monumento in Uzbechistan.* Acad. Sc. Rep. Soc. Sov. Uzb., Journ. Biol. Uzbechistan.

FANELLI C., 2001. *I funghi della carta.* In: *Giornata di studio conservazione e restauro delle opere d'arte su carta*, Accademia Nazionale delle Scienze detta dei XL, Roma: 152–165.

FANELLI C., FABBRI A.A., RICELLI A., CARPITA A., ROSSI R., 2001. *Effect of new synthetic antifungal compounds on the growth of fungi which are responsible for paper deterioration.* In: *Metodi Chimici, Fisici e Biologici per la Salvaguardia dei Beni Culturali,* AICAT & GICAT, Roma: 105–111.

FAVALI M.A., BARBIERI N., BASSI M., 1978. *A green alga growing on a plastic film used to protect archaeological remains.* International Biodeterioration Bulletin, *14* (3): 89–93.

FELLER R. L., 1994. *Accelerated aging. Photochemical and thermal aspects.* J. Paul Getty Trust. Edwards Bros., Ann Arbor, Michigan, USA.

FERONE C., PANSINI, M., MASCOLO, M.C., VITALE, A., 2000. *Preliminary study on the set up of mortars displaying biocidal activity.* In: FASSINA V. (Ed.), *Proceedings of the 9th International Congress on Deterioration and Conservation of Stone,* Venice. Elsevier, Amsterdam: 371–378.

FERRER M.R., QUEVEDO-SARMIENTO J., RIVADENEIRA M.A., BEJAR V., DELGADOAND R., RAMOS-CORMENZANA A., 1988. *Calcium carbonate precipitation by two groups of moderate halophilic microorganisms at different temperatures and salt concentrations.* Current Microbiology, *17*: 221–227.

FERRO G., FURNARI F., 1968. *Flora e vegetazione di Stromboli (Isole Eolie).* Arch. Bot. e Biogeogr. It., *44* (3): 59–85.

FILER K., SCHUTZ E., GEH S., 2001. *Detection of microbial volatile organic compounds (MVOCs) produced by moulds on various materials.* International Journal of Hygiene and Environmental Health, *204 (2–3)*: 111–121.

FILIPPI M., 1987. *Gli impianti nei musei.* Condizionamento dell'aria, riscaldamento e refrigerazione, Organo Ufficiale dell'AICARR, *8*: 965–984.

FILIPPI M., LOMBARDI C., SILVI C., 1994. *Conservazione di beni di interesse storico ed artistico: annotazioni a margine di una norma sulle condizioni ambientali.* Condizionamento dell'aria, riscaldamento e refrigerazione, *4*: 487–493.

FISHER G.G., 1972. *Weed damage to materials and structures.* International Biodeterioration Bulletin, *8 (3)*: 101–103.

FLETCHER R.L., 1988. *Brief review of the role of marine algae in biodeterioration.* International Biodeterioration, *24*: 141–152.

FLIEDER F., 1985. *Les agents de détérioration de l'image photographique et les moyens d'y remédier.* In: *Actes du Colloque Conservation et restauration du patrimoine photographique,* Paris Audiovisuel.

FLIEDER F., CAPDEROU C., 1999. *Sauvegarde des collections du Patrimoine. La lutte contre les détériorations biologiques,* CNRS Editions, Parigi.

FLIEDER F., LAVÉDRINE B., 1987. *Gli agenti di deterioramento delle immagini fotografiche e la protezione contro i loro danni.* In: MASETTI-BITELLI L., VLAHOV R. (Eds.), *La Fotografia. 1. Tecniche di conservazione e problemi di restauro,* Ed. Analisi, Bologna: 49–67.

FLIEDER F., RAKOTONIRAINY M., LEROY M., FOHRER F., 1995. *Disinfection of paper using gamma rays, electron beams and microwaves.* In: ARAYNAK C., SINGHASIRI C. (Eds), *Proceedings of the 3rd International Conference on Biodeterioration of Cultural Property,* Bangkok, Thailand: 174–182.

FLORENZANO G., 1986. *Fondamenti di microbiologia del terreno.* Ed. Reda, Roma.

FLORES M., LORENZO J., GOMEZ-ALARCON, G., 1997. *Algae and bacteria on historic monuments at Alcala de Henares, Spain.* International Biodeterioration & Biodegradation, *46*: 241–246.

FLORIAN M.L., 1988. *Deterioration of organic materials other than wood.* In: PEARSON C. (Ed.), *Conservation of marine archaeological objects,* Butterworth, London: 21–54.

FLORIAN M.L., 1993. *Conidial fungi (mould) activity on artifact material. A new look at prevention, control and eradication.* In: *ICOM Committee for Conservation, Proceedings of the 10th Triennial Meeting,* Washington DC, USA, *2*: 868–874.

FLORIAN M.L.E., 1994. *Conidial fungi (mould, mildew) biology: A basis for logical prevention, eradication and treatment for museum and archival collections.* Leather Conservation News, *10*: 1–26.

FLORIAN M.L., 1997. *Heritage eaters. Insects & fungi in heritage collections.* James & James, London.

FLORIAN M.L., 2000. *Fungal spots and others. Nomenclature, SEM Identification of Causative Fungi, and Effects on Paper.* In: CIFERRI O., TIANO P., MASTROMEI G. (Eds.), *Of Microbes and Art. The Role of Microbial Communities in the Degradation and Protection of Cultural Heritage,* Kluwer Academic/Plenum Publishers, New York: 135–151.

FLORIAN M.L., 2002. *Fungal facts. Solving fungal problems in heritage collections.* Archetype Publications Ltd., London.

FLORIAN M.L., GRIMSTAD K., 1987. *The effect on artifact materials of the fumigant ethylene oxide and the freezing used in insect control.* In: *ICOM Committee for Conservation, Proceedings of the 8th Triennial Meeting,* Sydney: 199–208.

FLORIAN M.L., MANNING L., 2000. *SEM analysis of irregular fungal fox spots in an 1854 book: Population dynamic and species identification.* International Biodeterioration & Biodegradation, *46*: 205–220.

FOGED N., 1983. *Diatoms in fountains, reservoirs and some other humid and dry localities in Rome, Italy.* Nova Hedwigia, *38*: 433–470.

FORD T., MITCHELL R., 1991. *The ecology of microbial corrosion.* In: MARSHALL K.C. (Ed.), *Advances in Microbial Ecology,* *11*: 231–262.

FORLANI G., SEVES A.M., CIFERRI O., 2000. *A bacterial extracellular proteinase degrading silk fibroin.* International Biodeterioration & Biodegradation, *46*: 271–275.

FORMICA CASTALDI V., 1988. *Criteri di intervento conservativo delle rocce del Parco Nazionale delle incisioni rupestri di Capo di Ponte.* In: *Il Parco delle incisioni rupestri di Grosio e la preistoria valtellinese, Atti I° Convegno archeologico provinciale,* Grosio 1985: 195–203.

FORNI U., 1866. *Manuale ragionato del pittore restauratore.* Le Monnier. Firenze.

FRANCESCHI V.R., HORNER H.T., 1980. *Calcium oxalate crystals in plants.* Botanical Review, *46*: 361–427.

FRIEDMANN I., 1956. *Beiträge zu Morphologie und Formwechsel der atmophytischen Bangioidee.* Phragmonema sordidum *Zopf.* Österr. Botan. Zeitschrift, *103 (5)*: 613–633.

FRY E. J., 1924. *A suggested explanation of the mechanical action of lithophytic lichens on rocks (shale).* Annals of Botany, *38*: 175–196.

GADD G.M., 1992. *Metals and microorganisms: A problem of definition.* FEMS Microbiol. Letters, *100*: 197–204.

GALLIEN J.P., GOUGET B., CARROT F., ORIAL G., BRUNET A., 2001. *Alteration of glasses by microorganisms.* Nuclear Instruments and Methods in Physics Research B, *181*: 610–615.

GALLO F., 1992. *Il biodeterioramento di libri e documenti.* Centro di studi per la conservazione della carta, ICCROM, Roma.

GALLO F., 1993. *Aerobiological research and problems in libraries.* Aerobiologia, *9*: 117–130.

GALLO F., PASQUARIELLO G., 1989. *Foxing, ipotesi sull'origine biologica.* Bollettino dell'Istituto Centrale per la Patologia del Libro, *43*: 136–176.

GALLO F., STRZELCZYK A., 1971. *Indagine preliminare sulle alterazioni microbiche della pergamena.* Bollet. Ist. Patol. Libro "Alfonso Gallo," *30*: 71–87.

GALLO F., VALENTI P., 1999. *Guida a un'indagine conoscitiva sullo stato di conservazione dei materiali librari.* In: *Conservazione dei materiali librari, archivistici e grafici.* U. Allemandi & C.: 121–128.

GALLO F., PASQUARIELLO G., VALENTI P., 1998. *Biblioteche e Archivi.* In: MANDRIOLI P., CANEVA G. (Eds), *Aerobiologia e Beni Culturali, metodologie e tecniche di misura,* Nardini, Firenze: 193–213.

GALLO F., MARCONI C., MONTANARI M., 1989. *Ricerche sperimentali sulla resistenza agli agenti biologici di materiali impiegati nel restauro dei libri.* Bollettino dell'Istituto Centrale per la Patologia del Libro, *43*: 106–120.

GALLO F., VALENTI P., COLAIZZI P., SCLOCCHI M.C., PASQUARIELLO G., SCORRANO M., MAGGI O., PERSIANI A.M., 1999a. *Research on the viability of fungal spores in relation to different microclimates and different materials.* In: FEDERICI C., MUNAFÒ P.F. (Eds.), *Conference on Conservation and Restoration of Archives and Library Materials,* Erice. Palumbo Editore, Palermo: 213–230.

GALLO F., VALENTI P., MAGGI O., PERSIANI A.M., VINCENZONI F., 1999b. *Modified starches: tests on biodegradability.* In: FEDERICI C., MUNAFÒ P.F. (Eds.), *International Conference on Conservation and Restoration of Archival and Library Materials,* Erice. Palumbo Editore, Palermo, Vol. *2*: 907–914.

GALLO F., PERSIANI A.M., MAGGI O., VALENTI P., SCLOCCHI C., PASQUARIELLO G., 2002. *La conservazione del patrimonio grafico e archivistico: le spore fungine deposte nella polvere e il rischio di biodeterioramento.* In: *Atti 97° Congresso della Società Botanica Italiana.* Ed. del Grillo, Lecce: 8.

GALLO L.M., PIERVITTORI R., 1993. *Lichenometry as a method for holocene dating: Limits in its applications and realibility.* Il Quaternario, *6*: 77–86.

GALUN M. (Ed.), 1988. *Handbook of Lichenology.* CRC Press, Boca Raton, Florida.

GAMBETTA A., ORLANDI E., 1982. *Sulla preservazione del legno messo in opera all'aperto.* Contributi scientifici pratici, Vol. 30. CNR Istituto per la Ricerca sul Legno, Firenze.

GAMPER U., BACCHETTA G., 2001. *La vegetazione dei muri di Venezia (NE-Italia).* Fitosociologia, *38 (2)*: 83–96.

GARCIA-VALLES M., URZÌ C., DE LEO F., SALAMONE P., VENDRELL-SAZ M., 2000. *Biological weathering and mineral deposits of the Belevi marble quarry (Ephesus,*

Turkey). International Biodeterioration & Biodegradation, *40 (2–4)*: 221–227.

GARG K.L., GARG N., MUKERJI K.G. (Eds.), 1994. *Recent Advances in Biodeterioration and Biodegradation,* Vol. 1. Naya Prokash, Calcutta, India.

GARG K.L., GARG N., MUKERJI K.G. (Eds.), 1994. *Biodeterioration & Biodegradation of Natural and Synthetic Products,* Vol. 2. Naya Prokash, Calcutta, India.

GARG K.L., MISHRA A.K., SINGH A., JAIN K.K., 1995. *Biodeterioration of cultural heritage: some case studies.* In: KAMLAKAR G., PANDIT RAO V. (Eds.), *Conservation, Preservation, and Restoration: Traditions, Trends, and Techniques*: 31–38.

GARTY J., 1990. *Influence of epilithic microorganisms on the surface temperature of building walls.* Canadian Journal of Botany, *68*: 1349–1353.

GARZA-VALDES L.A., STROSS B., 1992. *Rock varnish on a Pre-Columbian green jasper from the tropical rain forest (the Ahaw pectoral).* In: VANDIVER P.B., DRUZIK J.R., WHEELER G.S., FREESTONE I.C. (Eds.), *Materials Issues in Art and Archaeology III,* Symposium San Francisco, California. Materials Reserch Society, Pittsburgh, Pennsylvania: 891–900.

GAYLARDE P.M., GAYLARDE C.C., 2000. *Algae and cyanobacteria on painted buildings in Latin America.* International Biodeterioration & Biodegradation, *46 (2)*: 93–97.

GAYLARDE C., VIDELA H.A., 1987. *Localised corrosion induced by a marine vibrio.* International Biodeterioration, *23 (2)*: 91–104.

GAZZETTA UFFICIALE n. 229/L, 27 dicembre 1999. *Testo unico delle disposizioni legislative in materia di beni culturali e ambientali a norma dell'art.1 della legge n. 352 dell'8 Ottobre 1997 (DL 490, 29 Ottobre 1999).*

GAZZETTA UFFICIALE n. 244 (supplemento), fascicolo 238, 19 ottobre 2001. *Atto di indirizzo sui criteri tecnico-scientifici e sugli standard di funzionamento e sviluppo dei musei (DL 112/1998, art. 150, comma 6).*

GEHRMANN C.K., KRUMBEIN W.E., 1994. *Interaction between epilithic and endolithic lichens and carbonate rocks.* In: FASSINA V., OTT H., ZEZZA F. (Eds.), *Proceedings of the 3rd International Symposium on the Conservation of Monuments in the Mediterranean Basin,* Venice: 311–316.

GEHRMANN C.K., KRUMBEIN W.E., PETERSEN K., 1992. *Endolithic lichens and the corrosion of carbonate rocks – A study of biopitting.* Int. J. Mycol. Lichenol., *5 (1–2)*: 37–48.

GEROLA F.M., 1988. *Biologia Vegetale. Sistematica Filogenetica.* UTET, Torino.

GERWIN W., BAUMHAUER R., 2000. *Effect of soil parameters on the corrosion of archaelogical metal finds.* Geoderma, *96*: 63–80.

GIACCONE G., DI MARTINO V., 1999. *Biologia delle alghe e conservazione dei monumenti.* Bollettino dell' Accademia Gioenia di Scienze Naturali, Catania, *32*: 53–81.

GIACCONE G., ALONGI G., PIZZUTO F., COSSI A., 1994. *La vegetazione marina bentonica fotofila del Mediterraneo: II. Infralitorale e Circalitorale. Proposte di aggiornamento.* Bollettino dell' Accademia Gioenia di Scienze Naturaei, Catania, *27*: 1–47.

GIACOBINI C., 1974. *Prospettive di riconoscimento morfologico di alcuni tipi di alterazione dei monumenti.* In: *Atti del XXIX Congresso ATI*, Firenze: 129–132.

GIACOBINI C., LACERNA R., 1965. *Problemi di microbiologia nel settore degli affreschi.* Bollettino dell'Istituto Centrale del Restauro: 83–108.

GIACOBINI C., DE CICCO M.A., TIGLIÈ I., ACCARDO G., 1988. *Actinomycetes and biodeterioration in the field of fine art.* In: HOUGHTON D.R., SMITH R.N., EGGINS H.O.W. (Eds.), *Biodeterioration 7*, Elsevier, London & New York: 418–423

GIAMELLO M., PINNA D., PORCINAI S., SABATINI G., SIANO S., 2004. *Multidisciplinary study and laser cleaning tests of marble surfaces of Porta della Mandorla, Florence.* In: KWIATKOWSKI D., LÖFVENDAHL R. (Eds.), *Proceedings of the 10th International Congress on Deterioration and Conservation of Stone,* Stockholm: 841–848.

GIANI E., 2002. *La vetrina museale. Linee guida per la progettazione e la messa in opera.* Kermes, *48*: 37–48.

GIANI E., GIOVAGNOLI A., NUGARI M.P., 2002. *Il controllo ambientale nei musei: il caso della Galleria Doria Pamphilj a Roma.* In: *X Congresso Nazionale di Aerobiologia–Aria e Salute,* Riassunti, Bologna: 69.

GILBERT M., 1987. *Black spots on bronzes.* Newsletter ICOM Commitee for Conservation Metals Working Group, *3*: 12.

GILBERT M., 1989. *Inert atmosphere fumigation of museum objects.* Studies in Conservation, *34*: 80–84.

GILBERT M., 1991. *The Effects of Low Oxygen Atmospheres on Museum Pests.* Studies in Conservation, *36*: 93–98.

GILBERT O., 2000. *Lichens.* Harper Collins Publishers, London.

GILBERT O., 2003. *The lichen flora of unprotected soft sea cliffs and slopes.* Lichenologist, *35 (3)*: 245–254.

GIMINGHAM C.H., BIRSE E.M., 1957. *Ecological studies on growth-form in Bryophytes. I. Correlation between growth-form and habitat.* Journal of Ecology, *45*: 533–545.

GIORDANI P., MODENESI P., TRETIACH M., 2003. *Determinant factors for the formation of the calcium oxalate minerals, weddellite and whewellite, on the surface of foliose lichens.* Lichenologist, *35 (3)*: 255–270.

GIORDANO G., 1971. *Tecnologia del legno. Volume 1. La materia prima.* Unione Tipografico-Editrice Torinese. Torino: 621–629.

GIORDANO S., 1993. *Le basi morfologiche della conduzione dell'acqua nelle Briofite.* Informatore Botanico Italiano, *25*: 260–262.

GIULIANI M.R., NUGARI M.P., 1993. *A case of fungal biodeterioration on an ancient textile.* In: *Proceedings of the 10th Triennial ICOM Meeting,* Washington DC, ICOM Committee for Conservation, *1*: 305–307.

GOLUBIC S., 1967. *Algenvegetation der Felsen, eine ökologische Algenstudie im dinarischen Karstgebiet.* Binnengewässer, *23*: 1–183.

GOLUBIC S., PERKINS R.D., LUKAS K.J., 1975. *Boring microorganisms and microborings in carbonate substrates.* In: FREY R.W. (Ed.), *The Study of Trace Fossils,* Springer Verlag, New York: 229–259.

GOLUBIC S., FRIEDMAN E., SCHNEIDER J., 1981. *The lithobiontic ecological niche, with special reference to microorganisms.* J. Sediment. Petr., *51*: 475–478.

GOMEZ-ALARCON G., CILLEROS B., FLORES M., LORENZO J., 1995. *Microbial communities and alteration processes in monuments at Alcala de Henares, Spain.* The Science of the Total Environment, *167*: 231–239.

GOMEZ-ALARCON G., BLAZQUEZ A.B., LORENZO J., 1999. *Biocides used in the control of microorganisms on stone monuments.* In: *An International Conference on Microbiology and Conservation* (ICMC '99), *Of Microbes and Art. The Role of Microbial Communities in the Degradation and Protection of Cultural Heritage.* Florence, Italy: 228–232.

GONZALES J.M., 2003. *Overview of existing molecular techniques with potential interest in cultural heritage.* In: SAIZ-JIMENEZ C. (Ed.), *Molecular Biology and Cultural Heritage, Proceedings of International Congress on Molecular Biology and Cultural Heritage,* Seville. Balkema Publishers, Lisse (NL): 3–14.

GOODAY G.W., 1995a. *Cell Walls.* In: GOW N.A.R., GADD G.M. (Eds.), *The growing fungus.* Chapman & Hall, London: 43–62.

GORBUSHINA A.A., KRUMBEIN W.E., 1999. *The poikilo-trophic microorganism and its environment.* In: SECKBACH J. (Ed.), *Enigmatic microorganisms and life in extreme environments.* Kluwer Academic Publishers: 175–185.

GORBUSHINA A.A., KRUMBEIN W.E., 2000. *Patina (physical and chemical interactions of sub-aerial biofilms with objects of art).* In: CIFERRI O., TIANO P., MASTROMEI G. (Eds), *Of Microbes and Art. The Role of Microbial Communities in the Degradation and Protection of Cultural Heritage.* Kluwer Academic/Plenum Publishers, New York: 105–120.

GORBUSHINA A.A., PALINSKA K.A., 1999. *Biodeteriorative processes on glass: Experimental proof of the role of fungi and cyanobacteria.* Aerobiologia, *15*: 183–191.

GORBUSHINA A.A., KRUMBEIN W.E., HAMMAN C.H., PANINA L., SUOKHARJEVSKI S., WOLLENZIEN U., 1995. *Role of Black Fungi in Colour Change and Biodeterioration of Antique Marbles.* Geomicrobiology Journal, *11*: 205–222.

GORBUSHINA A.A., KRUMBEIN W.E., VLASOV D.Y., 1997. *The fungal microcosm of Mediterranean monuments and sites – Past, present and future.* In: MOROPOULU A., ZEZZA F., KOLLIAS E., PAPACHRISTODOULOU I. (Eds.), *Proceedings of the 4th International Symposium on the Conservation of Monuments in the Mediterranean,* Rhodes, Technical Chamber of Greece: 261–270.

GORBUSHINA A.A., DIAKUMAKU E., MULLER L., KRUMBEIN W.E., 2003. *Biocide treatment of rock and mural paintings: Problems of application, molecular techniques of control and environmental hazards.* In: SAIZ-JIMENEZ (Ed.), *Molecular Biology and Cultural Heritage, Proceedings of International Congress on Molecular Biology and Cultural Heritage,* Seville. Balkema Publishers, Lisse (NL): 61–71.

GORGONI C., LAZZARINI L., SALVADORI O., 1992. *Minero-geochemical transformations induced by lichens in the biocalcarenite of the Selinuntine monuments.* In: DELGADO RODRIGUEZ J., HENRIQUES F., TELMO JEREMIAS F. (Eds.), *Proceedings of the 7th International Congress on Deterioration and Conservation of Stone.* LNEC, Lisbon, Portugal: 531–539.

GOW N.A.R., ROBSON G.D., GADD G.M., 1999. *The fungal colony.* University Press, Cambridge.

GRANDIS E., 1985a. *Aspetti teorici della collatura in impasto.* Cellulosa e Carta, *XXXVI, 2*: 64–70.

GRANDIS E., 1985b. *Aspetti teorici della collatura in impasto.* Cellulosa e Carta, *XXXVI, 3*: 50–56.

GRANT C., BRAVERY A.F., 1985. *A new method for assessing the resistance of stone to algal disfigurement and the efficacy of chemical inhibitors.* In: FELIX G. (Ed.), *Proceedings of the 5th International Congress on Deterioration and Conservation of Stone.* Presses Polytechniques Romandes, Lausanne: 663–674.

GREGORY P.H., 1973. *The microbiology of the atmosphere.* John Wiley & Sons, New York.

GRICHKOVA A., PETOUCHKOVA Y., TREZVOV A., 1988. *Conservation of leather and textile objects using polymers designed to resist aging and biodeterioration.* In: *Procedings of the 6th International Restorer Seminar,* Veszprém: 311–316.

GRIFFIN D.H., 1996. *Fungal physiology.* Wiley Europe Publisher.

GROTH I., SAIZ-JIMENEZ C., 1999. *Actinomycetes in hypogean environments.* Geomicrobiology Journal, *16*: 1–8.

GU J.D., 2003. *Microbiological deterioration and degradation of synthetic polymeric materials: Recent research advances.* International Biodeterioration & Biodegradation, *52*: 69–91.

GUCCI P.M.B., 1989. *Presenza di alghe nell'acqua di rete: significato igienico sanitario. Microbiologia delle acque potabili,* Collana Quaderni di Tecniche di Protezione Ambientale, *6*: 61–71.

GUIAMET P.S., GOMEZ DE SARAVIA S.G., VIDELA H.A., 1998. *Biodeteriorating microorganisms of two archaeological buildings at the site of Uxmal, Mexico.* In: *Latincorr 198.* Proceedings, S11-01. NACE International, Huston, TX.

GUIDOBALDI F., LAWLOR P.O.P., 1990. *La Basilica e l'area archeologica di S. Clemente in Roma – Guida grafica ai tre livelli.* Adup S. Clemente. Roma.

GUIDOTTI G.R., 1985. *Ruolo dell'inquinamento atmosferico nel degrado delle opere d'arte.* Acqua-Aria, *8*: 775–783.

GUILLITTE O., 1995. *Bioreceptivity: a new concept for building ecological studies.* The Science of the Total Environment, *167*: 215–220.

GURTNER C., HEYRMAN J., PIÑAR G., LUBITZ W., SWINGS J., RÖLLEKE S., 2000. *Comparative analyses of the bacterial diversity on two different biodeteriorated wall painting by DGGE and 16S rRNA sequence analysis.* International Biodeterioration & Biodegradation, *46*: 229–239.

HADJIULCHEVA E., GESHEVA R., 1982. *Actinomycetes isolated from the Boyana Church mural paintings.* Comptes Rendus de l'Academie Bulgare des Sciences, *35*: 71–74.

HAGMAR L., SCHÜTZ A., HALLBERG T., SJÖHOLM A., 1990. *Health effect of exposure to endotoxins and organic dust in poultry slaughter-house workers.* Int. Arch. Environ. Health, *62*: 159–164.

HALE M.E., 1973. *Growth.* In: AHMADJIAN V., HALE M.E. (Eds.) *The lichens,* Academic Press, New York-London: 473–492.

HAMMOND P.M., 1995. *Described and estimated species numbers: An objective assessment of current knowledge.* In: ALLSOPP D., COLWELL R.R., HAWKSWORTH D.L. (Eds.), *Microbial diversity and ecosystem function,* CAB International, Wallingford, UK: 29–71.

HANUS J., 1985. *Gamma Radiation for Use in Archives and Libraries.* The Abbey Newsletter, *9*: 34–38.

HATAKKA A., 1994. *Lignin-modifying enzymes from selected white-rot fungi: Production and role in lignin degradation.* FEMS Microbiology Reviews, *13*: 125–135.

HAVIR E.A., ANAGNOSTAKIS S.L., 1983. *Oxaloacetate acetylhydrolase activity in virulent and hypovirulent strains of Endothia (Cryphonectria) parasitica.* Physiological Plant Pathology, *26*: 1–9.

HAWKSWORTH D.L. (Ed.), 1994. *Ascomycete Systematics. Problems and Perspectives in the Nineties.* NATO Advanced Science Institutes, Series 269. Plenum Press, London, New York.

HAWKSWORTH D.L., KIRK P.M., SUTTON B.C., PEGLER D.N. 1995. *Ainsworth & Bisby's Dictionary of the Fungi.* CAB International, Wallingford, UK.

HELMI F.M., 1988. *Deterioration phenomenon in the North Temple of Karanis, near Fayoum, Egypt.* In: *Proceedings of the 6th International Congress on Deterioration and Conservation of Stone.* Nicholas Copernicus University Press, Torun, Poland: 166–174.

HEMPEL K.B., 1978. *An improved method for the vacuum impregnation of stone.* Studies in Conservation, *21 (1)*: 40–43.

HENRY F., LOPES L., BENDJILALI C., MARTEAU S., 2001. *Encapsulation of antifungal essential oils for the protection of work of art.* In: *Fungi. A Threat for the People and Cultural Heritage through Micro-Organisms,* München: REMA: 33–34.

HERNANDEZ-MARINE M., CANALS T., 1994. Herpyzonema pulverulentum *(Mastigocladaceae), a new cavernicolous atmophytic and lime-incrusted cyanophyte.* Algological Studies, *75*: 123–136.

HERNANDEZ-MARINE M., ASENCIO A.D., CANALS A., ARIÑO X., ABOAL M., HOFFMANN L., 1999. *Discovery of population of the lime-incrusting genus* Loriella *(Stigonematales) in Spanish caves.* Arch. Hydrobiol., Algological Studies, *94*: 121–138.

HERNÁNDEZ-MARINE M., CLAVERO E., ROLDÁN M., 2003a. *Why there is such luxurious growth in the hypogean environments.* Arch. Hydrobiol., Algological Studies, *109*: 229–240.

HERNANDEZ-MARINE M., GONZALES-DEL VALLE M.A., ORTIZ-MARTINEZ A., LAIZ L., SAIZ-JIMENEZ C., 2003b. *Effect of Algophase on the cyanobacterium* Gloeothece membranacea *CCAP 1430/3.* In: SAIZ-JIMENEZ (Ed.), *Molecular Biology and Cultural Heritage, Proceedings of International Congress on Molecular Biology and Cultural Heritage,* Seville. Balkema Publishers, Lisse (NL): 195–200.

HESELMEYER K., FISCHER U., KRUMBEIN W.E., WARSCHEID T., 1991. *Application of Desulfovibrio vulgaris for the bioconversion of rock gypsum crusts into calcite.* BIOforum, *1/2*: 89.

HEYRMAN J., SWINGS J., 2001. *16S rDNA sequence analysis of bacterial isolates from biodeteriorated mural paintings in the Servilia tomb (necropolis of Carmona, Seville, Spain).* Systematic Applied Microbiology, *24*: 417–422.

HEYRMAN J., MERGAERT J., SWINGS J., 1999. *Diversity of bacterial heterotrophs present in biofilms on Roman and medieval mural paintings.* In: *An International Conference on Microbiology and Conservation (ICMC '99), Of Microbes and Art. The Role of Microbial*

Communities in the Degradation and Protection of Cultural Heritage, Florence, Italy: 96–101.

HEYRMAN J., BALCAEN A., DE VOS P., SCHUMANN P., SWINGS J., 2002. Brachybacterium fresconis *sp. nov.* and Brachybacterium sacelli *sp. nov., isolated from deteriorated parts of a medieval wall painting of the chapel of Castle Herberstein (Austria).* Int. J. Syst. Evol. Microbiol., *52*: 1641–1646.

HEYRMAN J., BALCAEN A., RODRIGUEZ-DIAZ M., LOGAN N.A., SWINGS J., DE VOS P., 2003. Bacillus decoloration is *sp. nov., a new species isolated from biodeteriorated parts of the mural paintings at the Servilia tomb (Roman necropolis of Carmona, Spain) and the Saint-Catherine chapel (castle Herberstein, Austria).* International Journal Systematic Evolution Microbiology.

HIGHLEY T. L., 1995. *Comparative Durability of Untreated Wood in Use Above Ground.* International Biodeterioration & Biodegradation, *35*: 409–419.

HIGUCHI T., CHANG H.M., 1980. *Lignin Biodegradation: Microbiology, Chemistry and Potential Applications.* KIRK KENT J. (Eds.), Vol. 1, CRC Press, Florida.

HIRSCH P., ECKHARDT F.E.W., PALMER R.J., 1995a. *Fungi active in weathering of rock and stone monuments.* Canadian Journal of Botany, *73*, suppl. *1*: 1384–1390.

HIRSCH P., ECKHARDT F.E.W., PALMER R.J., 1995b. *Methods for the study of rock-inhabiting microorganisms. A mini review.* Journal of Microbiological Methods, *23*: 143–167.

HOCKING A.D., 1990. *Responses of fungi to modified atmospheres.* In: CHAMP B.R., HIGHELY E., BANKS H.J. (Eds.), *Fumigation and controlled atmosphere storage of grain, Proceedings of an International Conference,* Singapore 1989. ACIAR, Canberra, Australia, *25*: 82.

HOFFMANN L., 1989. *Algae of Terrestrial Habitats,* The Botanical Review, *55 (2)*: 77–105.

HOFFMANN L., 2002. *Caves and other low-light environments: aerophitic photoautotrophic microorganisms.* In: BITTON G. (Ed.), *Encyclopedia of Environmental Microbiology,* John Wiley & Sons, New York: 835–843.

HONEGGER R., 1991. *Functional aspects of the lichen symbiosis.* Annual Review of Plant Physiology and Plant Molecular Biology, *42*: 553–578.

HONEGGER R., 1998. *The lichen symbiosis – what is so spectacular about it?* Lichenologist, *30 (3)*: 193–212.

HONEGGER R., 2001. *The Symbiotic Phenotype of Lichen-forming Ascomycetes.* In: HOCK (Ed.), *The Mycota IX Fungal Associations.* Springer Verlag, Berlin: 165–188.

HOOG DE G.S., 1987. *Taxonomic aims in the yeast-like fungi.* In: HOOG DE G.S., SMITH M.T., WEIJMAN A.C.M. (Eds.), *Proceedings of an International Symposium on the Perspectives of Taxonomy, Ecology and Phylogeny of Yeasts and Yeast-like Fungi,* Elsevier Science Publisher, Amsterdam: 13–16.

HOOG DE G.S., 1993. *Evolution of black yeasts: Possible adaption to the human host.* Antonie van Leeuwenhoek, *63*: 105–109.

HOOG DE G.S. (Ed.), 1999. *Ecology and evolution of black yeasts and their relatives.* Studies in Mycology, CBS, The Netherlands, *43:* 208.

HOOG DE G.S., ZALAR P., URZÌ C., DE LEO F., YURLOVA N.A., STERFLINGER K., 1999. *Relationships of dothide-aceous black yeasts and meristematic fungi based on 5.8S and ITS2 rDNA sequence comparison.* Studies in Mycology, *43:* 31–37.

HOPPERT M., SCHIEWECK O., 2004. *Microbial Biofilms on the Market Gate of Miletus – A case study.* In: KWIATKOWSKI D., LÖFVENDAHL R. (Eds.), *Proceedings of the 10th International Congress on Deterioration and Conservation of Stone,* Stockholm, *1*: 233–40.

HORBERT M., BLUME H.P., ELVERS H., SUKOPP H., 1980. *Ecological contributions to urban planning.* Urban Ecology, Blackwell, London: 255–275.

HORWARD G.T., BLAKE R.C., 1998. *Growth of Pseudomonas fluorescens on a polyester-polyurethane and the purification and characterization of a polyurethanase-protease enzyme.* International Biodeterioration & Biodegradation, *42*: 213–220.

HOYOS M., SOLER V., 1993. *La cueva de Nerja (Malaga): ejemplo de degradacion microambiental.* In: FORTEA F.J. (Ed.), *La proteccion y conservacion del arte rupestre Paleolitico,* Servicio de Publicaciones del Principado de Asturias, Oviedo: 95–107.

HRUSKA DELL'UOMO K., 1979. *Sur la végétation de la classe Parietarietea muralis* Riv. Mart. 1955 *dans les Marches (Italie centrale).* Documents Phytosociologique, n.s., *4*: 433–441.

HUECK H.J., 1965. *The biodeterioration of materials as a part of hylobiology.* Material und Organismen, *1 (1)*: 5–34.

HUECK H.J., 1968. *The biodeterioration of materials – An appraisal.* In: *Biodeterioration of Materials.* Elsevier, London: 6–12.

HUECK H.J., 1972. *Textiles Pests and Their Control.* In: *Textiles Conservation,* I.E. Leene, London: 76–97.

HUGHES S.J., 1982. *Penetration by rhizoids of the moss* Tortula muralis Hedw. *into well cemented oolitic limestone.* International Biodeterioration Bulletin, *18 (2)*: 43–46.

HUMMER C.W., SCUTHWELL C.R., ALEXANDER A.L., 1968. *Corrosion of metals in tropical environments. Copper and wrought copper alloys.* Materials Protection, *7 (1)*: 41–47.

HYVERT G., 1972a. *Ile de Pasque: les statues de Rapa Nui. Conservation et restauration.* Paris, Unesco: 40.

HYVERT G., 1972b. *The conservation of Borobudur Temple.* In: Unesco document no. RMO. RD/2646/CLP.

ICCROM-ICR, 1983. *Conservazione preventiva nei musei. Il controllo dell'illuminazione, il controllo del clima.* Roma.

ILIOPOULOU-GEORGOUDAKI J., PANTAZIDOU A., THEOULAKIS P., 1993. *An assessment of cleaning photoautotrophic microflora: The case of Perama cave,* Ioannina, Greece. Mem. Biospel., *20*: 117–120.

IMPAGLIAZZO G., RUGGIERO D., 2002. *Struttura e composizione della carta.* In: *Chimica e biologia applicata alla conservazione degli archivi,* Ministero per i Beni e le Attività Culturali, Direzione Generale per gli Archivi, Roma: 25–41.

IMPRESCIA U., 1995. *L'illuminazione scenografica delle grotte turistiche e problemi di salvaguardia ambientale.* In: *Atti Conv. Ass. It. di Illuminazione,* Ancona.

IMPRESCIA U., MUZI F., 1989. *Analisi dei risultati sperimentali riguardanti l'illuminazione di grotte turistiche.* Luce, *3*: 127–131.

INCERTI C., BLANCO-VARELA M.T., PUERTAS F., SAIZ-JIMENEZ C., 1997. *Halotolerant and halopilic bacteria associated to efflorescences in Jerez cathedral.* In: ZEZZA F. (Ed.), *Origin, mechanisms and effects of salts on degradation of monuments in marine and continental environments,* Protection and Conservation of the

European Cultural Heritage Research Report, *4*: 225–232.

INOUE M., KOYANO M., 1991. *Fungal contamination of oil paintings in Japan*. International Biodeterioration, *28 (1–4)*: 23–26.

IONITA I., 1971. *Contributions to the study of the biodeterioration of the works of art and of historic monuments. II. Species of fungi isolated from oil and tempera paintings*. Rev. Roum. Biol. Ser. Bot., *16*: 377–381.

ISO/DIS 11799, 2003. *Information and documentation. Document storage requirements for archive and library materials.*

ISO 18902, 1999. *Imaging materials. Processed photographic films, plates and papers – filing enclosures and storage containers.*

ISO 18911, 2000. *Imaging materials. Processed safety photographic films – storage practice.*

ISO 18918, 2000. *Imaging materials. Processed photographic plates – storage practice.*

ISO 18920, 2000. *Imaging materials. Processed photographic reflection prints – storage practice.*

IVERSON W.P., 1968. *Mechanisms of microbial corrosion*. In: WALTERS A.H., ELPHICK J.J. (Eds.), *Biodeterioration of materials. Proceedings of the 1st International Symposium*, Southampton. Elsevier Pub., Amsterdam: 28–43.

JACOB M.G., 1991. *Il dagherrotipo a colori: dalla tecnica alla conservazione*. Kermes, *4*: 51–54.

JAMES C., CORRIGAN C., ENSHAIAN M.C., GRECA M.R., 1991. *Manuale per la conservazione e il restauro di disegni e stampe antiche*. Olschki, Firenze.

JATON C., 1972. *Aspects microbiologiques des alterations des pierres de monuments*. In: *1er Colloque International sur la deterioration des pierres en oeuvre*, La Rochelle: 149–154.

JENNINGS D.H., LYSEK G.C., 1999. *Fungal Biology: Understanding the fungal lifestyle*. Bios Scientific Publishers, Springer.

JEYARAJ V., 1983. *Ethnological collection, conservation problems and their solution in Government Museum*. Madras. J Indian Museums, *39*: 123–128.

JOHN D.M., WHITTON B.A., BROOK A.J., 2003. *The Freshwater Algal Flora of the British Isles*. Cambridge Press.

JONES A.M., MOUZOURAS R., PITMAN A.J., POINTING S.B., 2003. *Preserving the timbers of the Tudor warship Mary Rose*. In: KOESTLER R.J., KOESTLER V.H., CHAROLA A.E., NIETO-FERNANDEZ F.E. (Eds.), *Art, Biology, and Conservation: Biodeterioration of works of art*. The Metropolitan Museum of Art, New York: 28–49.

JONES D., WILSON M.J., 1985. *Chemical activity of lichens on mineral surfaces. A review*. International Biodeterioration, *21*: 99–104.

JONES E.B., TURNER R.D., FURTADO S.E.J., KÜHNE H., 2001. *Marine biodeteriogenic organisms. I. Lignicolous fungi and bacteria and the wood-boring mollusca and crustacean*. International Biodeterioration & Biodegradation, *48*: 112–126.

JONES M.S., WAKEFIELD R.D., 2000. *A study of biologically decayed sandstone with respect to Ca and its distribution*. In: FASSINA V. (Ed.), *Proceedings of the 9th International Congress on Deterioration and Conservation of Stone*, Venice. Elsevier, Amsterdam: 473–481.

JORDAN B.A., 2001. *Site characteristics impacting the survival of historic waterlogged wood: A review*. International Biodeterioration & Biodegradation, *47 (1)*: 47–54.

KARBOWSKA-BERENT J., 2003. *Microbiodeterioration of mural paintings: A review*. In: KOESTLER R.J., KOESTLER V.H., CHAROLA A.E., NIETO-FERNANDEZ F.E. (Eds.), *Art, Biology, and Conservation: Biodeterioration of works of art*. The Metropolitan Museum of Art, New York: 266–301.

KARBOWSKA-BERENT J., STRZELCZYK A.B., 2000. *The Role of Streptomycetes in the Biodeterioration of Historic Parchment*. Wydawnictwo Uniwersytetu Mikolaja Kopernika, Torun.

KARPOVICH-TATE N., REBRIKOVA N.L., 1991. *Microbial communities on damaged frescoes and building materials in the cathedral of the Nativity of the Virgin in the Pafnutii-Borovskii Monastery, Russia*. International Biodeterioration, *27 (3)*: 281–296.

KELLER N.D., FREDERICKSON A.F., 1952. *The role of plants and colloid acids in the mechanisms of weathering*. American Journal of Sciences, *250*: 594–608.

KIRK P.M., CANNON P.F., STALPERS J.A. (Eds.), 2001. *Dictionary of Fungi*. Cabi Publishing, 9th edition.

KOCH V., 1993. *Composizione degli impasti per carta e fabbricazione della carta*. In: *Atti del Corso di formazione per tecnici di cartiera*, Fabriano: 3–6.

KOESTLER R.J., 2000. *Polymers and resins as food for microbes*. In: CIFERRI O., TIANO P., MASTROMEI G. (Eds.), *Of Microbes and Art. The Role of Microbial Communities in the Degradation and Protection of Cultural Heritage*, Kluwer Academic/Plenum Publisher, New York: 153–167.

KOESTLER R.J., SALVADORI O., 1996. *Methods of evaluating biocides for the conservation of porous building materials*. Science and Technology for Cultural Heritage 5 (1): 63–68.

KOESTLER R.J., SANTORO E.D., 1988. *Assessment of the Susceptibility to Biodeterioration of Selected Polymers and Resins*. Final Report, The Getty Conservation Institute, Los Angeles: 1–96.

KOESTLER R.J., CHAROLA A.E., WYPYSKI M., LEE J.J., 1985. *Microbiologically induced deterioration of dolomitic and calcitic stone as viewed by scanning electron microscopy*. In: FELIX G. (Ed.), *Proceedings of the 5th International Congress on Deterioration and Conservation of Stone*. Presses Polytechniques Romandes, Lausanne: 617–626.

KOESTLER. R.J., SANTORO E.D., PREUSSER F., RODARTE A., 1986. *A note on the reaction of methyl tri-methoxy silane to mixed cultures of microorganisms*. In: OREAR C.E., LEWELLYN G.C. (Eds.), *Biodeterioration Research I*, Plenum Press, New York: 317–321.

KOESTLER. R.J., SANTORO E.D., DRUZIK J., PREUSSER J., KOEPP L., DERRICK M., 1988. *Status report. Ongoing studies of the susceptibility of stone consolidants to microbiologically induced deterioration*. In: HOUGHTON D.R., SMITH R.N., EGGINS H.O.W. (Eds.), *Biodeterioration 7*, Elsevier Applied Science, New York: 441–448.

KOESTLER R.J., PARREIRA E., SANTORO E.D., NOBLE P., 1993. *Visual effects of selected biocides on easel painting materials*. Studies in Conservation 38 (4): 265–273.

KOESTLER R.J., WARSCHEID T., NIETO F., 1997. *Biodeterioration: risk factors and their management*. In:

BAER N.S., SNETHLAGE R. (Eds.), *Saving our architectural heritage: The conservation of historic stone structures*, John Wiley & Sons Ltd.: 25–36.

KOESTLER R.J, SARDJONO S., KOESTLER D.L., 2000. *Detection of insect infestation in museum objects by carbon dioxide measurement using FTIR*. International Biodeterioration & Biodegradation, 46(4): 285–292.

KOHLMEYER J., 1969. *Deterioration of wood by marine fungi in the deep sea*. In: *Materials Performance in the Deep Sea*. Special Technical, American Society for Testing Materials, 445: 20–30.

KÖPPEN W., 1936. *Das geographische system der klimate*. In: *Handbuch der Klimatologie*, Vol. 1.

KOVACIK L., 2000. *Cyanobacteria and algae as agents of biodeterioration of stone substrata of historical buildings and other cultural monuments*. In: CHOI S., SUH M., (Eds.), *New Millennium International Forum on Conservation of Cultural Property*, Institute of Conservation Science for Cultural Heritage, Kongiu, Korea: 44–58.

KOWALIK R., 1980a. *Microbiodeterioration of library materials. Part 1*. Restaurator, 4: 99–114.

KOWALIK R., 1980b. *Microbiodecomposition of basic organic library materials. Microbial deterioration of library materials. Part 2*. Restaurator, 4 (3–4): 135–219.

KRAMER G., KAINKA E., WILDFÜHR W., 1998. *Investigations according to mould fungi in the indoor area and at exponates of castle museum in Saxonia*. In: *Proceedings of the 6th International Congress on Aerobiology*, Perugia, Italy: 271.

KRUMBEIN W.E., 1983. *Microbiological Geochemistry*. Blackwell, Oxford.

KRUMBEIN W.E., 1988. *Biotransformations in monuments, a sociological study*. Durability of Building Materials, 5: 359–372.

KRUMBEIN W.E., 1992. *Colour change of building stone and their direct and indirect biological causes*. In: DELGADO RODRIGUEZ J., HENRIQUES F., TELMO JEREMIAS F. (Eds.), *Proceedings of the 7th International Congress on Deterioration and Conservation of Stone*. LNEC, Lisbon, Portugal: 443–452.

KRUMBEIN W.E., DIAKUMAKU E., 1996. *The role of fungi in the deterioration of stone*. In: *Interactive physical weathering and bioreceptivity study on building stones, monitored by Computerized X-Ray Tomography (CT) as a potential non-destructive research tool*. Protection and Conservation of the European Cultural Heritage, Research Report, 2: 140–170.

KRUMBEIN W.E., URZÌ C., GEHRMANN C., 1991. *Biocorrosion and biodeterioration of antique and medieval glass*. Geomicrobiol. Journal, 9: 139–165.

KRUMBEIN W.E., GORBUSHINA A., RUDOLPH C., URZÌ C., 1993. *Biological investigation on the question of organic and inorganic eutrophication-induced biocorrosions and biogenic deposition on late medieval church windows of the cathedral of Tours and St. Catharina in Oppenheim*. In: WELCK S., (Ed.), *Gemeinsames Erbe gemeinsam erhalten. 1*. Statuskolloquium des deutsch-fransösischen Forschungsprogramms zur Erhaltung von Baudenkmälern, Secretaire General du Programme Franco-Allemand de Recherche pour la Conservation des Monuments, Champs-sur-Marne: 269–275.

KRUMBEIN W.E., BRIMBECOMBLE P., COSGROVE D.E., STANIFORTH S. (Eds.), 1994. *Durability and change: the science, responsibility, and cost of sustaining cultural heritage*. John Wiley & Sons, Chichester, UK.

KUMAR R., KUMAR A.V., 1999. *Biodeterioration of stone in tropical environments. An overview*. The Getty Conservation Institute, Los Angeles: 1–95.

KUROCZIN J., KRUMBEIN W.E., 1987. *Studies on microbial degradation of ancient leather bookbinding. Part 1*. International Biodeterioration Bulletin, 23: 3–27.

KUTSCHERA L., LICHTENEGGER E., 1997. *Wurzeln. Bewurzelung von Pflanzen in verschiedenen Lebensräumen*. Stapfia, 49: 1–331.

KUTSCHERA L., LICHTENEGGER E., 2002. *Wurzelatlas-Mitteleuropaischer Waldbäume und Sträucher*. Leopold Stocker Verlag, Graz-Stuttgart.

KUTZNER H.J., 1981. *The Family Streptomycetaceae*. In: STARR M.P., STOLP H., TRÜPER H.G., BALOWS A., SCHLEGEL H.G. (Eds.), *The Prokaryotes. A handbook on habitats, isolation, and identification of bacteria*. Springer Verlag, Berlin, 2: 2028–2117.

LA CONO V., URZÌ C., 2003. *Fluorescent in situ hybridization (FISH) applied on samples taken with adhesive tape strips*. J. Microbiol. Methods, 55: 65–71.

LAIZ L., HERMOSIN B., CABALLERO B., SAIZ-JIMENEZ C., 2000a. *Bacteria isolated from the rocks supporting paintings in two shelters from Sierra de Cazorla, Jaen, Spain*. Aerobiologia, 16: 119–124.

LAIZ L., RECIO D., HERMOSIN B., SAIZ-JIMENEZ C., 2000b. *Microbial communities in salt efflorescences*. In: CIFERRI O., TIANO P., MASTROMEI G. (Eds.), *Of Microbes and Art. The Role of Microbial Communities in the Degradation and Protection of Cultural Heritage*. Kluwer Academic/Plenum Publishers, New York: 77–88.

LAIZ L., HERMOSIN B., SAIZ-JIMENEZ C., 2002. *Biodegradation of pollutants in urban environments*. In: GALÁN E., ZEZZA F. (Eds.), *Protection and Conservation of the Cultural Heritage of the Mediterranean Cities, Proceedings of the 5th International Symposium on the Conservation of Monuments in the Mediterranean Basin*, Seville: 179–182.

LAL B.B., 1978. *Weathering and preservation of stone monuments under tropical conditions: Some case histories*. In: *Altération et protection des monuments en pierre, Colloque International*, Unesco-Rilem, Paris: 7.8.

LAL GAURI K., CHOWDHURY A.N., KULSHRESHTHA N.P., PUNURU A.R., 1989. *The sulphation of marble and the treatment of gypsum crust*. Studies in Conservation, 34: 201–206.

LAL GAURI K., PARKS L., JAYNES J., ATLAS R., 1992. *Removal of sulphated-crust from marble using sulphate-reducing bacteria*. In: WEBSTER R.G.M. (Ed.), *Proceedings of International Conference on Stone Cleaning and the Nature, Soiling and Decay Mechanisms of Stone*, Donheadd, Edinburgh, UK: 160–165.

LALLEMANT R., DERUELLE S., 1978. *Presence de lichens sur les monuments en pierre: nuisance ou protection?* In: *Proceedings of the International Symposium on Alteration and Protection of Stone Monuments*, Unesco-Rilem, Paris, 2 (4.6): 1–6.

LAMENTI G., TIANO P., TOMASELLI L., 2000a. *Microbial communities dwelling on marble statues*. In: MONTE M. (Ed.), *Euromarble, Proceedings of the 8th Workshop*

EUROCARE- EUROMARBLE EU 496, CNR, Roma: 83–87.

LAMENTI G., TIANO P., TOMASELLI L., 2000b. *Biodeterioration of ornamental marble statues in the Boboli Garden (Florence, Italy).* Journal of Applied Phycology, *12*: 427–433.

LANDOLT A., 1977. *Okologische Zeigerwerte zur Schweizer Flora*, Veröffentlichungen des geobotanischen Institutes der eidg. Techn. Hochschule, Stiftung Rübel, Zurich, *64*: 46–63.

LARCHER W., 1975. *Physiological Plant Ecology.* Springer Verlag, Berlin-Heidelberg, New York.

LATTANZI E., TILIA A., 2004. *Area archeologica di Ostia Antica: analisi floristica preliminare e relativa valutazione della pericolosità.* In: *Atti del 40° Congresso della Società Italiana di Fitosociologia*, Roma: 19.

LAURENTI M.C., SALERNO C.S., FAZIO G., 1998. *La conservazione in situ.* In: PANDOLFI A.M., STELLA SPAMPINATO M.L. (Eds.), *Diagnosi e progetto per la conservazione dei materiali dell'architettura*, De Luca, Roma: 87–101.

LAURENTI M.C., 2000. *Strategie operative attuali per la conservazione delle aree archeologiche.* In: *Coperture per aree e strutture archeologiche: Repertorio di casi esemplificativi.* Arkos (Supplemento), I Grandi Restauri *1*: 6.

LAURENTI M.C. (Ed.), 2000. *Coperture delle aree archeologiche: Museo aperto. Progettazione di materiali/componenti e sistemi per la conservazione e la fruizione di siti archeologici.* Gangemi, Roma.

LAURENTI M.C., ALTIERI A., 2000. *Materiali e tecniche per la protezione dei mosaici pavimentali nelle aree archeologiche.* AISCOM, VI Coll.: 727–738.

LAURENTI TABASSO M., NUGARI M.P., PIETRINI A.M., SANTAMARIA U., 1991. *Sperimentazione per la scelta di un protettivo da applicare ai materiali costitutivi.* In: *Fontana di Trevi: la storia, il restauro.* Edizioni Carte Segrete, Roma: 167–173.

LAVÉDRINE B., 1990a. *Il dagherrotipo.* Kermes, *3*: 755–766.

LAVÉDRINE B., 1990b. *La conservation des photographies.* Presses du CNRS, Paris.

LAVÉDRINE B., GANDOLFO J.P., MONOD S., 2000. *Les collections photographiques: guide de conservation préventive.* Association pour la recherche scientifique sur les arts graphiques, Paris.

LAZAR G., SCHRÖDER F.R., 1992. *Degradation of lipids by fungi.* In: WINKELMANN G., WEINHEIM V.C.H. (Eds.), *Microbial degradation of natural products*: 267–291.

LAZZARINI L., SALVADORI O., 1989. *A reassessment of the formation of the patina called 'scialbatura.'* Studies in Conservation, *34*: 20–26.

LEE K.B., WEE Y.C., 1982. *Algae growing on walls around Singapore.* Malayan Nature Journal, *35*: 125–32.

LEEMING K., MOORE C.P., DENYER S.P., 2002. *The use of immobilised biocides for process water decontamination.* International Biodeterioration & Biodegradation, *49*: 39–43.

LEFEVRE M., 1974. *La "maladie verte" de Lascaux.* Studies in Conservation, *19*: 126–156.

LEHNINGER A.L., NELSON D.L., COX M.M., 1994. *Principi di biochimica.* Zanichelli, Bologna, 2nd ed.

LEITE MAGALHAES S., SEQUEIRA BRAGA M.A., 2000. *Biological colonization features on a granite monument from Braga (NW, Portugal).* In: FASSINA V. (Ed.), *Proceedings of the 9th International Congress on Deterioration and Conservation of Stone*, Venice. Elsevier, Amsterdam: 521–529.

LEONI M., 1984. *Elementi di metallurgia applicata al restauro delle opere d'arte.* OpusLibri, Firenze.

LEVI Y., ALBECK S., BRACK S., WEINER A., ADDADI L., 1998. *Control over aragonite crystal nucleation and growth: An* in vitro *study on biomineralization.* Chem. Eur. J., *4*: 389–396.

LEWIN S.Z., CHAROLA A.E., 1981. *Plant life on stone surfaces and its relation to stone conservation.* Scanning Electron Microscopy: 563–568.

LEWIS F., MAY E., BRAVERY A.F., 1985. *Isolation and enumeration of autotrophic and heterotrophic bacteria from decayed stone.* In: FELIX G. (Ed.), *Proceedings of the 5th International Congress on Deterioration and Conservation of Stone.* Presses Polytechniques Romandes, Lausanne: 633–642.

LEZNICKA S., 1992. *Antimicrobial protection of stone monuments with p-hydroxybenzoic acid esters and silicone resin.* In: DELGADO RODRIGUES J., HENRIQUES F., TELMO JEREMIAS F. (Eds.), *Proceedings of the 7th International Congress on Deterioration and Conservation of Stone*, LNEC, Lisbon, Portugal, *1*: 481–490.

LEZNICKA S., KUROCZKIN J., KRUMBEIN W.E., STRZELCZYK A.B., PETERSEN K., 1991. *Studies on the growth of selected fungal strains on limestones impregnated with silicone resins (Steinfestiger H and Elastosil E-41).* International Biodeterioration, *28*: 91–111.

LIGRONE R., 1993. *Le basi morfologiche della conduzione degli elaborati nelle Briofite.* Informatore Botanico Italiano, *25*: 256–259.

LIM G., TAN T.K., TOH A., 1989. *The fungal problem in buildings in the humid tropics.* International Biodeterioration, *25 (1–3)*: 27–38.

LISCI M., PACINI E., 1993. *Plant growing on the walls of Italian towns 2. Reproductive Ecology.* Giornale Botanico Italiano, *127*: 1053–1078.

LISCI M., MONTE M., PACINI E., 2003. *Lichens and higher plants on stone: A review.* International Biodeterioration & Biodegradation, *51*: 1–17.

LITTLE B., WAGNER P.A., LEWANDOWSKI Z., 1998. *The role of biomineralization in microbiologically influenced corrosion.* Corrosion 98, NACE Int., *294*: 1–18.

LORD B., LORD. G.D. (Eds.), 2002. *The Manual of Museum Exhibitions.* Altamira Press.

LOW G.A., YOUNG M.E., MARTIN P., PALFREYMAN J.W., 2000. *Assessing the relationship between the dry rot fungus* Serpula lacrymans *and selected forms of masonry.* International Biodeterioration & Biodegradation, *46*: 141–150.

LUGAUSKAS A., LEVINSKAITE L., PECIULYTE D., 2003. *Micromycetes as deterioration agents of polymeric materials.* International Biodeterioration & Biodegradation, *52*: 233–242.

LULL W.P., 1995. *Conservation Environment Guidelines for Libraries and Archives.* Canadian Council of Archives, Ottawa.

MACAULEY B.J., GRIFFIN D.M., 1969. *Effects of carbon dioxide and oxygen on the activity of some soil fungi.* Trans. Br. Mycol. Soc., *53*: 53–62.

MACQUEEN M., 1980. *Conservation of two Roman wheels with iron tyres from the Roman fort at Newstead.* The

Laboratories of the National Museum of Antiquities of Scotland, Edinburgh: 30–35.

MAGDEFRAU K., 1982. *Life forms of bryophytes.* In: SMITH A.J.F. (Ed.), *Bryophyte Ecology.* Chapman & Hall, London: 45–58.

MAGGI O., MANDRIOLI P., RANALLI G., SORLINI C., 1998a. *Il monitoraggio della componente biologica dell'aria.* In: MANDRIOLI P., CANEVA G. (Eds.), *Aerobiologia e beni culturali,* Nardini Editore, Fiesole: 121–123.

MAGGI O., MANDRIOLI P., RANALLI G., SORLINI C., 1998b. *Strategia delle tecniche di campionamento aerobiologico.* In: MANDRIOLI P., CANEVA G. (Eds.), *Aerobiologia e beni culturali,* Nardini Editore, Fiesole: 146–159.

MAGGI O., PIETRINI A.M., PIERVITTOI R., RICCI S., ROCCARDI A., 1998c. *L'aerobiologia applicata ai beni culturali.* In: MANDRIOLI P., CANEVA G. (Eds.), *Aerobiologia e beni culturali,* Nardini Editore, Fiesole: 32–33.

MAGGI O., PERSIANI A.M., GALLO F., VALENTI P., PASQUARIELLO G., SCLOCCHI M.C., SCORRANO M., 2000. *Airborne fungal spores in dust present in archives: Proposal for a detection method, new for archival materials.* Aerobiologia, 16: 429–434.

MAHOMED R.S., 1971. *Antibacterial and antifungal finishes.* In: MARK H., WOODING N.S., ATLAS S.M. (Eds.), *Chemical aftertreatment of textile.* Wiley Interscience New York, London: 507–552.

MAKES F., 1988. *Enzymatic consolidation of the portrait of Rudolf II as "Vertumnus" by Giuseppe Arcimboldo with a new multi-enzyme preparation isolated from Antarctic Krill* (Euphasia superba). Acta University Gotheburg, 23: 98–110.

MALAGODI M., NUGARI M.P., ALTIERI A., LONATI G., 2000. *Effects of combined application of biocides and protectives on marble.* In: FASSINA V. (Ed.), *Proceedings of the 9th International Congress on Deterioration and Conservation of Stone,* Venice. Elsevier, Amsterdam, 2: 225–233.

MAMBELLI R., RACAGNI B., DONATI F., FIORI C., 1989. *Influenza del trattamento con alcuni biocidi su tessere provenienti da mosaici della villa di Casignana (RC).* Quaderni IRTEC, Mosaico e restauro musivo, 2: 31–48.

MANDRIOLI P., 1998. *Principi di dispersione atmosferica delle particelle.* In: MANDRIOLI P., CANEVA G. (Eds.), *Aerobiologia e beni culturali,* Nardini Editore, Fiesole: 124–130.

MANDRIOLI P., CANEVA G. (Eds.), 1998. *Aerobiologia e beni culturali: metodologie e tecniche di misura.* Nardini Editore, Fiesole.

MANSCH R., BOCK E., 1998. *Biodeterioration of natural stone with special reference to nitrifying bacteria.* Biodegradation, 9 (1): 47–64.

MANSCH R., PINCK C., URZÌ C., DE LEO F., SALAMONE P., 1999. *Microbial colonisation of silicone treated mortar at Schloss Weissenstein in Pommersfelden, Germany.* In: *An International Conference on Microbiology and Conservation (ICMC '99). Of Microbes and Art. The Role of Microbial Communities in the Degradation and Protection of Cultural Heritage,* Florence, Italy: 200–205.

MANTOVANI O., 2002. *Degradazione del materiale cartaceo.* In: CENTRO DI FOTORIPRODUZIONE LEGATORIA E RESTAURO DEGLI ARCHIVI DI STATO (Ed.), *Chimica e biologia applicate alla conservazione degli archivi:* 297–320. Union Printing S.p.A., Roma.

MARA D.D., WILLIAMS D.J.A., 1972. *The mechanism of sulphide corrosion by sulphate-reducing bacteria.* Biodeterioration of Materials. In: *Atti Int. Biodet. Symp.,* Lunteren 1971, Applied Science Pub., London: 103–113.

MARAVELAKI P.V., ZAFIROPULOS V., KYLIKOGLOU V., KALAITZAKI M., FOTAKIS C., 1996. *Diagnostic techniques for laser cleaning of marble.* In: RIEDERER J. (Ed.), *Proceedings of the 8th International Congress on Deterioration and Conservation of Stone.* Berlin, Germany: 1395–1404.

MARCIANO P., DI LENNA P., MAGRO P., 1983. *Oxalic acid, cell wall-degrading enzymes and pH in pathogenesis and their significance in the virulence of two* Sclerotinia sclerotiorum *isolates on sunflower.* Physiological Plant Pathology, 22: 339–345.

MARCONE A.M., PARIS M., BUZZANCA G., LUCARELLI G., GERARDI G., GIOVAGNOLI A., IVONE A., NUGARI M.P., 2001. *Il progetto ICR di manutenzione e controllo della Galleria Doria Pamphilj: schedatura conservativa e monitoraggio ambientale.* Bollettino ICR. Nuova Serie, 2: 44–67.

MARESCOTTI P., 2002. *Ruolo dei microorganismi nei processi di alterazione dei vetri vulcanici basaltici: meccanismi, evidenze tessiturali e metodi di studio.* Plinius, 27: 195–202.

MARGALEF R., 1974. *Ecologia.* Omega, Barcellona.

MARKHAM P., BAZIN M.J., 1991. *Decomposition of cellulose by fungi.* In: ARORA D.K., RAI B., MUKERJI KNUDSEN G.R. (Eds.), *Handbook of Applied Mycology,* Marcel Dekker, New York, 1: 379–424.

MARKOVA N., WADSÖ L., 1998. *A microcalorimetric method of studying mould activity as a function of water activity.* International Biodeterioration & Biodegradation, 42: 25–28.

MARTINES G., 1983. *Marmo e restauro dei monumenti antichi: estetica delle rovine, degrado delle strutture all'aperto, una ipotesi di lavoro.* In: *Atti del Convegno "Marmo Restauro. Situazioni e Prospettive,"* Carrara: 83–92.

MARTINEZ P., CASTRO J., SANCHEZ A., 1994. *The biodeterioration of cultural property in Cuba.* In: GARG K.L., GARG N., MUKERJI K.J. (Eds.), *Recent advances in biodeterioration and biodegradation,* Naya Prokash, Calcutta, India, 1: 379–397.

MARTINEZ M., MARTINEZ P.C., LAVERDE P., GAYLARDE C.C., 2003. *Microbiological studies of biofilm present on stones from the National Museum building, Bogotà, Colombia.* In: SAIZ-JIMENEZ (Ed.), *Molecular Biology and Cultural Heritage, Proceedings of International Congress on Molecular Biology and Cultural Heritage,* Seville. Balkema Publishers, Lisse (NL): 259–262.

MARTINEZ-RAMIREZ S., PUERTAS F., BLANCO-VARELA M.T., THOMPSON G.E., 1998a. *Effect of dry deposition of pollutants on the degradation of lime mortars with sepiolite.* Cement and Concrete Research, 28 (1): 125–133.

MARTINEZ-RAMIREZ S., PUERTAS F., BLANCO-VARELA M.T., THOMPSON G.E., ALMENDROS P., 1998b.

Behavior of repair lime mortars by wet deposition process. Cement and Concrete Research, *28 (2)*: 221–229.

MASSARI G., MASSARI I., 1993. *Damp buildings, old and new.* ICCROM, Roma.

MATÈ D., 2002. *Il biodeterioramento dei supporti archivistici.* In: *Chimica e biologia applicate alla conservazione degli archivi.* Ministero per i Beni e le Attività Culturali, Direzione Generale degli Archivi di Stato, Istituto Poligrafico e Zecca dello Stato, Roma: 405–425.

MATÈ D., RESIDORI L., 2002. *La conservazione delle fotografie.* In: *Chimica e biologia applicate alla conservazione degli Archivi.* Centro di fotoriproduzione legatoria e restauro degli Archivi di Stato, Ministero per i Beni e le Attività Culturali. Direzione Generale degli Archivi di Stato, Roma: 475–487.

MATÈ D., RUSCHIONI E. (forthcoming). *Progetto di manutenzione per l'Archivio di Stato di Viterbo. Linee guida per un corretto intervento di spolveratura.* Pubblicazioni degli Archivi di Stato, Ministero per i Beni e le Attività Culturali.

MATÈ D., SCLOCCHI M.C. (forthcoming). *La ventilazione come fattore di controllo della crescita microbica su arredi metallici all'interno di ambienti confinati. Un'esperienza di lavoro nell'Archivio di Stato di Macerata.* In: *Atti della Conferenza Internazionale,* Dobbiaco 2002.

MATÈ D., SCLOCCHI M.C., RUGGIERO D., 2002. *I materiali fotografici e il loro deterioramento biologico.* Kermes, anno XV, *47*: 41–53.

MATTEINI M., MOLES A., 1986. *Le patine di ossalato sui manufatti in marmo.* O.P.D. Restauro (Num.spec.) Restauro del marmo/Opere e problemi: 65–73.

MATTEINI M., MOLES A., 1989. *La chimica nel restauro.* Nardini Editore, Firenze.

MATTIROLO O., 1928. *I licheni e la malattia delle vetrate antiche.* Riv. Archeol. della Provincia e Antica Diocesi di Como, fasc. *94–95*: 1–23.

MAY E., PAPIDA S., ABDULLA H., TAYLER S., 2000. *Comparative studies of microbial communities on stone monuments in temperate and semi-arid climates.* In: CIFERRI O., TIANO P., MASTROMEI G. (Eds.), *Of Microbes and Art. The Role of Microbial Communities in the Degradation and Protection of Cultural Heritage,* Kluwer Academic/Plenum Publishers, New York: 49–62.

MAY E., PAPIDA S., ABDULLA H., 2002. *Consequences of microbe-biofilm-salt interactions for stone integrity in monuments.* In: KOESTLER R.J., KOESTLER V.R., CHAROLA A.E., NIETO-FERNANDEZ F.E. (Eds.), *Art, Biology, and Conservation: Biodeterioration of Works of Art.* The Metropolitan Museum of Art, New York: 452–471.

MAZZOLENI S., RICCIARDI M., APRILE G., 1989. *Aspetti pionieri della vegetazione del Vesuvio.* Studi sul territorio (Annali di Botanica), *47 (6)*: 97–109.

MAZZONE B., 1999. *L'illuminazione quale parte integrante del progetto per il restauro finalizzato alla fruizione dei monumenti, esemplificata dalle case history delle installazioni per S. Pietro in Montorio e San Clemente a Roma.* Flare: 18–25.

McCARTHY B.J., 1987. *Rapid methods for the detection of biodeterioration in textiles.* International Biodeterioration, *23*: 357–364.

McLEOD I.D., RICHARDS V.L., 1996. *The impact of metal corrosion products on the degradation of waterlogged wood recovered from historic shipwreck.* In: HOFFMANN P., GRANT T., SPRIGGS J.A., DALEY T. (Eds.), *Proceedings of the 6th ICOM Group on Wet Organic Archaeological Materials Conference,* York: 331–349.

McNAMARA C., PERRY T.D., ZINN M., BREUKER M., MÜLLER R., HERNANDEZ-DUQUE G., MITCHELL R., 2003. *Microbial processes in the deterioration of Maya archaeological buildings in southern Mexico.* In: KOESTLER R.J., KOESTLER V.H., CHAROLA A.E., NIETO-FERNANDEZ F.E. (Eds.), *Art, Biology, and Conservation: Biodeterioration of Works of Art.* The Metropolitan Museum of Art, New York: 248–265.

McNEIL M.B., MOHR D.W., LITTLE B.J., 1991. *Correlation of laboratory results with observations on long-term corrosion of iron and copper alloys.* In: *2nd Symposium on Materials Issues in Art and Archaeology,* San Francisco, MRS: 753–759.

MELI P., ALONGI G., 1995. *Progetto di restauro del teatro greco di Eraclea Minoa.* Soprintendenza Beni Culturali e Ambientali di Agrigento, Agrigento.

MELLOR E., 1923. *Lichens and their action on the glass and leadings of church windows.* Nature, *112*: 299–300.

MELUCCO A., 1996. *Philosophies favoring in situ conservation.* In: *Proceedings of the 6th Conference of ICCM "Mosaics make a site. The conservation in situ of mosaics in archaeological sites,"* Nicosia, Cyprus: 17–22.

MENETREZ M., FOARDE K., 2002. *Microbial volatile organic compound emission rates and exposure model,* Indoor and Built Environment, *11 (4)*: 208–213.

MENICALI U., 1992. *I materiali nell'edilizia storica. Tecnologia e impiego dei materiali tradizionali.* La Nuova Italia Scientifica, Roma.

MENICALI U., 1993. *Tessuto o non tessuto?* Costruire, *127*: 93–97.

METRO B., GRZYWACZ C., 1992. *A showcase for preventive conservation.* In: *3eme Colloque international de ARAAFU,* Paris: 207–210.

MICHAELIDES D., 2001. *The International Committee for the conservation of mosaics: Profiles and strategies.* ICCM Newsletter, *11*: 8–14.

MINA A., MODICA G., 1987. *L'arte della fotografia. La stampa d'arte negli antichi procedimenti fotografici.* Hoepli Editore, Milano.

MINISTERO PER I BENI E LE ATTIVITÀ CULTURALI, 2000. *Atto di indirizzo sui criteri tecnico scientifici e sugli standard di funzionamento e sviluppo dei musei.* D.L. n. 112/98 art. 150 comma 6. Elaborati del gruppo di lavoro (D.M. 25/7/2000).

MIRAVALLE R., 1990. *Strategie per la gestione della vegetazione nella regione archeologica di Pompei.* In: MASTROROBERTO M. (Ed.), *Archeologia e Botanica, Atti Convegno di studi sul contributo della botanica alla conoscenza e conservazione delle aree archeologiche,* Pompei. L'Erma di Bretschneider, Roma: 85–94.

MISHRA A.K., GARG K.L., KAMAL K.J., 1995a. *Microbiological deterioration of stone – An overview.* Conservation, Preservation and Restoration: Tradition, Trends and Techniques: 217–228.

MISHRA A.K., JAIN K.K., GARG K.L., 1995b. *Role of higher plants in the deterioration of historic buildings.* The Science of the Total Environment, *167*: 375–392.

MITCHELL R., GU J-D., 2000. *Changes in the biofilm microflora of limestone caused by atmospheric pollutants.*

International Biodeterioration & Biodegradation, *46*: 299–303.

MODENESI P., LAJOLO L., 1988. *Microscopical investigation on a marble encrusting lichen.* Studia Geobotanica, *8*: 47–64.

MODENESI P., PIANA M., PINNA D., 1998. *Surface features in* Parmelia sulcata *(Lichenes) thalli growing in shaded or exposed habitats.* Nova Hedwigia, *66*: 535–547.

MONA-I-FAHD, 1994. *Biodeterioration of the mural paintings of the Tomb of Tutankhamum and its conservation.* Zeitschrift für Kunsttechnologie und Konservierung.

MONTACUTELLI R., TARSITANI G., MAGGI O., GABRIELLI N., 1992. *Metodi di controllo microbiologico degli affreschi: l'esempio del restauro della Cappella Sistina.* In: *Proceedings of the 3rd International Conference of Non-Destructive Testing, Microanalytical Methods and Environment Evaluation for Study and Conservation of Works of Art*, Viterbo, Italy: 1053–1064.

MONTACUTELLI R., FIORILLA M., FABIANI M., FLOCCIA F., TARSITANI G., 2000a. *Il problema polvere da un fondo librario antico. Studio microbiologico: l'esperienza della Biblioteca Lancisiana.* Annali di Igiene – Medicina preventiva e di comunità, *12 (1)*: 51–59.

MONTACUTELLI R., MAGGI O., TARSITANI G., GABRIELLI N., 2000b. *Aerobiological monitoring of the "Sistine Chapel": airborne bacteria and microfungi trends.* Aerobiologia, *21*: 67–84.

MONTE M., 2003. *Oxalate film formation in marble specimens caused by fungus.* Journal of Cultural Heritage, *4*: 255–258.

MONTE M., NICHI D., 1997. *Effects of two biocides in the elimination of lichens from stone monuments: preliminary findings.* Science and Technology for Cultural Heritage, *6 (2)*: 209–216.

MONTE M., FERRARI R., MASSA S., 1994. *Biodeterioration of Etruscan tombs: Aerobiology and microclimate.* In: *5th International Conference*, Bangalore, Agashe Ed. S.N.: 333–346.

MONTEGUT D., INDICATOR N., KOESTLER R.J., 1991. *Fungal deterioration of cellulosic textiles: A review.* International Biodeterioration, *28*: 209–226.

MONTEMARTINI CORTE A., FERRONI A., SALVO V.S., 2003. *Isolation of fungal species from test samples and maps damaged by foxing, and correlation between these species and the environment.* International Biodeterioration & Biodegradation, *51*: 167–173.

MORETTI C., SALERNO C.S., TOMMASI FERRONI S., 2004. *Ricette vetrarie muranesi: Gasparo Brunoro e il manoscritto di Danzica.* Nardini Editore, Firenze.

MORITA R.Y., 1980. *Calcite precipitation by marine bacteria.* Geomicrobiology Journal, *2*: 63–82.

MORTON L.H.G., SURMAN S.B., 1994. *Biofilms in biodeterioration – A review.* International Biodeterioration & Biodegradation, *34(3/4)*: 203–221.

MOTOC D., CONSTANTINESCU S., IONESCU A., 1968. *La corrosion microbiologique des metaux et les moyens de la prevenir.* In: *National Conference of General and Applied Microbiology*, Bucharest, Academy of the Socialist Rep. of Romania: 194–195.

MOUGA T.M., ALMEIDA M.T., 1997. *Neutralisation of herbicides. Effects on wall vegetation.* International Biodeterioration & Biodegradation, *40(2–4)*: 141–149.

MOUZOURAS R., 1986. *Patterns of timber decay caused by marine fungi.* In: Moss S.T. (Ed.), *The Biology of Marine Fungi,* Cambridge University Press, Cambridge: 341–354.

MUKERJI K.G., GARG K.L., MISHRA A.K., 1995. *Fungi in deterioration of museum objects.* In: ARAYNAK C., SINGHASIRI C. (Eds), *Proceedings of the 3rd International Conference on Biodeterioration of Cultural Property*, Bangkok, Thailand: 226–241.

MÜLLER E., DREWELLO U., DREWELLO R., WEISSMANN R., WUERTZ S., 2001. In situ *analysis of biofilms on historical window glass using confocal laser scanner microscopy.* Journal of Cultural Heritage, *2*: 31–42.

NARANJO S., 1991. *Agents of biodeterioration of ethnographical materials at the National Museum of The Philippines: a preliminary assessment.* In: AGRAWAL O.P., DHAWAN S. (Eds.), *Proceedings of the International Conference on Biodeterioration of Cultural Property*, Lucknow 1989. Rajkamal Eletric Press, New Delhi: 66–72.

NASH III T.H. (Ed.), 1996. *Lichen Biology.* Academic Press, Cambridge.

NASH III T.H., WIRTH V., 1988. *Lichens, Bryophytes and Air Quality.* Cramer, Berlin, Stuttgart.

NIES D.H., 1999. *Microbial heavy-metal resistance.* Appl. Microbiol. Biotechnol., *51*: 730–750.

NIETO F.E., LEE J.J., KOESTLER R.J., 1997a. *Assessing Biodeterioration in wood using ATP Photometry: Part I. Nucleotide extraction and wood interference.* International Biodeterioration & Biodegradation, *39*: 9–13.

NIETO F.E., LEE J.J., KOESTLER R.J., 1997b. *Assessing Biodeterioration in wood using ATP Photometry: Part II. Calculating a conversion factor for* Phanerochaete chrysosporium *using ATP and Adenylate Energy Charge,* International Biodeterioration & Biodegradation, *39*: 159–164.

NIETO-FERNANDEZ F.E., CENTENO S.A., WYPYSKI M.T., DI BONAVENTURA M.P., BALDWIN A.M., KOESTLER R.J., 2003. *Enzymatic approach to removal of fungal spots from drawings on paper.* In: KOESTLER R.J., KOESTLER V.H., CHAROLA A.E., NIETO-FERNANDEZ F.E. (Eds.), *Art, Biology, and Conservation: Biodeterioration of Works of Art.* The Metropolitan Museum of Art, New York: 110–127.

NIJLAND T.G., DUBELAAR C.W., van HEES R.P.J., van der LINDEN T.J.M., 2004. *Black weathering of Bentheim and Obernkirchen sandstone.* In: KWIATKOWSKI D., LÖFVENDAHL R. (Eds.), *Proceedings of the 10th International Congress on Deterioration and Conservation of Stone*, Stockholm, *1*: 27–42.

NIMIS P.L., 1987. *I Macrolicheni d'Italia. Chiavi analitiche per la determinazione.* Gortania, *8*: 101–220.

NIMIS P.L., 1990. *Air quality indicators and indices. The use of plants as indicators and biomonitors of air pollution.* In: COLOMBO A.G., PREMAZZI G. (Eds.), *Proceedings of the Workshop on Indicators and Indices*, JRS Ispra, EUR 13060, EN: 93–126.

NIMIS P.L., 1993. *The lichens of Italy.* Monografia XII. Museo Regionale di Scienze Naturali di Torino, Torino.

NIMIS P.L., 2003. *Checklist of the Lichens of Italy 3.0.* University of Trieste, Dept. of Biology, IN3.0/2 (http://dbiodbs.univ.trieste.it/).

NIMIS P.L., MARTELLOS S., 2001. *A New Information System on Italian Lichens.* Plant Ecology, *157 (2)*: 165–172.

NIMIS P.L., SALVADORI O, 1997. *La crescita dei licheni sui monumenti di un parco. Uno studio pilota a Villa Manin.* In: *Restauro delle sculture lapidee nel parco di Villa Manin a Passariano. Il viale delle Erme.* Quaderni di Studi e Ricerche del Centro Regionale di Restauro dei Beni Culturali, *4*: 109–142.

NIMIS P.L., MONTE M., TRETIACH M., 1987. *Flora e vegetazione di aree archeologiche del Lazio.* Studia Geobotanica, *7*: 3–162.

NIMIS P.L., PINNA D., SALVADORI O., 1992. *Licheni e conservazione dei monumenti.* CLUEB, Bologna.

NIMIS P.L., SCHEIDEGGER C, WOLSELEY P.A. (Eds.), 2002. *Monitoring with Lichens. Monitoring Lichens.* Kluwer, NATO Science Series, Earth and Envir. Ser. 7.

NORMAL 1/80, 1980. *Alterazioni macroscopiche dei materiali lapidei: Lessico.* CNR-ICR, Comas Grafica, Roma.

NORMAL 1/88, 1990. *Alterazioni macroscopiche dei materiali lapidei: Lessico.* CNR-ICR, Comas Grafica, Roma.

NORMAL 3/80, 1980. *Materiali lapidei: Campionamento.* CNR-ICR, Comas Grafica, Roma.

NORMAL 9/88, 1988. *Microflora autotrofa ed eterotrofa: Tecniche di isolamento in coltura.* CNR-ICR, Comas Grafica, Roma.

NORMAL 19/85, 1985. *Microflora autotrofa ed eterotrofa: Tecniche di indagine visiva.* CNR-ICR, Comas Grafica, Roma.

NORMAL 24/86, 1986. *Metodologia di rilevamento ed analisi della vegetazione.* CNR-ICR, Comas Grafica, Roma.

NORMAL 25/87, 1987. *Microflora autotrofa ed eterotrofa: Tecniche di isolamento e di mantenimento in coltura pura.* CNR-ICR, Comas Grafica, Roma.

NORMAL 38/93, 1993. *Valutazione sperimentale dell'efficacia dei biocidi.* CNR-ICR, Comas Grafica, Roma.

NORMAL 39/93, 1994. *Rilevamento della carica microbica dell'aria.* CNR-ICR, Comas Grafica, Roma.

NUGARI M.P., 1992. *The "Bandiera di San Giorgio," a silk and leather banner of the thirteenth-fourteenth century. Part 2: The silk elements. Appendix 2.* In: *Proceedings of International Symposium of the ICOM. Working Group of Leathercraft and Related Objects,* London: 16.

NUGARI M.P., 1995. *Indagini biologiche sul dipinto raffigurante la Crocifissione.* In: SONNINO E. (Ed.), *Due dipinti su tela di Tanzio da Varallo indagini e restauro,* Sopr. per i Beni Ambientali, Architettonici, Artistici e Storici per l'Abruzzo, L'Aquila: 37–38.

NUGARI M.P., 1999. *Interference of antimicrobial agents on stone.* In: *An International Conference on Microbiology and Conservation* (ICMC'99). *Of Microbes and Art. The Role of Microbial Communities in the Degradation and Protection of Cultural Heritage,* Florence, Italy: 211–214.

NUGARI M.P., 2003. *The aerobiology applied to the conservation of works of art.* Coalition. Newsletter n. 6, Special Issue: Coalition Advanced Course "Biological Problems in Cultural Heritage," 8–9 November 2002, Florence, Italy: 8–10.

NUGARI M.P., BARTOLINI M., 1997. *The protective substances for outside bronze works of art: An evaluation of resistance to biodeterioration.* In: MOROPOULOU A., ZEZZA F., KOLLIAS E., PAPACHRISTODOULOU I. (Eds.), *Proceedings of the 4th International Symposium on the Conservation of Monuments in the Mediterranean,* Rhodes, Technical Chamber of Greece, *3*: 257–264.

NUGARI M.P., PIETRINI A.M., 1997. *Trevi fountain: An evaluation of inhibition effect of water-repellents on Cyanobacteria and algae.* International Biodeterioration & Biodegradation, *40 (2):* 249–256.

NUGARI M.P., PRIORI G.F., 1985. *Resistance of acrylic polymers (Paraloid B72, Primal AC33) to microorganisms. first part.* In: FELIX G. (Ed.), *Proceedings of the 5th International Congress on Deterioration and Conservation.* Presses Polytechniques Romandes, Lausanne, *2*: 685–693.

NUGARI M.P., PRIORI G.F., 1993. *Relitto romano di Torre Flavia. Le analisi biologiche.* Bollettino di Archeologia Subacquea, *I (0):* 62–63.

NUGARI M.P., ROCCARDI A., 1996. *Sacro Monte di Varallo, Cappella della Crocefissione. Indagine aerobiologica.* In: *Atti del VII Congresso Nazionale dell'Associazione Italiana di Aerobiologia.* Firenze, Scientific Press: 171.

NUGARI M.P., ROCCARDI A., 2001. *Aerobiological investigations applied to the conservation of cultural heritage.* Aerobiologia, International Journal of Aerobiology, *17:* 215–223.

NUGARI M.P., ROCCARDI A., 2002. *Indagini aerobiologiche nella Tomba dell'Orco a Tarquinia.* In: *Congresso Nazionale di Aerobiologia,* Bologna. Associazione Italiana di Aerobiologia: 142.

NUGARI M.P., ROCCARDI A. (forthcoming). *Le analisi aerobiologiche per la valutazione dei rischi di degrado biologico nella Grotta della Basura.* Rivista di Studi Liguri, *LXVII.*

NUGARI M.P., SALVADORI O., 2002. *Biocides and treatment of stone: Limitations and future prospects.* In: *Art, Biology and Conservation 2002. Biodeterioration of Works of Art.* The Metropolitan Museum of Art, New York: 89–93.

NUGARI M.P., SALVADORI O., 2003a. *Biocides and treatment of stone: Limitations and future prospects.* In: KOESTLER R.J., KOESTLER V.H., CHAROLA A.E., NIETO-FERNANDEZ F.E. (Eds.), *Art, Biology, and Conservation: Biodeterioration of Works of Art.* The Metropolitan Museum of Art, New York: 518–535.

NUGARI M.P., SALVADORI O., 2003b. *Biodeterioration control in cultural heritage: Methods and products.* In: SAIZ-JIMENEZ (Ed.), *Molecular Biology and Cultural Heritage, Proceedings of International Congress on Molecular Biology and Cultural Heritage,* Seville. Balkema Publishers, Lisse (NL): 233–242.

NUGARI M.P., PRIORI G.F., MATÈ D., SCALA F., 1987. *Fungicides for use on textiles employed during the restoration of works of art.* International Biodeterioration, *23 (5):* 295–306.

NUGARI M.P., D'URBANO S., SALVADORI O., 1993a. *Test methods for comparative evaluation of biocide treatments.* In: THIEL M.J. (Ed.), *Proceedings of International UNESCO/RILEM Congress on Conservation of Stone and Other Materials,* E & FN Spon, London: 565–572.

NUGARI M.P., PALLECCHI P., PINNA D., 1993b. *Methodological evaluation of biocidal interference with stone materials – Preliminary laboratory tests.* In: THIEL M.J. (Ed.), *Proceedings of International UNESCO/RILEM Congress on Conservation of Stone and Other Materials,* Paris. E & FN Spon, London: 295–302.

NUGARI M.P., REALINI M., ROCCARDI A., 1993c. *Contamination of mural paintings by indoor airborne fungal spores.* Aerobiologia, *9 (2–3):* 131–139.

NUGARI M.P., RICCI S., ROCCARDI A., MONTE M., 1998. *Chiese ed ipogei*. In: MANDRIOLI P., CANEVA G. (Eds.), *Aerobiologia e beni culturali*, Nardini Editore, Fiesole: 229–246.

NUGARI M.P., GIULIANI M.R., BARTOLINI M., TARSITANI G.F., 2002a. *La conservazione dal punto di vista microbiologico dei manufatti presenti nel sarcofago.* In: ANDALORO M., VAROLI-PIAZZA R. (Eds.), *Il Sarcofago dell'Imperatore. Studi, ricerche e indagini sulla tomba di Federico II nella Cattedrale di Palermo. 1994–1999.* Regione Siciliana. Officine Grafiche Riunite, Palermo: 238–244.

NUGARI M.P., GIULIANI M.R., TARSITANI G.F., 2002b. *Problematiche microbiologiche.* In: ANDALORO M., VAROLI-PIAZZA R. (Eds.), *Il Sarcofago dell'Imperatore. Studi, ricerche e indagini sulla tomba di Federico II nella Cattedrale di Palermo. 1994–1999.* Regione Siciliana. Officine Grafiche Riunite, Palermo: 232–237.

NUGARI M.P., ROCCARDI A., CACACE C., 2003. *Analisi aerobiologiche: Biodeterioramento, agrobiologia e microclima.* In: BIANCHI A. (Ed.), *Il Restauro della Cripta di Anagni.* Artemide Edizioni, Roma: 221–228.

NYUKSHA Ju.P., 1994. *The biodeterioration of paper and books.* In: GARG K.L., GARG N., MUKERJI K.G. (Eds.), *Recent advances in biodeterioration and biodegradation.* Naya Prokash, Calcutta, India, Vol. 1: 1–88.

ODUM E.P., 1973. *Principi di ecologia.* Piccin Editore, Padova.

ODUM E.P., 1988. *Basi di ecologia.* Piccin Editore, Padova.

OGDEN S., 1999. *Preservation of Library & Archival Materials: A Manual.* Northeast Document Conservation Center, Andover, 3rd edition.

ONIONS A.H.S., ALLSOPP D., EGGINS H.O.W., 1981. *Smith's Introduction to Industrial Mycology.* Edward Arnold, Londra.

ORIAL G., 2002. *La pierre des monuments habitat pour bacteries nuisibles et utiles.* In: ROQUEBERT M.F. (Ed.), *Les contaminants biologiques des biens culturels,* Museum national d'histoire naturelle et Editions scientifiques et medicales, Elsevier: 13–33.

ORIAL G., MARIE-VICTOIRE E., 2002. *Les recouvrements biologiques sur beton. Essais comparatifs de traitements d'elimination.* In: ROQUEBERT M.F. (Ed.), *Les contaminants biologiques des biens culturels,* Muséum national d'histoire naturelle et Editions scientifiques et medicales, Elsevier: 363–376.

ORIAL G., CASTANIER S., LE METAYER G., LOUBIERE J.F., 1992. *The biomineralization: A new process to protect calcareous stone: Applied to historic monuments.* In: KTOISHI H., ARAI H., KENJO T., YAMANO K. (Eds.), *Proceedings of the 2nd International Conference on Biodeterioration of Cultural Property,* Pacifico Yokohama. International Communications Specialists, Tokyo: 98–116.

ORLITA A., 1977. *The occurrence of fungi on book leather bindings from the baroque period.* International Biodeterioration Bulletin, *13* (2): 45–47.

ORTEGA-CALVO J.J., SAIZ-JIMENEZ C., 1997. *Microbial degradation of phenanthrene in two European cathedrals.* FEMS Microbiology Ecology, *22*: 129–138.

ORTEGA-CALVO J.J., SANCHEZ-CASTILLO P.M., HERNANDEZ-MARINE M., SAIZ-JIMENEZ C., 1993. *Isolation and characterization of epilithic chlorophytes and cyanobacteria from two Spanish cathedrals (Salamanca and Toledo).* Nova Hedwigia, *57*: 239–253.

ORTEGA-CALVO J.J., ARIÑO X., HERNANDEZ-MARINE M., SAIZ-JIMENEZ C., 1995. *Factors affecting the weathering and colonisation of monuments by phototrophic microorganisms.* The Science of the Total Environment, *167*: 329–341.

OSTE C., 1988. *Polymerase chain reaction.* Biotechnology, *6*: 162–167.

OZENDA P., 1963. *Lichens.* In: ZIMMERMANN W., OZENDA P. (Eds.), *Handbuch der Pflanzenanatomie,* Borntraeger, Berlin, II ed.

PACINI E., 1991. *Dispersal in terrestrial plants.* Annali di Botanica, *40*: 169–174.

PALENI A., STAFFELDT E.E., CURRI S.B., 1978. *I cavalli di San Marco e la bioaggressione ai metalli.* In: *Arte e ambiente,* Ed. Poligrafico Artioli, Modena: 71–79.

PANTAZIDOU A., ROUSSOMOUSTAKAKI M., URZÌ C., 1997. *The micoflora of Milos catacombs.* In: SINCLAIR A., SLATER E., GOWLETT J. (Eds.), *Archaeological Sciences 1995, Proceedings of a conference on the application of scientific techniques to the study of archaeology,* Monography 64, Oxbooks, Oxford: 303 –312.

PAPIDA S., MURPHY W., MAY E., 2000. *Enhancement of physical weathering of building stones by microbial populations.* International Biodeterioration Bulletin, *46*: 305–317.

PARES Y., 1965. *Intervention des bactéries anaérobies dans le cycle de l'or.* Comptes rendus de l'Académie des Sciences de Paris, *260*: 2351–2352.

PASQUARIELLO G., 1992. *Problemi di conservazione dei negativi fotografici. Le lastre di vetro alla gelatina.* AFT, Rivista di Storia e Fotografia, Prato, *15*: 4–9.

PASQUARIELLO G., 1993. *Investigazione scientifica sulle lastre alla gelatina bromuro d'argento del fondo Peliti.* In: MIRAGLIA M. (Ed.), *Federico Peliti (1844–1914). Un fotografo piemontese in India al tempo della regina Vittoria,* Peliti Associati Ed., Torino: 289–293.

PASQUARIELLO G., 2000. *L'Aerobiologia per la salvaguardia dei beni culturali.* In: *IX Congresso Nazionale di Aerobiologia "Ambiente e Prevenzione,"* Riassunti: 189.

PASQUARIELLO G., 2001. *Analisi dei fattori di biodeterioramento delle opere d'arte su carta.* In GRANITI A. (Ed.), *Atti Giornata di Studio Conservazione e Salvaguardia delle opere d'arte su carta: Stampe e disegni.* Accademia Nazionale delle Scienze detta dei XL, Roma: 129–150.

PASQUARIELLO G., MAGGI O., 1998. *Musei.* In: MANDRIOLI P., CANEVA G. (Eds.), *Aerobiologia e beni culturali.* Nardini Editore, Firenze: 215–227.

PASQUARIELLO G., MAGGI O., 2003. *Museums.* In: MANDRIOLI P., CANEVA G., SABBIONI C. (Eds.), *Cultural Heritage and Aerobiology,* Kluwer Academic Publishers, Dordrecht/Boston/London: 195–206.

PASQUARIELLO G., TALARICO F., COLADONATO M., FAVETTI S., 1996. *Problems in conservation of historical glass plate negatives: state of researches and possibile developments.* In: *International Conference on Conservation and Restoration of Archive and Library Materials,* Erice. Palumbo Editore, Palermo: 241–260.

PASQUARIELLO G., MISSORI M., PINZARI F., CARUSO G., DE MICO A., 2003. *Foxing di origine biologica: Riproduzione* in vitro, *analisi SEM-EDS e spettroscopia.*

In: *Convegno "Biologia e Beni culturali,"* Como, Preprints: 64.

PAULUS W., 1990. *Microbicides for protection of materials: Yesterday, today and tomorrow.* In: ROSSMORE H.W. (Ed.), *Biodeterioration and Biodegradation 8,* Elsevier, New York: 35–52.

PEDELÌ C., PULGA S., 2002. *Pratiche conservative sullo scavo archeologico. Principi e metodi.* All'Insegna del Giglio, Florence.

PELTOLA J., SALKINOJA-SALONEN M.S., 2003. *Biodeterioration of miniature paintings from the 18th and 19th centuries.* In: SAIZ-JIMENEZ C. (Ed.), *Molecular Biology and Cultural Heritage, Proceedings of International Congress on Molecular Biology and Cultural Heritage,* Seville. Balkema Publishers, Lisse (NL): 79–83.

PETERSEN K., HAMMER I., 1992. *Biodeterioration of Romanesque Wall Paintings under Salt Stress in the Nonnberg Abbey, Salzburg, Austria.* In: TOISHI K., ARAI H., KENJO T., YAMANO K. (Eds.), *Proceedings of the 2nd International Conference on Biodeterioration of Cultural Property,* Pacifico Yokohama. International Communications Specialists, Tokyo: 263–277.

PETERSEN K., KRUMBEIN W.E., HAFNER N., LUX E., MIETH A., 1993. *Aspects of biocide application on wall-paintings – Report on Eurocare project EU 489, Biodecay.* In: THIEL M.J. (Ed.), *Conservation of Stone and Other Materials, Proceedings of the International RILEM/UNESCO Congress,* E & FN Spon, London: 295–302.

PETERSEN K., YUN Y., KRUMBEIN W.E., 1995. *On the occurrence of alkalitolerant and alkaliphilic microorganisms on wall paintings and their interaction in restoration/consolidation.* In: ARAYNAK C., SINGHASIRI C. (Eds), *Proceedings of the 3rd International Conference on Biodeterioration of Cultural Property,* Bangkok, Thailand: 371–385.

PETUSHKOVA J.P., GRISHKOVA A.F., 1990. *Bacterial degradation of limestone treated with polymers.* In: *9th Triennal Meeting ICOM Committee for Conservation,* Dresda: 347–349.

PETUSHKOVA J.P., LYALIKOVA N., 1986. *Microbiological degradation of lead-containing pigments in mural paintings.* Studies in Conservation, *31:* 65–69.

PETUSHKOVA J.P., LYALIKOVA N., 1989. *Microbiological transformation of metals in architectural monuments.* In: *Conservation of Metals, Proceedings of International Restorer Seminar,* Veszprém: 157–158.

PETUSHKOVA J.P., LYALIKOVA N., 1994. *The microbial deterioration of historical buildings and mural paintings.* In: GARG K.L., GARG N., MUKERJI K.G. (Eds.), *Recent Advances in biodeterioration and biodegradation,* Naya Prokash, Calcutta, India, Vol. 1: 145–171.

PETUSHKOVA J.P., LYALIKOVA N., NICHIPOROV F.G., 1988. *Effect of ionizing radiation on monument deteriorating organisms.* J. Radioanalytical and Nuclear Chem. Articles, *125 (2):* 367–371.

PIERANTONELLI L., MATÉ D., SCALA F., 1984–85. *Il restauro di documenti d'archivio in seta. Prove di resistenza all'attacco biologico di vari adesivi,* Bollettino dell'Istituto Centrale della Patologia del Libro, *39:* 119–136.

PIERVITTORI R., 2003. *Lichens and biodeterioration of stoneworks: The Italian experience.* In: ST. CLAIR L.,

SEAWARD M.R.D. (Eds.), *Biodeterioration of Stone Surfaces.* Kluwer Academic Publishers, Dordrecht: 45–68.

PIERVITTORI R., CARAMIELLO R., 2001. *Importance of biological elements in conservation of stonework: A case study on a Romanesque church (Cortazzone, N. Italy).* In: *Science and Technology for the Safeguard of Cultural Heritage in the Mediterranean Basin, Proceedings 3rd Congress* (A. Guarino, ed.), *II:* 891–894.

PIERVITTORI R., LACCISAGLIA A., 1993. *Lichens as biodeterioration agents and biomonitors.* Aerobiologia, *9:* 181–186.

PIERVITTORI R., ROCCARDI A., 1998. *Licheni.* In: MANDRIOLI P., CANEVA G. (Eds.), *Aerobiologia e Beni culturali,* Nardini Editore, Firenze: 179–183.

PIERVITTORI R., ROCCARDI A., 2002. *Indagini aerobiologiche in ambienti esterni: Valutazione della componente lichenica.* X Congresso Nazionale di Aerobiologia, Aria e Salute. Sezione Beni Culturali, Bologna, Associazione Italiana di Agrobiologia: 70.

PIERVITTORI R., SALVADORI O., LACCISAGLIA A., 1994. *Literature on lichens and biodeterioration of stonework. I.* Lichenologist, *26 (2):* 171–192.

PIERVITTORI R., SALVADORI O., LACCISAGLIA A., 1996. *Literature on lichens and biodeterioration of stonework. II.* Lichenologist, *28 (5):* 471–483.

PIERVITTORI R., SALVADORI O., ISOCRONO D., 1998. *Literature on lichens and biodeterioration of stonework. III.* Lichenologist, *30 (3):* 263–277.

PIERVITTORI R., ROCCARDI A., ISOCRONO D., 2002. *Aspetti del controllo della colonizzazione lichenica sui monumenti.* Notiziario della Società Lichenologica Italiana, *15:* 71.

PIERVITTORI R., SALVADORI O., ISOCRONO D., 2004. *Literature on lichens and biodeterioration of stonework. IV.* Lichenologist, *36 (2):* 145–157.

PIETRINI A.M., 1988. *Studi ed interventi di carattere biologico.* In: CARDILLI L. (Ed.), *Il Tritone restaurato.* Ed. Quasar, Roma: 57.

PIETRINI A.M., 1991. *Un' indagine conoscitiva sulla microflora.* In: CARDILLI L. (Ed.), *Fontana di Trevi. La storia, il restauro.* Carte Segrete Ed., Roma: 187–189.

PIETRINI A.M., 1995. *Aspetti del biodeterioramento delle fontane.* In: LIO A. (Ed.), *Restauri in Piazza – La fontana di Piazza Colonna.* Bonsignori Ed., Roma: 85–86.

PIETRINI A.M., RICCI S., 1991. *Laboratory and field trials of algicidal biocides for the treatment of mural paintings.* In AGRAWAL O.P., DHAWAN S. (Eds.), *Proceedings of the International Conference on Biodeterioration of Cultural Property,* Lucknow 1989. Rajkamal Eletric Press, New Delhi: 353–358.

PIETRINI A.M., RICCI S., 1993. *Occurrence of a calcareous blue-green alga, Scytonema julianum (Kuetz.) Meneghini, on the frescoes of a church carved from the rock in Matera, Italy.* Crypt. Bot., *3:* 290–295.

PIETRINI A.M., RICCI S., BARTOLINI M., GIULIANI M.R., 1985. *A reddish alteration caused by algae on stoneworks. Preliminary studies.* In: FELIX G. (Ed.), *Proceedings of the 5th International Congress on Deterioration and Conservation of Stone.* Presses Polytechniques Romandes, Lausanne, *2:* 653–662.

PIETRINI A.M., RICCI S., BARTOLINI M., 1999. *Long-term evaluation of biocide efficacy on algal growth.* In: *An*

International Conference on Microbiology and Conservation (ICMC '99), Of Microbes and Art. The Role of Microbial Communities in the Degradation and Protection of Cultural Heritage, Florence, Italy: 238–245.

PIETRINI A.M., ALTIERI A., RICCI S., 2002. Il Colombario degli Scipioni (Roma): I biodeteriogeni quali bioindicatori. In: Atti del 97° Congresso della Società Botanica Italiana, Lecce, Ed. Del Grifo: 6.

PIGNATTI S. (Ed.), 1995. Ecologia vegetale. UTET, Torino.

PILEGGI C., BERSELLI S., GIULIANI M.R., NUGARI M.P., SCARAMELLA L., 1996. Il fondo di lastre gelatina bromuro d'argento dell'I.C.R.: Problemi di conservazione. In: Proceedings of the International Conference on Conservation and Restoration of Archive and Library Materials, Erice. Palumbo Editore, Palermo: 695–707.

PIÑAR G., LUBITZ W., RÖLLEKE S., GURTNER C., 1999. Identification of Archaea in deteriorated ancient wall paintings by molecular means. In: An International Conference on Microbiology and Conservation (ICMC '99), Of Microbes and Art. The Role of Microbial Communities in the Degradation and Protection of Cultural Heritage. Florence, Italy: 46–50.

PIÑAR G., RAMOS C., RÖLLEKE S., SCHABEREITER-GURTNER C., VYBIRAL D., LUBITZ W., DENNER E.B.M., 2001. Detection of indigenous Halobacillus populations in damaged ancient wall paintings and building materials: Molecular monitoring and cultivation. Applied and Environmental Microbiology, 67: 4891–4895.

PIÑAR G., SCHABEREITER-GURTNER C., LUBITZ W., 2003. Analysis of the microbial diversity present on the wall paintings of Castle of Herberstein by molecular techniques. In: SAIZ-JIMENEZ C. (Ed.), Molecular Biology and Cultural Heritage, Proceedings of International Congress on Molecular Biology and Cultural Heritage, Seville. Balkema Publishers, Lisse (NL), 35–45.

PINCK C., BALZAROTTI-KAMMLEIN R., MANSCH R., 2002. Biocidal efficacy of Algophase® against nitrifying bacteria. In: GALÁN E., ZEZZA F. (Eds.), Protection and Conservation of the Cultural Heritage of the Mediterranean Cities. Proceeding of the 5th International Symposium on the Conservation of Monuments in the Mediterranean Basin, Seville: 449–453.

PINI R., SIANO S., SALIMBENI R., PAQUINUCCI M., MICCIO M., 2000. Tests of laser cleaning on archaeological metal artefacts. J. Cultural Heritage, 1: 129–137.

PINNA D., 1993. Fungal physiology and the formation of calcium oxalate films on stone monuments. Aerobiologia, 9 (2–3): 157–167.

PINNA D., SALVADORI O., 1999. Biological growth on Italian monuments restored with organic or carbonatic compounds. In: An International Conference on Microbiology and Conservation (ICMC '99), Of Microbes and Art. The Role of Microbial Communities in the Degradation and Protection of Cultural Heritage, Florence, Italy: 149–154.

PINNA D., SALVADORI O., 2000. Endolithic lichens and conservation: An underestimated question. In: FASSINA V. (Ed.), Proceedings of the 9th International Congress on Deterioration and Conservation of Stone, Venice. Elsevier, Amsterdam: 513–519.

PINNA D., SALVADORI O., TRETIACH M., 1998. An anatomical investigation of endolithic lichens from the Trieste Karst (NE Italy). Plant Biosystems, 132: 183–195.

PIROLA A., MONTANARI C., CREDARO V., 1980. Valutazione speditiva del grado di protezione del mantello vegetale contro l'azione delle acque cadenti e dilavanti. Collana del programma Finalizzato "Promozione della qualità dell'ambiente" C.N.R., AQ/1/75: 1–20.

PITT J.I., 1975. Xerophylic fungi and the spoilage of food of plant origin. In: DUCKWORTH R.B. (Ed.), Water Relations of Food, London, Academic Press: 273–307.

PITT J.I., HOCKING A.D., 1985. Fungi and food spoilage. Aspen Publisher, Inc. Gaithersburg, Maryland.

PITZURRA L., GIRALDI M., SBARAGLIA G., BISTONI F., 2003. Microbial environmental monitoring of stone cultural heritage. In: FASSINA V. (Ed.), Proceedings of the 9th International Congress on Deterioration and Conservation of Stone, Venice. Elsevier, Amsterdam: 483–491.

PITZURRA L., 1997. Il Monitoraggio Ambientale Microbiologico (MAM). Ann. lg., 9: 439–454 .

PIZZUTO F., 1999. On the structure, typology and periodism of a Cystoseira brachycarpa J. Agardh emend. Giaccone community and of a Cystoseira crinita Duby community from the eastern coast of Sicily (Mediterranean Sea). Plant Biosystems, 133 (1): 15–35.

PIZZUTO F., SERIO D., 1995. A study on the qualitative minimal area of two Cystoseira (Cystoseiraceae, Fucophyceae) communities from Eastern Sicilian coast (Ionian Sea). Giornale Botanico Italiano, 128 (6): 1092–1095.

PLOSSI ZAPPALÀ M., 1999. Modified starches for library and archives conservation: Effect on paper. In: FEDERICI C., MUNAFÒ P.F. (Eds.), International Conference on Conservation and Restoration of Archival and Library Materials, Erice. Palumbo Editore, Palermo, Vol. 2.

POCHON J., COPPIER O., 1950. Role des bactéries sulfato-reductrices dans l'altération biologique des pierres des monuments. C.R. de l'Académie des Sciences: 1584–1585.

POCHON J., JATON C., 1971. Facteurs biologiques de l'altération des pierres. In: WALKERS H., HUECK VAN DER PLAS E.H. (Eds.), Biodeterioration of Materials 1, London, Applied Science Publisher: 358–368.

POHL W., SCHNEIDER J., 2002. Impact of endolithic biofilms on carbonate rock surfaces. In: SIEGESMUND S., WEISS T., VOLLBRECHT A. (Eds.), Natural stone, weathering phenomena, conservation strategies and case studies. Geological Society, London, Special Publication 205: 177–194.

POINDEXTER J.S., 1981. Oligotrophy, feast and famine existence. Adv. Microbial. Ecol., 5: 63–89.

POINTING S.B., JONES E.B.G., JONES A.M., 1998. Decay prevention in waterlogged archaeological wood using gamma irradiation. International Biodeterioration & Biodegradation, 42: 17–24.

POLDINI L., 1989. La vegetazione del Carso isontino e triestino. Lint Editore, Trieste.

POLI MARCHESE E., GRILLO M., LO GIUDICE E., 1995. Aspetti del dinamismo della vegetazione sulla colata lavica del 1651 del versante orientale dell'Etna. Coll. Phytosoc., 24: 241–264.

POLI MARCHESE E., DI BENEDETTO L., LUCIANI F., RAZZARA S., GRILLO M., STAGNO F., AURICCHIA A., 1997. Biodeteriogeni vegetali di monumenti del centro storico della città di Noto. Arch. Geobot., 3: 71–80.

Poli Marchese E., Luciani F., Razzara S., Grillo M., Auricchia A., Stagno F., Giaccone G., Di Geronimo R., Di Martino V., 1998. *Biodeteriorating plants on sandstone monuments in the historical city centre of Catania.* In: *Proceedings of the 1st International Congress on Sciences and Technology for the Safeguard of Cultural Heritage in the Mediterranean Basin:* 1195–1203.

Polsinelli M., de Felice M., Galizzi A., Galli E., Mastromei G., Mazza P., Viale G., 1993. *Microbiologia.* Bollati Boringhieri, Torino, nuova edizione.

Prieto B., Rivas M.T., Silva B.M., 1996. *Effectiveness of biocide treatments on granite.* In: *Degradation and conservation of granitic rocks in monuments, Proceedings of EC workshop,* Santiago de Compostela, 1994: 361–366.

Proctor M.C.F., 1984. *Structure and ecological adaptation.* In: Dyer A.F., Duckett J.G. (Eds.), *The experimental biology of Bryophytes,* Academic Press, London: 9–30.

Prosperi Porta C., 1996. *Arslantepe (Malatya) – Una esperienza in corso. Protezione delle strutture e musealizzazione del sito.* Scienza e Beni Culturali, *12:* 549–597.

Punja Z.K., Huang J.S., Jenkins S.F., 1985. *Relationship of mycelial growth and production of oxalic acid and cell wall degrading enzymes to virulence in* Sclerotium rolfsii. Can. J. Plant Pathol., *7 (2):* 109–117.

Purvis O.W., 1984. *The occurence of copper oxalate in lichens growing on copper sulphide-bearing rocks in Scandinavia.* Lichenologist, *16 (2):* 197–204.

Putt N., Menegazzi C., 1999. *ICCROM preventive conservation experiences in Europe.* In: *Pre-prints of ICOM Commitee for conservation, 12th Triennial Meeting,* Lyon, James & James, London, *1:* 93–99.

Quaresima R., Baccante A., Volpe R., Corain B., 1997. *Realisation and possibility of polymeric metallo-organic matrixes with biocides activity.* In: Moropoulou A., Zezza F., Kollias E., Papachristodoulou I. (Eds.), *Proceedings of the 4th International Symposium on the Conservation of Monuments in the Mediterranean Basin,* Rhodes, Technical Chamber of Greece, 3: 323–335.

Raffellini G., Cellai C., 1997. *La distribuzione dell'aria.* In: Alfano G., Filippi M., Sacchi E. (Eds.), *Impianti di climatizzazione per l'edilizia. Dal progetto al collaudo,* Masson, Milano: 521–544.

Rakotonirainy M.S., Flieder F., 1994. *L'assainissement des aires de stockage par thermonébulisation.* Nouvelles de l'ARSAG, *10:* 16–17.

Rakotonirainy M.S., Raisson M.A., Flieder F., 1998. *Evaluation of the fungistatic and fungicidal activity of six essential oils and their related compounds.* In: *Pre-prints of the Jubilee Symposium of the School of Conservation,* Copenhagen, The Royal Danish Academy of Fine Arts: 121–130.

Rakotonirainy M.S., Flieder F., Fohrer F., leroy M., 1999a. *La desinfection des papiers par les faiseaux d'electrons et les micro-ondes.* In: *Les Documents Graphiques et photographiques: Analyse et conservation.* Travaux du CRCDG, 1994–1998, La documentation francaise, Paris: 159–172.

Rakotonirainy M.S., Fohrer F., Flieder F., 1999b. *Research on fungicides for aerial disinfections by thermal fogging in libraries and archives.* International Biodeterioration & Biodegradation, *44:* 133–139.

Rambelli A., Pasqualetti M., 1996. *Nuovi fondamenti di Micologia.* Jaca Book, Milano.

Ranalli G., 2000. *Rapid diagnosis of microbial growth and biocide treatments on stone materials by bioluminescent low-light imaging tecnique.* In: Fassina V. (Ed.), *Proceedings of the 9th International Congress on Deterioration and Conservation of Stone,* Venice. Elsevier, Amsterdam: 499–505.

Ranalli G., Coppola R., Sorlini C., 1995. *Preliminary investigations on air-borne microorganisms in indoor environment of artistic interest.* In: Araynak C., Singhasiri C. (Eds), *Proceedings of the 3rd International Conference on Biodeterioration of Cultural Property,* Bangkok, Thailand: 268–271.

Ranalli G., Chiavarini M., Guidetti V., Marsala F., Matteini M., Zanardini E., Sorlini C., 1996. *The use of microorganisms for the removal of nitrates and organic substances on artistic stoneworks.* In: Riederer J. (Ed.), *Proceedings of the 8th International Congress on Deterioration and Conservation of Stone.* Berlin, Germany: 1421–1427.

Ranalli G., Chiavarini M., Guidetti V., Marsala F., Matteini M., Cardini E., Sorlini C., 1997. *The use of microorganisms for the removal of sulphates on artistic stoneworks.* International Biodeterioration & Biodegradation, *40:* 255–261.

Ranalli G., Pasini P., Roda A., 1998. *Bioluminescent low-light imaging technique as a rapid method to detect spatial distribution and activity of biodeteriogen agents in cultural heritages.* In: Roda A., Pazzagli M., Kricka L.J., Stanley P.E. (Eds.), *Bioluminescence and Chemiluminescence: Perspectives for the 21th Century,* Wiley, Bologna: 153–156.

Ranalli G., Matteini M., Tosini I., Zanardini E., Sorlini C., 2000. *Bioremediation of cultural heritage: Removal of sulphates, nitrates and organic substances.* In: Ciferri O., Tiano P., Mastromei G. (Eds.), *Of Microbes and Art. The Role of Microbial Communities in the Degradation and Protection of Cultural Heritage.* Kluwer Academic/Plenum Publishers, New York: 231–245.

Ranalli G., Belli C., Baracchini C., Caponi G., Pacini P., Zanardini E., Sorlini C., 2003. *Deterioration and bioremediation of fresco: A case study.* In: Saiz-Jimenez C. (Ed.), *Molecular Biology and Cultural Heritage, Proceedings of International Congress on Molecular Biology and Cultural Heritage,* Seville. Balkema Publishers, Lisse (NL): 243–246.

Raper K.B., Fennell D.I., 1965. *The genus* Aspergillus. The Williams & Wilkins Company, Baltimore.

Rasmussen L., Johnsen I., 1976. *Uptake of minerals, particularly metals, by epiphytic* Hypnum cupressiforme. Oikos, *27:* 483–487.

Rautela G.S., Cowling E.B., 1966. *Simple cultural test for relative cellulolityc activity of fungi.* Applied Environmental Microbiology, *14:* 892–898.

Raven P.H., Evert R.F., Eichhorn S.E., 2002. *Biologia delle piante.* Zanichelli, Bologna.

Raychaudhuri M.R., Brimblecombe P., 2000. *Formaldehyde oxidation and lead corrosion.* Studies in Conservation, *45:* 226–232.

RAYNER A.D.M, M. RAMSDALE, Z.R. WATKINS. 1995. *Origins and significance of genetic and epigenetic instability in mycelial systems*. Canadian Journal of Botany 73 (Suppl. 1): 1241–1248.

RAZDAN M., CHATPAR H.S., 1991. *Biological aspects of biodeterioration on leather and its control*. In: AGRAWAL O.P., DHAWAN S. (Eds.), *Proceedings of the International Conference on Biodeterioration of Cultural Property*, Lucknow 1989. Rajkamal Eletric Press, New Delhi: 160–167.

REALINI M., TONIOLO L. (Eds.), 1996. *Proceedings of the II Symposium "The oxalate films in the conservation of works of art."* Editeam, Bologna.

REALINI M., SORLINI C., BASSI M., 1985. *The Certosa of Pavia: A case of biodeterioration*. In: FELIX G. (Ed.), *Proceedings of 5th International Congress on Deterioration and Conservation of Stone*. Presses Polytechniques Romandes, Lausanne, 2: 627–629.

REBRIKOVA N.L., AGEEVA E.N., 1995. *An evaluation of biocide treatments on the rock art of Baical*. In: *International Colloquium on Methods of Evaluating Products for the Conservation of Porous Building Materials in Monuments*. ICCROM, Rome: 69–74.

REBRIKOVA N.L., MANTUROVSKAYA N.V., 2000. *Foxing – A new approach to an old problem*. Restaurator, 21: 85–100.

REILLY J.M., 1980. *The Albumen and Salted Paper Book, the history and practice of photographic printing: 1840–1985*. Light Impressions, Rochester.

REILLY J.M., 1986. *Care and identification of 19th Century photographic prints*. Eastman Kodak Company, Rochester.

RESIDORI L., 2002a. *I materiali fotografici, cenni di storia, fabbricazione e manifattura*. In: *Chimica e biologia applicate alla conservazione degli Archivi*. Centro di fotoriproduzione legatoria e restauro degli Archivi di Stato, Ministero per i Beni e le Attività Culturali, Direzione Generale degli Archivi, Roma: 163–215.

RESIDORI L., 2002b. *Struttura e composizione dei materiali fotografici*. In: *Chimica e biologia applicate alla conservazione degli Archivi*. Centro di fotoriproduzione legatoria e restauro degli Archivi di Stato, Ministero per i Beni e le Attività Culturali, Direzione Generale degli Archivi, Roma. 217–270.

RICCI S., 2003. *La flora algale*. In: CANEVA G., TRAVAGLINI C.M. (Eds.), *Atlante Storico-ambientale di Anzio e Nettuno*, Croma, Ed. De Luca, Roma: 78–79.

RICCI S., PIETRINI A.M., 1994a. *Caratterizzazione della microflora algale presente sulla Fontana dei Quattro Fiumi, Roma*. In: FASSINA V., OTT H., ZEZZA F. (Eds.), *Atti del 3° Simposio Internazionale "La Conservazione dei Monumenti nel Bacino del Mediterraneo,"* Venice: 353–357.

RICCI S., PIETRINI A.M., 1994b. *Rinvenimento di* Spelaeopogon lucifugus Borzì *(Stigonemataceae) su affreschi della Basilica Inferiore di San Clemente, Roma*. Giornale Botanico Italiano, 128 (1): 215.

RICCI S., PIETRINI A.M., 2004. *Il Colombario degli Scipioni a Roma. Caratterizzazione delle alterazioni di natura biologica*. Kermes, 54: 61–66.

RICCI S., PIETRINI A.M., GIULIANI M.R., 1985. *Il ruolo delle microalghe nel degrado biologico degli intonaci*. In: *L'intonaco: Storia, cultura e tecnologia, Atti Convegno di Studi*, Bressanone: 53–61.

RICELLI A., FABBRI A.A., BRASINI S., FANELLI C., 1995. *Effect of some antifungals on the prevention of paper biodeterioration caused by fungi*. In: *Proceeding of the 1st International Congress on Science and Technology for the Safeguard of Cultural Heritage in the Mediterranean Basin*.

RICELLI A., FABBRI A.A., FANELLI C., MENICAGLI R., SAMARITANI S., PINI D., RAPACCINI S.M., SALVATORI P., 1999. *Fungal growth on samples of paper: Inhibition by new antifungals*. 20: 97–107.

RICHARDS B.N., 1974. *Introduction to the soil Ecosystem*. Longman, Harlow, Essex.

RIEDERER J., 1984. *The restoration of archeological monuments in the tropical climate*. In: *ICOM Committee for Conservation Preprints, 7th Triennial Meeting*, Copenhagen: 21–22.

RIKKINEN J., 1995. *What's Behind the Pretty Colors: A Study on the Photobiology of Lichens*. Bryobrothera, 4: 1–239.

RIVADENEYRA M.A., PEREZ-GARCÌA SALMERÒN I., RAMOS-CORMENZANA A., 1985. *Bacterial precipitation of calcium carbonate in presence of phosphate*. Soil. Biol. Biochem., 17: 171–172.

RIVADENEYRA M.A., DELGADO R., QUESADA E., RAMOS-CORMENZANA A., 1991. *Precipitation of calcium carbonate by* Deleya halophila *in media containing NaCl as sole salt*. Current Microbiology, 22: 185–190.

RIVADENEYRA M.A., DELGADO R., DEL MORAL A., FERRER M.R., RAMOS-CORMENZANA A., 1994. *Precipitation of calcium carbonate by* Vibrio spp. *from an inland saltern*. FEMS Microbiol Ecology, 13: 197–204.

RIZZI LONGO L., POLDINI L., GOIA F., 1980. *La microflora algale delle pareti calcaree del Friuli-Venezia Giulia (Italia nord-orientale)*. Studia Geobotanica, 1(1): 231–263.

ROCCARDI A., 2003. *Il Santuario della Madonna di Cibona alle Allumiere, tutela e valorizzazione di un monumento: Indagini sulla flora lichenica*. Ministero per i Beni e le Attività Culturali. Gangemi Editore, Roma. 287–294.

ROCCARDI A., BIANCHETTI P.L., 1989. *The distribution of lichens in some stoneworks in the surroundings of Rome*. Studia Geobotanica, 8: 89–97.

RODRIGUEZ-NAVARRO C., RODRIGUEZ-GALLEGO M., BEN CHEKROUN K., GONZALES-MUNOZ M.T., 2000. *Carbonate production by* Mixococcus xantus: *A possible application to protect/consolidate calcareous stones*. In: *Quarry, Laboratory, Monument*, Pavia: 493–498.

ROLDÁN M., CLAVERO E., HERNÁNDEZ-MARINÉ M., 2003. *Aerophytic biofilms in dim habitats*. In: SAIZ JIMÉNEZ C. (Ed.), *Molecular Biology and Cultural Heritage, Proceedings of International Congress*, Seville. Balkema Publishers, Lisse (NL): 163–169.

RÖLLEKE S., MUYZER G., WAWER C., WANNER G., LUBITZ W., 1996. *Identification of bacteria in a biodegradated wall painted by denaturing gradient gel electrophoresis of PCR-amplified gene fragments coding for 16S rRNA*. Applied and Environmental Microbiology, 62: 2059–2065.

RÖLLEKE S., WITTE A., WANNER, G., LUBITZ W., 1998. *Medieval wall paintings: A habitat for archea – Identification of archea by denaturing gradient gel electrophoresis (DGGE) of PCR-amplified gene fragments coding for 16S rRNA in medieval wall painting*. International Biodeterioration & Biodegradation, 41: 85–92.

RÖLLEKE S., GURTNER C., DREWELLO U., DREWELLO R., LUBITZ W., WEISSMANN R., 1999. *Analysis of bacterial communities on historical glass by denaturing gel electrophoresis of PCR-amplified gene fragments coding for 16S rRNA.* Journal of Microbiological Methods, *36*: 107–114.

RÖLLEKE S., GURTNER C., PINAR G., LUBTITZ W., 2000. *Molecular approaches for the assessment of microbial deterioration of objects of art.* In: CIFERRI O., TIANO P., MASTROMEI G. (Eds.), *Of Microbes and Art, The Role of Microbial Communities in the Degradation and Protection of Cultural Heritage.* Kluwer Academic/Plenum Publishers, New York: 39–47.

ROSENQUIST I.T., 1961. *Subsoil corrosion of steel.* Norwegian Geotechnical Institute, Oslo.

ROSSI R., BELLINA F., CIUCCI D., CARPITA A., FANELLI C., 1998. *A new synthesis of fungicidal methyl(E)-3 methoxypropenoates.* Tetrahedron, *54*: 7595–7614.

ROSSI R., CARPITA A., BELLINA F., FABBRI A.A., MONTI S., FANELLI C., 1999. *Effect of synthetic analogues of strobilurin A on the growth of fungi which damage papery materials.* Modern Fungicides and Antifungal Compounds: 77–84.

ROSSMOORE H.W., ROSSMOORE K., SONDOSSI M., KOESTLER R.J., 2002. *Results of a novel germicidal lamp system for reduction of airborne microbial spores in museum collection.* In: *Art, Biology, and Conservation 2002. Biodeterioration of Works of Art,* The Metropolitan Museum of Art, New York: 121.

ROTHAMEL U., 1998. *The textiles coming from medioeval tombs: A hundred years of experience, today's knowledge, and unsoved problems.* In: *Interim Meeting ICOM-CC "Interdisciplinary Approach about Studies and Conservation of Medieval Textiles,"* Palermo: 71–76.

RUHLIMG A., TYLER G., 1970. *Sorption and retention of heavy metals in the woodland moss Hylocomium splendens (Hedw.) Br. et Sch.* Oikos, *21*: 92–97.

SAARELA M., ALAKOMI H-L., SUIHKO M-L., MAUNUKSELA L., RAASKA L., MATTILA-SANDHOLM T., 2004. *Heterotrophic microorganisms in air and biofilm samples from Roman catacombs, with special emphasis on actinobacteria and fungi.* International Biodeterioration & Biodegradation, *54*: 27–37.

SADIRIN H., 1988. *The deterioration and conservation of stone historical monuments in Indonesia.* In: *Proceedings of the 6th International Congress on Deterioration and Conservation of Stone.* Nicholas Copernicus University Press, Torun, Poland: 722–731.

SAGAR B.F., 1988. *Biodeterioration of Textile Materials and Textile Preservation.* In: HOUGHTON D.R., SMITH R.N., EGGINS H.O.W. (Eds.), *Biodeterioration 7,* Elsevier, New York: 683–702.

SAIZ-JIMENEZ C., 1990. *Deposition of airborne organic pollutants on historic buildings.* Atmos. Environ., *27b*: 77–85.

SAIZ-JIMENEZ C., 1995a. *Deposition of anthropogenic compounds on monuments and their effects on airborne microorganisms.* Aerobiologia, *11*: 161–175.

SAIZ-JIMENEZ C., 1995b. *Microbial melanins in stone monuments.* The Science of the Total Environment, *167*: 273–286.

SAIZ-JIMENEZ C., 1997. *Biodeterioration vs biodegradation: The role of microorganisms in the removal of pollutants deposited on to historic buildings.* International Biodeterioration & Biodegradation, *40*: 225–232.

SAIZ-JIMENEZ C., 1999. *Biogeochemistry of weathering processes in monuments.* Geomicrobiology Journal, *16*: 27–37.

SAIZ-JIMENEZ C., LAIZ L., 2000. *Occurrence of halotolerant and halophilic bacterial communities in deteriorated monuments.* International Biodeterioration & Biodegradation, *46*: 319–326.

SAIZ-JIMENEZ C., SAMSON R.A., 1981a. *Biodegradation de obras de arte. Hongos implicados en la degradacion de los frescos del monasterio de la Rabida (Huelva).* Botanica Macaronesica, *8–9*: 255–264.

SAIZ-JIMENEZ C., SAMSON R.A., 1981b. *Microorganisms and environmental pollution as deteriorating agents of the frescoes of the Monastery of Santa Maria de la Rabida.* In: *Proceedings of the 6th Triennial Meeting ICOM Committee for Conservation,* Huelva, Spain, *81 (15)*: 5–14.

SALKINOJA-SALONEN M.S., PELTOLA J., ANDERSSON M.A., SAIZ-JIMENEZ C., 2003. *Microbial toxins in moisture damage indoor environment and cultural assets.* In: SAIZ-JIMENEZ C. (Ed.), *Molecular Biology and Cultural Heritage, Proceedings of International Congress on Molecular Biology and Cultural Heritage,* Seville. Balkema Publishers, Lisse (NL): 93–105.

SALVADORI O., 2000. *Characterisation of endolithic communities of stone monuments and natural outcrops.* In: CIFERRI O., TIANO P., MASTROMEI G. (Eds.), *Of Microbes and Art. The Role of Microbial Communities in the Degradation and Protection of Cultural Heritage.* Kluwer Academic/Plenum Publishers, New York: 89–101.

SALVADORI O., LAZZARINI L., 1991. *Lichens deterioration on stones of Aquileian monuments.* Botanika Chronika, *10*: 961–968.

SALVADORI O., NUGARI M.P., 1988. *The effect of microbial growth on synthetic polymers used on works of art.* In: HOUGHTON D.R., SMITH R.N., EGGINS H.O.W. (Eds.), *Biodeterioration 7,* Elsevier Applied Science, New York: 424–427.

SALVADORI O., REALINI M., 1996. *Characterization of biogenic oxalate films.* In: REALINI M., TONIOLO L. (Eds.), *The Oxalate Films in the Conservation of Works of Art.* EDITEAM, Bologna: 337–351.

SALVADORI O., TRETIACH M., 2002. *Thallus-substratum relationships of silicicolous lichens occurring on carbonatic rocks of the Mediterranean region.* In: LUMBSCH H.T., OTT S. (Eds.), *Progress and Problems in Lichenology at the Turn of the Millenium.* IAL 4, X. Llimona, Bibliotheca Lichenologica, *82*: 57–64.

SALVADORI O., SORLINI C., ZANARDINI E., 1994. *Microbiological and biochemical investigations on stone of the Ca' d'Oro facade (Venice).* In: FASSINA V., OTT H., ZEZZA F. (Eds.), *La conservazione dei monumenti nel bacino del mediterraneo. Atti del 3° Simposio Internazionale,* Venice: 343–346.

SAMENO PUERTO M., VILLEGAS SANCHEZ R., GARCIA ROWE J., 1996. *Inventario de la vegetacion y estudio de la interferencia biocidi con los materiales petreos del yacimiento del cerro de la plaza de armas de Puente Tablas (Jaen).* PH Boletin del Istituto Andaluz del Patrimonio Historico IV, *14*: 67–74.

SAMPÒ S., LUPPI MOSCA A.M., 1989. *A study of the fungi occurring on 15th century frescoes in Florence, Italy.* International Biodeterioration 25: 343–353.

SAND W., 1997. *Microbial mechanisms of deterioration of inorganic substrates – A general mechanistic overview.* International Biodeterioration & Biodegradation, 40: 183–190.

SAND W., GEHRKE T., JOZSA P.G., SCHIPPERS A., 2001. *Biochemistry of bacterial leaching – Direct vs. indirect bioleaching.* Hydrometallurgy, 59: 159–175.

SANTORO E.D., KOESTLER R.J., 1991. *A methodology for biodeterioration testing of polymers and resins.* International Biodeterioration, 28: 81–92.

SANTORO S., SANTOPUOLI N., 2000. *La protezione delle aree archeologiche: ricerca e prassi operativa.* In: *Coperture per aree e strutture archeologiche: Repertorio di casi esemplificativi.* Arkos (Supplemento), I Grandi Restauri, UTET Periodici, 1.

SAXENA V.K., JAIN K.K., SINGH T., 1991. *Mechanisms of biologically induced damage to stone materials.* In: AGRAWAL O. P., DHAWAN SHASHI (Eds.), *Proceedings of the International Conference on Biodeterioration of Cultural Property*, Lucknow 1989. Rajkamal Eletric Press, New Delhi: 249–258.

SBARAGLIA G., BELLEZZA T., BON DI VALSASSINA C., GARIBALDI V., GIRÁDI M., PITZURRA L., BISTONI F., 1999. *Microbial environmental monitoring of museums.* In: *An International Conference on Microbiology and Conservation* (ICMC '99), *Of Microbes and Art. The Role of Microbial Communities in the Degradation and Protection of Cultural Heritage*, Florence, Italy: 102–108.

SCARAMELLA L., 1999. *Alcune tecniche fotografiche del XIX secolo. Platinotipia, Kallitipia, Cianotipia.* In: *Il patrimonio fotografico, tutela e valorizzazione.* Centro per il Restauro e la Conservazione della Fotografia, Berselli, Milano.

SCHABEREITER-GURTNER C., PINAR G., LUBITZ W., ROLLEKE S., 2001. *Analysis of fungal communities on historical church window glass by denaturing gradient gel electrophoresis and phylogenetic 18S rDNA sequence analysis.* Journal of Microbiological Methods, 47: 345–354.

SCHNABEL L., 1991. *The treatment of biological growth on stone: A conservator's view point.* International Biodeterioration, 28: 125–131.

SCHOSTAK V., KRUMBEIN W.E., 1992. *Occurrence of extremely halotolerant and moderate halophilic bacteria on salt impaired wall paintings.* In: DELGADO RODRIGUEZ J., HENRIQUES F., TELMO JEREMIAS F. (Eds.), *Proceedings of the 7th International Congress on Deterioration and Conservation of Stone.* LNEC, Lisbon, Portugal: 551–560.

SCHROETER B., SANCHO G., 1996. *Lichens growing on glass in Antartica.* Lichenologist, 28 (4): 385–390.

SCHULZ B., 1994. *Zum Umgang mit Pastellbildern. Konservatorische Maßnahmen an drei Porträts.* Restauro, 3: 174–177.

SCLOCCHI C. 2002. *La disinfezione e la disinfestazione dei supporti archivistici.* In: *Chimica e biologia applicate alla conservazione degli archivi*, Ministero per i Beni e le Attività Culturali, Direzione Generale degli Archivi di Stato. Union Printing, Roma: 557–575.

SCLOCCHI M.C., MAGGI O., PERSIANI A.M., VALENTI P., PASQUARIELLO G., GALLO F., 2002. *Le aerospore fungine deposte nella polvere ed il rischio di biodeterioramento del patrimonio librario.* X Congresso Nazionale di Aerobiologia – Aria e Salute, Riassunti, Bologna: 71.

SCLOCCHI M.C., MATÈ D., RUGGIERO D., 2003. *I costituenti organici dei materiali fotografici: Loro suscettibilità ad attacco biologico. AFT, 35.* Rivista di Storia e Fotografia, Prato.

SCOTT G., 1994. *Moisture, ventilation and mould growth.* In: ROY A., SMITH P. (Eds.), *Preventive Conservation Practice, Theory and Research, ICC Preprints of the Contribution to the Ottawa Congress.* The International Institute for Conservation of History and Artistic Works, London: 149–153.

SCOTT P.J.B., DAVIES M., 1994. *Biodeterioration of materials in marine environments.* Material World, 2 (4): 189–191.

SEAWARD M.R.D., GIACOBINI C., 1988. *Lichen-induced biodeterioration of Italian monuments, frescoes and other archaeological materials.* Studia Geobotanica, 8: 3–11.

SEAWARD M.R.D., GIACOBINI C., GIULIANI M.R., ROCCARDI A., 1989. *The role of lichens in the biodeterioration of ancient monuments with particular reference to Central Italy.* International Biodeterioration Bulletin, 25: 49–55.

SEAWARD M.R.D., GIACOBINI C., ROCCARDI A., 1990. *I licheni a Villa Cordellina.* In: *Tiepolo e il Settecento Vicentino. Le sculture del giardino. Vicende e problemi di conservazione* (catalogo). Electa, Milano: 327–328.

SECCO SUARDO G., 1866. *Manuale ragionato per la parte meccanica dell'arte del restauratore dei dipinti.* Milano.

SEITZ L.M., SAUER D.B., BURROUGHS R., MOHR H.E., HUBBARD J.D., 1979. *Ergosterol as a measure of fungal growth.* Phytopathology, 69: 1202–1203.

SEVES A.M., SORA S., CIFERRI O., 1996. *The microbial colonization of oil paintings. A laboratory investigation.* International Biodeterioration & Biodegradation, 37: 215–224.

SEVES A.M., ROMANÒ M., MAIFRENI T., SORA S., CIFERRI O., 1998. *The microbial degradation of silk: A laboratory investigation.* International Biodeterioration & Biodegradation, 42: 203–211.

SEVES A.M., ROMANÒ M., MAIFRENI T., SEVES A., SCICOLONE G., SORA S., CIFERRI O., 2000. *A laboratory investigation on the microbial degradation of cultural heritage.* In: CIFERRI O., TIANO P., MASTROMEI G. (Eds.), *Of Microbes and Art. The Role of Microbial Communities in the Degradation and Protection of Cultural Heritage*, Kluwer Academic/Plenum Publishers, New York: 121–133.

SHAH R.P., SHAH N.R., 1992–93. *Growth of plants on monuments.* Studies in Museology, 26: 29–34.

SHARMA K.D., 1991. *Testing of chrome tanned leather against biodeterioration by fungi under high relative humidity.* In: AGRAWAL P.O., DHAWAN S. (Eds.), *Proceedings of the International Conference on Biodeterioration of Cultural Property*, Lucknow 1989. Rajkamal Eletric Press, New Delhi: 153–159.

SHARMA O.P., SHARMA K.D., 1980. *Application of fungicides in the control of fungal deterioration of finished leather in India.* International Biodeterioration Bulletin, 16 (49): 107–112.

SIGNORINI M.A., 1995. *Lo studio e il controllo della vegetazione infestante nei siti archeologici. Una proposta metodologica*. In: Marino L., Nenci C. (Eds.), *L'area archeologica di Fiesole. Rilievi e ricerche per la conservazione*. Alinea Ed., Firenze: 41–46.

Signorini M.A., 1996. *L'indice di pericolosità: Un contributo del botanico al controllo della vegetazione infestante nelle aree monumentali*. Inf. Bot. Ital., *28 (1)*: 7–14.

Singh A.P., Butcher J.A., 1991. *Bacterial degradation of wood cells: A review of degradation patterns*. Journal of the Institute of Wood Science, *12*: 143–157.

Singh A.P., Nilsson T., Daniel G.F., 1990. *Bacterial attack of* Pinus sylvestris *wood under near-anaerobic conditions*. Journal of the Institute of Wood Science, *11*: 237–249.

Singh A.P., Meenakshi G., Sing A.B., 1995. *Fungal spores are an important component of library air*. Aerobiologia, *11*: 231–237.

Singh J., 1993. *Nature and extent of deterioration in buildings due to fungi*. In: Singh J. (Ed.), *Building Mycology*, Chapman & Hall, London: 34–53.

Sjöström E., 1993. *Wood Chemistry: Fundamentals and Applications*. San Diego, Academic Press, Inc.

Smith A.J.E., 1993. *The Moss Flora of Britain and Ireland*. Cambridge University Press, Cambridge.

Smith V.L., Punja Z.K., Jenkins S.F., 1986. *A histological study of infection of host tissue by* Sclerotium rolfsii. Phytopathology, *76*: 755–759.

Somavilla J.F., Khayyat N., Arroyo V., 1978. *A comparative study of the microorganisms present in the Altamira and La Pasiega caves*. International Biodeterioration Bulletin, *14*: 103–109.

Sorlini C., Sacchi M., Falappi D., 1983. *I microrganismi che concorrono al deterioramento degli affreschi della casa romana dell'Orto di S. Giulia in Brescia*. Notiziario dell'Ecologia, *3*: 17–19.

Sorlini C., Sacchi M., Ferrari A., 1987. *Microbiological deterioration of Gambara's frescoes exposed to open air in Brescia, Italy*. International Biodeterioration, *23*: 167–179.

Sorlini C., Falappi D., Ranalli G., 1991. *Biodeterioration Preliminary Test on Samples of Serena Stone Treated with Resin*. Ann. Microbiol., *41*: 71–79.

Sorlini C., Zanardini E., Albo S., Praderio G., Cariati F., Bruni S., 1994. *Research on Chromatic Alterations of Marbles from the Fountain of Villa Litta (Linate, Milan)*. International Biodeterioration & Biodegradation, *33*: 152–164.

Staffeldt E.E., Kohler D.A., 1973. *Assessment of corrosion products removed from "La Fortuna" punta della dogana da mar, Venezia*. In: *Atti del Congresso Petrolio e Ambiente*, Ed. Poligrafico Artioli, Roma: 166–170.

Stainer R.Y., Doudoroff M., Adelberg E.A., 1970. *The microbial world*, Prentice Hall, Inc, Englewood Cliffs, NJ.

Stanley P.E., 1988. *Rapid microbiology: the use of luminescence and adenosine triphosphate (ATP) for enumerating and checking effectiveness of biocides: present status and future prospects*. In: Houghton D.R., Smith R.N., Eggins H.O.W. (Eds.), *Biodeterioration 7*. Elsevier Applied Science, London and New York: 664–668.

Sterflinger K., 1998. *Temperature and NaCl tolerance of rock-inhabiting meristematic fungi*. Antonie van Leeuwenhoek, *74*: 271–281.

Sterflinger K., Krumbein W.E., 1995. *Multiple stress factors affecting growth of rock-inhabiting black fungi*. Botanica Acta, *108*: 490–496.

Sterflinger K., Krumbein W.E., 1997. *Dematiaceous fungi as a major agent for biopitting on Mediterranean marbles and limestones*. Geomicrobiology Journal, *14 (219)*: 21–230.

Sterflinger K., De Baere R., de Hoog G.S., De Wachter R., Krumbein W.E., Haase G., 1997. Coniosporium perforans *and* C. apollinis, *two new rock-inhabiting fungi isolated from marble in the Sanctuary of Delos (Cyclades, Greece)*. Antonie van Leeuwenhoek, *72*: 349–363.

Sterflinger K., de Hoog G.S., Haase G., 1999. *Phylogeny and ecology of meristematic ascomycetes*. Studies in Mycology, *43*: 5–22.

Stolow N., 1987. *Conservation and exhibitions: Packing, transport, storage and environmental considerations*. Butterworths, London.

Stone H.E., Armentrout V.N., 1985. *Production of oxalic acid by* Sclerotium ceviporum *during infection of onion*. Mycologia, *77*: 526–530.

Stroebel L., Compton J., Current L., Zakia R., 1993. *Fondamenti di fotografia, Materiali e processi*, Zanichelli, Bologna.

Strzelczyk A.B., 1981. *Painting and sculpture*. In: Rose A.H. (Ed.), Microbial Biodeterioration. Economic Microbiology. Academic Press, London, *6*: 203–234.

Strzelczyk A.B., Kuroczkin J., Krumbein W.E., 1987. *Studies on the microbial degradation of ancient leather bookbindings. Part I*. International Biodeterioration, *23 (1)*: 3–27.

Strzelczyk A.B., Kuroczkin J., Krumbein W.E., 1989. *Studies on the microbial degradation of ancient leather bookbindings. Part 2*. International Biodeterioration, *25 (13)*: 39–47.

Strzelczyk A.B., Bannach L., Kurowska A., 1997. *Biodeterioration of Archeological Leather*. International Biodeterioration & Biodegradation., *39 (4)*: 301–309.

Sukopp H., 1990. *Urban ecology and its application in Europe*. In: Sukopp H. et al. (Eds.), *Urban ecology*, SPB, Acad. Publ., The Hague: 1–22.

Suzuki J., Koestler R.J., 2003. *Visual assessment of biocide effects on Japanese paint materials*. In: Koestler R.J., Koestler V.H., Charola A.E., Nieto-Fernandez F.E. (Eds.), *Art, Biology, and Conservation: Biodeterioration of Works of Art*. The Metropolitan Museum of Art, New York: 410–425.

Szczepanowska H., Cavaliere A.R., 2000. *Fungal deterioration of 18th and 19th century documents: A case study of the Tilghman family collection, Wye House, Easton, Jour*. International Biodeterioration & Biodegradation, *46*: 245–249.

Szczepanowska H.M., Cavaliere A.R., 2003. *Artworks, drawings, prints, and documents – Fungi eat them all!* In: Koestler R.J., Koestler V.H., Charola A.E., Nieto-Fernandez F.E. (Eds.), *Art, Biology, and Conservation: Biodeterioration of Works of Art*. The Metropolitan Museum of Art, New York: 128–151.

Szczepanowska H., Lovett C.M., 1992. *A study of the removal and prevention of fungal stains on paper*. Journal of the American Institute for Conservation, *2 (1)*: 147–160.

Talarico F., Caldi C., Valenzuela M., Zaccheo C., Zampa A., Nugari M.P., 2001. *Applicazioni dei gel come supportanti nel restauro.* Bollettino ICR Nuova Serie, *3*: 101–118.

TANASI M.T., 2002. *Il deterioramento di natura chimica della pergamena.* In: *Chimica e biologia applicate alla conservazione degli archivi.* Ministero per i Beni e le Attività Culturali. Direzione Generale degli Archivi di Stato, Istituto Poligrafico e Zecca dello Stato: 321–330.

TARSITANI G., TRAMA A., 2002. *I rischi di salute per gli operatori dei beni culturali.* In: *X Congresso Nazionale di Aerobiologia – Aria e Salute*, Bologna: 72.

TARSITANI G.F., NUGARI M.P., GIULIANI M.R., 1998. *A method for microbiologic study prior to the opening of a tomb.* In: *Interim Meeting ICOM-CC "Interdisciplinary Approach about Studies and Conservation of Medieval Textiles,"* Palermo: 77–79.

TAVZES C., POHLEVEN F., KOESTLER R.J., 2001. *Effect of anoxic conditions on wood-decay fungi treated with argon or nitrogen.* International Biodeterioration & Biodegradation, *47*: 225–231.

TAVZES C., POHLEVEN J., POHLEVEN F., KOESTLER R.J., 2003. *Anoxic eradication of fungi in wooden objects.* In: KOESTLER R.J., KOESTLER V.H., CHAROLA A.E., NIETO-FERNANDEZ F.E. (Eds.), *Art, Biology, and Conservation: Biodeterioration of Works of Art.* The Metropolitan Museum of Art, New York: 426–439.

TAYLER S., MAY E., 1991. *The seasonality of heterotrophic bacteria on sandstones of ancient monuments.* International Biodeterioration, *28*: 48–64.

TAYLER S., MAY E., 1995. *A comparison of methods for the measurement of microbial activity on stone.* Studies in Conservation, *40*: 163–170.

TÉTREAULT J., 1999. *Coatings for display and storage in museums.* Canadian Conservation Institute Technical Bulletin, *21*: 1–46.

THOMPSON J.M.A., 1994. *Manual of curatorship: A guide to museum pratice.* Butterworths, London, 2nd edition.

THOMSON G., 1986. *The museum environment.* Butterworths, London.

TIANO P., 1995. *Stone reinforcement by calcite crystals precipitation induced by organic matrix macromolecules.* Studies in Conservation, *40*: 171–176.

TIANO P., 1998. *Biodeterioration of monumental rocks: Decay mechanisms and control methods.* Science and Technology for Cultural Heritage, *7 (2)*: 19–38.

TIANO P., 2003. *Biomediated calcite treatments for stone conservation.* In: SAIZ-JIMENEZ C. (Ed.), *Molecular Biology and Cultural Heritage, Proceedings of International Congress on Molecular Biology and Cultural Heritage*, Seville. Balkema Publishers, Lisse (NL): 201–208.

TIANO P., TOMASELLI L., 1989. *Un caso di biodeterioramento del marmo.* Arkos, *6*: 12–18.

TIANO P., BIANCHI R., GARGANI G., VANNUCCI S., 1975. *Research on the presence of sulphur-cycle bacteria in the stone of some historical buildings in Florence.* Plant and Soil, *43*: 211–217.

TIANO P., TOMASELLI L., ORLANDO C., 1989. *The ATP-bioluminescence method for a rapid evaluation of the microbial activity in the stone materials of monuments.* Journal of Bioluminescence and Chemoluminescence, *3*: 213–216.

TIANO P., ADDADI L., WEINER S., 1992. *Stone reinforcement by calcite crystals using organic matrix*

macromolecules: feasibility study. In: DELGADO RODRIGUES J., HENRIQUES F., TELMO JEREMIAS F. (Eds.), *Proceedings of the 7th International Congress on Deterioration and Conservation of Stone.* LNEC, Lisbon, Portugal: 1317–1326.

TIANO P., ACCOLLA P., TOMASELLI L., 1994. *Biocidal Treatments on Algal Biocenosis. Control of the Lasting Activity.* Science and Technology for Cultural Heritage, *3*: 89–94.

TIANO P., ACCOLLA P., TOMASELLI L., 1995a. *Phototrophic biodeteriogens on lithoid surfaces: An ecological study.* Microbial Ecology, *29*: 299–309.

TIANO P., CAMAITI M., ACCOLLA P., 1995b. *Methods for evaluation of products against algal biocenosis of monumental fountains.* In: *International Colloquium on Methods of Evaluating Products for the Conservation of Porous Building Materials in Monuments.* ICCROM, Rome: 75–86.

TIANO P., TOSINI I., RIZZI M., TSAKOMA M., 1996. *Calcium oxalate decomposing microorganisms: A biological approach to the oxalate patinas elimination.* In: REALINI M., TONIOLO L. (Eds.), *Proceedings of the 2nd International Symposium: The Oxalate Films in the Conservation of Works of Art.* Editeam, Milano: 25–27.

TIANO P., TOSINI I., RIZZI M., 1997. *Antimould action of some biocides mixed with glues used in painting conservation.* Science and Technology of Cultural Heritage, *6 (2)*: 129–134.

TIANO P., BIAGIOTTI L., BRACCI S., 2000. *Biodegradability of products used in monuments' conservation.* In: CIFERRI O., TIANO P., MASTROMEI G. (Eds.), *Of Microbes and Art. The Role of Microbial Communities in the Degradation and Protection of Cultural Heritage.* Kluwer Academic/Plenum Publishers, New York: 169–181.

TILAK S.T., 1989. *Airborne fungal spores.* In: *Airborne pollen and fungal spores*, Vaijayanti Prakashan, Aurangabad, India: 125–141.

TILAK S.T., 1991. *Biodeterioration of paintings at Ajanta.* In: AGRAWAL O.P., DHAWAN S. (Eds.), *Proceedings of the International Conference on Biodeterioration of Cultural Property*, Lucknow 1989. Rajkamal Eletric Press, New Delhi: 305–312.

TOISHI K., ARAI H., KENJO T., YAMANO K. (Eds.), 1992. *Proceedings of the 2nd International Conference on Biodeterioration of Cultural Property*, Pacifico Yokohama. International Communications Specialists, Tokyo.

TOMASELLI L., MARGHERI M. C., FLORENZANO G., 1979. *Indagine sperimentale sul ruolo dei cianobatteri e delle microalghe nel deterioramento di monumenti ed affreschi.* In: BADAN B. (Ed.), *Atti del 3° Congresso Internazionale sul Deterioramento e la Conservazione della Pietra, Venezia.* Litografia La Photograph, Padova: 313–325.

TOMASELLI L., LAMENTI G., BOSCO M., TIANO P., 2000a. *Biodiversity of photosynthetic micro-organisms dwelling on stone monuments.* International Biodeterioration & Biodegradation, *46*: 251–258.

TOMASELLI L., LAMENTI G., TIANO P., 2000b. *Relationships between stone-dwelling cyanobacteria and damage.* In: GUARINO A. (Ed.), *Proceedings of the 2nd International Congress, Science and Technology for the Safeguard of Cultural Heritage in the Mediterranean Basin*, Paris, France, *1*: 841–842.

TOMASELLI L., TIANO P., LAMENTI G., 2000c. *Occurrence and fluctuation in photosynthetic biocoenosis dwelling on stone monuments*. In: CIFERRI O., TIANO P., MASTROMEI G. (Eds.), *Of Microbes and Art. The Role of Microbial Communities in the Degradation and Protection of Cultural Heritage*. Kluwer Academic/Plenum Publishers, New York: 63–76.

TOMASELLI L., LAMENTI G., TIANO P., 2003. *Diagnostic tools for monitoring phototrophic biodeteriogens*. In: SAIZ-JIMENEZ (Ed.), *Molecular Biology and Cultural Heritage, Proceedings of International Congress on Molecular Biology and Cultural Heritage*, Seville. Balkema Publishers, Lisse (NL): 247–251.

TOMASELLI R., 1955. *Relazione sulla nomenclatura botanica speleologica*. Archivio Botanico, *31 (4–15)*: 193–211.

TOSCO U., 1970. *La vegetazione delle grotte di Castellana (Bari)*. Le Grotte d'Italia, *4 (2)*: 193–202.

TRETIACH M., 1996. *La coevoluzione di funghi con alghe e cianobatteri all'interno della simbiosi lichenica*. In: *Atti dell'Accademia Nazionale dei Lincei, XXIII Seminario sulla Evoluzione Biologica e i grandi problemi della Biologia, Coevoluzione e Coadattamento*: 121–142.

TRETIACH M., PINNA D., GRUBE M., 2003. Caloplaca erodens *[sect. Pyrenodesmia]*, *a new lichen species from Italy with a unusual thallus type*. Mycological Progress *2 (2)*: 127–136.

TUDOR P.B., MATERO F.G., KOESTLER R.J., 1990. *A case study of the biocidal compatibility of biocidal cleaning and consolidation in the restoration of a marble statue*. In: *Biodeterioration Research 3*, Plenum Publishing Corp., New York.

TYLECOTE R.F., 1979. *The effect of soil conditions on the long-term corrosion of buried tin bronzes and copper*. Journal Archaeological Science, *6 (4)*: 345–368.

UCHIDA E., OGAWA Y., MAEDA N., NAKAGAWA T., 1999. *Deterioration of stone materials in the Angkor monuments, Cambodia*. Engineering Geology, *55*: 101–112.

UNGER A., SCHNIEWIND A.P., UNGER W., 2001. *Conservation of wood artifacts. A handbook*. Springer Verlag, Berlin Heidelberg.

UNI 10380, 1994. *Illuminotecnica. Illuminazione di interni con luce artificiale*. UNI, Milano.

UNI 10586, 1997. *Documentazione. Condizioni climatiche per ambienti di conservazione di documenti grafici e caratteristiche degli alloggiamenti*. UNI, Milano.

UNI 10813, 1999. *Beni Culturali. Materiali lapidei naturali ed artificiali. Verifica della presenza di microrganismi fotosintetici su materiali lapidei mediante determinazione spettrofotometrica UV/Vis delle clorofille a, b, e c*. UNI, Milano.

UNI 10829, 1999. *Beni di interesse storico artistico. Condizioni ambientali di conservazione. Misurazioni ed analisi*. UNI, Milano.

UNI 10922, 2001. *Beni Culturali. Materiali lapidei naturali ed artificiali. Allestimento di sezioni sottili e sezioni lucide di materiali lapidei colonizzati da biodeteriogeni*. UNI, Milano.

UNI 10923, 2001. *Beni Culturali. Materiali lapidei naturali ed artificiali. Allestimento di preparati biologici per l'osservazione al microscopio ottico*. UNI, Milano.

UNI 10969, 2002. *Principi generali per la scelta e il controllo del microclima per la conservazione dei beni culturali in ambienti interni*. UNI, Milano.

UNI EN 335–1, 1992. *Durability of wood and wood-based products. Definition of hazard classes of biological attack. Part 1: General*.

UNI EN 335–2, 1992. *Durability of wood and wood-based products. Definition of hazard classes of biological attack. Part 2: Application to solid wood*.

URZÌ C., ALBERTANO P., 2001. *Studying phototrophic and heterotrophic microbial communities on stone monuments* In: DOYLE R.J. (Ed.), *Methods in Enzymology*, Academic Press, San Diego, CA: *336*: 340–355.

URZÌ C., KRUMBEIN W.E., 1994. *Microbiological impacts on the cultural heritage*. In: KRUMBEIN W.E., BRIMBLECOMBE P., COSGROVE D.E., STAINFORTH S. (Eds.), *Durability and Changes: The Science, Responsibility, and Cost of Sustaining Cultural Heritage*, John Wiley & Sons, Chichester, UK: 107–135.

URZÌ C., REALINI M., 1998. *Colour changes of Noto's Calcareous Sandstone as Related to its Colonisation by Microorganisms*. International Biodeterioration & Biodegradation, *42*: 45–54.

URZÌ C., LISI S., CRISEO G., PERNICE A., 1991. *Adhesion to and degradation of marble by a* Micrococcus *strain isolated from it*. Geomicrobiology Journal, *9*: 81–90.

URZÌ C., KRUMBEIN W.E., WARSCHEID TH., 1992. *On the question of biogenic colour changes of Mediterranean monuments (coating. crust. microstromatolite. patina. scialbatura. skin. rock varnish)*. In: DECROUEZ D., CHAMAY J., ZEZZA F. (Eds.), *Proceedings of the 2nd International Symposium, Musée d'art et d'histoire*, Geneve: 397–420.

URZÌ C., KRUMBEIN W.E., LYALIKOVA N., PETUSHKOVA J., WOLLENZIEN U., ZAGARI M., 1994. *Microbiological investigations of marbles exposed to natural and anthropogenic influences in northern and southern climates*. In: FASSINA V., OTT H., ZEZZA F. (Eds.), *Proceedings of the 3rd International Symposium on the Conservation of Monuments in the Mediterranean Basin*, Venice: 297–304.

URZÌ C., DE LEO F., DE HOOG S., STERFLINGER K., 2000a. *Recent advances in the molecular biology and ecophysiology of meristematic stone-inhabiting fungi*. In: CIFERRI O., TIANO P., MASTROMEI G. (Eds.), *Of Microbes and Art. The Role of Microbial Communities in the Degradation and Protection of Cultural Heritage*. Kluwer Academic/Plenum Publishers, New York: 3–19.

URZÌ C., DE LEO F., GALLETTA M., SALAMONE P., 2000b. *Efficiency of biocide in "in situ" and "in vitro" treatment. Study case of the "Templete de Mudejar," Guadalupe, Spain*. In: FASSINA V. (Ed.), *Proceedings of the 9th International Congress on Deterioration and Conservation of Stone*, Venice. Elsevier, Amsterdam: 531–539.

URZÌ C., DONATO P., LO PASSO C., ALBERTANO P., 2002. *Occurrence and biodiversity of* Streptomyces *strains isolated from Roman Hypogea*. In: GALAN E., ZEZZA F. (Eds.), *Protection and Conservation of the Cultural Heritage of the Mediterranean Cities. Proceeding of the 5th International Symposium on the Conservation of Monuments in the Mediterranean Basin*, Seville: 269–272.

URZÌ C., LA CONO V., DE LEO F., DONATO P., 2003a. *Fluorescent In Situ Hybridization (FISH) to Study Biodeterioration*. In: SAIZ-JIMENEZ C. (Ed.), *Molecular*

Biology and Cultural Heritage, Proceedings of International Congress on Molecular Biology and Cultural Heritage, Seville. Balkema Publishers, Lisse (NL): 55–60.

Urzì C., De Leo F., Donato P., La Cono V., 2003b. *Multiple approaches to study the structure and diversity of microbial communities colonizing artistic surfaces*. In: Saiz-Jimenez C. (Ed.), *Molecular Biology and Cultural Heritage, Proceedings of International Congress on Molecular Biology and Cultural Heritage*, Seville. Balkema Publishers, Lisse (NL): 187–194.

Urzì C., De Leo F., Donato P., La Cono V., 2003c. *Study of microbial communities colonizing hypogean monuments surfaces using destructive and non-destructive sampling methods*. In: Koestler R.J., Koestler V.R., Charola A.E., Nieto-Fernandez F.E. (Eds.), *Art, Biology, and Conservation: Biodeterioration of Works of Art*. The Metropolitan Museum of Art, New York: 316–327.

Urzì C., La Cono, V., Stackebrandt, E., 2004. *Design and application of two oligonucleotide probes for the identification of* Geodermatophilaceae *strains using Fluorescence In Situ Hybridization (FISH)*. Environmental Microbiology, 6(7): 678–685.

Valcarce M.P., Busalmen J.P., de Sànchez S.R., 2002. *The influence of the surface condition on the adhesion of* Pseudomans fluorescens *(ATCC 17552) to copper and aluminium brass*. International Biodeterioration & Biodegradation, 50: 61–66.

Valentin N., 2003. *Microbial contamination in museum collection: organic materials*. In: Saiz-Jimenez C. (Ed.), *Molecular Biology and Cultural Heritage, Proceedings of International Congress on Molecular Biology and Cultural Heritage*, Seville. Balkema Publishers, Lisse (NL): 85–91.

Van den Hoek C., Mann D.G., Jahns H.M. (Eds.), 1995. *Algae. An introduction to phycology*. Cambridge University Press, Cambridge.

Van Der Molen L.J., Garty B.W., Aardema W., Krumbein W., 1980. *Growth control of algae and cyanobacteria on historical monuments by a mobile UV unit (MUVU)*. Studies in Conservation, 25 (2): 71–77.

Vasella A., Davies G.J., Bohm M., 2002. *Glycosidase mechanisms*. Curr. Opin. Chem. Biol., 6: 619–629.

Vedovello S., 2000. *Lo stato di conservazione*. In: Capponi G., Vedovello S. (Eds.), *Il restauro della Torre di Pisa. Il cantiere di progetto per la conservazione delle superfici*. Litografica Iride, Roma: 73–83.

Videla H.A., Saiz-Jimenez C., 2002. *Biodeterioro de monumentos de Iberoamerica*. Programma CITED.

Videla H.A., Guiamet P.S., Gomez de Saravia S., 2000. *Biodeterioration of Mayan archaeological sites in the Yucatan Peninsula, Mexico*. International Biodeterioration & Biodegradation, 46: 335–341.

Vigini, G., 1985. *Glossario di biblioteconomia e scienza dell'informazione*. Editrice Bibliografica, Milano.

Vigo T.L., 1977. *Preservation of Natural Textile Fibers – Historical Perspectives*. In: Williams J.C. (Ed.), *Preservation of Paper and Textiles of Historic and Artistic Value*, Advances in Chemistry, American Chemical Society, Washington, 164: 189–207.

Vigo T.L., 1980. *Protection of textiles from biodeterioration*. In: *Conservazione e restauro dei tessili, Convegno internazionale*, Como: 18–26.

Vittori O., Mestitz A., 1976. *Four golden horses in the sun*. International Fund for Monuments, New York.

Vlad Borrelli L., 1954. *Distacco di due frammenti dalla tomba del colle*. Boll. Ist. Centrale Restauro: 33–37.

Von Endt D.W., Jessup W.C., 1986. *The deterioration of protein materials in museum*. In: Barry S., Houghton D.R. (Eds.), *6th International Biodeterioration Symposium, Washington, DC, August 1984*. Biodeterioration, 6: 332–335.

Von Plehwe-Leisen E., Warscheid T., Leisen H., 1996. *Studies of long-term behaviour of conservation agents and microbiological contamination on twenty years exposed treated sanstone cubes*. In: Riederer J. (Ed.), *Proceedings of the 8th International Congress on Deterioration and Conservation of Stone*. Berlin, Germany 2: 1029–1037.

Voronina L.I., Nazarova O.N., Petushkova U.P., Rebrikova N.L., 1981. *Damage of parchment and leather caused by microbes*. In: Grattan D.W. (Ed.), *Proceedings of the ICOM Waterlogged Wood Working Group Conference, Ottawa*. ICOM, Ottawa: 19/3.1–19/3.11.

Wadum J., 1998. *Microclimate Boxes for Panel Paintings*. In: Dardes K., Rothe A.N. (Eds.), *Proceedings of Symposium "The Structural Conservation of Panel Paintings."* J. Paul Getty Museum, 1995. Getty Conservation Institute, Los Angeles: 497–524.

Wakefield R.D., Jones M.S., 1996. *Some effects of masonry biocides on intact and decayed stone*. In: Riederer J. (Ed.), *Proceedings of the 8th International Congress on Deterioration and Conservation of Stone*, Berlin, Germany: 703–716.

Walsh J.H., Stewart C.S., 1971. *Effect of temperature, oxygen and carbon dioxide on cellulolytic activity of some fungi*. Trans. Br. Mycol. Soc., 57: 75–84.

Walter H., 1994. *Vegetation of the Earth and Ecological System of the Geo-biosphere*. Heidelberg Science Library, Springer Verlag, Berlin.

Warscheid Th., 2000. *Integrated concepts for the protection of cultural artifacts against biodeterioration*. In: Ciferri O., Tiano P., Mastromei G. (Eds.), *Of Microbes and Art. The Role of Microbial Communities in the Degradation and Protection of Cultural Heritage*. Kluwer Academic/Plenum Publishers, New York: 185–201.

Warscheid Th., 2003. *The evaluation of biodeterioration processes on cultural objects and approaches for their effective control*. In: Koestler R.J., Koestler V.H., Charola A.E., Nieto-Fernandez F.E. (Eds.), *Art, Biology, and Conservation: Biodeterioration of Works of Art*. The Metropolitan Museum of Art, New York: 14–27.

Warscheid Th., Braams J., 2000. *Biodeterioration of Stone: A review*. International Biodeterioration & Biodegradation, 46 (4): 343–368.

Warscheid Th., Krumbein W.E., 1996. *Biodeterioration of inorganic materials – General aspects and selected case*. In: Heitz H., Sand W., Flemming H.C. (Eds.), *Microbially induced corrosion of materials*, Springer, Berlin: 273–295.

Warscheid Th., Petersen K., Krumbein W.E., 1990. *A rapid method to demonstrate and evaluate microbial activity on decaying sandstone*. Studies in Conservation, 35: 137–147.

WEE Y.C., LEE K.B., 1980. *Proliferation of algae on surfaces of buildings in Singapore.* International Biodeterioration & Biodegradation, *16*: 113–117.

WEED S.B., DAVEY C.B., COOK M.G., 1969. *Weathering of mica by fungi.* Soil Sci. Soc. Amer. Proceedings, *33*: 813–814.

WEISSMANN R., DREWELLO R., 1996. *Attack on glass.* In: HEITZ E., FLEMMING H. C., SAND W. (Eds.), *Microbially Influenced Corrosion of Materials,* Springer Verlag, Berlin: 339–352.

WELLHEISER J.G., 1992. *IFLA Publication 60. Nonchemical Treatment Processes for Disinfestation of Insects and Fungi.* Library Collections, K.G. Saur Publ.

WENDLER E., PRASARTSET C., 1999. *Lichen growth on old Khmer-style sandstone monuments in Thailand: Damage factor or shelter?* In: *Preprints of the 12th Triennial Meeting, Lyon, ICOM Committee for Conservation.* James & James, London, *2*: 750–754.

WESSEL D.P., 2003. *The use of metallic oxides in control of biological growth on outdoor monuments.* In: KOESTLER R.J., KOESTLER V.H., CHAROLA A.E., NIETO-FERNANDEZ F.E. (Eds.), *Art, Biology, and Conservation: Biodeterioration of Works of Art.* The Metropolitan Museum of Art, New York: 536–551.

WESSELS D., WESSELS L., 1995. *Biogenic weathering and microclimate of Clarens sandstone in South Africa.* Cryptogamic Botany, *5*: 288–298.

WESSELS J.G.H., 1994. *Developmental regulation of fungal cell wall formation.* Ann. Rev. Phytophatol., *32*: 413–417.

WHEELER A.P., SIKES C.S., 1989. *Matrix-crystal interaction in CaCO₃ biomineralization.* In: MANN S., WEBB J., WILLIAMS R.J.P. (Eds.), *Biomineralization: Chemical and Biochemical Perspectives.* VCH, Friburg: 95–131.

WHIPPS J.M., 1987. *Method for estimation of chitin content of mycelium of ectomycorrhizal fungi grown on solid substrates.* Transaction of the British Mycological Society, *89*: 199–203.

WHITERHOUSE H.L.K., 1980. *The production of protonemal gemmae by mosses growing in deep shade.* Journal of Bryology, *11*: 133–138.

WICKLOW D.T., CARROL G.C., 1981. *The fungal comunity. Its organization and role in the ecosystem.* Marcel Dekker, New York.

WILIMZIG M., FAHRIG N., BOCK E., 1992. *Biologically influenced corrosion of stones by nitrifying bacteria.* In: DELGADO RODRIGUEZ J., HENRIQUES F., TELMO JEREMIAS F. (Eds.), *Proceedings of the 7th International Congress on Deterioration and Conservation of Stone.* LNEC, Lisbon, Portugal: 459–469.

WILSON M.J., JONES D., McHARDY W.J., 1981. *The weathering of serpentinite by Lecanora atra.* Lichenologist, *13*: 167–176.

WIRTH V., 1995. *Flechtenflora.* UTB-Verlag, Ulmer Verlag, Stuttgart.

WOESE C.R., 2000. *Interpreting the universal phylogenetic tree.* In: *Proceedings of the National Academy of Science,* USA *97(15)*: 8392–8396.

WOLBERS R., 2000. *Cleaning painted surface: Aqueous Methods.* Archetype, London.

WOLLENZIEN U., DE HOOG G.S., KRUMBEIN W.E., URZÌ C., 1995. *On the isolation of microcolonial fungi occurring on and in marble and other calcareous rocks.* The Science of Total Environment, *167*: 287–294.

WOLLENZIEN U., DE HOOG G.S., KRUMBEIN W.E., UIJTHOF J.M.J., 1997. Sarcinomyces petricola, *a new microcolonial fungus from marble in the Mediterranean basin.* Antonie van Leeuwenhoek, *71*: 281–288.

WORLD HEALTH ORGANIZATION (WHO), 1990. *Indoor air quality: Biological contaminants.* WHO Regional Publications, European Series, *31.*

YOUNG G.S., WAINWRIGHT I.N.M., 1995. *The control of algal biodeterioration of a marble petroglyph site.* Studies in Conservation, *40*: 82–95.

YOUNG M.E., WAKEFIELD R., URQUHART D.C.M., NICHOLSON K., TONG K., 1995. *Assessment in a field setting of various biocides on sandstone.* In: *International Colloquium on Methods of Evaluating Products for the Conservation of Porous Building Materials in Monuments,* ICCROM, Rome: 93–99.

YOUNG V., URQUHART D.C.M., 1996. *Access to the Built Heritage: Advice on the provision of access for people with disabilities to historic sites open to the public.* Historic Scotland Technical Advice Note 7. The Stationery Office, Edinburgh: 1–51.

ZAINAL A.S., GHANNOUM M.A., SALLAL A.K., 1983. *Microbial biodeterioration of leather-containing exhibit in Kuwait National Museum.* Biodeterioration, *5*: 416–426.

ZALAR P., DE HOOG G.S., GUNDE-CIMERMAN N., 1999. *Ecology of halotolerant dothideaceous black yeasts.* Studies in Mycology, *43*: 38–48.

ZANARDINI E., ANDREONI V., BORIN S., CAPPITELLI F., DAFFONCHIO D., TALOTTA P., SORLINI C., RANALLI G., BRUNI S., CARIATI F., 1997. *Lead-resistant Microorganisms from Red Stains of Marble of the Certosa of Pavia, Italy and Use of Nucleic Acid-based Techniques for their Detection.* International Biodeterioration & Biodegradation, *40 (2–4)*: 171–182.

ZANARDINI E., ABBRUSCATO P., GHEDINI N., REALINI M., SORLINI C., 2000. *Influence of atmospheric pollutants on the biodeterioration of stone.* International Biodeterioration & Biodegradation, *45*: 35–42.

ZANARDINI E., ABBRUSCATO P., SCARAMELLI L., ONELLI E., REALINI M., PATRIGNANI G., SORLINI C., 2003. *Red stains on Carrara marble: A study case of Certosa of Pavia (Italy).* In: KOESTLER R.J., KOESTLER V.H., CHAROLA A.E., NIETO-FERNANDEZ F.E. (Eds.), *Art, Biology and Conservation,* The Metropolitan Museum of Art, New York: 226–247.

ZANOTTI CENSONI A.L., MANDRIOLI P., 1979. *Aerobiological investigation in Scrovegni Chapel (Padua, Italy).* In: *Proceedings of the 3rd International Congress on Deterioration and Preservation of Stone, Venezia.* Litografia La Photograph, Padova: 699–703.

ZUNINO M., ZULLINI A., 1995. *Biogeografia: la dimensione spaziale dell'evoluzione.* Ed. Ambrosiana, Milano.

ZYSKA B., 1997. *Fungi isolated from library materials: A review of the literature.* International Biodeterioration & Biodegradation, *40 (1)*: 43–51.

CONTRIBUTORS

Patrizia B. ALBERTANO, Department of Biology, Università degli Studi di Roma "Tor Vergata," Via della Ricerca Scientifica, 00133 Roma. *albertano@uniroma2.it*

Antonella ALTIERI, Laboratory for Biological Analyses, Istituto Centrale del Restauro (Central Conservation Institute), Piazza San Francesco di Paola 9, 00184 Roma. *aaltieri.icr@beniculturali.it*

Stefano BERTI, IVALSA/Istituto per la Valorizzazione del Legno e delle Specie Arboree (Trees and Timber Institute) of the CNR/Consiglio Nazionale delle Ricerche (National Research Council), Via Madonna del Piano, 50019 Sesto Fiorentino (FI). *berti@ivalsa.cnr.it*

Giulia CANEVA, Department of Biology, Università degli Studi Roma Tre, Viale G. Marconi 446, 00146 Roma. *caneva@uniroma3.it*

Francesca CAPPITELLI, Department of Microbiology and Food Science and Technology (DISTAM), Sezione MAAE, Università degli Studi di Milano, Via Celoria 2, 20133 Milano. *francesca.cappitelli@unimi.it*

Simona CESCHIN, Department of Biology, Università degli Studi Roma Tre, Viale G. Marconi 446, 00146 Roma. *ceschin@uniroma3.it*

Filomena De LEO, Department of Microbiological, Genetic, and Molecular Sciences, Università degli Studi di Messina, Salita Sperone 31, 98166 Messina. *deleo@chem.unime.it*

Corrado FANELLI, Department of Plant Biology, Università degli Studi di Roma "La Sapienza," Largo Cristina di Svezia 24, 00165 Roma. *corrado.fanelli@uniroma1.it*

Elisabetta GIANI, Laboratory for Biological Analyses, Istituto Centrale del Restauro (Central Conservation Institute), Via di S. Michele 23, 00153 Roma. *egiani.icr@beniculturali.it*

Oriana MAGGI, Department of Plant Biology, Università degli Studi di Roma "La Sapienza," Piazzale Aldo Moro 5, 00185 Roma. *oriana.maggi@uniroma1.it*

Paolo MANDRIOLI, Istituto di Scienze dell'Atmosfera e del Clima (Atmosphere and Climate Institute), Aerobiology, CNR/Consiglio Nazionale delle Ricerche (National Research Council), Via Gobetti 101, 40129 Bologna. *p.mandrioli@isac.cnr.it*

Donatella MATÈ Centro di Fotoriproduzione Legatoria e Restauro degli Archivi di Stato (Center for Photo Reproduction, Bookbinding, and Conservation of the State Archives), Via C. Baudana Vaccolini 14, 00153 Roma. *cflr.biol@archivi.beniculturali.it*

Pierluigi NIMIS, Department of Biology, Università di Trieste, Via L. Giorgieri 10, 34127 Trieste. *nimis@univ.trieste.it*

Maria Pia NUGARI, Istituto Centrale del Restauro (Central Conservation Institute), Piazza San Francesco di Paola 9, 00184 Roma. *mpnugari.icr@beniculturali.it*

Alessandra PACINI, Department of Biology, Università degli Studi Roma Tre, Viale G. Marconi 446, 00146 Roma. *paciniale@libero.it*

Ettore PACINI, Department of Environmental Sciences "Giacomino Sarfatti," Università di Siena, Via P.A. Mattioli 4, 53100 Siena. *pacini@unisi.it*

Sabrina PALANTI, IVALSA/Istituto per la Valorizzazione del Legno e delle Specie Arboree (Trees and Timber Institute) of the CNR/Consiglio Nazionale delle Ricerche (National Research Council), Via Madonna del Piano, 50019 Sesto Fiorentino (FI). *s.palanti@ivalsa.cnr.it*

Giovanna PASQUARIELLO, Istituto Nazionale per la Grafica (National Graphic Arts Institute), Via della Lungara 230, 00165 Roma. *giovannapasquariello@virgilio.it*

Anna Maria PERSIANI, Department of Plant Biology, Università degli Studi di Roma "La Sapienza," Piazzale Aldo Moro 5, 00185 Roma. *annamaria.persiani@uniroma1.it*

Rosanna PIERVITTORI, Department of Biology and CEBIOVEM/Centro di Eccellenza per la Biosensoristica Tramite l'Utilizzo di Organismi Vegetali e Microbici (Center for Excellence for the Biosensoristic Approach with the Use of Plant and Microbial Organisms), Università degli Studi di Torino, Viale Mattioli 25, 10125 Torino. *rosanna.piervittori@unito.it*

Anna Maria PIETRINI, Istituto Centrale del Restauro (Central Conservation Institute), Piazza San Francesco di Paola 9, 00184 Roma. *ampietrini.icr@beniculturali.it*

Daniela PINNA, Opificio delle Pietre Dure (Firenze) (Precious Stones Institute) e Soprintendenza per il Patrimonio Storico, Artistico ed Etnoantropologico

(Bologna) (Administrative Body for Historical, Artistic, and Ethnoanthropological Heritage). *daniela.pinna@unibo.it*

Flavia PINZARI, Istituto Centrale per la Patologia del Libro (Central Institute for Book Pathology), Biology Laboratory, Via Milano 76, 00184 Roma. *flavia.pinzari@quipo.it*

Giancarlo RANALLI, Department of Agriculture and Food, Environmental and Microbiological Sciences and Technologies, Università degli Studi del Molise, Via De Sanctis 46, 86100 Campobasso. *ranalli@unimol.it*

Sandra RICCI, Istituto Centrale del Restauro (Central Conservation Institute), Piazza San Francesco di Paola 9, 00184 Roma. *sricci.icr@beniculturali.it*

Ada ROCCARDI, Istituto Centrale del Restauro (Central Conservation Institute), Piazza San Francesco di Paola 9, 00184 Roma. *aroccardi.icr@beniculturali.it*

Ornella SALVADORI, Soprintendenza Speciale per il Polo Museale Veneziano (Special Administration for the Association of Venetian Museums), Scientific Laboratory, Cannaregio 3553, 30131 Venezia. *osalvadori@arti.beniculturali.it*

Maria Carla SCLOCCHI, Centro di Fotoriproduzione Legatoria e Restauro degli Archivi di Stato (Center for Photo Reproduction, Bookbinding, and Conservation of the State Archives), Via C. Baudana Vaccolini 14, 00153 Roma. *cflr.biol@archivi.beniculturali.it*

Maria Adele SIGNORINI, Department of Plant Biology, Università degli Studi di Firenze. Piazzale delle Cascine 28, 50144 Firenze. *msignorini@unifi.it*

Claudia SORLINI, Department of Microbiology and Food Science and Technology (DISTAM), Sezione MAAE, Università degli Studi di Milano, Via Celoria 2, 20133 Milano. *claudia.sorlini@unimi.it*

Gianfranco TARSITANI, Department of Public Health Sciences "Edificio G. Sanarelli" ex-Igiene, Università degli Studi di Roma "La Sapienza," Piazzale A. Moro 5, 00185 Roma. *gianfranco.tarsitani@uniroma1.it*

Piero TIANO, ICVBC/Istituto per la Conservazione e la Valorizzazione dei Beni Culturali (Institute for the Conservation and Exploitation of Cultural Heritage) of the CNR/Consiglio Nazionale delle Ricerche (National Research Council), Via Madonna del Piano 10, Edificio C, 50019 Sesto Fiorentino (FI). *p.tiano@icvbc.cnr.it*

Maria Luisa TOMASELLI, ISE/Istituto per lo Studio degli Ecosistemi (Institute for the Study of Ecosystems) of the CNR/Consiglio Nazionale delle Ricerche (National Research Council), Via Madonna del Piano, 50019 Sesto Fiorentino (FI). *tomaselli@ise.cnr.it*

Mauro TRETIACH, Department of Biology, Università degli Studi di Trieste, Via L. Giorgieri 10, 34127 Trieste. *tretiach@univ.trieste.it*

Clara URZÌ, Department of Microbiological, Genetic, and Molecular Sciences, Universita' degli Studi di Messina, Salita Sperone 31, 98166 Messina. *urzicl@unime.it*

Paola VALENTI, Istituto Centrale per la Patologia del Libro (Central Institute for Book Pathology), Biology Laboratory, Via Milano 76, 00184 Roma. *icplbio@tin.it*

Elisabetta ZANARDINI, Department of Chemical and Environmental Sciences (Como campus), Università degli Studi Insubria, Via Valleggio 11, 22100 Como. *elisabetta.zanardini@uninsubria.it*

INDEX